Series in BioEngineering

The Series in Bioengineering serves as an information source for a professional audience in science and technology as well as for advanced students. It covers all applications of the physical sciences and technology to medicine and the life sciences. Its scope ranges from bioengineering, biomedical and clinical engineering to biophysics, biomechanics, biomaterials, and bioinformatics.

Indexed by WTI Frankfurt eG, zbMATH.

More information about this series at http://www.springer.com/series/10358

Stephen J. Beebe · Ravi Joshi ·
Karl H. Schoenbach · Shu Xiao

Ultrashort Electric Pulse Effects in Biology and Medicine

Springer

Stephen J. Beebe
Frank Reidy Research Center for Bioelectrics
Old Dominion University
Norfolk, VA, USA

Karl H. Schoenbach
Frank Reidy Research Center for Bioelectrics
Old Dominion University
Norfolk, VA, USA

Ravi Joshi
Department of Electrical and Computer Engineering
Texas Tech University
Lubbock, TX, USA

Shu Xiao
Department of Electrical and Computer Engineering
Old Dominion University
Norfolk, VA, USA

ISSN 2196-8861 ISSN 2196-887X (electronic)
Series in BioEngineering
ISBN 978-981-10-5112-8 ISBN 978-981-10-5113-5 (eBook)
https://doi.org/10.1007/978-981-10-5113-5

This Springer imprint is published by the registered company Springer Nature Singapore Pte Ltd.
The registered company address is: 152 Beach Road, #21-01/04 Gateway East, Singapore 189721, Singapore

Contents

1 **Introduction** 1
Karl H. Schoenbach
1.1 Bioelectrics—The Beginning 1
1.2 How Electrical Pulses Affect the Membrane Potentials of Cells 2
1.3 Electroporation—"Electrical Breakdown" of the Plasma Membrane 5
1.4 From Classical Electroporation Towards Nanoporation 9
1.5 Modeling Intracellular Electroporation 15
1.6 Intracellular Electromanipulation/Biological Studies 17
1.7 Towards Medical Treatments 19
1.8 Scaling of Nanosecond Pulse Effects 22
1.9 From Nanosecond to Picosecond Pulses 23
1.10 New Research Directions 25
1.11 Concluding Remarks 26
References 26

2 **Effects of usEPs on Plasma Membranes—Pores, Channels, and Repair** 33
Stephen J. Beebe
2.1 Some Terms About Pulses 33
2.2 Introduction 37
2.3 Plasma Membrane Structure and Function 38
2.4 Plasma Membrane Channels and Pores in Proteins 42
2.5 Studies with usEP Effects on Plasma Membranes 43
2.6 usEPs Induces Plasma Membrane Nanopores 45
2.7 Effects of usEPs on Plasma Membrane Channels 49
2.8 usEPs Activate Plasma Membrane Cell Repair Mechanisms 56
2.9 Effects of usEPs on Plasma Membrane Redox Systems (PMRS) 62
2.10 Summary 68

References ... 69

3 Simulations of Membrane Effects of Cells After Exposure to Ultrashort Pulses ... 77
Ravi Joshi
3.1 Introduction ... 77
3.2 Computational Aspects ... 79
3.2.1 Molecular Dynamics Methods ... 79
3.2.2 Some Molecular Dynamics Examples in the Nanosecond Pulsing Context ... 82
3.2.3 Analytic Methods for Transmembrane Potentials ... 92
3.2.4 Distributed Network Analysis for Transmembrane Potentials ... 96
3.3 Flows Through Membrane Nanopores ... 102
References ... 103

4 Comparison Between Monopolar and Bipolar Pulses for Effective Nanoporation ... 109
Ravi Joshi
4.1 Introduction ... 109
4.2 Simulation Results ... 111
References ... 122

5 usEP Effects on the Endoplasmic Reticulum (ER) ... 127
Stephen J. Beebe
5.1 Introduction—ER Structure and Function-Interactions with the Nucleus and Mitochondria ... 128
5.2 usEPs Induce Ca^{2+} Release ... 130
References ... 138

6 Intra-cellular Calcium Release Dynamics Due to Nanosecond Electric Pulsing ... 143
Ravi Joshi
6.1 Introduction ... 143
6.2 Intra-cellular Calcium Release Dynamics Due to High-Intensity, Short Electric Pulse ... 146
References ... 155

7 Effects of usEPs on DNA, Nuclear, and Subnuclear Compartments ... 159
Stephen J. Beebe
7.1 Introduction ... 160
7.2 Structure of DNA and the Nucleus ... 160
7.3 Effects of usEPs on the Nucleus and DNA ... 162
7.4 Effects of usEPs on Nuclear Substructures ... 168
References ... 172

8 Mitochondria as usEP Sensors 175
Stephen J. Beebe
8.1 Introduction 176
8.2 From Prokaryotes to Eukaryotes—The Origin of Mitochondria 177
8.3 Up Close and Personal with Mitochondria 180
8.4 Effects of usEP on Mitochondria 192
8.5 usEPs Upregulate Genes in the Electron Transport Chain and ATP Synthase 214
References 219

9 usEP Induce Regulated Cell Death Mechanisms 227
Stephen J. Beebe
9.1 Introduction 227
9.1.1 The Meaning of Necrosis 229
9.1.2 Pyroptosis and Necroptosis Are Called Regulated Necrosis 230
9.2 Cell Death Can Be Programmed or Regulated 230
9.3 Cell Stress Responses to usEPs 234
9.4 Regulated Cell Death (RCD) Responses to usEPs 235
9.5 Examples of usEP-Induced RCD 242
9.6 Use of Jurkat Cell Mutants to Demonstrate usEP-Induced Caspase-Dependent Apoptosis and Caspase-Independent RCD 244
9.7 usEPs Induce RCD in a Cell Type-Dependent Manner 249
References 258

10 Model Rate Equation Evaluation of an Extrinsic Apoptotic Pathway 265
Ravi Joshi
10.1 Introduction 265
10.2 Single Cell Simulation Results 268
10.3 Predictions of Ensemble Population Dynamics 270
References 273

11 Thermal Effects in Bioelectrics 275
Karl H. Schoenbach
11.1 Introduction 275
11.2 Heating Mechanisms in Bioelectrics 276
11.2.1 Joule Heating 276
11.2.2 Heating Due to Dielectric Relaxation 278
11.2.3 External Heating 282
11.3 Thermal Effects on Cells 283
11.3.1 Thermal Effects on Cell Membranes 286
11.4 How Heating Affects Electroporation 288
11.4.1 Modeling Results 288

11.4.2 Experimental Results—Joule Heating 289
11.4.3 Experimental Results—External Heating or Cooling Based on Thermal Conduction 290
11.4.4 Experimental Results—External Heating Including Radiative Sources 291
11.5 Summary 297
References 298

12 Synergy Between Electric Pulse and Thermal Effects 301
Ravi Joshi
12.1 Introduction 301
12.2 Simulation Results with Heating and Electric Pulsing 303
12.3 Role of Thermal Gradients 308
References 314

13 Synergy Between Electric Pulsing and Shock Waves for Cell Poration 317
Ravi Joshi
13.1 Introduction 317
13.2 Molecular Dynamics Simulation Results 318
References 324

14 Probing Potential for Cellular Stimulation by Time-Varying Magnetic Fields 327
Ravi Joshi
14.1 Introduction 327
14.2 Simple Analysis 328
14.3 Some Simulation Results 331
14.4 Brief Discussion Relating to Electromagnetic Bio-stimulation 333
References 337

15 Pulsed Power Generators 339
Shu Xiao
15.1 Introduction 339
15.2 High Voltage Switches 340
15.2.1 Gas Spark Gap Switches 341
15.2.2 Liquid Switches 344
15.2.3 MOSFET Switches 345
15.2.4 Switches for Subnanosecond Pulse Generators 346
15.3 Basic Pulsed Power Circuits 349
15.3.1 Combination of a Capacitor and an Inductor 349
15.3.2 Pulse Forming Line and Pulse Forming Network 350
15.3.3 Blumlein Line Generator 353
15.3.4 Pulse Generators with MOSFET Switches 356
15.4 High Voltage Boosters 365

15.4.1 Transformer 365
15.4.2 Marx Generator 367
References 370

16 Pulse Delivery and Exposure Systems 373
Shu Xiao
16.1 Introduction 373
16.2 Tissue Properties and Electric Field Determination 374
16.3 Pulse Distortion 375
16.4 Electrodes 377
16.4.1 Cylindrical Electrodes 377
16.4.2 Parallel-Plate Cuvette 380
References 384

17 Pulse Voltage Measurement 385
Shu Xiao
17.1 Introduction 385
17.2 Resistor Voltage Divider 386
17.3 Coaxial Attenuators 388
References 390

18 usEPs in Pre-clinical Cancer Treatment 391
Stephen J. Beebe
18.1 Introduction 391
18.2 NPS for Cancer Therapy 393
18.3 An Initial Study for the Effects of usEPs on Fibrosarcoma Tumors in Mice 394
18.4 Murine Melanomas Were Shown to Self-destruct as usEPs Effectively Targeted Cancer 396
18.5 usEPs Eliminate Melanoma and Its Blood Supply 397
18.6 Optimization of usEP for Treating Melanoma 398
18.7 Additional Melanoma Studies with usEP 400
18.8 usEP Eliminate Hepatocellular Carcinomas (HCCs) 401
18.9 Moderate Heat Enhances usEP Ablation of Ectopic Tumors 403
18.10 usEPs Demonstrate Efficacy Against Spontaneous Canine Osteosarcoma 404
18.11 usEPs Treatments of Human Cancer Xenografts 405
18.11.1 Human Melanomas 406
18.11.2 Human Hepatocellular Carcinomas 407
18.11.3 Human Breast Cancer 408
18.11.4 Pancreatic Cancer 409
18.11.5 Human Squamous Carcinomas 409
18.11.6 Human Glioma Xenograft in a Vascularized Avian Chorioallanttoic Membrane (CAM) 410

18.11.7 Human Triple Negative Breast Cancer (TNBC) Patient-Derived Xenograft (PDX) in NOD-Scid Gamma (NGS) and NSGM3 Mice 410
18.11.8 Summary of usEPs in tumor treatment 413
References 415

19 usEPs as a Possible Immunotherapy 419
Stephen J. Beebe
19.1 Introduction 420
19.2 usEPs Induce Immunity 421
19.2.1 usEP Treatment Induces an Immune-Mediated Vaccine Effect in Orthotopic N1-S1 HCC Tumors 423
19.2.2 N1-S1 Tumor Microenvironment Exhibits Time-Dependent Increases in Granzyme B After usEP Treatment 424
19.2.3 Increased Numbers of CD4+ and CD8+ Tumor-Infiltrating Lymphocytes (TILs) Permeate the N1-S1 HCC Tumor Microenvironment (TME) 425
19.2.4 usEP Treatment Enriches the Presence of Memory T-Cells in the Spleens of Rats with N1-S1 Regressing Tumors 426
19.2.5 usEPs Decrease the Percentages of Tregs in the N1-S1 HCC TME 428
19.2.6 usEPs Activate the Innate Immune System 429
19.2.7 usEP Treatment Vaccinates Mice Against N1-S1 HCC 432
19.2.8 Splenic CD8+ Lymphocytes Cytotoxicity After usEP Treatment Demonstrates Active Immunity 435
19.2.9 usEPs Induce Immune Responses, Eradicating Breast Cancer and Reducing Distant Metastases 437
19.2.10 Other Evidence that usEPs Induce Immune Responses 440
References 446

Chapter 1
Introduction

Karl H. Schoenbach

Abstract The effects of intense electrical pulses on cell membranes have been studied extensively since the second half of the twentieth century, and have found multiple applications, ranging from bacterial decontamination to medical therapies. The first studies on this effect, named electroporation, rarely extended beyond microsecond pulse effects on cells. In the past two decades, however, the use of nanosecond and even shorter pulses gained interest, since electrical circuit models of biological cells indicated that not only the plasma membrane, but also the subcellular structures of mammalian cells could be affected by such extremely short pulses. The first experimental study published in 2001 confirmed this hypothesis. It was followed by a large number of publications which showed that such ultrashort, high electric field, but low electrical energy pulses, affect cell functions, such as programmed cell death, and a lower intensity, calcium mobilization from intracellular structures. This chapter, after a short introduction to electroporation, provides an overview of the progress of basic studies on nano- and picosecond pulse effects on cells, tissues, and organisms over the past two decades.

1.1 Bioelectrics—The Beginning

Bioelectrical effects and applications cover a wide range of electrical parameters and the targets of electric fields include biological cells, tissues, and organisms. Studies on bioelectric effect, or more general bioelectromagnetic effects, range from exploring the effect of cell phone radiation on cells to their use in bioelectromagnetic medicine. The reader who is looking for literature on the various topics in this area can easily find it on the internet, or even better in the journals of the Bioelectromagnetics Society (BEMS) and the European Bio-Electromagnetic Association (EBEA). In this book we will focus on such bioelectric effects in medical applications as they are observed when intense, short electrical pulses are applied to biological systems. Starting out with a brief history of pulsed electric field effects on cells, we will focus on bioelectric effects of nanosecond and shorter pulse with pulse amplitudes in the range of thousands of Volts per centimeter. But the history of bioelectric effects begins with much longer pulses. Pulsed electric effects on humans were generated by

S. J. Beebe et al., *Ultrashort Electric Pulse Effects in Biology and Medicine*,
Series in BioEngineering, https://doi.org/10.1007/978-981-10-5113-5_1

"electric fishes" such as the Mediterranean Torpedo or the Nile Catfish, which have been recorded and explored for over 2000 years. Electric shocks by such fish were considered to be helpful in of exorcising spirits (Kellaway 1946). Electroexorcism, by other means of electric pulse sources, was practiced even up to the twentieth century (Wickland 1968). A history of these early bioelectric applications, which were strongly related to theology, can be found in reference (Stilling 2004).

The use of electrical fish shocks in medical applications, such as treatment of headache, arthritis, and other ailments, already began in the second century AD as reported by Largus and Dioscorides (Largus 1528; Dioscorides 512). Dawud Al-Antaki used electrical pulses for the treatment of Epilepsy in the sixteenth century (Leibowitz 1957). Research on bioelectric effects took off in the late eighteenth and the nineteenth century after the invention of the Leyden Jar and the voltaic batteries. It became possible to perform more controlled studies on physiological effects on humans. Quoting Faraday "Wonderful as are the laws and phenomena of electricity when made evident to us in inorganic and or dead matter, their interest can bear scarcely any comparison with that which attaches to the same force when connected with the nervous system and with life." In the nineteenth century and early twentieth century, the studies on intense electrical pulse effect focused on the response of humans to electrical stimulation, whereas the second half of the twentieth century brought us into an area of increasing interest on electrical effects on biological cells.

1.2 How Electrical Pulses Affect the Membrane Potentials of Cells

The effects of intense electrical pulses on mammalian cells has been a topic of research since the 1950s (Staempfli 1958). "Intense" means that the amplitude of the electric field is such that it causes temporary or permanent changes in the structure of cell membranes or even affects the structure of proteins. That the membranes, particularly the outer membrane or plasma membrane was mainly affected by such intense pulses can be understood by considering the electrical properties of cells.

A cross-section of a generalized mammalian cell is shown in Fig. 1.1. The cell is surrounded by an about 5 nm thick plasma membrane (Fig. 1.2) which consists of a lipid bilayer with embedded proteins. From the electrical point of view, the plasma membrane can be considered a high resistance layer with a resistivity of 10^7 Ωcm. Because of its large resistivity compared to that of the cytoplasm, which is approximately 100 Ωcm, it is often considered a perfect dielectric. It capacitance is rather large due to its small thickness. It has been measured as approximately 1 μF/cm^2. When a monopolar voltage pulse is applied across such a cell, the membrane is charged as schematically shown in Fig. 1.3, where only the contours of the plasma membrane and of some membranes across intracellular structures, such as the nucleus, are shown.

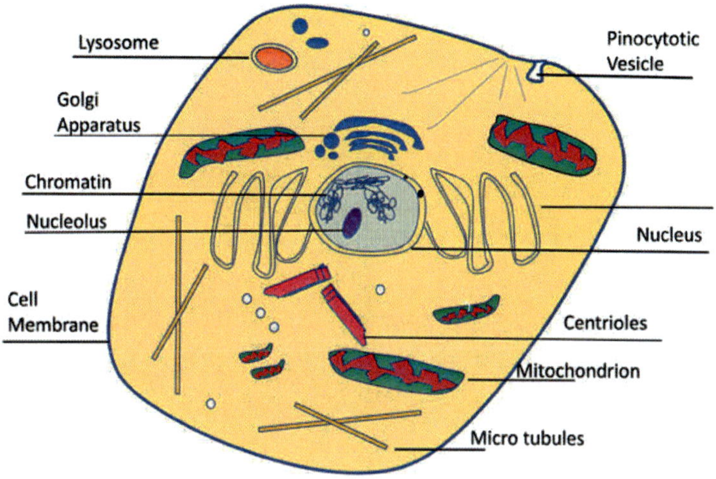

Fig. 1.1 Cross-section of a generalized mammalian cell

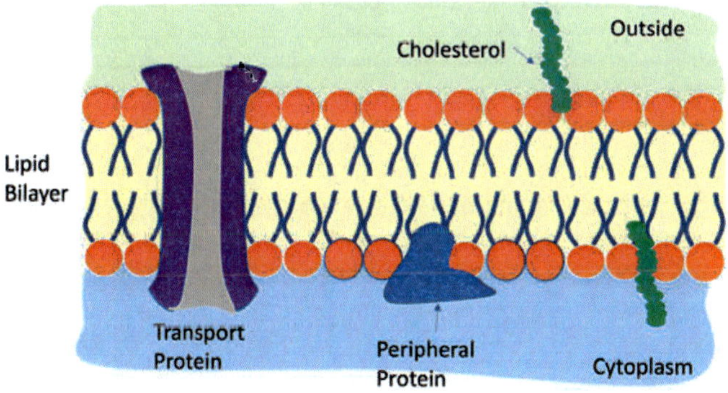

Fig. 1.2 Cross-section and top view of a generalized plasma membrane

The charging time constant of the plasma membrane is for spherical cells at a volume fraction, V, in a suspension given as (Cole 1937):

$$\tau_{om} = \left[\frac{1+2V}{1-V}\frac{\rho_1}{2} + \rho_2\right]C_m D \tag{1.1}$$

where ρ_c is the resistivity of the cytoplasm and ρ_m, that of the medium. C is the plasma membrane capacitance and D is the cell diameter. The charging time constant is the time required to charge the membrane to (1 − 1/e) which is approximately 63% of its final voltage value when a step pulse is applied. With a cell diameter of 10 μm, typical for many mammalian cells, such as HL60, a membrane capacitance of 1 μF/cm^2, and a resistivity of the cytoplasm of 100 Ωcm, a medium resistivity

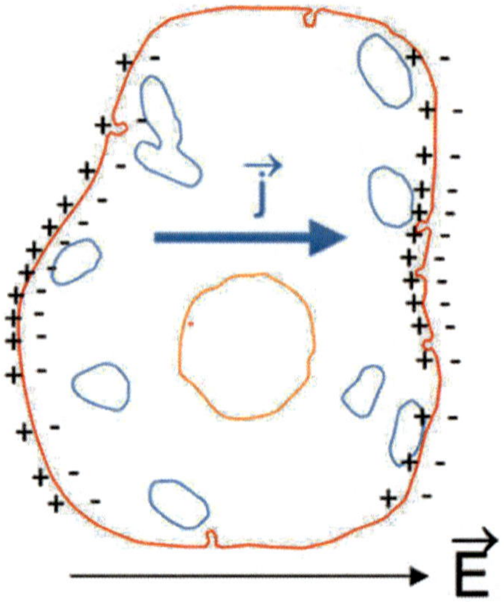

Fig. 1.3 Cross-section of the generalized mammalian cell, as shown in Fig. 1.1, but emphasizing the plasma membrane and selected subcellular membranes. It demonstrates the direction of the ion current density, J, inside the cell, and the corresponding charging of the plasma membrane, when an electric field, E, is applied. For pulsed electric fields with rise-time and duration long compared to the plasma membrane charging time constant, the electric field has only a very minor charging effect on the subcellular membranes

the same as that of the cytoplasm, and volume fractions very small compared to 1, the membrane charging time constant is 75 ns. Since most early studies were done with pulse durations large compared to microseconds, and generally with pulse risetimes far above the charging time constant for the plasma membrane, the charging of the membrane occurred much longer than the charging time constant, and did not require high current density levels. This "slow" charging had only minor effects on the membranes of subcellular structures—the electric field inside which is related to the current density by Ohm's law was not sufficient to cause considerable charging of the membranes of the subcellular structure.

However, charging of the plasma membrane at low applied electric fields may even strongly affect the membrane structure. With the membrane fully charged, the membrane voltage is given as:

$$V_m = f\,E\,a\cos\delta \tag{1.2}$$

where f is a constant, which for spherical cells is 1.5, E is the applied electric field (if the cell is positioned in a medium between two plates, the average electric field is the applied voltage divided by the plate distance), and δ is the angle with respect to the direction of the electric field (cosδ is ± 1 at the poles of the cell). Even for a low electric field of 100 V/cm, the induced voltage across the membrane at the poles of a cell with a 10 μm diameter is about 50 mV above the resting potential. Considering the extremely small thickness of the membrane (5 nm), this increase in voltage corresponds to an increase in the average electric field in the membrane by 10^5 V/cm. Likewise, an electric field of 1 kV/cm causes the membrane electric field to reach a value of 1 MV/cm. At electric fields on that level, instant modifications of the cell membrane structure can be expected.

1.3 Electroporation—"Electrical Breakdown" of the Plasma Membrane

First systematic studies on the effects of pulsed electric fields of the magnitude (1 MV/cm) revealed permeability changes in the membrane of such a magnitude that it caused the release of cellular components, such as calcium ions, and catecholamines, and chromogranine proteins from isolated chromaffin granules of bovine adrenal medullae (Neumann and Rosenheck 1972). Later studies demonstrated the uptake of macro molecules, such as DNA, through the electro-permeated membrane (Neumann et al. 1982; Wong and Neumann 1982).

The electrically induced permeability of the membrane is consistent with the results of the membrane conductance measurements. The rapid increase in conductance during the application of intense electrical pulses led some authors of the first papers on this topic to name this effect "electrical breakdown", because of its similarity to the electric breakdown of dielectric layers (Benz and Zimmermann 1980a). Measurements of the changes in the electrical characteristics of lipid bilayers were initially performed using a pulse charge method (Benz and Zimmermann 1980b). In this method, a lipid bilayer membrane, which has a capacitance C_m, is charged to a voltage V_m by injecting a charge Q. V_m is given as:

$$V_m = Q/C_m \tag{1.3}$$

By varying both, the injected charge at identical length, and the pulse lengths at identical charge the breakdown voltage for lipid bilayer membranes could be obtained. It was found that it is not a constant, but dependent on the time it took reach the breakdown voltage. In the following we will use the expression "pulse duration" for this time. The results of these measurements on lipid bilayer membranes made from oxidized cholesterol/n-decane at temperatures of 25 and 40 °C are shown in Fig. 1.4. They show that the membrane breakdown voltage is approximately 0.5 V at pulse durations in the range from 10 to 1 μs for a temperature of 40 °C. For shorter duration it increases to values of 1 V (Fig. 1.4, left). For lower temperatures, 25 °C, the breakdown voltage is higher. It reaches 1.2 V for submicrosecond pulses (Fig. 1.4, right).

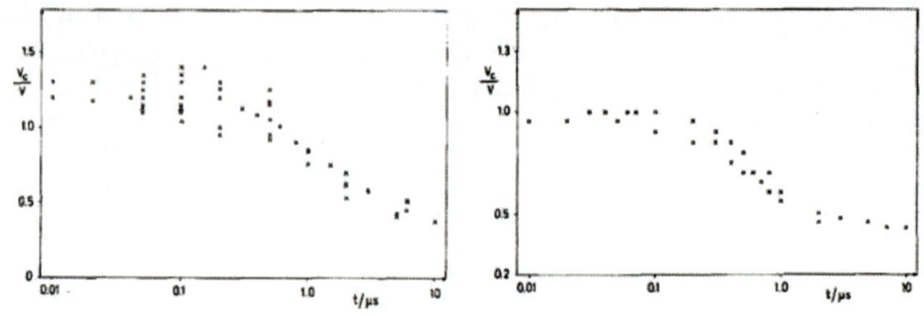

Fig. 1.4 The breakdown voltage, V_c (in Volts), of lipid bilayer membranes is dependent on the charging time, t, (in microseconds) at a temperature of 25 °C (left), and 40 °C (left) (Benz and Zimmermann 1980a)

Although the reduced membrane conductivity and the uptake and/or release of macromolecules through the permeated cell can be described in general terms by homogeneous changes in the structure of the membrane, it was held that the increase in membrane conductance corresponds to the formation of local openings in the membrane, i.e., electro-pores (Kinosita and Tsong 1977). The permeabilization of the plasma membrane by applying high voltage pulses has therefore been named "electroporation", short for electric pore formation (Neumann et al. 1982). For pulses at high electric fields and/or long pulse application, these pores can grow to such a large size so that they lead to cell death, a process called "irreversible electroporation". For the electric field/ pulse duration combination below that required for cell death, the pores are transient, a process called "reversible electroporation". The temporal development of such pores was considered to occur in two stages (Weaver et al. 1984; Glaser et al. 1988). The initial state of the pores was described as short-living hydrophobic pores, water-filled defects, with a radius of 0.3–0.5 nm. With increasing time the hydrophobic pores change into long-living hydrophilic pores with diameters of 0.6 to more than 1 nm by the reorientation of the lipid molecules (Fig. 1.5).

The studies with electric fields of 100 s of V/cm for millisecond pulses to kV/cm for microsecond pulses led to a host of novel medical and environmental applications,

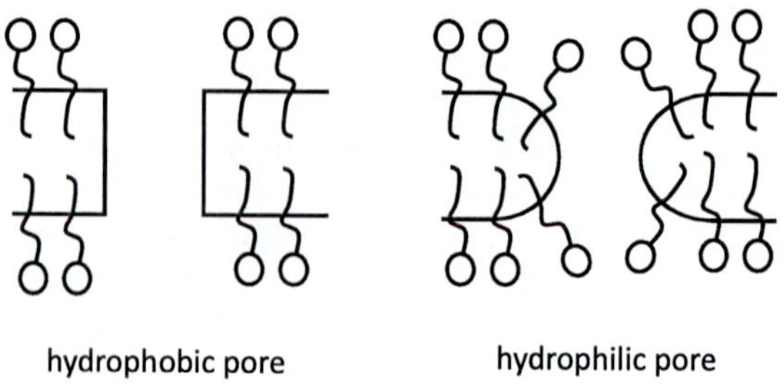

Fig. 1.5 Schematics of the lipid bilayer arrangement of a hydrophobic (left) and hydrophilic pore (right)

as well as applications in food and biomaterials processing. In medical applications, cancer treatment emerged as the major application. It included electrochemotherapy where drugs are inserted through the permeabilized membrane (Okino and Mohri 1987; Mir et al. 1991; Sersa et al. 1995; Heller et al. 1997; Frandsen et al. 2012) and gene therapy where DNA plasmid is delivered into the cell (Titomirov et al. 1991; Heller et al. 1996; Daud et al. 2008). Whereas these methods are based on reversible electroporation, in which the pores reseal after some time, irreversible electroporation, where microsecond pulses are used to eliminate tumor cells by inhibiting pore closure, has added to the electrical methods in cancer treatment (Davalos et al. 2005; Jiang et al. 2015).

Another area which also relies on irreversible electroporation is food processing, which deals with plant cells and microbes rather than mammalian cells. The use of pulsed electric fields for liquid food preservation through microbial killing and to extract intracellular components has been demonstrated successfully in juice extraction from grapes, sugar beets, and other plants (Frey et al. 2017).

In order to determine the magnitude of an applied pulsed electric fields required to achieve electroporation, a cell can be modeled as a spherical conducting body surrounded by a thin layer of insulating material and immersed in a conducting medium (Fig. 1.1). Applying a step function electric field, E, to the cell, the voltage across the cell membrane, V_{cm}, increases with time, t, as

$$V_{om}(t) = fE\frac{D}{2}\cos\theta(1 - \exp(-t/\tau_{om})) \tag{1.4}$$

For electroporation to occur for a given pulse duration, τ, the electric field at the poles of the cell (θ = 0 or π), E_{cr}, must reach the critical value, V_{cr}:

$$E_{cr} = \frac{V_{cr}}{f\frac{D}{2}(1 - \exp(-\tau/\tau_{om}))} \tag{1.5}$$

Figure 1.6 shows the normalized critical field for electroporation depending on the pulse duration, τ. With an assumed critical voltage for electroporation of 1 V, consistent with measured values at pulse durations shorter than 100 ns at temperatures of 20 ^{0}C (Fig. 1.4a), the electric field required for electroporation at a pulse duration of 75 ns ($\tau/\tau_c = 1$) is approximately 3 kV/cm. It increases almost linearly with reduced pulse duration. For 7.7 ns pulses it would reach 30 kV/cm, and for 750 ps pulses about 300 kV/cm.

This simple calculation of the critical electric fields does not take into account the increasing critical membrane voltage (V_{cr}) with shorter pulses. This is obvious from the membrane breakdown voltage measurement results shown in Fig. 1.4. A stochastic model for electric field-induced membrane pores shows a similar tendency. As shown in Fig. 1.7, the transmembrane voltage V, required to generate pores with a lower limit size for the conduction of alkali metal ions, increases with reduced average time interval (Sugar and Neumann 1984). A comparison of theoretical data with experimental show a reasonable fit. Interestingly, for pulses with "durations" as

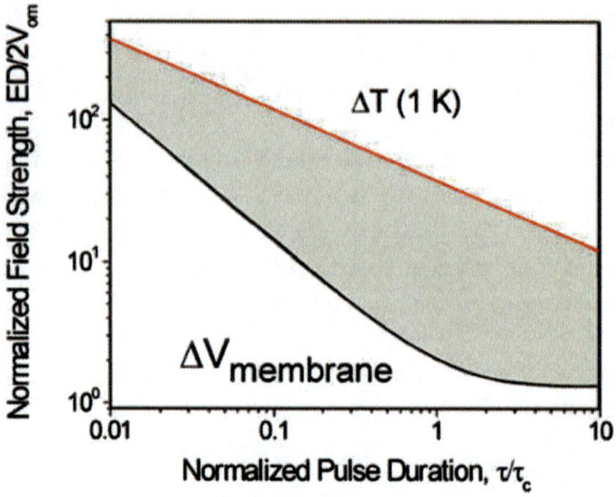

Fig. 1.6 Using Eq. 1.5, the normalized critical electric field, $ED/2V_{cr}$, for the onset of electroporation is plotted versus the normalized pulse duration τ/τ_c,. In this graph it is assumed that the critical voltage, V_{cr}, is independent of the pulse duration. The normalized critical field approaches a constant value for pulses longer than τ_c, and increases linearly for pulse durations shorter than τ_c. Taking the results of measurements of the breakdown of liquid bilayer membranes (Fig. 1.4) into account will result in a more than linear increase in critical voltage for pulse durations less than τ_c,. The effect of Joule heating caused by the electrical pulse sets the range of nonthermal effects. In this example, assuming that the upper limit of temperature increase is 1 K, the shaded area between the two curves depicts the operational range of E and τ

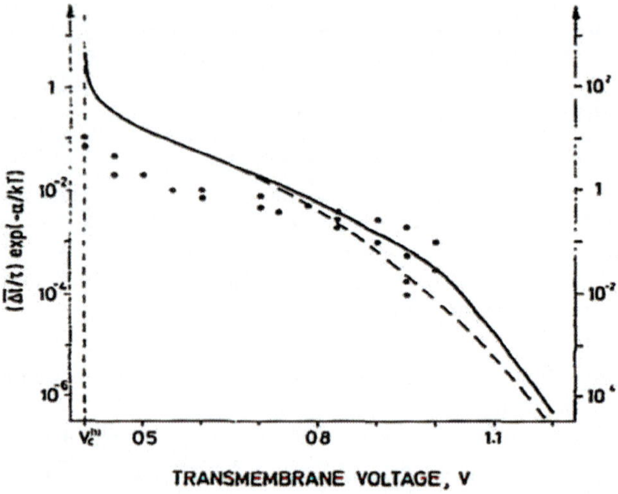

Fig. 1.7 The normalized interval Δt of building up of the lower limit pore size for the conduction of alkali metal ions, as a function of the transmembrane voltage, V, at two threshold pore sizes (solid curve and dashed curve). For a given scaling factor the ordinate on the right-hand side marks the absolute time scale of Δt (Sugar and Neumann 1984). The dots show the experimental data of Benz and Zimmermann as depicted in Fig. 1.4

short as 10^{-4} μs (100 ps), there is a sharp increase in breakdown voltage. The critical membrane voltage also depends on the temperature (Fig. 1.4). For temperatures of 40 °C, it increases from 1 to 1.2 V. It is reasonable to assume that for lower temperatures, e.g. 25 °C, the voltage for subnanosecond pulses increases by about the same factor, from 1.2 (Fig. 1.4b) to ~ 1.5 V.

The electrical pulse for electroporation causes an increase in temperature, ΔT, which depends on E and τ, the conductivity, σ, of the cell suspension, the density, ρ, and the specific heat, c, of the suspension:

$$\Delta T = \sigma E^2 \tau / \rho c \quad (1.6)$$

Using the same cell parameters which lead to a charging time constant of 75 ns, and assuming a medium density of 1 g/cm^3, a specific heat capacity of 1 cal/g °K, and a conductivity of 10 mS/cm, typical for many in vitro studies in cuvettes, an increase in temperature by 1 °K is dependent on electric field and pulse duration (Fig. 1.6). In order to limit the temperature increase to values below 1 °K, the electric field and pulse duration should always be kept below the temperature curve in Fig. 1.6. That criterion determines a crescent shaped area, shown in Fig. 1.6 as shaded area between the two curves, where “nonthermal” electroporation can be performed. As the nonthermal range of electroporation will depend on the type of studies performed, the diagram only serves to show tendencies. One of them is the narrowing of the operational E and τ range for shorter pulses. It indicates that for nanosecond pulses and shorter, care needs to be taken to avoid thermal processes due to Joule heating.

That for nanosecond pulses the “breakdown voltage” exceeds 1 V is generally in agreement with the results of optical measurements which were performed on living cells, rather than planar lipid bilayers. The change in the membrane potential of Jurkat cells was studied for pulses with a duration of 60 ns (Frey et al. 2006). The membranes were stained with a fast voltage-sensitive dye, ANNINE-6, which has a subnanosecond voltage response time. A temporal resolution of 5 ns was achieved by the excitation of this dye with a tunable laser pulse, synchronized with the applied electric field to record images at times before, during, and after exposure. When exposing the Jurkat cells to a pulse, the voltage across the membrane at the anodic pole of the cell reached values of 1.6 V after 15 ns as shown in Fig. 1.8. Voltages across the membrane on the side facing the cathode reached values of only 0.6 V in the same time period, indicating a strong asymmetry in conduction mechanisms in the membranes of the two opposite cell hemispheres.

1.4 From Classical Electroporation Towards Nanoporation

It is obvious from these relative simple calculations of the electric field dependence on the pulse duration as shown in Fig. 1.6 that reducing the pulse duration to values on the order of and less than the charging time constant of the plasma membrane requires pulses with drastically increased electric field amplitudes. In order to reach

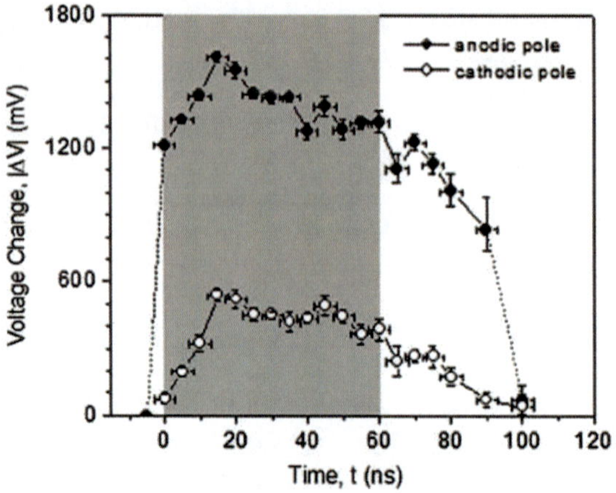

Fig. 1.8 Temporal development of the membrane voltage at the cathodic and anodic pole. The data represent the mean membrane voltages ± SE in three to five experiments. The shaded area depicts the time when the electric field was applied (Frey et al. 2006)

the threshold for electroporation pulses with durations in the single nanosecond regime require amplitudes of ten to hundred kV/cm. This was obviously at least one of the reasons that initial studies focused on the modifications of the plasma membrane with monopolar electrical pulses of tens of microseconds and longer duration, for which much lower fields were needed. An exception were studies by Benz and Zimmermann, where already in the late 1970s and early 1980s, studies of electroporation were performed with pulses as short as 10 ns. However, rather than mammalian cells, these studies focused on electrical breakdown (Benz and Zimmermann 1980a, b) and membrane recovery (Benz and Zimmermann 1981) of artificial bilayer membranes, and of membranes of giant algal cells. The studies resulted in important information on the temporal development of pore opening and pore closure in bilayer structures. However, due to the limited voltage of the pulse generators available at that time, the effects on intracellular structures in the temporal range of biological cells couldn't be explored.

Pulse power technology was the enabling technology which allowed us to study the effect of nanosecond and shorter pulses on biological cells and tissues. Pulse power technology deals with the physics and applications of high voltage pulse generators, which were originally developed for military applications. The concept of such pulse generators was increasingly applied to bioelectric studies in the 1990s (Schoenbach et al. 1997a). Most of these nanosecond pulse generators which were used for biological studies are so-called line-type pulse generators (Kolb et al. 2006). They allow us to generate square wave electrical pulses with voltages in the tens of kV range and pulse durations as short as single digit nanoseconds. The concept of such pulse generators is based on the use of transmission lines, such as coaxial cables or strip-lines, consisting of two parallel conductors separated by a dielectric layer. Such lossless transmission lines can be considered as a series of capacitors, C′, and

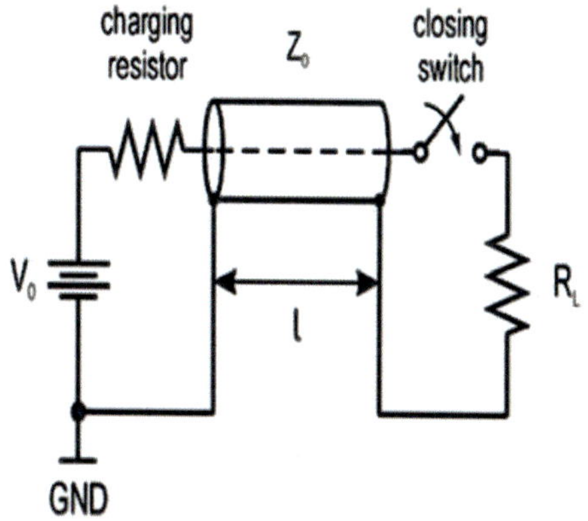

Fig. 1.9 Schematics of a line-type pulse generator. By closing a switch with a closing time of much less than the pulse length, a square wave pulse with duration of twice the travelling time of the electromagnetic wave, which is equal to the length of the transmission line divided by the propagation velocity of the electromagnetic wave along the line, and an amplitude of half the applied voltage is generated across the load for a matched load ($R_L = Z_0$)

inductors, L′, per length, with an impedance, Z_0, given as $(L'/C')^{1/2}$. The impedance for cables is usually about 50 Ω. With strip-lines, impedances in the 10 Ω range and less can be obtained.

In the most basic circuit of a high power monopolar pulse generator, the transmission line is charged to the desired voltage, and then discharged into the load, R_L, by means of a fast closing switch (Fig. 1.9). The closure time of the switch is small compare to the pulse duration, causing the generation of a square wave pulse across a matched load—that means a load which has a resistance equal the transmission line impedance. The duration is twice the time it takes for the pulse to finish a round trip of the transmission line. For a coaxial cable with the relative permittivity of the dielectric material between the two coaxial conductors being about 2, a length of one meter produces a pulse with a duration of 10 ns. Shortening or increasing the cable length allows to change the pulse duration. The voltage across the load for this circuit is half the applied voltage.

It needs to be mentioned that by modifying the circuit shown in Fig. 1.9 into a "Blumlein" circuit it is possible to generate ultrashort electrical pulse of the same amplitude as the applied voltage, rather than only half of the applied voltage (Kolb et al. 2006). For most of the bioelectric studies, such Blumlein pulse generators, which are described in more detail in Chap. 16, have been used. An example of a 10 ns, 35 kV pulse generated by a Blumlein pulse generator with an impedance of 10 Ω is shown in Fig. 1.10. Although the electrical energy of the pulse deposited in a 10 Ω load is just about 1.2 J, the electrical power is 120 Million Watts.

Efforts to explore the bioelectric effects of such ultrashort pulses, defined as pulses with submicrosecond duration, began in the mid-1990s, not initially through basic studies on their effect on cells, but as a means for more efficient bacterial

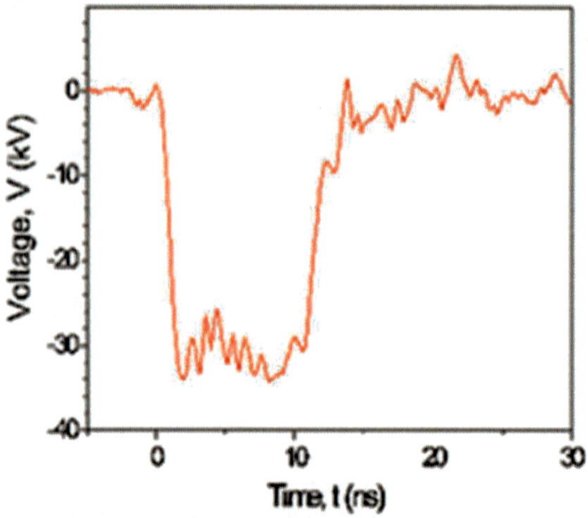

Fig. 1.10 A 10 ns, 35 kV pulse, generated by means of a 10 Ω transmission line Blumlein type pulse generator

decontamination and to prevent biofouling (Schoenbach et al. 1997a, Ghazala and Schoenbach 2000). Shortly after the initial studies on the effect of ultrashort pulses on aquatic nuisance species, first measurements on the effect of nanosecond pulses on cancer cells were performed and compared to those obtained with microsecond pulses (Schoenbach et al. 1997b). Shorter pulses (50 ns) caused damage to the nucleus, but did not seem to affect the outer membrane. On the other hand 5 μs long pulses didn't seem to affect the nucleus, but did not seem to cause damage to the outer membrane.

The explanation for the observed intracellular effects with nanosecond pulses was based on the following considerations: When a step function electrical pulse is applied to a cell, the temporal increase in the voltage across the outer membrane is determined by the charging time constant of this membrane. For mammalian cells, typical values for this time constant are in the submicrosecond range. When a high-voltage pulse with a rise-time short compared to the charging time constant is applied to the cell, charging the plasma membrane requires a large ionic current density during the initial phase of the pulse, and, according to Ohm's law, a large electric field inside the cell. For the initial nanoseconds, or tens of nanoseconds, this internal electric field can reach values which are comparable to the applied electric field. Provided that the applied electric field is large enough, the internal electric field can cause electroporative effects on organelle membranes. This led to the conclusion that subcellular, membrane-bound structures may be affected if the pulse duration is reduced to values on the order of and less than the charging time constant of the plasma membrane. This case is consistent with the well-known fact that cell membranes become transparent for high frequencies, in the MHz range (Foster and Schwan 1996), as can be shown through a Fourier analysis of ultrashort pulses.

These considerations indicate that it should be possible to reach electric fields inside the cell which are of sufficient magnitude to cause portion of subcellular membranes., The use of an electrical equivalent circuit for cells with a substructure, first presented in a paper in 1997 (Schoenbach et al. 1997b), allowed us to obtain more

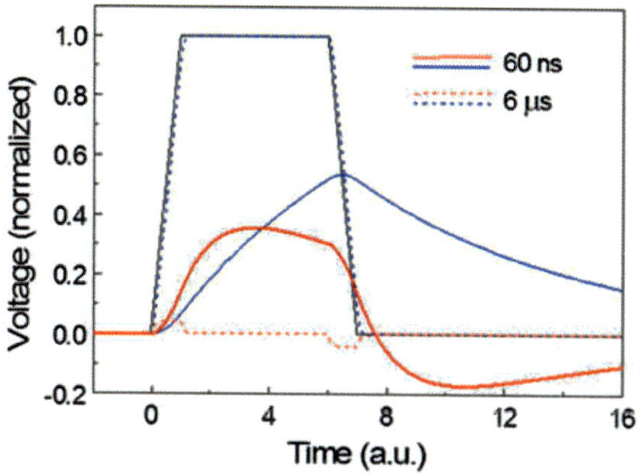

Fig. 1.11 The computed temporal development of the voltage across the surface membrane (dashed blue line for the 6 μs pulse, solid blue line for the 60 ns pulse), and the voltage across a double intracellular structure membrane (dashed red line for the 6 μs pulse, solid red line for the 60 ns pulse). The applied trapezoidal voltage pulse is solid black line. The abscissa has two different time scales. For the 60 ns pulse the scale is given in units of 10^{-8} s, for the 6 μs pulse it is given in units of 10^{-6} s. With the application of a 60 ns pulse, voltages across both surface and intracellular membrane are of the same magnitude. With longer (6 μs) electric field application, the voltage across the intracellular membrane is almost negligible, while the voltage across the surface membrane approximates the magnitude of the applied electrical pulse itself (Schoenbach et al. 2001)

quantitative information on the required pulse parameters for intracellular electromanipulation (Schoenbach et al. 2001). In such a model, with the equivalent circuit of a cell shown in Fig. 1.1, the plasma membrane and intracellular membranes—in this case the nuclear membrane—are represented by capacitors, and the cytoplasm, nucleoplasm and medium, by resistors. The voltages across the membranes were compared for the case of a trapezoidal 6 μs pulse, with that of a 60 ns pulse (Fig. 1.11). For the 6 μs pulse (dashed blue line), the voltage across the plasma membrane closely follows the applied pulse voltage (solid black line). The voltage across the nuclear membrane (dashed red line) shows a small positive value during the rise of the 6 μs pulse and a small negative voltage during the fall time. For the 60 ns pulse the amplitudes of the voltages across plasma membrane and nuclear membrane are on the same order of magnitude. However, the voltage across the plasma membrane increases up to the end of the pulse, and then decays exponentially, whereas the voltage across the nuclear membrane exceeds that of the plasma membrane for a short time, then decays before the end of the pulse, and turns briefly negative after the pulse. It needs to be noted that in this cell model, the capacitance of the nuclear membrane was set to half of the value because two lipid bilayer membranes comprise the nuclear envelope. That explains why the voltage across the nuclear membrane for short pulses exceeds that of the plasma membrane.

Reducing the pulse duration into the hundred-nanosecond range—which implies a sub-hundred nanosecond rise-time—and increasing the pulse amplitude to values which allow us to raise the voltage across subcellular membranes to values

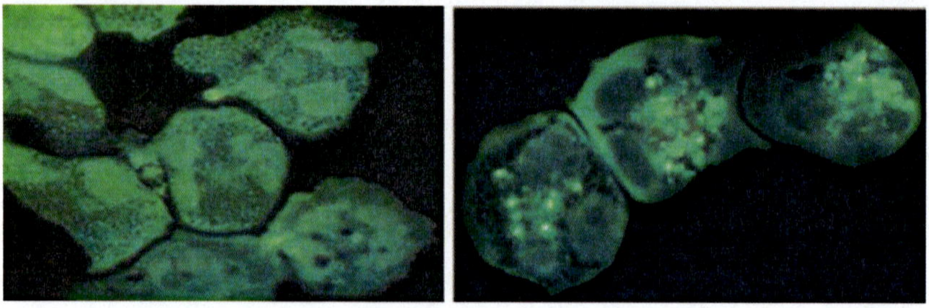

Fig. 1.12 Images of eosinophils under a microscope before (left) and after (right) the application of five 60 ns pulses with electric fields of 60 kV/cm. The photograph on the right shows that inner structures have been opened and taken up dyes—shown as sparklers (Schoenbach et al. 2001)

comparable to those required for electroporation of the plasma membrane, was expected to cause substantial subcellular effects (Schoenbach et al. 1997b, 1999). Modeling and experimental results showed this "intracellular electromanipulation" requires nanosecond pulses to generate electric field in the cells ranging from kV/cm to hundreds of kV/cm. This is orders of magnitude higher than needed for electroporation with millisecond pulses (Schoenbach et al. 2002, 2004, 2007).

The first experimental study that proved the hypothesis that nanosecond pulses affect intracellular structures was performed in 2000 and published in 2001 in the Bioelectromagnetics Journal (Schoenbach et al. 2001). Human eosinophils were loaded with calcein-AM (calcein-acetoxymethylester), an anionic fluorochrome that enters cells freely and becomes trapped in the cytoplasm by an intact cell surface membrane following removal of the AM group. The eosinophil granules stay unlabeled because the cytosolic calcein is impermeant to the granular membranes. When 60-ns pulses with an electric field amplitude of 60 kV/cm were applied to the eosinophils suspended in Hanks Balanced Salt Solution, granules, which were dark before pulsing, began to fluoresce brightly (Fig. 1.12). This was a strong evidence for the breaching of the granule membranes and ionic binding of free calcein from the cytosol to the cationic granule components. On the other hand, the retention of the cytoplasmic calcein staining seemed to indicate that the plasma membrane was not electroporated.

That the plasma membrane was not electroporated by nanosecond pulses seemed to be confirmed by studies on COS-7 cells where it was shown that endocytosed vacuoles were permeabilized with 100, 50 ns, 7.5 kV/cm pulses applied at a repetition rate of 20 Hz (Tekle et al. 2005). The permeabilization of the inner vacuoles occurred without a detectable permeabilization of the outer membrane.

However, in a later study on the permeabilization of endocytic vesicles in B16 F1 mouse melanoma cells, where the effect of one to twenty, 60 ns, 50 kV/cm pulses, applied with a repetition rate of 1 kHz, was explored, different results were obtained. It was found that permeabilization of both the vesicles, but also permeabilization of the plasma membrane occurred (Napotnik et al. 2010). That the plasma membrane is also affected by nanosecond pulses had earlier been confirmed by studies based on differential uptake of fluorescent markers (Pakhomov et al. 2007) and using patch clamp (Pakhomov et al. 2009). With nanosecond pulse application, small pores, so-called nanopores, are formed in the plasma membrane. They have diameters of less than 1.5 nm, too small to allow transfer of calcein through the membrane. The concept of nanoporation, or supra electroporation, or the creation of high density "nanopores" in the plasma membrane through application of nanosecond pulses had been introduced by the Joshi group (Hu et al. 2005a) and by the Weaver group (Gowrishankar et al. 2006; Vasilkoski et al. 2006), respectively.

Following the 2001 publication on intracellular electromanipulation, numerous in-depth studies on this topic were performed in cooperation with scientists at the Eastern Virginia Medical School. Most of the studies were done at the then new Center for Bioelectrics (now: Frank Reidy Research Center for Bioelectrics), established in 2002 by Old Dominion University. One of those studies provided further proof that nanosecond pulses affect intracellular structures of cells, without necessarily electroporating the plasma membrane in the same way as millisecond and microsecond pulses (Chen et al. 2004). The change in the integrity of the plasma membrane and the nucleus of HL-60 cells induced by 10 and 60 ns long pulses of about the same energy were recorded by fluorescence changes with propidium iodide (PI) and acridine orange (AO) using confocal microscopy. A single 10 ns pulse at an amplitude of 65 kV/cm didn't cause PI uptake over the time of observation (30 min) whereas the nucleus in the AO stained HL-60 cell became irregularly shaped with decreasing fluorescence over time. The observed diffusion of the DNA stained by acridine orange to the cytoplasm after nanosecond pulse application suggests changes in the nuclear membrane integrity.

1.5 Modeling Intracellular Electroporation

Numerical modeling of the effects of ultrashort pulses on cells provide a much more accurate description of membrane electroporation effects of both the plasma membrane and subcellular membranes compared to the simple equivalent circuit models. A detailed discussion of bioelectric modeling methods and results can be found in Chaps. 3, 4, 6, and 10. This section serves only as a brief introduction into the efforts to describe and understand bioelectric effects through modeling.

Several "active" cell models have been developed by Joshi's group at Old Dominion University and Weaver's group at the Massachusetts Institute of Technology. Weaver's group has used a lattice model (Gowrishankar and Weaver 2003). The results of such a model are shown in Fig. 1.13 where the equipotential lines at and inside a spherical cell, exposed to a 60 ns pulse are compared to that of a 100 μs pulse. It clearly shows that for nanosecond pulses the electric field reaches into the cell interior and electroporates both the plasma membrane as well as subcellular membranes, whereas for the 100 μs pulse the subcellular membranes are not electroporated (Gowrishankar et al. 2006). Joshi et al. have used a distributed circuit model with current continuity, and a coupled Smoluchowski equation for pore development

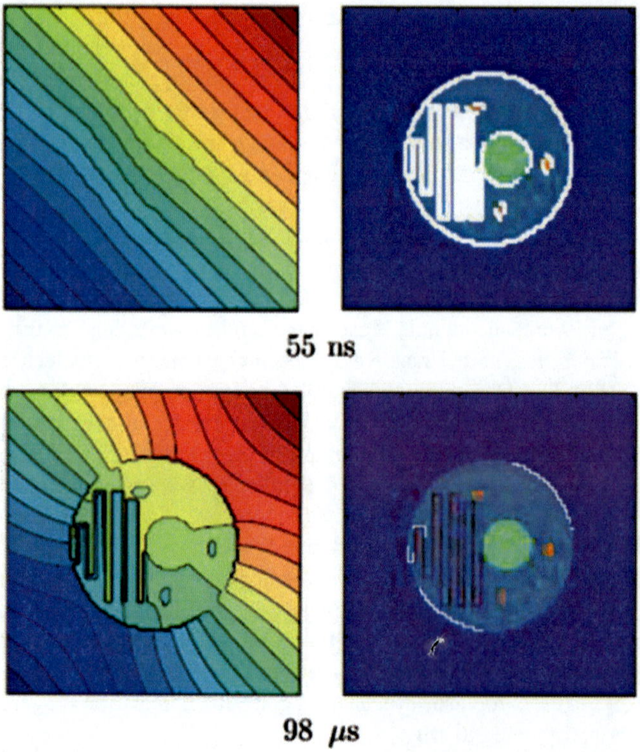

Fig. 1.13 The effect of a 60 ns pulse with an electric field amplitude of 60 kV/cm (top) and that with a 100 μs long pulse with 1.1 kV/cm field amplitude (bottom) on a 10 μm radius cell. The cell model includes representations for the plasma membrane, the nuclear membrane, the endoplasmic reticulum membrane, and three mitochondria with inner and outer membranes. Shown is the equipotential distribution close to the end of the two pulses, at 55 ns for the 60 ns pulse, and at 98 μs for the 100 μs pulse (left). The electroporated sites are shown in white (right) (Gowrishankar et al. 2006)

(Joshi et al. 2004). The model, which provided information on the temporary development of the membrane potentials of plasma membrane and subcellular membranes, also yields time dependent distributions of the membrane pore population.

The distributed parameter cell models, which allowed us to obtain information on the spatial distribution and temporal development of electropores for single cells and assemblies of cells, does not provide information on the development of single pores. This gap in knowledge can be closed in part by using molecular dynamics simulations. They provide the most basic and fundamental approaches for the modeling of effects of electric fields on cell membranes by computing the interaction of charged subgroups which characterize the structure of biological membranes with applied electric fields (Tieleman et al. 2003). Because it is computationally intensive, the method is restricted to the modeling of small patches of the membrane for limited pulse durations. This works well for nanosecond pulses. First molecular dynamics studies focusing on the development of nanopores have been performed by the group of Joshi (Hu et al. 2005a, b) and by the group of Vernier (Vernier et al. 2006). A discussion of the modeling procedures and results would be beyond the scope of this manuscript. However, the reader can find more information on molecular dynamics simulations of pore formation in membranes in chapters of the book on "Advanced Electroporation Techniques in Biology and Medicine" (Vernier 2010; Tarek and Delemotte 2010), and in references (Bennett et al. 2014; Son et al. 2014; Rems et al. 2016).

Whereas the modeling studies were able to shed light on the physical effects of nanosecond pulses on the cell membranes, a number of important subsequent physical and biological effects have been explored using numerous experimental techniques. Some of these effects turned out to be significant for their potential use in novel medical therapies.

1.6 Intracellular Electromanipulation/Biological Studies

In the years following the first studies on intracellular electroporation, or "intracellular electromanipulation", as it was named in the early publications (Schoenbach et al. 2001; Beebe et al. 2002), efforts were made to explore the physical processes in the cells caused by nanosecond pulses. Comparisons of the effects of microsecond pulses with nanosecond pulse effects showed a relatively long delay in uptake of PI in Jurkat cells for the nanosecond pulses compared to microsecond pulses (Deng et al. 2003). For pulses of approximately the same energy, pulses with a duration of 10 μs were found to cause immediate uptake of PI, whereas for 60 ns pulses the median delay time for PI uptake was 15 min. This indicated that the permeabilization of the plasma membrane with nanosecond pulses is likely determined by biological processes, whereas for microsecond pulses (and longer pulses) it seems to be an instant physical effect.

Another striking difference between the effects of nanosecond versus microsecond pulses was that for long pulses the location of the uptake showed anodic preference, as expected for long duration pulses (Gabriel and Teissie 1997), whereas for nanosecond pulses the loss of membrane integrity was not found to have any discernable electrode preference. In a follow-up study, under similar conditions as described above (Deng et al. 2003), these results were not only confirmed, but it was also found that increasing the pulse duration had a profound effect on the swelling of cells (Buescher and Schoenbach 2003). For long pulses cellular swelling was observed, whereas for short, 60 ns pulses, no swelling was seen. All these results—delayed PI uptake, lack of anodic preference, absence of cellular swelling for nanosecond pulses—indicated that such nanosecond pulse effects are due to intracellular effects rather than direct effects on the plasma membrane.

An important feature of nsPEF effects on cells is its specificity. That means that different cell types are differently affected by nsPEF. The changes in the membrane potential, using $DiOC_5(3)$, a slow-response membrane potential sensitive dye, and the uptake of PI were studied for various types of cells after applying one to five, 60 and 300 ns pulses at electric field intensities between 15 and 60 kV/cm (Hair et al. 2003). The cell types included Jurkat cells, polymorphonuclear leukocytes, fresh and aged human mononuclear cells, human trunk skin cells, and human macrophages in various mixtures of two different cell types. The studies showed different responses for the different cell types in the PI uptake characteristics and the effects on the membrane potential, but they were expressed mainly at the lowest pulse numbers. With increasing pulse numbers, the heterogeneous nsPEF effects on cells tend to become homogeneous, suggesting that cumulative effects will occur with an increased pulse number. Interestingly, the nsPEF effects on cells for low pulse numbers were found to be independent of cell size. This is quite different from the plasma membrane electroporation-effects of multi-microsecond and millisecond pulses, which have been used to discriminate between cells of different sizes (Eppich et al. 2000).

Early research showed that the application of nanosecond pulses at high electric fields causes cell death. It was Stephen Beebe with his team at Eastern Virginia Medical School, in cooperation with researchers at Old Dominion University, who explored nanosecond induced cell death thoroughly and published a number of highly cited papers on this topic (Beebe et al. 2002, 2003a, b, 2004). The observation that cells which were exposed to nanosecond electrical pulses with pulse parameters above certain duration, amplitude, and pulse number underwent apoptosis was an exciting discovery. Apoptosis is a regulated cell death characterized by nuclear condensation, cell shrinkage, membrane blebbing, and DNA fragmentation. It is not an instant cell death but occurs over hours if initiated by UV light or by toxic chemicals. It is nature's primary means of getting rid of cells no longer needed.

Using nanosecond pulses it was shown to be possible to control the progression of apoptosis by the choice of nanosecond pulse parameters. Caspases, a family of cysteine proteases, are the central regulators of apoptosis. Higher electric fields cause faster progression. Figure 1.14 shows the increase in the number of Jurkat cells which are undergoing caspase activation after a single 300 ns. In this case caspase activity

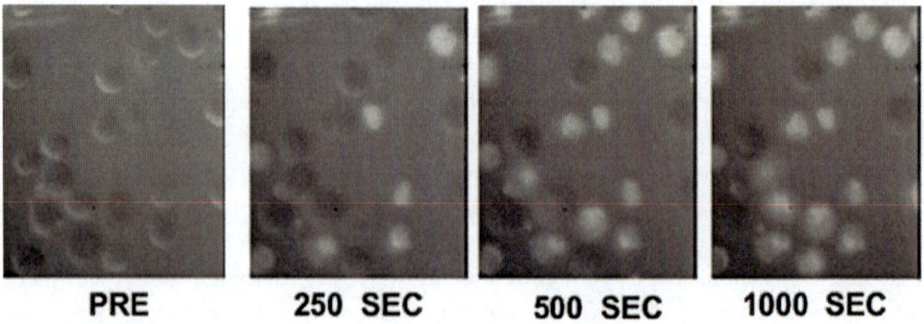

Fig. 1.14 Caspase Activation in Jurkat Cells after a 300 ns, 60 kV/cm pulse (Courtesy of S. Buescher, EVMS)

is induced on a time scale of minutes, but can be longer at lower electric fields. Once caspases are activated, cell death ensues relatively rapidly.

Since the first report (Beebe et al. 2002), and studies by the group at USC that focused on phosphadityl-serine externalization as an indicator of apoptotic cells (Vernier et al. 2003a), more than forty in vitro studies focusing on apoptosis induction by nanosecond puled electric fields have been published before February 2016 (Napotnik et al. 2016). The results of these studies indicate that nanosecond pulsed electric fields act differently on apoptotic pathways in different cells. In those studies, the major cell lines were Jurkat and HL60 cells. It could be clearly shown that the cells responded to nanosecond pulsed electric fields by apoptosis induction. For other cells, different regulated cell mechanisms need to be considered (Galluzi et al. 2012).

One of the nanosecond pulse effects, observed at moderate electric field strengths is calcium release. First publications on this topic appeared in 2003 (Vernier et al. 2003b; Buescher and Schoenbach (2003); Beebe et al. 2003b). Calcium is known as an important molecule which regulates a number of responses including secretion of neurotransmitters and apoptosis induction. It is mainly stored in the endoplasmic reticulum (ER) and in alpha granules for platelets. At lower levels it is stored in the mitochondria. One of the earliest studies, carried out on human polymorphonuclear leukocytes (PMNs) moving along the surface of a coverslip of a microscope, showed a sudden spike in calcium release when a 300 ns long pulse of 15 kV/cm amplitude was applied (Buescher et al. 2004). This spike was associated with a loss of mobilization which lasted several minutes. The origin of this calcium spike was not determined at that time, but was later shown to be due to the release of calcium from subcellular stores as well as influx from the medium through a nanoporated plasma membrane. As expected from physical models, the effect on subcellular structures was stronger with shorter pulses (Semenov et al. 2013).

1.7 Towards Medical Treatments

The discovery of apoptosis induction through nanosecond pulses led to studies on their use as cancer treatment therapies. Since apoptosis is suppressed in tumors, stimulating apoptosis can therefore be considered a means to suppress tumor growth and cancer development. A paper by Beebe et al. in 2003 (Beebe et al. 2003a) showed ex vivo indications of controlled destruction of tumor tissue through nanosecond pulses. In 2005 the first in vivo studies confirmed that the nanosecond pulse technology might lead to a purely physical, apoptosis-based **cancer therapy**. The results were published in 2006 (Nuccitelli et al. 2006; Schoenbach et al. 2006). First surface photographs of the melanoma tumor regression following nanosecond pulse treatment in mice are shown in Fig. 1.15. Following the first publications on tumor treatment with nanosecond pulses in 2006, more than 60 papers have been published on this topic. An important result was that only a single treatment of melanoma tumors in mice with 50, 100 ns pulses at 30 kV/cm was required to obtain complete regression of the tumor (Nuccitelli et al. 2010).

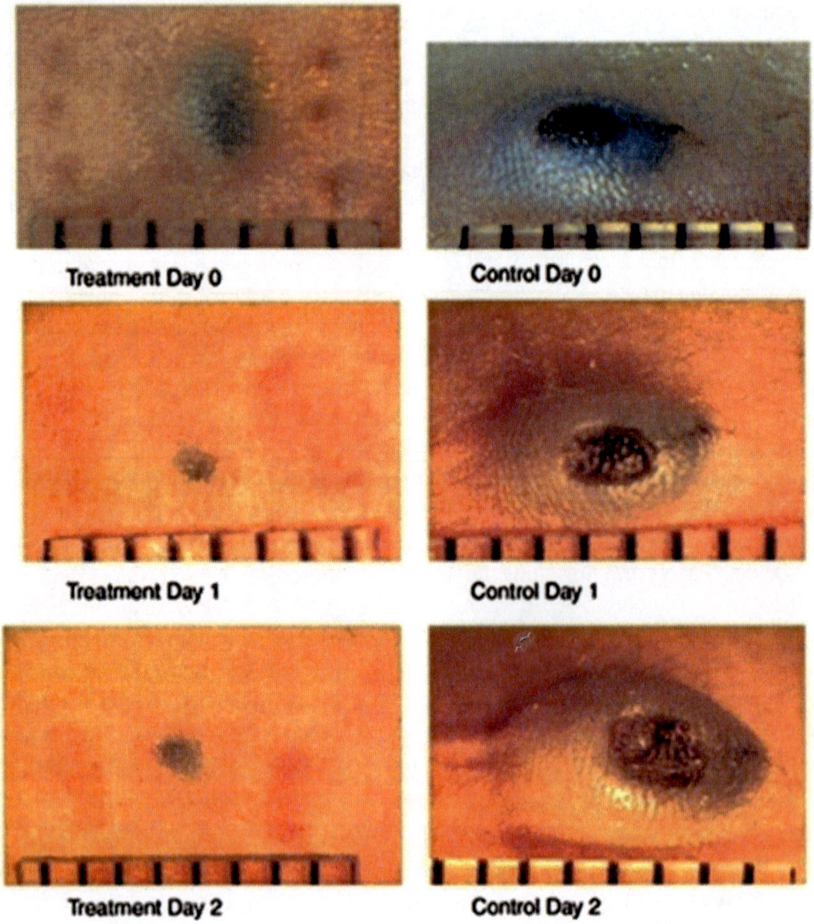

Fig. 1.15 Surface photographs of the tumor development in a mouse over three days. The left row of photographs shows the response of a melanoma to 1000, 300 ns, 10 kV pulses at three locations. The pulses were applied with needle electrodes. The three locations are visible in the top left photograph. After two days the tumor regressed dramatically, whereas the control tumor, shown in the right row of photographs, grew continuously (Schoenbach 2018)

Over the past years research on nanosecond pulse cancer treatment has led to a very exciting discovery: that nanosecond pulsed electric fields don't only just ablate tumors, but they also induce a protective, vaccine-like effect. The first observation of this effect dates back to 2011 (Beebe et al. 2011). When mice carrying Hepa1-6 subcutaneous liver tumors were treated with nanosecond pulsed electric fields and the tumors were eliminated by this treatment, the mice did not develop tumors when they were challenged with a second injection of cancer cells. This was the first indication that nsPEFs could induce an immune response. Other suggestive evidence for an immune response was shown when nsPEF treatment was superior to tumor excision at accelerating secondary tumor rejection in these mice (Nuccitelli et al. 2012). Another study suggested an immune response was CD8-dependent (Nuccitelli et al. 2015).

In a larger study following the early 2011 paper (Beebe et al. 2011), nsPEFs induced a vaccine-like effect was shown again when nsPEF-ablated, tumor-free rats did not develop tumors after being challenged with second injections of the same cancer cells (Chen et al. 2014). More recently, nsPEF treatment of mouse breast cancer tumors have demonstrated the first direct evidence for nsPEF-induced immune responses by specifically defining several immune mechanisms (Guo et al. 2017).

Wound healing is another important application of nanosecond pulsed electric field effects. It is based on platelet aggregation. Both intracellular calcium released from the stores and calcium influx through the nanoporated membrane were found to cause platelet activation and aggregation (Fig. 1.16). This effect was successfully used for wound healing. It must be noted that as an additional effect, growth factors were released from alpha granules in platelets after pulsing (Zhang et al. 2008).

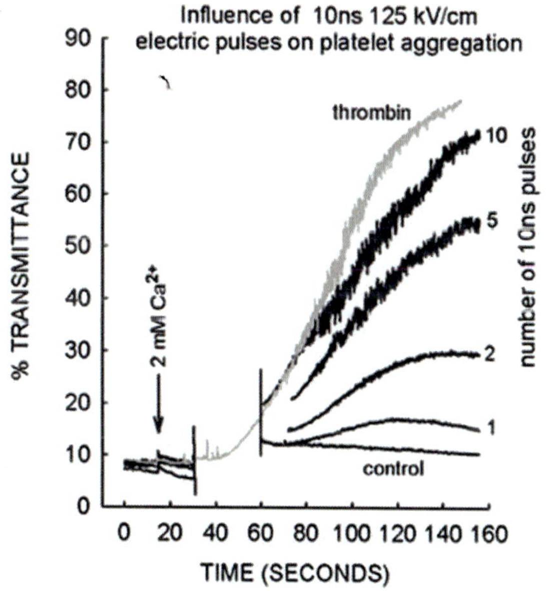

Fig. 1.16 Influence of 10 ns, 125 kV/cm electrical pulses on platelet aggregation (parameter is the pulse number) and comparison to the effect of thrombin. For aggregation studies 0.6 ml of washed platelets were added to a glass aggregometer. One minute before aggregation measurements were performed 2.0 mM Ca^{2+} was added to the platelet suspension, a baseline optical measurement was collected, then after 15 s the entire contents of the aggregation cuvette was removed with a glass Pasteur pipette into a 0.2 mm pulsing cuvette. The pulsing cuvette was then pulsed once or multiple times. The platelets were then removed from the pulsing cuvette back into the aggregation cuvette and optical measurements continued (Schoenbach 2018)

Besides using platelet-rich plasma activated by nsPEF for accelerated healing of skin lesions, more recent studies have shown cardioprotective effects of nanosecond pulse activated platelet rich plasma (nsPRP) on the ischemic heart during reperfusion (Hargrave et al. 2016). Studies on rabbit and mouse hearts have shown improvement in left ventricular pumping. Also, it could be shown that nsPRP injection restores electrical activity even before reperfusion begins.

Another important emerging application of nsPEF induced cell death is **treatment of cardiac arrhythmia**. Most common treatments involve surgical ablation of cardiac tissue during open heart surgery. The goal is to create nonconducting lesions which intercept pathways of recurring arrhythmias or isolate sources of arrhythmia. Current ablation technologies are either based on heating (RF) technologies or on cryoablation. Both methods suffer from substantial recurrence rates, thermal side effects, and long procedure times. nsPEF, without heating the tissue, creates more uniform ablation and allows us to reduce the time of treatment from minutes to seconds (Xie et al. 2015).

1.8 Scaling of Nanosecond Pulse Effects

The effects of nanosecond and shorter pulses on biological cells have been found to be dependent on a number of pulse parameters: pulse duration, pulse amplitude, pulse number, and pulse frequency. For moderate pulse frequencies, defined as frequencies where a sequence of pulses cause only minor increases in temperature through Joule heating, the last parameter is likely irrelevant. Measurements of the temperature increase in mouse skin, showed an increase in temperature in the mouse skin by 4 ^{0}C when 18 pulses of 300 ns duration, at an average electric field of 17 kV/cm were applied at a repetition frequency of 0.5 Hz (Pliquett and Nuccitelli 2014). Although Joule heating in in vitro studies, due to better thermal conductance of the medium, likely plays a lesser role, it is important to consider thermal effects at increased repetition rates.

On the other hand, if thermal effects can be excluded by limiting the number of pulses, varying the repetition rate of nanosecond pulses might allow us to explore and possibly control the uptake of molecules. This has been shown by recording the uptake of Propidium Iodide, FM 1-43, and YO-PRO-1 in CHO-K1 cells over a range of pulse repetition frequency from 5 Hz to 500 kHz (Steelman et al. 2016). However, even without taking the effect of pulse frequency into account, the electrical parameters of the pulse, duration, rise time and amplitude, and the pulse number cover a wide range.

Studies of nanosecond pulse effects varying the three parameters, pulsed electric field, E, pulse duration, τ, and pulse number, n, seem to indicate that the total number of ions passing through the cell membrane is the determining parameter for nanosecond pulse effects. In other words, the magnitude of bioeffects depends on the product of the current density (which, according to Ohm's law, for constant conductivity is linearly dependent on the electric field intensity) and the pulse duration.

Instead of scaling with energy density, it scales with electrical charge. Nanosecond bioelectric effects seems therefore to scale with the product of electric field, E, and pulse duration, τ. In addition, the dependence on the number of pulses seem to follow a square root law (Schoenbach et al. 2009). The square root dependence on the pulse number in in vitro studies can be considered the result of the thermally initiated statistical rotation of cells with respect to the electric field direction.

The scaling law allows us to vary pulse parameters for a certain biological application. As long as the square root of the product of electric field intensity and pulse duration is identical, identical or close in vitro results could be expected for the same cell type. It needs to be noted that this scaling law for nanosecond pulses has been derived for monopolar pulses. For pulses of a different shape, e.g. for bipolar pulses (Pakhomov et al. 2014), it needs to be modified.

1.9 From Nanosecond to Picosecond Pulses

Generating ever higher amplitude electrical pulses when moving further into the ultrashort pulse regime has in the past been challenging. However, modern pulsed power technology is now so advanced that pulses into the range of 100 ps duration and amplitudes of tens of kV can be produced (Schoenbach and Xiao 2016). The use of such picosecond pulses allows us to enter a range of electrical field-cellular membrane coupling with new bioelectric applications. By applying such ultrashort pulses it also becomes possible to use wideband antennas, rather than direct contact electrodes, to deliver the pulses to tissue.

When the pulse duration is reduced from nanoseconds to picoseconds, increasingly the dielectric properties, rather than the resistive characteristics of the media determine the electric field distribution (Schoenbach et al. 2008). Generally, in modeling the effect of pulsed electric fields on cells, the capacitive component of cytoplasm and medium are neglected. These assumptions are only valid for a pulse duration that is long compared to the dielectric relaxation time (ε/σ) of cytoplasm and medium, where ε is the permittivity and σ the conductivity. For pulses short compared to the dielectric relaxation time of the cytoplasm and the medium, the coupling of the electric field with the cell is defined by the permittivity of the cell components and the cell environment. For a membrane with a relative dielectric constant of 8, the electric field in the membrane is approximately 10 times higher than the electric field in the medium, which has a dielectric constant of about 80. The electric field then acts directly on the lipid bilayer and the embedded membrane proteins, rather than causing charging of the membrane. If sufficiently strong, it can cause direct and instant conformational changes in membrane proteins.

Another bioelectric effect which becomes relevant when the pulse duration is reduced into the lower nanosecond and subnanosecond range is heating of the membranes through dielectric relaxation. Dielectric spectroscopy on multilamellar liposomes has shown dispersion frequencies related to the diffuse thermal rotational motion of the phosphatidylcholine headgroup and to bound water (Kloesgen et al.

1996). Dielectric relaxation in the frequency range of 10–100 MHz consequently leads to increased energy deposition in the plasma membrane in this frequency range (Kotnik and Miklavcic 2000). Considering the frequency distribution of square wave pulses, such dielectric relaxation effects are expected for nanosecond, and even more for picosecond pulses. This energy deposition leads to a spike in membrane temperature and possibly to changes in the lipid bilayer (Croce et al. 2010).

Biological studies with picosecond-range pulses have focused less on inducing cell death, but more on neural stimulation. Studies on primary rat hippocampus neurons exposed to subnanosecond pulses were performed using a patch clamp method, where a constant current was injected to the neurons before the pulses were applied. It could be shown that pulsed electric fields of just 10 kV/cm (100 pulses at a repetition rate of 500 Hz) caused membrane depolarization, which culminated in an action potential at injection currents of − 80 pA (Xiao et al. 2013). More recent studies on different cell types indicate that even a single pulse, at an electrical field far below that required for electroporation, can cause an increase in cytosolic calcium. The results indicate that such picosecond pulses result in long-lasting opening of voltage gated calcium channels (Semenov et al. 2015).

A practical reason for entering the sub-nanosecond pulse range is the possible use of antennas as pulse delivery systems instead of electrodes. These antenna systems initially studied at the Frank Reidy Research Center for Bioelectrics have been based on the use of a prolate spheroidal reflector, where the pulse is launched from one focal point and reflected into a second focal point (Xiao et al. 2010). The concept and a photograph of such a focusing antenna used in our studies is shown in Fig. 1.17. The reason to use pulses with a risetime of picoseconds, which provide electric field pulses of the same duration as the risetime at the target, is the size of the focal volume, i.e., the small spatial resolution achievable with such an antenna. The shorter the pulse

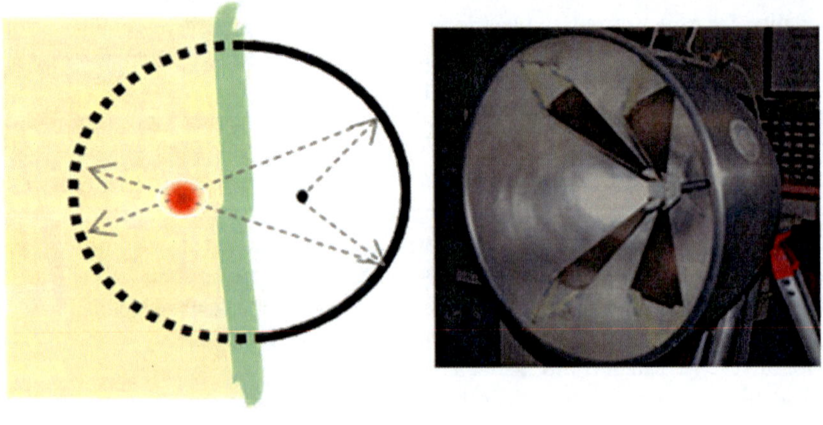

Fig. 1.17 Left: Concept of a focusing antenna using a prolate-spheroidal reflector. The reflector focuses the electromagnetic radiation into the tissue. The focal volume in the tissue is shown in red. Right: photograph of an antenna with a prolate-spheroidal reflector

duration and the higher the relative permittivity of the medium, the smaller is the spot size, and consequently the higher is the spatial resolution. For a medium with relative permittivity of 80 such as water, the pulse duration needs to be in the one hundred picosecond range to obtain a spatial resolution of less than one centimeter.

Whereas the field which deals with picosecond pulse effects is still in a nascent state, an increasing number of studies on biological effects and new developments in reducing the size of the antenna into the centimeter range (Petrella et al. 2017; Xiao et al. 2020) promise to open up a new area of studies on electric field–cell interactions. What is also intriguing in using picosecond electric fields is the possibility to apply extremely large electric fields and not to cause thermal effects because of the lower energy of the short pulses. A pulse generator developed at the Frank Reidy Research Center for Bioelectrics provides 700 ps pulses at voltages of up to 250 kV (Heeren et al. 2007). I It is possible to generate electric fields in excess of 1 MV/cm in a 2 mm cuvette without generating electrical sparks. This might allow the study of the effect of such pulses on molecular structures, opening a new field of research in bioelectrics.

1.10 New Research Directions

Whereas most of the studies with nanosecond pulses focused on the effects of the plasma membrane and intracellular membranes, only a few studies (Stacey et al. 2003) addressed the question of the effects on macromolecules. Data obtained using the comet assay (Lehmann et al. 1998) show clearly the induction of DNA damage in Jurkat and HL-60 cells exposed to 1, 5, and 10 pulses of 60 ns at 60 kV/cm. Interestingly, the application of multiple pulses (5 and 10) caused little extra damage compared to that observed with a single pulse. Measurements of the nuclear speckle change following the application of 10 ns, 150 kV/cm pulses indicated that such pulses directly influence nuclear processes, such as changes in the nuclear RNA–protein complexes (Chen et al. 2007). More recent measurements focused on the direct impact of even shorter pulses (200 ps) on amyloid fibrils, which are associated with neurodegenerative diseases, such as Parkinson's and Alzheimer's. It was shown by using electron microscopy that bursts of 16,000, 200 ps long pulses at electric field intensities between 40 and 120 kV/cm cause fragmenting of the amyloid-beta fibrils (Schoenbach and Greene 2015).

Thermal effects, although generally avoided in bioelectric studies, have their place in bioelectrics, even for pulses short compared to the dielectric relaxation time of the membrane. Simultaneously pulsing and heating cells, either by utilizing the Joule heat generated through high repetition rate pulsing (Xiao et al. 2011) or using pulse-independent methods such as microwave heating, allow us to reduce the electrical energy required for inducing cell death by a considerable amount. This has been shown recently in a study where 200-ps long pulses were applied to a slightly heated cell suspension (Camp et al. 2012). Cell death, as recorded by means of trypan blue uptake, increased considerably for heating above physiological temperature. The results, which are in more detail described in Chaps. 12 and 13, indicate that thermal assistance of electro-technologies, such as tumor ablation and gene therapy, will allow us to reduce the electrical energy (pulse amplitude) and consequently reduce tissue damage.

1.11 Concluding Remarks

According to a review on the effects of high-voltage nanosecond and picosecond electrical pulses on eukaryotic cells, published in 2016 (Napotnik et al. 2016), at least 203 papers on this topic have been published. Several of them are review papers on the ultrashort pulse effects on cells and tissue (Schoenbach et al. 1997a, 2002, 2004, 2007; Schoenbach 2010; Chopinet and Rols 2015; Napotnik et al. 2016). There are also two books which in part cover important aspects of the basics and application of bioelectric effects of ultrashort pulses (Pakhomov et al. 2010; Akiyama and Heller 2017). This book, however, focuses solely on the history, the discoveries, the efforts in understanding and modeling of the ultrashort pulse effects, and the exciting applications which have emerged from these studies. It is written by four of the scientists and engineers who were working in this field from its very beginning in the late 90s and still are involved in this particular bioelectrics research. We hope that it provides not just an inside in the physics, biology, and clinical applications of ultrashort pulse effects in bioelectrics, but also demonstrates how this field developed over the past two decades.

References

Akiyama H, Heller R (eds) (2017) Bioelectrics. Springer Nature, Japan

Beebe SJ, Fox PM, Rec LJ, Somers K, Stark RH, Schoenbach KH (2002) Nanosecond pulsed electric field (nsPEF) effects on cells and tissues: apoptosis induction and tumor growth inhibition. IEEE Trans Plasma Sci 30:286–292

Beebe SJ, White J, Blackmore PF, Deng Y, Somers K, Schoenbach KH (2003) Diverse effects of nanosecond pulsed electric fields on cells and tissues. DNA Cell Biol 22:785–796

Beebe SJ, Fox PM, Rec LJ, Willis LK, Schoenbach KH (2003) Nanosecond, high intensity pulsed electric fields induce apoptosis in human cells. FASEB J 17:1493–1495

Beebe SJ, Blackmore PF, White J, Joshi RP, Schoenbach KH (2004) Nanosecond pulsed electric fields modulate cell functions through intracellular signal transduction mechanisms. Physiol Meas 25:1077–1093

Beebe SJ, Ford WE, Ren W, Chen X (2011) Pulse power ablation of melanoma with nanosecond pulsed electric fields. In: Morton R (ed) Tech croatia, treatment of metastatic melanoma, pp 231–268. ISBN 978-953-307-574-7

Bennet WFD, Sapay N, Tieleman DP (2014) Atomistic simulations of pore formation and closure in lipid bilayers. Biophys J 106:210–219

Benz R, Zimmermann U (1980) Relaxation studies on cell membranes and lipid bilayers in the high electric field range. Bioelectrochem Bioenerg 7:723–739

Benz R, Zimmermann U (1980) Pulse-length dependence of the electrical breakdown of lipid bilayer membranes. Biochim Biophys Acta 597:637–642
Benz R, Zimmermann U (1981) The resealing process of lipid bilayers after reversible electrical breakdown. Biochim Biophys Acta 640:169–178
Buescher ES, Schoenbach KH (2003) The effects of submicrosecond, high intensity pulsed electric fields on living cells—intracellular electromanipulation. IEEE Trans Dielectr Elec Insul 10:788–794
Buescher ES, Smith RR, Schoenbach KH (2004) Submicrosecond intense pulsed electric field effects on intracellular free calcium: mechanisms and effects. IEEE Trans Plasma Sci 32:1563–1572
Camp JT, Jing Y, Zhuang J, Zhanrgen K, Beebe SJ, Song J, Joshi RP, Xiao S, Schoenbach KH (2012) Cell death induced by subnanosecond pulsed electric fields at elevated temperatures. IEEE Trans Plasma Sci 40(10):2334–2347
Chen N, Schoenbach KH, Kolb JF, Swanson RJ, Garner AL, Yang J, Joshi RP, Beebe SJ (2004) Leukemic cell intracellular responses to nanosecond electric fields. Biochem Biophys Res Commun 317:421–427
Chen N, Garner AL, Chen G, Jing Y, Deng Y, Swanson RJ, Kolb JF, Beebe SJ, Joshi RP, Schoenbach KH (2007) Nanosecond electric pulses penetrate the nucleus and enhance speckle formation. Biochem Biophys Res Commun 364:220–225
Chen R, Sain NM, Harlow KT, Chen YJ, Shires PK, Heller R, Beebe SJ (2014) A protective effect after clearance of orthotopic rat hepatocellular carcinoma by nanosecond pulsed electric fields. Eur J Cancer 50:2705–2713
Chopinet L, Rols MP (2015) Nanosecond electric pulses: a mini-review of the present state of the art. Bioelectrochemistry 103:2–6
Cole KS (1937) Electric impedance of marine egg membranes. Trans Faraday Soc 23:966–972
Croce RP, De Vita A, Pierro V, Pinto IM (2010) A thermal model for pulsed EM field exposure effects in cells at nonthermal levels. IEEE Trans Plasma Sci 38:149–155
Daud AI, DeConti RC, Andrews S, Urbas P, Riker AI, Sondak VK, Munster PN, Sullivan DM, Ugen KE, Messina JL, Heller R (2008) Phase I trial of interleukin-12 plasmid electroporation in patients with metastatic melanoma. J Clin Oncol 26:5896–5903
Davalos RV, Mir L, Rubinski B (2005) Tissue ablation with irreversible electroporation. Ann Biomed Eng 33:223–231
Deng J, Schoenbach KH, Buescher ES, Hair PS, Fox PM, Beebe SJ (2003) The effects of intense submicrosecond electrical pulses on cells. Biophys J 84:2709–2714
Dioscorides (512) Dioscorides, pedanius, of anazarbos: the greek herbal of dioscorides, illustrated by a Byzantine AD 512. Translated by John Goodyer AD 1655. Hafner, London 1968, Facsimile of 1934 edition
Eppich HM, Foxall R, Gaynor K, Dombkowski D, Miura N, Cheng T, Silva-Arrieta S, Evans RH, Mangano JA, Preffer FI, Scadden DT (2000) Pulsed electric fields for selection of hematopoietic cells and depletion of tumor cell contaminants. Nat Biotechnol 18:882–887
Foster KR, Schwan HP (1996) Dielectric properties of tissues. In: Polk C, Postow E (eds) Handbook of biological effects of electromagnetic fields. CRC Press, pp 25–102
Frandsen SK, Gissel H, Hojman P, Tramm T, Eriksen J, Gehl J (2012) Direct therapeutic applications of calcium electroporation to effectively induce tumor necrosis. Cancer Res 72:1336–1341
Frey W, White JA, Price RO, Blackmore PF, Joshi RP, Nuccitelli R, Beebe SJ, Schoenbach KH, Kolb JF (2006) Plasma membrane voltage changes during nanosecond pulsed electric field exposure. Biophys J 90:3608–3615
Frey W, Gusbeth C, Sakugawa T, Sack M, Mueller G, Sigler J, Vorobiev E, Lebovka N, Alvarez I, Raso J, Heller LC, Malik MA, Eing C, Teissie J (2017) Environmental applications, food and biomass processing by pulsed electric fields. In: Akiyama H, Heller R (eds) "Bioelectrics", chap 6. Springer Nature, Japan, pp 389–476
Gabriel B, Teissie J (1997) Direct observations in the millisecond time range of fluorescent molecular asymmetrical interaction with the electropermeabilized cell membrane. Biophys J 73:2630–2637

Galluzi L, Vitale I, Abrams JM, Alnemri ES, Baehrecke EH, Blagosklonny MV, Dawson TM, Dawson VL, El-Deiry WS, Fulda S, Gottlieb E, Green DR, Hengartner MO, Kepp O, Knight RA, Kumar S, Lipton SA, Lu X, Madeo F, Malorni W, Mehlen P, Nunez G, Peter ME, Piacentini M, Rubinsztein DC, Shi Y, Simon HU, Vandenabeele P, White E, Yuan J, Zhivotovsky B, Melino G, Kroemer G (2012) Molecular definitions of cell death subroutines: recommendations of the nomenclature committee on cell death. Cell Death Differ 19:107–120

Ghazala A, Schoenbach KH (2000) Biofouling prevention with pulsed electric fields. IEEE Trans Plasma Sci 28:115–121

Glaser RW, Leikin SL, Chernomordik LV, Pastushenko VF, Sokirko AI (1988) Reversible electrical breakdown of lipid bilayers: formation and evaluation of pores. Biochim Biophys Acta 940:275–287

Gowrishankar TR, Weaver JC (2003) An approach to electrical modeling of single and multiple cells. Proc Nat Acad Sci 100:32033208

Gowrishankar TR, Esser AT, Vasilkoski Z, Smith KC, Weaver JC (2006) Microdosimetry for conventional and supra-electroporation in cells with organelles. Biochem Biophys Res Commun 341:1266–1276

Guo S, Jing Y, Burcus NI, Lassiter BP, Tanaz R, Heller R, Beebe SJ (2017) Nano-pulse stimulation induces potent immune responses, eradicating local breast cancer while reducing distant metastases. Int J Cancer. https://doi.org/10.1002/ijc.31071

Hair PS, Schoenbach KH, Buescher ES (2003) Sub-microsecond, Intense pulsed electric field applications to cells show specificity of effects. J Bioelectrochem 61:65–72

Hargrave B, Varghese F, Barabutis N, Catravas J, Zemlin CW (2016) Nanosecond pulsed platelet-rich plasma (nsPRP) improves mechanical and electrical cardiac function following myocardial reperfusion injury. Physiol Rep 4:12710

Heeren T, Camp JT, Kolb JF, Schoenbach KH, Katsuki S, Akiyama H (2007) 250 kV subnanosecond pulse generator with adjustable pulsewidth. IEEE Trans Dielectric Electric Insul 14:884–888

Heller R, Jaroszeski M, Atkin A, Moradpour D, Gilbert R, Wands J (1996) Nicolau, C: in vivo gene electroinjection and expression in rat liver. FEBS Lett 389:225–228

Heller R, Jaroszeski M, Perrott R, Messina J, Gilbert R (1997) Effective treatment of B16 melanoma by direct delivery of bleomycin using electrochemotherapy. Melanoma Res 7:10–18

Hu Q, Viswanadham S, Joshi RP, Schoenbach KH, Beebe SJ, Blackmore PF (2005) Simulations of transient membrane behavior in cells subjected to a high-intensity ultrashort electrical pulse. Phys Rev E 71:031914-1–31919

Hu Q, Joshi RP, Schoenbach KH (2005a) Simulations of nanopore formation and phosphatidylserine externalization in lipid membranes subjected to a high-intensity, ultrashort electric pulse. Phys Rev E 72:031902-1–031902-10

Jiang C, Davalos RV, Bischof JC (2015) A review of basic to clinical studies of irreversible electroporation therapy. IEEE Trans Biomed Eng 62:4–20

Joshi RP, Hu Q, Schoenbach KH (2004) Modeling studies of cell response to ultrashort high-intensity electric fields—implications for intracellular manipulation. IEEE Trans Plasma Sci 32:1677–1686

Kellaway P (1946) The part played by electric fish in the early history of bioelectricity and electrotherapy. Bull Hist Med 20:112

Kinosita K, Tsong TY (1977) Voltage-induced pore formation and hemolysis of human Eerythrocytes. Biochim Biophys Acta 471:227–242

Kloesgen B, Reichle C, Kohlsmann S, Kramer KD (1996) Dielectric spectroscopy as a sensor of membrane headgroup mobility and hydration. Biophys J 71:3251–3260

Kolb JF, Kono S, Schoenbach KH (2006) Nanosecond pulse electric field generators for the study of subcellular effects. Bioelectromagnetics 27:172–187

Kotnik T, Miklavcic D (2000) Theoretical evaluation of the distributed power dissipation in biologicalcells exposed to electric fields. Bioelectromagnetics 21:385–394

Largus K (1528) Scribonius: De compositionibus medicamentorum. Liber unius, antehac nusquarn excusus: Joanne Ruellio, Paris

Lehmann J, Pollet D, Peker S, Steinkraus V, Hoppe U (1998) Kinetics of DNA strand breaks and protection by antioxidants in UVA- or UVB-irradiated HaCaT keratinocytes using the single cell electrophoresis assay. Mutat Res 407:97–108

Leibowitz JO (1957) Electroshock therapy in Ibn-Sina's Canon. J Hist Med 12:71

Mir LM, Belehradek M, Domenge C, Orlowski S, Poddevin B, Belehradek J Jr, Schwaab G, Luboinski B, Paoletti C (1991) Electrochemotherapy, a new antitumor treatment: first clinical trial. C R Acad Sci III 313:613–618

Napotnik TB, Rebersek M, Kotnik T, Lebrasseur E, Cabodevila G, Miklavcic D (2010) Electropermeabilization of endocytic vesicles in B16 F1 mouse melanoma cells. Med Biol Eng Comput 48:407–413

Napotnik TB, Rebersec M, Vernier PT, Mali B, Miklavcic D (2016) Effects of high voltage electric pulses on eukaryotic cell (in vitro): a systematic review. Bioelectrochemistry 110:1–12

Neumann E, Rosenheck K (1972) Permeability changes induced by electric impulses in vesicular membranes. J Membr Biol 10:279–290

Neumann E, Schaefer-Ridder M, Wang Y, Hofschneider PH (1982) Gene transfer into mouse lyoma cells by electroporation in high electric fields. EMBO J 1:841–845

Nuccitelli R, Pliquett U, Chen X, Ford W, Swanson RJ, Beebe SJ, Kolb JF, Schoenbach KH (2006) Nanosecond pulsed electric fields cause melanomas to self-destruct. Biochem Biophys Re Commun 343:351–360

Nuccitelli R, Tran K, Sheikh S, Athos B, Kreis M, Nuccitelli P (2010) Optimized nanosecond pulsed electric field therapy can cause murine malignant melanomas to self-destruct with a single treatment. Int J Cancer 127:1727–1736

Nuccitelli R, Tran K, Lui K, Huynh J, Athos B, Kreis M, Nuccitelli P, De Fabio EC (2012) Non-thermal nanoelectroablation of UV-induced murine melanomas stimulates an immune response. Pigment Cell Melanoma Res 25:618–629

Nuccitelli R, Berridge JC, Mallon Z, Kreis M, Athos B, Nuccitelli P (2015) Nanoelectroablation of murine tumors triggers a CD8-dependent inhibition of secondary tumor growth. PLoS One 10(7):e0134364

Okino M, Mohri H (1987) Effects of a high-voltage electrical impulse and an anticancer drug on in vivo growing tumors. Jap J Cancer Res 78:1319–1321

Pakhomov AG, Shevin R, White J, Kolb JF, Pakhomova ON, Joshi RP, Schoenbach KH (2007) Membrane permeabilization and cell damage by ultrashort electric field shocks. Arch Biochem Biophys 465:109–118

Pakhomov AG, Bowman AM, Ibey BI, Andre FM, Pakhomova ON, Schoenbach KH (2009) Lipid nanopores can form a stable, ion channel-like conduction pathway in cell membrane. Biochem Biophys Commun 385:181–186

Pakhomov AG, Miklavcic D, Markov MS (eds) (2010) Advanced electroporation techniques in biology and medicine. CRC Press, Taylor and Francis, Bota Racon, FL

Pakhomov AG, Semenov I, Xiao S, Pakhomova ON, Gregory B, Schoenbach KH, Ullery JC, Beier HT, Rajulapati SR, Ibey BL (2014) Cancellation of cellular responses to nanoelectroporation by reversing the stimulus polarity. Cell Mol Life Sci 71:4431–4441

Petrella R, Schoenbach KH, Xiao S (2017) A dielectric rod antenna driven by pulsed power system for stimulating biological tissues. IEEE Trans Dielectric Electric Insul 24:2157–2163

Pliquett U, Nuccitelli R (2014) Measurement and simulation of Joule heating during treatment of B-16 melanoma tumors in mice with nanosecond pulsed electric fields. Bioelectrochemistry 100:62–68

Rems E, Tarek M, Casciola M, Miklavcic D (2016) Properties of lipid electropores II: comparison of continuum-level modeling of pore conductance to molecular dynamics simulations. Bioelectrochemistry 112:112–124

Schoenbach KH, Greene L (2015) Method and device for treatment of conditions aggravated by amyloid fibrils. US Patent, No. 8,948,878 B2

Schoenbach KH (2010) Bioelectric effects of intense nanosecond pulses. In: Pakhomov AG, Miklavcic D, Markov MS (eds) Advanced electroporation techniques in biology and medicine. CRC Press, Taylor and Francis Group, Bota Racon, FL, pp 19–49
Schoenbach KH (2018) From the basic science of biological effects of ultrashort electrical pulses to medical therapies. Bioelectromagnetics 39(4):257–276
Schoenbach KH, Xiao S (2016) Nanosecond pulses and beyond—towards antenna applications. In: Jarm T, Kramar P (eds) 1st world congress on electroporation and pulsed electric fields in biology, medicine and food and environmental technologies. IFMBE Proceedings, vol 53. Springer, Singapore
Schoenbach KH, Peterkin FE, Alden RW, Beebe SJ (1997a) The effect of pulsed electric fields on biological cells: experiments and applications. Trans Plasma Sci 25:284–292
Schoenbach KH, Abou-Ghazala A, Vithoulkas T, Alden RW, Turner R, Beebe SJ (1997b) The effect of pulsed electric fields on biological cells. In: Cooperstein C, Vitkovitsky I (eds) Proceedings of 11th IEEE international pulsed power conference, Baltimore, ML, pp 73–78
Schoenbach KH, Beebe JS, Buescher ES (1999) Biological effects of high power, microsecond and submicrosecond electrical pulses. ElectroMed99, Norfolk, VA, Symposium Record Abstracts, pp 35
Schoenbach KH, Beebe SJ, Buescher ES (2001) Intracellular effect of ultrashort electrical pulses. Bioelectromagnetics 22:440–448
Schoenbach KH, Katsuki S, Stark RH, Beebe S, Buescher S (2002) Bioelectrics—new applications for pulsed power technology. IEEE Trans Plasma Sci 30:293–300
Schoenbach KH, Joshi RP, Kolb JF, Chen N, Stacey M, Blackmore PF, Buescher ES, Beebe SJ (2004) Ultrahort electrical pulses open a new gateway into biological cells. Proc IEEE 92:1122–1137
Schoenbach KH, Nuccitelli R, Beebe SJ (2006) ZAP: Extreme voltage could be a surprisingly delicate tool in the fight against cancer. IEEE Spectr 43:20–26
Schoenbach KH, Hargrave B, Joshi RP, Kolb JF, Nuccitelli R, Osgood C, Pakhomov A, Stacey M, Swanson RJ, White JA, Xiao S, Zhang J, Beebe SJ, Blackmore PF, Buescher ES (2007) Bioelectric effects of intense nanosecond pulses. IEEE Dielectric Electric Insul 14:1088–1109
Schoenbach KH, Xiao S, Joshi RP, Camp JT, Heeren T, Kolb JF, Beebe SJ (2008) The effect of intense subnanosecond electrical pulses on biological cells. IEEE Trans Plasma Sci 36:414–424
Schoenbach KH, Baum CE, Joshi RP, Beebe SJ (2009) A scaling law for membrane permeabilization with nanopulses. IEEE Trans Dielectric Electric Insul 16:1224–1235
Semenov I, Xiao S, Pakhomova ON, Pakhomov AG (2013) Recruitment of the intracellular Ca2+ by ultrashort electric stimuli: the impact of pulse duration. Cell Calcium 54:145–150
Semenov I, Xiao S, Kang D, Schoenbach KH, Pakhomov A (2015) Cell stimulation and calcium mobilization by picosecond electrical pulses. Bioelectrochemistry 105:65–71
Sersa G, Cemazar M, Miklavcic D (1995) Antitumor effectiveness of electrochemotherapy with cis-diamminedichloroplatinum(II) in mice. Cancer Res 55:3450–3455
Son RS, Smith KC, Gowrishankar TR, Vernier PT, Weaver JC (2014) Basic features of a cell electroporation model: illustrative behavior for two very different pulses. J Membrane Biol 247:1209–1228
Stacey M, Stickley J, Fox P, Statler V, Schoenbach K, Beebe SJ, Buescher S (2003) Differential effects in cells exposed to ultra-short, high intensity electric fields: cell survival, DNA damage, and cell cycle analysis. Mutat Res 542:65–75
Staempfli R (1958) Reversable breakdown of excitable membrane of a Ranvier node. Ann Acad Bras Ciencias 30:57
Steelman ZA, Tolstykh GP, Beier HT, Ibey BL (2016) Cellular response to high pulse repetition rate nanosecond pulses varies with fluorescent marker identity. BBRC 478:1261–1267
Stillings D (2004) The theology of electricity: electricity, alchemy, and the unconscious. In: Rosch PJ, Markov MS (eds) Bioelectromagnetic medicine. Marcel decker Inc., New York, pp 61–75
Sugar IP, Neumann E (1984) Stochastic model for electric field-induced membrane pores electroporation. Biophys Chem 19:211–225

Tarek M, Delemotte L (2010) Electroporation of lipid membranes: insights from molecular dynamics simulations. In: Pakhomov A, Miklavcic D, Markov MS (eds) Chapter II-7 Advanced electroporation techniques in biology and medicine. CRC Press, Taylor and Francis Group, Bota Racon, FL

Tekle E, Oubrahim H, Dzekunov SM, Kolb JF, Schoenbach KH, Chock PB (2005) Selective field effects on intracellular vacuoles and vesicle membranes with nanosecond electric pulses. Biophys J 89:274–284

Tieleman DP, Leontiadou H, Mark AE, Marrink SJ (2003) Simulation of pore formation in lipid bilayers by mechanical stress and electric fields. J Am Chem Soc 125:6382–6383

Titomirov AV, Sukharev S, Kistanova E (1991) In vivo electroporation and stable transformation of skin cells of newborn mice by plasmid DNA. Biochim Biophys Acta 1088:131–134

Vasilkoski Z, Esser AT, Gowrishankar TR, Weaver JC (2006) Membrane electroporation: the absolute rate equation and nanosecond time scale pore creation. Phys Rev E 74:021904–1–21912

Vernier PT (2010) Nanoscale restructuring of lipid bilayers in nanosecond electric fields. In: Pakhomov AG, Miklavcic D, Markov MS, (eds) Chapter II- 8 in advanced electroporation techniques in biology and medicine. CRC Press, Taylor and Francis Group, Bota Racon, FL

Vernier PT, Sun Y, Marcu L, Salemi S, Craft CM, Gundersen MA (2003) Calcium bursts induced by nanosecond electrical pulses. Biochem Biophys Commun 310:286–295

Vernier PT, Li A, Marcu L, Craft CM, Gundersen MA (2003) Ultrashort pulsed electric fields induce membrane phospholipid translocation and caspase activation: differential sensitivities of Jurkat T lymphoblasts and rat glioma C6 cells. IEEE Trans Dielectric Electric Insul 10:795–803

Vernier PT, Ziegler MJ, Sun Y, Gundersen MA, Tielemann DP (2006) Nanopore-facilitated, voltage-driven translocation in lipid bilayers—in cells and in silico. Phys Biol 3:233–247

Weaver JC, Powell KT, Mintzer RA, Sloan SR, Ling H (1984) The diffuse permeability of bilayer membranes: the contribution of transient aqueous pores. Bioelectrochem Bioenerg 12:405–412

Wickland CA (1968) Thirty years among the dead. Spiritual Press, London

Wong TK, Neumann E (1982) Electric field mediated gene transfer. Biophys Biochem Res Commun 107:548–587

Xiao S, Altunc S, Kumar P, Baum CE, Schoenbach KH (2010) A reflector antenna for focusing in the near field. IEEE Antennas Wirel Propag Lett 9:12–15

Xiao S, Guo S, Vasyl N, Heller R, Schoenbach KH (2011) Subnanosecond electrical pulses cause membrane permeabilization and cell death. IEEE Trans Biomed Eng 58:1239–1245

Xiao S, Pakhomov A, Guo F, Polisetty S, Schoenbach KH (2013) Neurostimulation using subnanosecond electrical pulses. SPIE Proc 8585:0M

Xiao S, Zou X, Huynh K, Yamada R, Petrella R, Hani M, Beebe S (2020) A high power dielectric biconical antenna for treatment of subcutaneous targets. Bioelectromagnetics 41:413–424

Xie F, Varghese F, Pakhomov AG, Semenov I, Xiao S, Philpott J, Zemlin C (2015) Ablation of myocardial tissue with nanosecond pulsed electric fields. PLoS ONE 10:e0144833

Zhang J, Blackmore PF, Hargrave BY, Xiao S, Beebe SJ, Schoenbach KH (2008) Nanosecond pulse electric field (nanopulse): A novel non-ligand agonist for platelet activation. Arch Biochem Biophys 471:240–248

Chapter 2
Effects of usEPs on Plasma Membranes—Pores, Channels, and Repair

Stephen J. Beebe

Abstract Receptors, channels, glycoprotein, and transport proteins, among other integral proteins, load the plasma membrane lipid bilayer, which receives reinforcement from cytoskeletal filaments as the boundaries of cellular life. Early studies suggested that usEPs did not induce plasma membrane permeabilization until relatively high amplitude conditions. As will be discussed here and elsewhere, this was because the usEP-induced plasma membrane pores were so small that they did not allow entry of propidium iodide, a commonly used marker for permeability, or calcein exit from the cell. The plasma membrane contained nanoelectropores or nanopores, so-called because they were ~1 nm in diameter. These lipid nanopore structures exhibited complex conductances similar to classic pores, which are common in protein structures. Another finding presented some confounding information regarding plasma membrane phosphatidylserine (PS) externalization, commonly used as a marker for apoptosis. usEPs "pulled" PS through these nanopores. Now PS externalization was an ambiguous marker for apoptosis. Although plasma membrane lipids were clear usEP targets, plasma membrane channels were also targets for usEPs. As will be discussed, these effects could be due indirectly to nanopore formation or possibly directly on the channels themselves, although more evidence for direct effects is necessary. In any event, patch-clamp techniques provided new, exciting, and valuable information about usEP effects on plasma membranes. As might be expected, cells have mechanisms that confront nanopore formation and other possible usEP-induced injuries, which will be presented. Finally, in relatively uncharted territory, usEPs were also shown to affect redox systems in plasma membranes as electron carriers. Although usEPs are unique for their effects on intracellular structures and functions, their impact on the plasma membrane has provided a wealth of information about how they can be used as tools in biology and medicine.

2.1 Some Terms About Pulses

What is in a name? Terms used for Ultrashort Electric Pulses—A variety of terms have been used to denote ultrashort electric pulses. The term "ultrashort" implies electric fields with pulse durations less than 1 microsecond (μs). This then

S. J. Beebe et al., *Ultrashort Electric Pulse Effects in Biology and Medicine*, Series in BioEngineering, https://doi.org/10.1007/978-981-10-5113-5_2

indicates pulses with durations in the nanosecond (ns) and picosecond (ps) ranges. The greatest body of literature at the present time concerns nanosecond pulses, with fewer studies on picosecond pulses. In addition to ultrashort electric pulses (usEPs), these pulses have been called nanosecond pulsed electric fields (nsPEFs), nanosecond electric pulses (nsEP), intense ultrashort pulses, intense submicrosecond electrical pulses, submicrosecond intense electrical pulses, nanoelectropulses, nanoelectroablation and most recently, nano-pulsed stimulation (NPS). This latest term conveys the concept that ultrashort electric pulses cause cell and tissue stimulation that can include stimulation of programmed cell death mechanisms, ablation and elimination of cancer cells or other unwanted tissues as well as activation of cell signaling mechanisms such as mobilization of calcium from the endoplasmic reticulum that promotes platelet activation or activation of bone-derived dendritic cells. Another aspect to the pulse of the commonly used term "pulses" is that in some studies a single exposure to an electric field were used. However, in the most common application to date, that is cancer treatment, multiple pulses, even hundreds of pulses, are needed to completely ablate tumors and to induce an immune response after treatment, which have been demonstrated in mice or rats. Nevertheless, nano-pulse stimulation is a fitting term.

It is about the ultrashort pulse duration with short rise-fall time and the high power—There are also further distinctions based on basic physics principles of pulsed power and principles of frequency content from pulse durations. Pulsed power, which is the basis of ultrashort electric pulses, concentrates electrical energy and releases it is short bursts of high power low energy electric fields. This technology has been used in radar, particle accelerators, high powered pulsed lasers, and fusion research, among others. Sandia National Laboratories have an advanced pulsed power program that they say produces "boundless energy in an instant" with a goal to produce fusion energy in the laboratory.

Both conventional electroporation and ultrashort electric pulses compress electrical energy and release it is short bursts that produce power measured in watts. The difference is the amount of power released. If a joule of energy is released in one second, it produces 1 W. If it is released in one microsecond it produces a megawatt. If the pulses are ultrashort, they are released in one nanosecond producing a gigawatt and or released in a picosecond producing a terawatt. While it is relatively easy to produce milli- and micro-second pulses with low electric fields, it is not trivial to produce nano- and pico-second pulses with short (fast) rise-fall times and high electric fields. Applying ultrashort electric pulses brings these physics concepts to fields of biology and medicine. So, in experiments or cancer treatments that use ultrashort pulses of 100 ns and 50 kV/cm compared to using irreversible electroporation (IRE) pulses of 100 μs and 4 kV/cm, the power is 1000 times greater and the amplitude more than 10 times greater for the usEPs. Herein lies some of the differences in the basic technology and the biological consequences between these two technologies.

A primary hypothesis states that shorter pulse durations with shorter rise-fall times may are better for intracellular effects and may be better for usEP-induced immunity—Through the biology chapters, the comparison of usEP conditions are based on charging effects quantified as Volt seconds / centimeter (Vs/cm). This unit is determined as Eτn where E = electric fields in volts, τ = pulse duration in seconds and n = pulse number. The use of Vs/cm is based on usEP-induced cell/tissue charging effects as a hypothetical concept for ablation and now also considered a factor for immune induction. Not surprisingly, there is a Vs/cm threshold for ablation based on the electric field and/or pulse number; however, less is known about optimal conditions or a usEP threshold for immunity. The charging effect as Vs/cm is used instead of energy density (E2τn) because usEPs are less energy-dependent and more related to electric field, power and charging effects (Schoenbach et al. 2009). This theory may explain why usEPs as 100 ns with 10 ns shorter rise-fall time appear better than 200 ns with 40–50 ns long (slow) rise-fall time for immunity. This concept is based on the plasma membrane charging time constant, which is considered to be responsible for unique intracellular effects of usEPs. Since immunity is enhanced by exposure/release of danger associated molecular patterns (DAMPs) such as translocation of calreticulin from the ER to the plasm membrane and release of intracellular ATP and HMGB1, it is reasonable to consider the induction of intracellular effects would be important for a usEP-induced RCD mechanism(s) that is/are immunogenic and enhance immunity (see Chap. 9). Basically, the theory states that if the pulse duration is shorter than the charging time constant of the cell membrane, intracellular effects occur during the rise and fall times of the pulse, while plasma membrane effects occur during the pulse plateau. For cells in suspension, the charging time constant is about 75 ns; however, it is much longer for cells in tumors, perhaps 1 μs. In either case, the plasma membrane becomes more transparent for electric field changing in a time frame shorter than the charging time constant (Schwan 1985). Thus, the concept that short pulse durations and short rise-fall times can be considered better for immunity can be based on frequency considerations as they have broader high frequency contents which favors electric field penetration through the cell's capacitive branch in the equivalent circuit. Shorter pulse durations were better for release of Ca^{2+} from the endoplasmic reticulum [10 ns > 60 ns > 300 ns (Semenov et al. 2013b, (see Chap. 5)] and a short rise-fall time of 15 ns was better than longer rise-fall time of 150 ns for dissipation of the ΔΨm and induction of death (Beebe et al. 2012, 2013), [see Chap. 8]. These two studies provide strong evidence that shorter pulses with shorter rise-fall times are more effective for intracellular effects. Evidence suggest that the same short pulse conditions are optimal for induction of RCD mechanisms that are immunogenic and therefore for induction of immunity. These later suggestions are under study and require more data to support them.

A few words about differences between Ultrashort Electric Pulses and Conventional Electroporation—In attempts to be most specific and descriptive, there are more scientifically precise and unambiguous terms for ultrashort electric pulses (usEPs) with nanosecond and picosecond durations than conventional electroporation with microsecond to millisecond durations. These differences include not only the obvious duration difference, but also differences in the frequency content, energy density, power, and cellular effect.

Pores and intracellular structures—Two other major differences at the biological cell level are effects on pore formation in cell plasma membranes and effects on intracellular membranes and possibly other intracellular structures. Convention EP causes rapid rearrangements of plasma membrane lipids that cause formation of transient pores in the membrane structure, which allows transport of relatively large molecules, such as DNA, across the plasma membrane of cells with the specific purpose of maintaining viability (Neumann et al. 1982). Thus, effects of conventional electroporation are primarily on the plasma membrane while intracellular contents are shielded from the electric fields. This provides a means to transform bacteria or transfect cells to express genes of interest, for example. Electrochemotherapy uses conventional electroporation pulses to transport impermeable chemotherapeutic drugs into cells to eliminate cells. IRE uses conventional electroporation pulses with sufficiently high electric field amplitudes to specifically prevent viability. So, the fundamental feature of conventional electroporation is that it causes a relatively low density of relatively large pores on the plasma membrane, with only minor effect on the ER adjacent to the plasma membrane. Also see Chap. 9.

In contrast to conventional EP, usEP produce *nanopores* in plasma membranes (Vernier et al. 2006a; Pakhomov et al. 2009). In addition, usEP induce intracellular effects (Schoenbach et al. 2001). These effects are often, but not exclusive, considered to be due to what is called intracellular electroporation. This was called supraelectroporation (Stewart et al. 2004; Gowrishankar et al. 2006). These nanopores were shown to cover much more of the plasma membranes surface than conventional EP; these are high density nanopores. In addition, nanosecond pulses induce formation of high density nanopores in all cell membranes including intracellular membranes. Supraelectroporation pulses go through cells, while conventional electroporation pulses go around cells (Gowrishankar et al. 2006). These simulation studies were experimentally varied at the plasma membrane level in cardiac myocytes (Semenov et al. 2015b).

In general, these two observations—the formation of nanopores and intracellular effects—and their biological consequences are believed to be the primary mechanistic basis of how usEPs affect biological cells and their systematic regulation of biological function. Most broadly, primary effects have been studied at the level of plasma membrane and organelle membranes, especially the endoplasmic reticulum, mitochondria, nucleus, and cytoskeletal membranes. Nevertheless, although there are increases in intracellular calcium, there are other effects on mitochondria and other organelles that is are unrelated to plasma membrane permeabilization. For example, there is some evidence that usEPs affect proteins, such as the cAMP-dependent protein kinase in vitro (Beebe 2015), some membrane channels (Semenov et al. 2015a) and non-membranous mitochondrial components (Beebe et al. 2013). If usEP biological targets are considered, lipid membranes are primary targets since they surround and protect cells and their organelles. It is also reasonable to consider other targets. Given that electric fields charge cells, charging membranes causing dielectric breakdown is clear. Other targets are less obvious, but proteins, RNA and DNA are structures with charges (see Chap. 7), and these can possibly be modified by electric fields.

UsEP versus IRE—There can also be confusion with the use of the term IRE. IRE is correctly applied to the use of conventional electroporation pulses in the microsecond range that cause irreversible loss of cell viability by using high electric field. These cells are sufficiently damaged to prevent programmed cellular repair functions; cells die by blunt necrosis. This is sometimes called accidental death (ACD) in general cell death literature, except here it is not accidental. So conventional electroporation can be designed to be reversible for the expression of genes or other molecular effects in cells that survive the electric fields, or it can be used for irreversible effects in ablation with IRE.

Nanosecond pulses or usEPs can also be used for reversible and irreversible effects. However, another major loss of specificity and distinction can occur when nanosecond pulse-induce ablation is referred to as IRE. When ablative nanosecond pulses are applied to cells or tumors there is a nanoporation of cell membranes that results in the irreversible loss of viability; however, the poration effect itself is different and more importantly, the cell response mechanisms are different. Death resulting from nanosecond pulses is not due to blunt necrosis. There are a large number of studies that have shown that cells do not die instantly, but die by programmed or regulated cell death (PCD, RCD) that can include apoptosis and other RCD mechanisms that are less well defined for usEPs. Thus, calling cell death by usEPs as IRE loses intrinsic specificities and includes discrepancies about the applied pulse duration, electric field amplitudes, and the biological response to them.

UsEPs can stimulate—When non-lethal ultrashort pulses are used, they can stimulate or modulate cell functions similar to hormones, cytokines or drugs that act through receptors. Ultrashort pulses appear to be less specific and affect the entire plasma membrane, intracellular membranes and/or other cellular structures. Thus, in a sense ultrashort pulses are non-ligand agonists (Hall et al. 2007; Zhang et al. 2008; Beebe 2013). For this reason, nanosecond pulses have been called nano-pulse-stimulation (NPS). Such stimulation have been shown to mobilize calcium from the endoplasmic reticulum (ER) (White et al. 2004; Semenov et al. 2013a, b). Most notably, NPS with pulses as short as 500 picoseconds causes long lasting opening of voltage gated calcium channels (VGCC) by a mechanism(s) that do not involve conventional electroporation, heating, or membrane depolarization by opening of VG Na(+) channels (Semenov et al. 2015b).

2.2 Introduction

When cells are exposed to usEPs, they are sensing an impact that cells have not experienced in the evolutionary history of life on earth. These ultrashort pulses present enormously high power for exceedingly short periods of time. usEPs have significant effects on the plasma membrane like that observed for milli- and micro-second (long) pulses; however, because usEPs are exceptionally brief, high power and low energy pulses, they are more likely than longer pulses to have intracellular effects such as modifying intracellular organelles like mitochondria, endoplasmic reticulum, nucleus, lysosomes, and the cytoskeleton. These intracellular effects are hypothesized to be a unique feature of usEPs. Simulations and modelling studies suggest that micro- and milli-second pulses can have intracellular effect (Gowrishankar et al. 2006; Esser et al. 2009, 2010); however, there is not enough experimental evidence to support this concept; in silico data have not been verified with in vitro or in vivo experimental data. In contrast, there are scores of studies that demonstrate intracellular effects of usEPs on structures and functions of intracellular organelles.

2.3 Plasma Membrane Structure and Function

It can be expected that when outer plasma membranes of cells are exposed to usEPs, there are cellular reckonings to consider to manage the consequencies. usEPs are considered physical stimuli and they challenge multiple systems that respond to such impact. While most reports of usEP effects on cells define specific cellular mechanisms that are affected, these are generally downstream modules that are part of a response to "danger" signals. At the plasma membrane and beyond, the response depends on the usEP intensity, which is determined by the pulse duration, electric fields, and pulse numbers as well as repetition rates. Repetition rates have generally not been carefully studied as a component of intensity, until recently. In general, high repetition rates have the potential to generate heat, which is a variable that "contaminates" pure electric field effects. Thus, repetition rates are favorably low for most studies so that heat is not a variable. Nevertheless, heat has been used to enhance usEP effects (Edelblute et al. 2018), as well as enhancing gene delivery by electroporation (Donate et al. 2016; Edelblute et al. 2017; see Chaps. 11 and 12). In addition, rapid repetition rates have also been used in studies with excitable tissues as will be discussed below.

Upon first consideration from the electric fields' perspective, the plasma membrane is a lipid bilayer that can be charged as a dielectric, rearranging its well-organized structure to allow permeabilization, and thereby altering the ionic equilibrium of the cell and presenting the cell with a potentially catastrophic problem if not quickly resolved. Prolonged loss of membrane integrity would extinguish cellular life. The plasma membrane is the cell's barrier to the outside world and as such defines the limits of life itself. This outer membrane protects the elaborate and often mystifying sophistication of intracellular molecular mechanisms that defines and maintains life. Yet the plasma membrane is very much a part of this living system's energy. This double layered lipid enclosure separates these living systems from the outer world's environment.

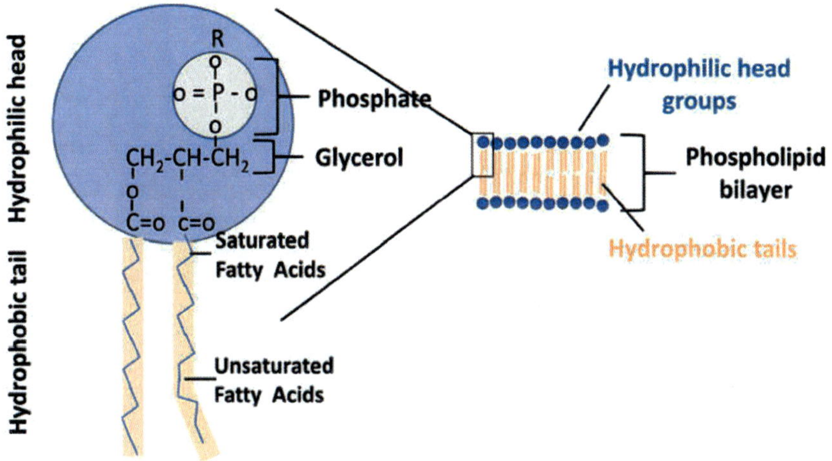

Fig. 2.1 Structure of phospholipids in the phospholipid bilayer

The plasma membrane is made of two layers of lipids. Lipids consist of hydrophilic (water loving or attracting) phosphate head groups, a phosphate attached to a glycerol backbone with two attached long hydrophobic (water fearing or repelling) tail groups (acyl chains). Figure 2.1 shows that the bilayer is formed with the hydrophobic tails facing the interior and the hydrophilic head groups facing the inside and outside of the structure in contact with aqueous medias. Sheets of these phospholipid bilayers surround the living interior of the cell. All lipid groups with choline moieties, such as phosphatidylcholine and sphingomyelin, are on the outer leaflet; lipids that include terminal primary amino groups such as phosphatidylserine are in the inner leaflet of the bilayer (Op den Kamp 1979).

The fluidity of the plasma membrane is determined by several factors. The longer the lipid tails, the more molecular interactions occur. Long lipid tails make membranes with more rigidity and less fluid. Conversely shorter lipid tails make membranes more fluid. The phospholipid tails can be either saturated, meaning they have no double bonds, or unsaturated with double bonds present. Without double bonds, saturated lipids are straight and have greater interactions. Thus, membranes with saturated lipids are less fluid. The double bonds in unsaturated lipids make them more bent and crimped. Since these lipid tails have fewer molecular interactions, they give the membrane more fluidity.

Another factor that affects membrane fluidity is cholesterol. Cholesterol makes significant contributions to membrane structures (Fig. 2.2). Molecule per molecule, it generally makes up about half of the structure. However, because it is smaller, it is proportionally about 20% of the structure (Alberts et al. 2014). Like phospholipids, cholesterol is an *amphipathic molecule*, meaning it contains a hydrophobic and a hydrophilic segment. Its structure includes a four carbon steroid rings between a hydroxyl group and a hydrocarbon chain segment. Cholesterol's hydroxyl (OH)

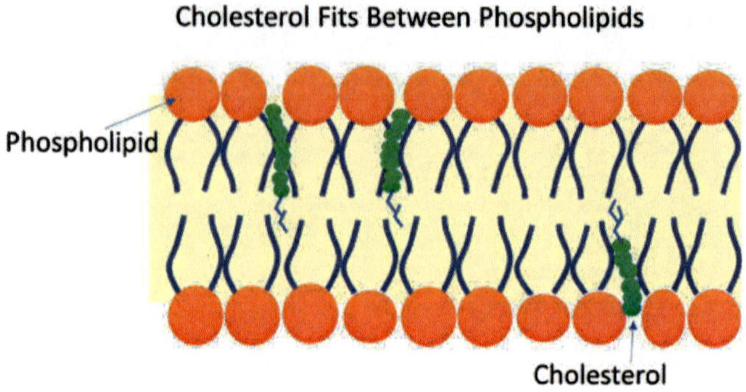

Fig. 2.2 Cholesterol in the lipid bilayer

group aligns with the phosphate heads of the phospholipids. The remaining portion of it tucks into the fatty acid portion of the membrane. This provides some mobility to the outer surface of the membrane and makes it less soluble to smaller water-soluble molecules. While cholesterol adds strength and rigidity to the plasma membrane and prevent it from becoming overly fluid, keeps it from collapsing on itself, yet also helps maintain its fluidity. Due to its position in the membrane, cholesterol helps separate the phospholipids so that the fatty acid chains cannot come together and crystallize. So, cholesterol helps prevent extremes by acting like a buffer that keeps the membrane not too rigid and not to fluid. So, it is a bidirectional regulator of membrane fluidity by stabilizing the membrane at higher temperatures and preventing them from crowding together and hardening at low temperatures.

While the plasma membranes are lipid bilayers, they consist of much more than lipids (Fig. 2.3). The plasma membrane is the purveyor of news from the outside

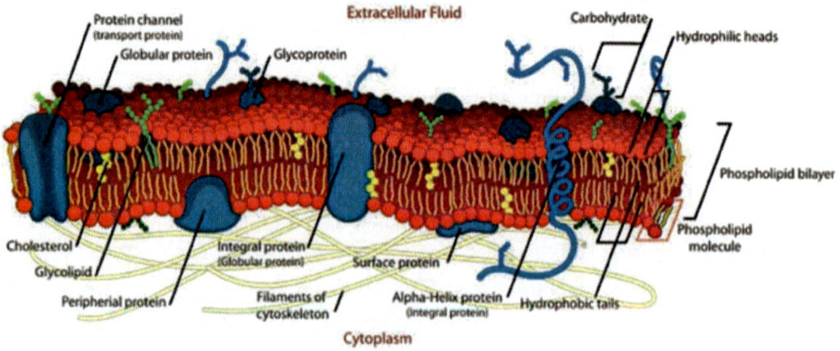

Fig. 2.3 Plasma membrane with all of its molecular components.(from Mariana Ruiz Villarreal (User: LadyofHats/Wikimedia Commons); License-Public. *Source* https://commons.wikimedia.org/wiki/File:Cell_membrane_detailed_diagram_en.svg

environment to the intracellular milieu. Thus, the plasma membrane is the communication director of external affairs, receiving messages from abroad, next door and also from itself. These messages can arrive to receptors that are specific to signaling molecules sent from distant cells in the body (hormone effects), from neighboring cells (paracrine effects) or messages from the cells themselves (autocrine effects). This allows cells to respond to signals for coordinating activities with other cells, with their immediate cellular environment as well as to respond to their own signals so they can function effectively and communally within a tissue, an organ, and/or an organism. These signaling receptors are often integral proteins and some are transmembrane proteins extend from the inner leaflet to the outer leaflet of the lipid bilayer. However, some signaling mechanisms occur through molecules that are peripheral proteins or proteins that are surface proteins anchored by lipids. Plasma membranes also express transporters and channels that traffic ions and inorganic and organic molecules that maintain homeostasis and respond to the changing needs of the cell and its environment. Plasma membranes are decorated with glycoproteins, glycolipids, and carbohydrates. As discussed below, usEP may be direct effects on protein channels. Because usEPs have such a short duration and rise-fall times, they too can act as signals to intracellular structures as they pass through the plasms membrane.

In order to organize cellular activities, the plasma membrane contains signaling platforms called lipid rafts (not shown here). These membrane stages provide localization and distribution of signaling molecules (Bubb et al. 2017; Nordzieke and Medraño-Fernandez 2018). These structures are rich in cholesterol and sphingolipids, which have long and saturated fatty acids that aggregate providing some rigidity that is enhanced by the presence of cholesterol. These lipid rafts make the membrane thicker. The cholesterol also helps stabilize proteins in lipid rafts and supports their functions. This provides an excellent environment for protein to be "soluble" in membranes (Simons and Toomre 2000). Caveolae are a subset of lipid rafts that form invaginations in the plasma membrane by polymerizing proteins called caveolins, which are integral proteins that tightly bind free cholesterol. Caveolae participate in signal transduction, but do not appear to be required. They are important in signaling from growth factors, protein kinases A and C as well as nitric oxide synthase, but are not in all cells types. They are plentiful in adipocytes, fibroblasts, and endothelial cells, sparse in hepatocytes and absent in lymphocytes and non-adherent monocytes (Fielding and Fielding 2000). During stress in cardiac myocytes, caveolae form physical interactions with mitochondria that may specifically target mitochondria for post-translational modification (Sun et al. 2015). There have not been any studies showing effects of usEP on lipid rafts. However, as discussed below, the presence of cholesterol changes the impact of usEP on plasma membranes.

The plasma membrane can also act as a capacitor for regulating energy and metabolism and communication in connection with intracellular components (Ray et al. 2016). In this context, cholesterol in the plasma membrane serves as stored cellular substrates that can be utilized under conditions of stress. For example, the plentiful levels of cholesterol in caveolae in the lung are involved in the uptake and exchange of oxygen, which is upregulated in response to hypoxia (Botto et al. 2006). The membrane lipids serve as capacitors themselves as they are now known to exhibit bioactivity. Some lipids are stored in inactive components in membranes and then released and activated to serve roles in signal transduction. As an example, inactive phospholipids can be converted into oxidized phospholipid signaling molecules by action of peroxidases on membranes and lipoproteins (Davies and Guo 2014). Triacylglycerols (TAG) stores as intracellular cytoplasmic lipid droplets as a main constituent of body fat that can be hydrolyzed to produce a wide variety of bioactive molecules that can act as co-factors of transcription or enzymes that regulate various cellular processes (Papackova and Cahova 2015).

2.4 Plasma Membrane Channels and Pores in Proteins

During the physiological activation of plasma membrane transport systems, "pores" are formed through protein structures that transiently connect the intracellular and extracellular milieu. Pores through transport proteins are most often channels that are nanoscale structures with diameters of 0.2 − 4.0 nm. However, there can also be maligned movements of molecules across plasma membranes. Staphylococcus aureus protein α-toxin and Streptococcus pyogenes protein streptolysin O create pores that allow unsolicited molecular transfer across membranes that disrupts vital functions. Cytotoxic T-cells permeabilize invading cell membranes with perforin, creating pores through which granzymes can induce apoptosis in invading cells. These pores have a lumen of 5–30 nm with the majority at 13–20 nm (Law et al. 2010).

During some regulated cell death (RCD) mechanisms pores or channels are produced in membranes from the inside out. These "pores" are designed to end in RCD by lysing cells. These mechanisms are sometimes referred to as programmed or regulated necrosis or secondary necrosis (Zhang et al. 2018). Two relatively well-studied regulated necrosis mechanisms are necroptosis and pyroptosis (see Chap. 9). Necroptosis is known to occur in several pathological conditions including inflammation, autoimmunity, or immunodeficiency (Chan et al. 2015). When necroptosis is initiated, a protein kinase called receptor-interacting protein kinase 3 (RIP-3) is activated by RIP-1, forming what is called the necrosome that phosphorylates and recruits a pseudokinase called MLKL (mixed-lineage kinase domain-like protein) to the plasma membrane. MLKL forms octamers, eight subunit proteins, which span the membrane to form a channel. MLKL forms unique cation channels that exhibit preferential permeability to Mg^{2+}, but not Ca^{2+}, in the presence of Na^{+} and K^{+}. (Xia et al. 2016; Huang et al. 2017) Transient potential receptor (TRP) channels TRPM7 may mediate post-MLKL channels resulting in Ca^{2+} influx into cells that thereby die by cell lysis (Cai et al. 2014). Nevertheless, pore-forming activity of MLKL is required for necroptosis (Xia et al. 2016; Huang et al. 2017).

Pyroptosis is another form of RCD that ends in cell lysis (Aglietti et al. 2016; Ding et al. 2016; Liu et al. 2016; Sborgi et al. 2016). This mechanism depends on inflammasome-mediated mechanisms that activate caspase-1 and/or 11 in mice (caspase-4/-5 in humans) and induce secretion of IL-1β and IL-18 (Denes et al. 2012).

Inflammasomes like the NLRP3 (nucleotide-binding domain, leucine rich repeat containing receptor protein 3) are large multiprotein complexes that are activated by DAMPs (danger-associated molecular patterns) and PAMPs (pathogen-associated molecular patterns) as well as UV (ultraviolet) radiation. Like necroptosis, pyroptosis also uses oligomerization and translocation of intracellular structures to the plasma membrane to form permeabilizing structures from the inside out, in this case forming pores instead of channels. However, instead of the phosphorylation mechanism in necroptosis, pyroptosis forms lytic structures by proteolytic cleavage of gasdermin D (GSDMD) by caspase-1/-11 (-4/-5 in human). The N-terminal fragment of GSDMD translocates to the plasma membrane and forms non-selective pores with an inside diameter of 10–14 nm (Ding et al. 2016). There are many similarities between RCD by necroptosis and pyroptosis, but there are also many differences (Chen et al. 2016; Shi et al. 2017). These mechanisms of RCD by necroptosis and pyroptosis will be discussed in Chap. 9.

2.5 Studies with usEP Effects on Plasma Membranes

The plasma membrane is essentially a leaky dielectric and when electric fields are applied with voltage increased to about 1 V, the membranes become electroporated and leaky to water and small ions. Conventional electroporation uses pulse durations of microseconds (μs) and milliseconds (ms) (Neumann et al. 1999). These conditions lead to relatively large pores in the plasma membrane (Weaver and Chizmadzhev 1996) that allow entry of large molecules such as DNA or impermeable chemotherapeutic drugs. Shorter pulses in the nanosecond range cause pore formation in the plasma membrane that are smaller allowing only small ions to enter the cell (Deng et al. 2003; Chen et al. 2004; Hu et al. 2005b). When the rise-fall time of the pulses are faster than the charging time of the plasma membrane, the electric fields pass through the cells, into the cytoplasm and small pores are formed in intracellular membranes (Schoenbach et al. 2001) and in all cell membranes (Gowrishankar et al. 2006).

So, in contrast to conventional EP, usEPs were hypothesized to bypass the resistance of the plasma membrane and affect intracellular structures and functions. This concept was based on a simple electrical model of biological cells that predicted increasing likelihood for electric field effects on intracellular structures when the pulse durations were in the sub-microsecond range (Schoenbach et al. 2001). When the duration is short enough or more accurately, when the pulse rise-fall times are short enough, compared to the charging time constant of the plasma membrane dielectric, the electric field occurs across the intracellular milieu while the charge across the external membranes may be relatively small enough not to cause poration (Schoenbach et al. 2004). The charging time constant is a measure of the time that the cell interior is exposed to the applied electric field or about 75 ns (Schoenbach et al. 2001). From a different perspective, the plasma membrane becomes more transparent for oscillating electric fields when the angular frequency of the oscillation is greater than a value given by the inverse of the charging time (Schwan 1985). A 75 ns charging time constant is determined for cells in vitro, but would be much longer for tissues or tumors in vivo. However, these arguments for significance of the charging time constant can be made equally for in vitro and in vivo effect.

Early experiments with usEPs supported a hypothesis of selective intracellular effects with usEPs because propidium iodide (PI), a DNA binding dye that is impermeable to cells with intact plasma membranes and commonly used as a permeability marker, did not readily enter cells after usEPs (Schoenbach et al. 2001; Deng et al. 2003; Beebe et al. 2003a, b, Vernier et al. 2004a). For example, studies with Jurkat cells exposed to pulse durations between 100 μs and 60 ns and electric fields 3–150 kV/cm found that shorter pulses exhibited temporally delayed PI uptake with heterogeneous staining magnitudes and minimal cell swelling (60 ns more delayed, less swelling than 300 ns). In contrast, pulses with 10 and 100 μs durations showed rapid and homogeneous permeability changes with rapid cell swelling (Deng et al. 2003). These studies, like others (Beebe et al. 2002, 2003a, b), demonstrated distinct differences between usEPs and conventional EP pulses, yet when electric fields were sufficiently intense, plasma membranes became permeable to usEPs defined by PI uptake.

This theory was further supported by the selective permeability of intracellular granules to calcein in human eosinophils, referred to as eosinophil sparklers, without apparent effects on the plasma membrane (Schoenbach et al. 2001). Additional support for these ideas was found in several studies, showing that calcium was mobilized from intracellular stores in cells that were incubated in media containing EGTA to chelate extracellular calcium (Vernier et al. 2003; Beebe et al. 2003b; Buescher et al. 2004; White et al. 2004). In addition, selective permeability was observed in intracellular vacuoles and vesicle membranes (Tekle et al. 2005). In time, a number of biological responses to usEPs were considered to be due to intracellular effects with minimal effects on the plasma membrane. One of the most interesting findings, given the recent understanding of how cells could die on their own accord, was the observation that usEPs induced cytochrome *c* release from mitochondria into the cytoplasm and activated caspases, indicating intrinsic apoptosis (Beebe et al. 2003a; Ren et al. 2012). Intrinsic apoptosis is due to generation of signals from intracellular origin, further supporting the hypothesis that usEPs induced intracellular mechanisms including apoptotic cell death (see Chap. 9).

In further support for intracellular charging effects, a most noteworthy study by Frey and co-workers (2006) examined voltage changes in the membrane potential of Jurkat cells in response to usEPs for pulses with a duration of 60 ns and maximum field strengths of ~100 kV/cm (100 V/cell diameter). The study loaded cells with the fast voltage sensitive dye ANNINE-6 with a voltage sensitive temporal resolution of 5 ns achieved by the excitation of this dye with a tunable laser pulse, which was synchronized with the applied electric field to record images at times before, during, and after exposure. When a pulse was applied to the Jurkat cells the anodic voltage reached values of 1.6 V within 15 ns. This was greater than the voltage required for electroporation. Voltage across the anodic side reached values of only 0.6 V with the same time range. So, this strong asymmetry in conduction mechanisms in the membranes of the two opposite cell hemispheres, indicates that during the pulse, almost all the voltage across the cell appears across the interior of the cell after conduction pores through the membrane have formed. This means that the cell interior sees approximately the same electric field as the entire cell—nearly 100 kV/cm (6×10^{-3} Vs/cm). It is reasonable to assume that these high electric fields charge subcellular membranes and likely other structures in the same way as external electric fields charge the plasma membrane, and indeed, this "electroporation effect" on subcellular membranes has been observed (Schoenbach et al. 2001). As indicated throughout this book, these usEPs have been the focus of a significant number of studies. Many of such studies have been reviewed (Schoenbach et al. 2004) and enumerated (Batista Napotnik et al. 2016).

2.6 usEPs Induces Plasma Membrane Nanopores

This novel idea of selective intracellular effects led to several additional studies investigating usEP effects on biological cells. Of course, a number of these studies focused on effects of usEPs on the plasma membrane. However, there were several simulation studies that contradicted this hypothesis. Analytic models (Foster 2000; Joshi et al. 2002; Gowrishankar and Weaver 2006; Kotnik and Miklavcic 2006; Smyth et al. 2006) and numerical simulations (Tieleman 2004; Hu et al. 2005b; Tarek 2005) predicted that usEPs would induce poration in plasma membranes. Thus, while considerable attention was focused on intracellular effects, the plasma membrane became a focus of a many studies to resolve the apparent discrepancies between modeling and experimental analysis.

As indicated earlier in this subsection on plasma membrane structure and function, it was noted that the plasma membrane is a well-organized structure with phosphatidylcholine and sphingomyelin in the outer leaflet and phosphatidylserine (PS) in the inner leaflet of the bilayer (Op den Kamp 1979; Alberts et al. 2014). During apoptosis, PS flips from the intracellular to the extracellular leaflet, likely by aminophospholipid translocase (flippase) and scramblase enzymes. Apoptosis induces increases in calcium signaling that inhibits translocases, which transport PS from the outer to the inner leaf of the plasma membrane lipid bilayer. Simultaneously, increases in intracellular calcium activates scramblases, which randomly translate lipids bidirectionally resulting in loss of membrane asymmetry (Bailey et al. 2009; Segawa et al. 2014). Thus, during apoptosis, the plasma membrane symmetry is lost, and PS appears on the outer membrane leaflet as an "eat me" signaling for phagocytosis. Considerations of PS externalization will be revisited in a specific section of usEPs induce RCD mechanisms (see Chap. 9).

One of the many intriguing observation for effects of usEPs on biological cells was immediate, within milliseconds, externalization of PS after the application of a single usEP (30 ns, 25 kV/cm, 7.5×10^{-4} Vs/cm) in human Jurkat cells (Vernier et al. 2003, 2004a, b, 2006a). Given the lipid structure of the plasm a membrane, the appearance of PS on the outer leaflet using FMI-43 indicated that the electric field had caused a rearrangement of the lipids in the plasma membrane, but only on the anodal side of the cell. This was consistent with specific electric field interactions with lipids in the plasma membrane; however, this was observed in the absence of specific plasma membrane poration effects or membrane damage using PI and trypan blue as a conventional indicator of plasma membrane integrity. The rapid appearance of PS on the cell exterior in response to usEPs was not consistent with apoptosis-induced PS externalization, because it appeared too quickly. Furthermore, the appearance of PS only on the anodic side of cells was telling. It was hypothesized that the negatively charged head group of PS was driven toward the positive electrode with the electric field providing the energy for traversing the plasma membrane (Vernier et al. 2003, 2004a, b). This suggested that PS externalization was an electric field-driven event without the appearance of membrane permeabilization using the classical membrane integrity markers PI or trypan blue.

Additional studies tracking PS with FMI-43 and using YO-PRO-1 as a permeability marker in plasma membranes provided several findings supporting the hypothesis that usEPs provided the means for PS translocation to the outer layer of the membrane bilayer, but provided an alternative mechanism for how the translocation occurs (Vernier et al. 2006a, b). Not only did PS appear on the anodic side with a unipolar pulse, but a bipolar pulse induced PS on both the anodic and cathodic side of cells. This provided additional support that PS externalization was an electric field-driven phenomena. Thus, usEP-induced PS externalization occurred as a result of a direct interaction of the electric field on the lipid components of the plasma membrane. This study also found that trains of shorter pulses required higher electric fields to induce PS externalization compared to PS induced by trains of longer pulse duration. For equivalent pulse magnitudes, thirty, 4 ns pulses produced weaker responses than four, 30 ns pulses. This provided additional support that PS externalization was due to charging the plasma membrane. Each of these studies suggested direct electric field effects on the plasma membrane as opposed to secondary biological events that were subsequent to those triggered by usEPs. Because usEPs had been shown to induce release of intracellular calcium (Vernier et al. 2003; Beebe et al. 2003b; White et al. 2004; Buescher et al. 2004), it was still possible that usEP-induced release of intracellular calcium could inactivate translocases and activate scramblases causing PS externalization; however, the speed of PS externalization argued against such a slower enzymatic response. The demonstration that PS externalization was calcium-independent, ruled out the possibility that calcium-dependent enzyme activities could be responsible for PS externalization (Vernier et al. 2006a, b).

Another finding in this study was that under conditions that did not show permeability to PI, Jurkat cells were permeable to YO-PRO-1 (Vernier et al. 2006a). At higher electric fields, permeability to YO-PRO-1 increased and permeability to PI also became evident. Greater permeability to YO-PRO-1 than to PI, which are nearly the same size and same double-positive charge (at physiological pH), suggested that pores on the scale of a nanometer were induced by usEPs with durations of 3–30 ns. However, YO-PRO-1 has a smaller cross-section than PI, suggesting it can traverse smaller pores (Romeo et al. 2011). Nevertheless, while it was previously proposed that PS's negative head groups could be driven by electric field energy across the lipid bilayer (Vernier et al. 2003, 2004a, b), it was now possible to envision PS with a net negative charge following a low energy pathway from the inside to the outside layer of the plasma membrane through a hydrophilic pore (Vernier et al. 2006a, b). So, it was possible that electric fields could charge the membrane dielectric forming nanometer-sized pores and that anionic PS could be electrophoretically pulled through the pore to the outside of the membrane. This indicated that, unlike previous concepts that usEPs only had intracellular effects, now usEPs could be predicted to induce pores in plasma membranes that were too small to allow transport of PI, calcein, or trypan blue (TB).

In a series of studies with GH3 and HeLa cells, long-last increases in passive electrical conductance of cell membranes were observed with usEPs (five 60 ns, 12 kV/cm pulses, 1.61×10^{-3} Vs/cm), suggesting the opening of stable conductance pores described as nano-electropores (Pakhomov et al. 2007a). These nanopores were impermeable to PI and TB, but were non-selective with permeability to small cations and anions, remained open for minutes and passively resealed. That is, they slowly close without cellular controlled energy sources. When cells were exposed to 200 of these pulses (1.02×10^{-2} Vs/cm), plasma membranes remained impermeable to PI and TB, but cells swelled, blebbed and cytoplasm granulated with 20–30 s after exposure, suggesting rapid ACD, often referred to as necrotic cell death. HeLa cells were must less sensitive than GH3 cells. The conductance with small pulse numbers and necrosis with long pulse trains were blocked by Lanthanide (La^{3+}) and gadolinium (Gd^{3+}) ions. While these nanopores characteristics were consistent with de novo formed small electropores, contribution of usEP-activated endogenous ion conductance could not be ruled out.

A subsequent study applied whole patch-clamp techniques for the first time for analyzing effects on usEP-exposed cell membranes (Pakhomov et al. 2007b). Using GH3, Jurkat and HeLa cells, usEPs induced long-lasting decreases in plasma membrane resistance accompanied by a decrease in the membrane potential. As in the previous study, the gradual transition from open to closed state and the long-lasting decreases in plasma membrane resistance and membrane potential were again unexpected, based on other studies that predicted such electropores would last in the nanosecond (Vernier et al. 2003, 2004a, 2006a, b) to millisecond range (Gowrishankar and Weaver 2006).

Further analyses of these nanopore in GH3 and CHO-K1 cells demonstrated that they were voltage-sensitive, opening at high negative voltages, and inwardly rectifying pores. They were a about ~ 1 nm, based on their exclusion of PI, which is about 1 nm in diameter (Pakhomov et al. 2009). It took minutes for the pores to seal having significant effects on electrolyte and water balance. In some cells, the ion channel-like properties evaporated as nanopores appeared to break into larger, PI permeable, conventional EP pores without inward rectification. Although these nanopores exhibited complex characteristic observed in protein-containing pores, nanopores were shown to lipid-based pores through with PS externalization within seconds after exposure, differentiating this from apoptosis-mediated PS externalization, confirming previous conclusions (Vernier et al. 2006a, b). These nanopores were not a creation of or affected by electrophysiological methodology since they could be detected by using (thallium) Tl^+ and a Tl^+-sensitive intracellular fluorophore. Increases in Tl^+ fluorescence was evident in < 100 ms and was insensitive to the presence or absence of Ca^{2+} or the K^+-channel blockers TEA and 4-AP, respectively (Bowman et al. 2010). Since Tl^+ was not present within cells, this approach provided a sensitive indicator of nanopore formation that did not depend on membrane permeabilization to Ca^{2+} or Na^+ or activation of voltage-gated channels. Since CHO-K1 cells are devoid of voltage-gated channels, ion transport through such pores in these cells was ruled out. Overall, these studies demonstrated that usEP-induced nanopores were distinct from membrane defects induced by conventional plasma membrane EP and were distinct from classic ion channels (Vernier et al. 2006a; Pakhomov et al. 2007a, b, 2009; Pakhomov and Pakhomova 2010).

The authors suggested that usEP-induced nanopores in plasma membranes were not artifacts or anomalies of usEPs but may represent naturally occurring phenomena that were enhanced by the presence of usEPs. That these nanopores have complex conductive properties that were not unlike protein channels, suggested that these nanopores may be endogenous channels and not mere abnormalities or glitches in nature that randomly appear in plasma membranes. Cells become permeable to water movements periodically despite the tight barrier formed by the lipid bilayer and nanopores may be responsible for these events or other events. Nevertheless, roles these nanopores may play in cell function aside from responses to usEPs remains to be determined.

The lipid content of the plasma membrane is expected to have an impact on how cell membrane "sense" usEPs. As indicated earlier, cholesterol is a major component of plasma membranes and is known to determine many of its physical properties by increasing the fluid-phase of the membrane affecting its structural and function mechanics. However, cholesterol is a major component of caveolae (Fielding and Fielding 2000) and lipid rafts (Simons and Toomre 2000), which are major signaling platforms that regulate signal transductions (Bubb et al. 2017; Nordzieke and Medraño-Fernandez 2018). So, changes in plasma membrane cholesterol level have physical effects on fluidity as well as signaling effects from signaling stations in lipid rafts.

Molecular dynamics (MD) analyses addressed the question of the role cholesterol in electric field effects on membranes including permeability such as pore formation and structural properties such as PS externalization. One MD study using DOPC (1,2-dipalmitoyl-sn-glycero-3-phosphatidylcholine) membranes increasing percentages of cholesterol caused a near linear increase in electric field from 32.5 kV/cm with no cholesterol to 75 kV/cm with 40 mol% of cholesterol for pore formation (Fernández et al. 2010). The study suggested that cholesterol increases the cohesive natural stability of the cholesterol-containing membranes, condensing it and thereby requiring higher electric fields for pore formation. Hydrophilic pores formed more slowing and remained open for longer periods of time (nanoseconds vs hundreds of picoseconds). A second study using similar MD simulation that mimicked applications of a high intensity nanosecond pulses came to the same conclusions (Casciola et al. 2014). Using pulses that simulated conventional low intensity longer pulses, the authors showed similar increases in cholesterol concentrations also required higher transmembrane voltages to produce pores in the plasma membrane. However, the nature of the conventional pores suggested different molecular and ionic transport that would be expected affect pore closure and lifetime.

These MD simulations were confirmed by experimental data in CHO-K1 and Jurkat cells (Cantu et al. 2016). When cholesterol was depleted from membranes using methyl-β-cyclodextrin (MβCD), cells were more sensitive to usEPs. That is, with fewer pulses or lower energy levels, cholesterol-depleted cells exhibited greater permeability to PI, greater Ca^{2+} influx and greater PS externalization. Likewise, both CHO-K1 and Jurkat cells exhibited greater toxicity to usEPs when cholesterol was depleted from their membranes. Not surprisingly, cholesterol levels are not the primary determinant of cell responses to usEPs. It was previously shown that viability of Jurkat cells was more sensitive to usEPs than U937 cells, at least in part because usEPs induced Fas-ligand secretion in Jurkat cells and usEPs induced U937 cells expressed the anti-apoptotic protein c-FLIP (Estlack et al. 2014). Jurkat cell secretion of Fas-ligand is likely to play a role in having greater sensitivity to usEPs in this study as well. As the authors point out and as indicated above, cholesterol is also a major constituent in lipid rafts and caveolae (Fielding and Fielding 2000; Simons and Toomre 2000; Bubb et al. 2017; Nordzieke and Medraño-Fernandez 2018), so cholesterol depletion is also likely playing a role in altering these signaling platforms in ways that were not specifically analyzed. This group additionally demonstrated that cellular "stiffness" plays a role in bio-effects of usEPs by showing that disrupting the actin cortex decreased rigidity in CHO cells making them more vulnerable to usEPs (Thompson et al. 2014).

2.7 Effects of usEPs on Plasma Membrane Channels

The great body of evidence for effects of usEPs on cells clearly demonstrates effects on lipids in cellular membranes. Consistent with this, the first evidence for an effect of usEPs on a channel as indicated above was a non-conventional lipid channel called nanoelectropores or plasma membranes nanopore. Also discussed above, these nanopores were distinct from conventional electroporation pores in smaller size, greater density, and presence in intracellular membranes as well as plasma membranes(see section RCD mechanisms, Chap. 9). These lipid nanopore structures exhibit complex conductances similar to classic pores, which are protein structures. Demonstrating effects on classical membrane protein channels presented interesting problems. The primary task was to differentiate ion transport across plasma membranes caused by nanopores or nanoporation from transport through voltage-gated or other protein channels; both could transport common ions such as Na^+, Ca^{2+}, K^+ or Cl^-. This differentiation between lipid nanopores and protein channels was accomplished in several ways. One approach was to use specific inhibitors of selective channels; however, the key here was a common pharmacologic problem of specificity. Nevertheless, this was manageable given the depth of data on specific channel pharmacology. Another strategy was to change ions in the media by chelating Ca^{2+}, changing ion concentrations or ion substitutions, such as NMDG + or choline + for Na^{+}. A third approach was to differentiate nanoporation versus activation of voltage-gated channels based on their different sensitivities to usEPs. As it turned out, there was a fine distinction because differences in sensitivities for nanoporation, intracellular Ca^{2+} transients and effects on ion channels, which were subtle and sometimes overlapping. These tactics was aided by using specific inhibitors and by distinguishing the kinetics of ion transport or currents across the plasma membrane. One kinetic approach was based on the time required for a voltage sensor in a voltage-gated channel to respond to an electrical signal. Yet another method was to use different cell types that were known to have or not have specific types of channels.

Defining these differences was accomplished using two major approaches, including whole cell patch-clamp methodologies and optical membrane imaging. Patch-clamp techniques are especially useful in excitable cells such as neurons, cardiomyocytes, chromaffin cells, or excitable cultured cells. Here it was possible to determine currents of specific single ion channel molecules and to voltage clamp cells to determine currents of various membrane potentials. However, there were problems to maintain the gigaseal of the patch in the presence of usEPs. The gigaseal was important to electronically isolate currents across the membrane as they were measured to prevent "noise" and provide physical stability during the recording. Another issue with patch clamp recording was a possible problem of "usEP pick-up" artifacts—these are effects pulse generation can have on recording equipment. While many of these problems have been resolved, there are alternative approaches. An alternative to avoid potential problems of patch clamp techniques is to use fluorescent probes. One of these approaches used Tl^{+} to distinguish nanoporation, since Tl^{+} is small enough to cross nanopores and is not transported by any channel because it is not present in cells (Pakhomov et al. 2009; Bowman et al. 2010). Another approach was to measure optical membrane potentials with fluorescent molecules such as FluoVolt (Pakhomov et al. 2017).

Another issue with demonstrating effects on channels, was that they were proteins, and there was a dearth of experimental evidence for direct effects of usEPs on proteins. There is experimental evidence that usEPs could inactivate the catalytic subunit of protein kinase A (Beebe 2015); however, while there were antioxidants in the media, there was not a concentrated effort to include the presence and absence of antioxidants to rule out effects of oxidation as a means of kinase inactivation as opposed to a direct electric field effects on protein structure that inactivated its function. However, it is possible that direct usEP effects could alter the conformation of the kinase itself. While it was arguable that an usEP charging effect can affect protein structures, during phosphorylation the attachment of a negatively charged phosphoryl group can significantly affect the structure and function of the phosphorylated protein (Graves and Krebs 1999; Cohen 2000; Pawson and Scott 2005; Taylor et al. 2012). There are a number of modeling studies that predict usEP effects can alter protein conformation (Marracino et al. 2013; Reale et al. 2013). However, effects of usEPs on classical voltage-gated channels may be more easily envisioned since these channels are in membrane structure affected by usEPs, have voltage-sensors that are well-characterized, and could possibly be triggered by a direct usEP charging effect on the channel voltage sensor themselves. It is possible that some usEPs are too short to affect the voltage sensor (Semenov et al. 2015a, b) as will be discussed below. A combination of experimental approaches in several studies from different groups, appear to have ultimately determined that usEPs can have direct effects on specific voltage-gated channels. However, this took a decade of intense investigation.

While some earlier studies investigated non-lethal effects of usEPs on biological functions of mammalian cells, the finding that usEP could kill cancer cells, especially by apoptosis as a "safe" death (Beebe et al. 2002, 2003a), set the stage for a number of early studies to investigate usEP-induced mechanisms of cell death and what cellular target usEPs used to kill cancer cells (see section on RCD mechanisms, Chap. 9). These studies had evolved from experiements that investigated how energy efficient usEPs were for killing bacteria (Schoenbach et al. 1997, 2000). Thus, in the early phases of studies with usEPs, decontamination and cancer therapy were investigative strategies. While not all of these early studies investigated lethal actions of usEPs, studies with usEP effects on ion channels provided a general shift in strategies for defining more fine-tuned effects of usEPs on cell structures and functions as a means to activate or excite cells and tissues. Without a doubt, greater therapeutic applications of usEPs on excitable cells such as neurosecretory cells, neurons and cardiac myocytes could be achieved through excitation of modulation of function rather than eliminating their functions.

One of the earliest attempts to determine if usEPs could activate plasma membrane channels exploited bovine adrenal chromaffin cells (Vernier et al. 2008). This was one of the first studies for analyzing effects of usEPs on excitable cells. These cells are triggered to release epinephrine and norepinephrine by exocytosis by increasing Ca^{2+} influx after activation of voltage-gated Ca^{2+} channels (VGCC) (Craviso et al. 2009, 2010). usEPs with a 4 ns duration and 80 kV/cm (3.2×10^{-4} Vs/cm) induced a rapid increase in intracellular Ca^{2+}. This increase was insensitive to depletion of intracellular Ca^{2+} with caffeine, which empties Ca^{2+} from ryanodine-sensitive endoplasmic reticulum (ER) Ca^{2+} stores, and was insensitive to thapsigargin, which inhibits Ca^{2+} uptake into ER stores by inhibiting Ca^{2+} ATPase. In contrast, uptake of calcium was prevented by chelating extracellular Ca^{2+} with EGTA. These studies indicated that Ca^{2+} entered the cell through the plasma membrane from the extracellular media. Increases in Ca^{2+} were completely inhibited by nitrendipine, a selective antagonists of L-type VGCC. Since the plasma membrane was not permeable to YO-PRO-1, suggesting the plasma membrane was intact and sensitive to nitrendipine, these results suggested that usEPs directly affected L-type Ca^{2+} in the plasma membrane. The Ca^{2+} increase was similar to Ca^{2+} influx in response to a nicotinic receptor agonist. This was particularly interesting because the results were in contrast to usEP-induce Ca^{2+} release from intracellular stores in Jurkat and HL-60 cells (Vernier et al. 2003; Beebe et al. 2003b; White et al. 2004; Buescher et al. 2004) and the natural response for catecholamine release from chromaffin cells is by activation of VGCC. These same pulses were in fact shown to release epinephrine and norepinephrine through VGCC (Craviso et al. 2009).

A later study shed new light onto these earlier finding of usEP effects on Ca^{2+} entry into chromaffin cells (Craviso et al. 2010). The finding that usEPs affected L-type channels was correct, but the mechanisms of a direct effect was not. This finding was based on complete inhibition Ca^{2+} influx with the L-type VGCC blocker nifedipine, which was used at a relative high concentration. When inhibitors were used for VGCC of the N-type (ω-conotoxin GVIA), P-type (ω-agatoxin IVA low concentration), Q-type (ω-agatoxin IVA high concentration) and N- and P/Q-type (ω-conotoxin MVIIC), all had inhibitory effects on usEP-induced Ca^{2+} influx. Thus, consistent with the various type of VGCC channels in chromaffin cells, usEPs affected L-, N-, and P/Q-type VGCC. Another consideration was the possibility that membrane depolarization was sufficient to activate VGCC (Frey et al. 2006). So, it was possible that usEPs activated VGCC by depolarizing the membrane potential. When the extracellular concentration of K^+ was reduced to depolarize the cell, thus lowing the threshold for opening VGCC and when Na^+ was excluded from the extracellular media, effects of usEPs on Ca^{2+} entry was attenuated, suggesting that Na^+ entry preceded activation of VGCC. The lack of an effect of tetrodotoxin (TTX), a voltage gated Na^+ channel (VGSC) blocker, had no effect, indicating that Na^+ entry through VGSC was not involved in Ca^{2+} influx. While the entry of Na^+ through some non-voltage-gated channel was not ruled out, it was also not ruled out, but possibly Na^+ could enter through small pores that were responsible for phosphatidylserine externalization (Hu et al. 2005a; Vernier et al. 2006a). These same channels were likely the same as those defined as long lasting nanopores, permeable to ions like Na^+, but not to YO-PRO-1 (Pakhomov et al. 2007a, b, 2009), and the same as those used previously to suggest that VGCC allowed Ca^{2+} entry without permeabilizing the plasma membrane (Vernier et al. 2008; Wang et al. 2009a).

Another study with usEP effects on excitable cells used rat cardiac myocytes (Wang et al. 2009a). usEPs were shown to induce cardiac myocyte action potentials, but by a new, indirect mechanism for cardiac excitation. The study used 4 ns pulses with electric fields from 10 to 80 kV/cm, (4×10^{-5} to 3.2×10^{-4}). Treatment of cardiac myocytes with conventional electroporation (EP) [1 ms, 0.5–2.4 kV/cm, (0.5–2.4 Vs/cm)] caused asymmetrical excitation with depolarization at the anodal side of the cells with an inward Na^+ current and a hyperpolarization at the cathodal side of the cell as an outward K + current, which causes a secondary depolarization (Cheng et al. 1999; Sharma and Tung 2002). Thus, the EP-induced excitation response is due to voltage-gated channels. In contrast, usEP induced cardia myocyte excitation was due to nanopore formation in the plasma membrane. Several pieces of evidence pointed to this mechanism. First, the response was not all-or-none, which is a definitive characteristic of a voltage-gate channels. Instead, the response to usEPs was electric field-dependent and graded (not sure what "graded" means). In addition, an increase in the pulse repetition rate increased the excitation. Pulses were resistant to verapamil, a VGCC inhibitor, but Ca^{2+} levels were higher than with EP-induced Ca^{2+} entry, indicating additional Ca^{2+} release from the sarcoplasmic reticulum as a Ca^{2+}-induced Ca^{2+} release. UsEP repetition overcame inhibition by tetrodotoxin with anodal Ca^{2+} entry and Ca^{2+} waves. Finally, blockade of Na^+ / Ca^{2+} exchanger decreased responsiveness to usEP indicating that the Na^+/Ca^{2+} exchanger was involved in usEP excitation of cardiac myocytes. So, usEP-induced action potentials in cardiac myocytes are multifaceted, but most likely are due to nanoporation of the plasma membrane with non-selective cation entry, secondary activation of VCSC, which leads to activation of Na^+/Ca^{2+} exchanger.

The modulation of excitable cells can occur by permeabilization of the plasma membrane by nanopores, which are well characterized by different procedures including electrophysiological methods (Pakhomov et al. 2007a; Nesin 2012a; Ibey et al. 2010; Ibey et al. 2009), detection of optical dyes and ion mobilization (Frey et al. 2006; Cantu et al. 2016; Semenov et al. 2016) as well as simulation using molecular dynamics (Vernier et al. 2006b; Sridhara and Joshi 2014; Joshi et al. 2005) and mathematical analyses (Batista Napotnik et al. 2016; Joshi et al. 2005). However, this modulation of membrane potential occurs in all cells. Although nanopores exhibit channel-like properties such as sensitivities to voltage and current, ion selectivity and inward rectification (Pakhomov et al. 2007a; Nesin et al. 2012a, b) nanoporation occurs independently of events that naturally modulate excitable cells such as voltage gated (VG) channels for Ca^{2+} (VGCC) and Na^+ (VGSC). Investigations of VG channels by usEPs using electrophysiological measurements are complicated due to usEP-created "noise" that affects cellular voltage and current-clamp signals, so only delayed usEP effects have been analyzed. Consequently, the most common approaches determine indirect effects of usEP responses from VGCC and VGSC using optimal measurements of cytosolic Ca^{2+} in the presence and absence of channel inhibits such as verapamil for VGCC and tetrodotoxin (TTX) for VGSC.

Roth and colleagues (2013) determine nanoporation thresholds for NG108 neuroblastoma cells and primary hippocampal neurons (PHN) determining membrane asymmetry changes using Annexin-V and FM1-43 and Ca^{2+} Green to determine Ca^{2+} mobilization. Using 600 ns pulses, they found that ED50 values for the uptake of Ca^{2+} for NG108 cells was 1.76 kV/cm (1.1×10^{-3} Vs/cm) and for PHN was 0.84 kV/cm (5.0×10^{-4} Vs/cm). With 16.2 kV/cm the EC50 for Ca^{2+} uptake was 95 ns (1.5×10^{-3} Vs/cm) for both cell types. The effects were not blocked by the non-specific Ca^{2+} channel blocker cadmium indicating these effects were due to nanoporation.

Experiments with cardiac myocytes (Wang et al. 2009a) using fluorescence in cardiac myocytes demonstrated Ca^{2+} transients with 4 ns pulses between 10 and 80 kV/cm (4×10^{-5} to 3.2×10^{-4} Vs/cm) that were suggestive of occurring through non-selective ion channels. In experiments with bovine chromaffin cells, usEPs with 5 ns durations and electric field amplitudes of 50 kV/cm (2.5×10^{-4} Vs/cm) mobilized Ca^{2+} by opening L-type VGCC that were indirect likely induced by nanoporation (Craviso et al. 2010).

Pakhomov et al. (2017), whose laboratory has produced the greatest number of studies with usEPs effects on plasma membranes and ion channels, explored effects of 200 ns usEPs using optical membrane potential (OMP) changes with FluoVolt found that single action potential were induced in 40% of primary hippocampal neurons by 200 ns pulses with a threshold of about 3 kV/cm (6×10^{-4} Vs/cm) while the nanoporation threshold was slightly lower at about 1.7 kV/cm (3.4×10^{-4} Vs/cm). The exclusion of extracellular Ca^{2+} had no effect on reducing the depolarization, so contributions from VGCC were unlikely. While the thresholds for nanoporation and AP overlapped, the authors did not necessarily suggest causal relationships, but did conclude that usEPs were less effective for inducing APs than conventional electric stimulation.

Another study was reported with usEP effects on murine cardiac myocytes (Azarov et al. 2018). This study used much longer pulses with 200 ns and lower electric fields (~1 to 3 kV/cm) and FluoVolt optical action potentials (APs). In these cells the action potential were at least in part responsive to slow sustained depolarization (SSD). However, there were other contributing factors. Like the study with GH3 and NG108 cells with longer pulses (300 ns and 1.6–5.3 kV/cm, 4.8×10^{-4} to 1.6×10^{-3} Vs/cm) (Nesin 2012a), the various ion transport mechanisms were difficult to differentiate. For example, in mouse cardiac myocytes, the thresholds for the observed SSD, action potential and VGC-insensitive Ca^{2+} transients were 1.26 (2.5×10^{-4} Vs/cm), 1.34 (2.68×10^{-4} Vs/cm) and 1.40 kV/cm (2.8×10^{-4} Vs/cm), respectively. Nevertheless, usEP-induced TTX-sensitive action potentials were coincident or preceded by SSD and most often transient after depolarization waves. The SSD was a possible result of several different mechanisms. These included nanoporation of the plasma membrane, opening of selective and/or non-selective cation channels, such as TRPCs (transient receptor potential channels) or the forward mode of the Na^+/Ca^{2+} exchanger. Because the SSD was present when VGSC and VGCC were inhibited by TTX and verapamil, respectively, the SDD were not due to VG channels. Nanoporation was suspected based on time courses and the drug sensitivity. However, there were several issues that suggested active mechanisms were involved. The removal of extracellular Ca^{2+} and verapamil-induced attenuation of the SDD suggested some Ca^{2+}-dependent mechanism. Elevation of intracellular Ca^{2+} could be due to Ca^{2+} influx or from nanoporation of the sarcoplasmic reticulum (SR); however, SR leakage was not likely because Ca^{2+} transients were absent in the presence of EGTA. Another candidate was the Na^+/Ca^{2+} exchanger, which is an antiporter that uses energy stored in the inward Na^+ gradient to remove Ca^{2+} (3 Na^+ in, 1 Ca^{2+} out), which causes depolarization. Inhibiting this antiporter with SEA0400 attenuated the SSD amplitude and increased the action potential. Finally, the generation of the after depolarizations, a common characteristic in many usEP-treated cardiac myocytes, are related to the L-type Ca^{2+} channel, implicating this channel in the action potential. Overall, it is likely that the usEP induced action potential is due to Ca^{2+} entry, allied with SSD that results in an action potential. While Ca^{2+} entry was not dependent on VGCCs, the SSD was most likely supported by the Na^+/Ca^{2+} exchanger and L-type Ca^{2+} channels.

A further exploration of effects of usEPs on VGCC used pulses with durations of 0.5 ns (500 picoseconds, ps) with electric fields of 190 kV/cm (0.9×10^{-4} Vs/cm) (Semenov et al. 2015a). Fura2-ratiometric imaging was used to determine cytosolic Ca^{2+} imaging in GH3 and NG108 cell, which have multiple types of VGCCs, and CHO cells, which do not express VGCCs. In CHO cells, as many as 100 pulses at 190 kV/cm failed to elevate intracellular Ca^{2+}. This indicated that this condition did not cause electroporation of the plasma membrane. In GH3 cells, these same pulses induced increases in intracellular Ca^{2+} levels that were inhibited 95% by the

L-type VGCC inhibitor verapamil. This clearly implicated the L-type VGCCs for the increase in intracellular Ca^{2+}. In NG108 cells these pulses increased intracellular Ca^{2+}, which were partially inhibited by verapamil, but inhibited 85–100% by a cocktail of verapamil and ω-conotoxin, which indicate multiple types of VGCCs were activated by usEPs in the picosecond range. To rule out activation of VGCCs by membrane depolarization, replacing Na^+ with NMGD did not cause a decrease in intracellular Ca^{2+}, indicating that VGSCs were not responsible for the Ca^{2+} increases like that in other studies (Wang et al. 2009a). These studies demonstrated in all these cell types that usEPs can have effects on protein membrane channels, especially on VGCCs when the plasma membrane is affected by nanoporation. This is intriguing because the pulse durations used are shorter than the time required for responses through the voltage sensors in the VGCC, which is 5–6 times faster than the activation time for the VGCC or VGSC (Armstrong and Hille 1998). The authors suggested that usEP effects on VGCC were either non-conventional electroporation, such as failure to detect short-lived pore formation, or a non-conventional mechanism for opening VGCCs that bypasses the voltage sensor mechanisms and acts directly on the channel gate or by some other unknown mechanism. These results are interesting based on other theoretical considerations. Given that the pulse duration is shorter than the membrane charging time constant (~100 ns under these conditions) and shorter than the dielectric relaxation time of the cytoplasm (~700 ps under these conditions), the possibility that the effects on the membrane are determined by membrane charging process through charging currents may not be correct. Under such circumstances of usEPs less than 1 ns, the electric field interactions could be better explained by the cytoplasm and media permittivity rather than conductivity.

In additional studies with GH3 and NG108 cells with 300 and 600 ns pulses at and above 1.5–2 kV/cm, it was found that there were overlaps in currents with prolonged inhibition of VGCCs and VGSCs and so-called non-inactivating "leak" currents, were presumably due to nanopore formation in the plasma membranes (Nesin et al. 2012a). The leak currents were abrupt, were not coincident with PI uptake, increased at high negative membrane potentials, and were inward rectifying. These are all typical characteristics of nanopores. In NG108 cells, but not in GH3 cells, inhibition of the VGSC currents could be isolated by subtracting the leak current. The combined current (VGSC and leak) was measured as a negative peak in current–voltage (I–V) traces. As the voltage steps increased at the beginning of the trace, where the leak current was greatest, it could be subtracted from the VG channels and the two currents could be isolated independently. Since the leak current started and ended earlier than the VG currents, it was considered whether increases in the leak current could cause or affect inhibition of VG channels. However, the magnitude and kinetic relationship between the two currents appeared to be rather weak, suggesting that one was not caused by the other. While both currents increased with higher electric fields, the leak current often increased dramatically, while inhibition of the VGSC current changed less or did not change. From another perspective, inhibition of the VG channels could remain, while the leak current had dissipated. These observations suggested, but did not prove, that the leak currents either caused or modulated the VC currents.

So, the increase in the leak current could be due to Na^+ and/or Ca^{2+} coming through nanopores and/or Ca^{2+} could be released from nanoporation of the endoplasmic reticulum (White et al. 2004; Semenov et al. 2013a, b). In a subsequent study (and Pakhomov 2012b), the Na^+ question was resolved by increasing Na^+ in the pipette in patch clamp approaches, which did not cause inhibition of the VGSC. The Ca^{2+} question was addressed by showing that BAPTA chelation from the pipette did not prevent or reduce inhibition of the VGSC. Thus, while nanopores brought cation influxes, inhibition of the VGSC was shown to be Ca^{2+}- and Na^+-independent. These results suggested a Ca^{2+}-independent local downregulation of the VGSC channels or a direct effect of the usEPs on the channels themselves. This was suggestive evidence that usEPs could directly affect proteins.

2.8 usEPs Activate Plasma Membrane Cell Repair Mechanisms

When cells are exposed to mechanical, chemical, or pathological injuries, and the membrane is breached, the security of the membrane bound intracellular living systems are threatened. Immediate responses are initiated by multiple mechanisms to ensure rapid membrane repair and that the normal ionic equilibrium can be maintained. These will clearly be operational when cell membranes are affected or wounded by usEPs, such as in nanoporation. There are several steps that are essential to successfully respond to plasma membrane wounding. These include sensing the wound, resealing the membrane by closing the wound and remodeling the plasma membrane and the underlying cortical cytoskeleton. The classical cellular membrane repair routine is like a well-coordinated "fire drill" for cell survival. These protocols have been highly conserved through evolution, so they are well-rehearsed and synchronized when activated. A typical membrane repair scenario includes initial clotting mechanisms with Ca^{2+} binding protein aggregation to create a new barrier that patches the wound. Membrane internalization by endocytosis to remove wounded lipids, endocytosis of adjacent vesicles or membrane fusion with intracellular vesicles serve as sources of recycled membranes for larger wounds. Exocytosis or membrane shedding is another mechanism to remove "holes". Another cellular approach is to invoke the actin cytoskeleton at the site of the wound and actomyosin contraction responses as F-actin assembly and disassembly at the leading and trailing edge, respectively, of the closing wound. These contractile mechanisms are downstream of the RhoA GTPase regulators for wound repair (Horn and Jaiswal 2018; Nakamura et al. 2018).

Many of these response mechanisms, or variations of them, have been observed in response to usEPs. The universal signal for membrane repair is an influx of Ca^{2+} across the plasma membrane. Thus, it is not surprising that most of the response for membrane repair are Ca^{2+}-dependent. When cells are exposed to usEPs, the supra-electroporation of the cell has this outer barrier receiving the electric field electric field impact as it passes into the cell interior. As already indicated, short pulses and short pulse rise-fall times have greater effects on intracellular structures and functions than pulses with longer durations and long rise-fall time; however, under these shorter pulse conditions, all cell membranes experience supraelectroporation such that high density nanopores are created in all cell membranes throughout the cell (Stewart et al. 2004; Gowrishankar et al. 2006; see section RCD mechanisms, Chap. 9). The injury associated with usEPs are most likely sensed as a mechanical stress since electric fields charge the lipid bilayer, its associated structures, and intracellular structures. These dense yet nanoscale-sized breaches in the plasma membrane set in motion Ca^{2+} mobilization, which drives down a 10,000-fold electrochemical gradient into the cell so that intracellular Ca^{2+}-sensitive binding proteins discern this Ca^{2+} rush and rapidly respond to close these membrane breaches. However, nanopores can remain open for relatively long time periods. Thus, Ca^{2+} and other ions affect the cell membrane potential and requires voltage-gated and other ion channels to maintain the membrane potential and avoid cell death. This also requires changes in the cytoskeleton that participate in wound remodeling during repair and modulates gene transcription events mediated by several signaling mechanisms including protein kinase C and p38 MAPK (Togo 2004). There are also responses to reestablish the ionic equilibrium with a regulatory volume decrease (RVD) response, which includes K^+ and Cl^- efflux mechanisms. Swelling responses and increases in intracellular Ca^{2+} activates efflux from intermediate-conductance Ca^{2+}-dependent K^+ (IK) channels. Swelling also causes release of ATP, which stimulates purinergic ATP (P2Y2) receptors, thereby inducing phospholipase C-mediated Ca^{2+} mobilization that augments the IK channels. Another G protein-coupled Ca^{2+}-sensing receptor increases efflux activity of volume-sensitive outwardly rectifying Cl^- channels (Okada et al. 2001). Thus, increases in intracellular Ca^{2+}, swelling-induced ATP release, and G-protein-coupled receptors including phosphatidylinositol 4,5-bisphosphate (PIP2) and phospholipase C (PLC) activation coordinate ionic mechanisms of RVD that attempts to correct the swelling induced by nanoporation from usEPs.

One of the most important, early clogging mechanisms of membrane repair involves the Ca^{2+}-binding annexin protein family. While annexin activities have not been specifically evaluated in response to usEPs, phosphatidylserines (PSs) are known to be released in response to usEPs either as an apoptosis-mediated response, which is relatively late, or a response from specific responses to electric fields "pulling PS from the inner to the outer leaflet of the plasma membrane through nanopores", which is immediate. In addition, PS is an important participant in the repair response. The annexin family has a broad range of Ca^{2+} sensitivities with a diverse range of interacting partners with roles for clotting, membrane trafficking, recruiting, and cytoskeletal reorganizations. Annexins accumulate at the site of injury and facilitate repair by forming 2-dimensional protein arrays that stabilize the wound and limit its expansion. Annexin-V binds to PS that accumulates at the wound site. While PS is translocated to the outer leaflet of the plasma membrane during apoptosis as an "eat me" signal for phagocytes and serves to suppress inflammation, it also binds to mitsugumin 53 (MG53/TRIM72), which is known to be an important component of the membrane repair machinery in several cell types (Cai et al. 2009). MG53 is recruited and binds to PS as a means to associate with intracellular vesicles on the cytoplasmic side of the plasma membrane. In association with other proteins such as dysferlin, and caveolin, this allows vesicles to fuse with the disrupted membrane. Wound constriction is mediated by caveolae that gather and fuse around larger wounds, leading to wound constriction and intracellular fusion of caveolar endosomes (Blazek et al. 2015). Although these mechanisms have not been analyzed in response to usEPs, they are likely to play roles as nanopores fuse to make larger pores under higher intensity usEP conditions. In any event, while the appearance of PS on the externalized plasma membrane signals apoptosis, PS in the cytoplasmic side of the membrane participates in wound repair.

In addition to Ca^{2+} and PS, PIP2 and its downstream signaling molecules acts as critical regulators of actin polymerization and cytoskeleton/membrane linkages, modulators of adhesion of the actin-based cortical cytoskeleton to the plasma membrane and crucial regulators of membrane organization during repair. As already mentioned, PIP2 and PLC are activated in response to purinergic P2Y2 receptors activation. Some evidence suggests that binding of PIP2 to cytoskeletal proteins can modify their lateral movements to affect lipid symmetry during membrane repair (Hayes et al. 2004). PIP2 binds with high affinity to cholesterol and annexin A2, one of the many annexin family proteins, which binds to actin and lipids in a Ca2 + -dependent manner providing for tight control of repair processes (Demonbreun and McNally 2016). Plasma membrane injury-induced increase in intracellular Ca^{2+} activates phospholipase C (PLC) generating IP3 and diacylglycerol (DAG). A signaling platform is established at the injury site for the downstream activation of protein kinase C (PKC) and Rho GTPases. It is not presently clear how these signaling mechanisms translate to membrane repair (Horn and Jaiswal 2018).

It is clear that the plasma membrane does not act alone in these repair processes since it is not isolated from the contractile cortex that lies beneath the lipid bilayer composed of cytoskeleton, which is rich in actin filaments, myosin II and actin binding proteins that form a layer beneath the membrane about a micron or more thick. It is these structures that determine how cells respond to mechanical stresses, such as usEPs. The lipid bilayer has limited flexibility and does not readily expand when exposed to mechanical forces. The cortical layer is a mesh-like shell that is more resistant to mechanical stresses and deformation and can produce a force by actin-polymerization or contraction induced by myosin II. This latter mechanism can induce motility. Rho A, a small GTPase, is the principal contraction regulator of the cortex assembly that activates both actin polymerization and myosin II-mediated recruitment for chemotaxis and cytokinesis (Charras et al. 2006).

Hydrolysis of PIP2 and PLC activity have been reported in response to usEPs. These activities are also reported to be responsible for remodeling of the actin cytoskeleton and membrane blebbing (Tolstykh et al. 2016, 2017). Blebbing is vesicular protrusions of the plasma membrane above points of detachment of the underlying cortex. Blebs are filled and inflated with flowing cytoplasm from intracellular pressure, but are devoid of cortex and associated proteins. When expansion slows, the actin cortex is reassembled. Actin liner proteins, actin, actin-associated proteins and myosin motor proteins are recruited to the base of the bleb. The myosin motor proteins retract the bleb (Charras et al. 2006). Blebbing in response to usEPs had been reported earlier (Pakhomov et al. 2007a; Tekle et al. 2008) and described and defined in detail in a somewhat unique cell-type specific response to usEPs (Rassokhin and Pakhomov 2012, 2014), as will be presented later.

While the PIP2 hydrolysis and PLC activation were described in response to usEPs, the authors were more focused on these responses as mediated through G-protein coupled muscarinic and Angiotensin II (AngII) receptors as opposed to responses to cell injury (Tolstykh et al. 2016, 2017). Given what is known about responses to plasma membrane injury, these findings are clearly consistent with cells responding to nanoporation of plasma membranes. usEPs (1 and 20 pulses, 600 ns and 16.2 kV/cm, $9.7x10^{-3}$ -$4.3x10^{-2}$ Vs/cm; relatively intense pulses) and oxotremorine, a muscarinic agonist, induced a time-dependent hydrolysis of PIP2 and like AngII, activated PLC as shown by a translocation from the membrane to the cytoplasm that correlated with dissociation of the actin cytoskeleton and an increase in plasma membrane blebbing. These were inhibited by the PLC inhibitor edelfosine, demonstrating that usEPs, muscarinic agonists and AngII initiate PLC-induced hydrolysis of PIP2, which resulted in actin cytoskeletal remodeling and plasma membrane blebbing. Given that nsEPs induce cell injury by plasma membrane nanoporation and cell responses to plasma membrane injury involved PLC-induced PIP2 hydrolysis and actin-cytoskeleton remodeling and membrane blebbing, these cell responses are quite typical of downstream events expected in response to usEP-induced plasma membrane injury.

Another change that occurs at the plasma membrane in response to usEPs, as well as in response to other agonists, is externalization of PS (PSE), which can also be associated with plasma membrane blebbing. PSE is generally associated with apoptosis, but this is not always the case. For example, autophagy-deficient cells undergoing apoptosis do not show PSE (Qu et al. 2007). Also, human neutrophils exposed to usEPs, which induce PSE in Jurkat cells, did not show PSE (Beebe, unpublished). In contrast, early features of cells undergoing regulated cell death by parthanatos and netosis also show PSE (Verhoven et al. 1995; Wang et al. 2009b; Mihalache et al. 2011). For treating cells with usEPs there were also some considerations that were found to be required when evaluating PSE as an indicator for apoptosis. As discussed above, it was determined that PSE occurring on the anodic side of the cell during nanopore formation indicated that was not a molecular translocation across the plasma membrane, but was a nanopore facilitated event as negatively charged PS head groups were pulled through nanopores toward the positively charged electrode providing the energy as a direct electric field event across the plasma membrane.

Thus, nanopore formation and PSE occurred simultaneously as the anionic PS head group was electrophoretically transported through the hydrophilic pore while the pore was still being formed. This occurred much quicker than required for PSE in response to apoptosis. Thus, given these usEP findings as well as caveats with other confounders with PSE in regulated cell death, caution should be exercised when using PSE as an apoptosis marker. When evaluating PSE in response to usEPs, about 30 min should pass so re-internalization of usEP-induced PS can occur following usEP treatments (Batista Napotnik et al. 2012). As a more scientifically sound approach, determination of apoptosis or any RCD mechanisms should always use multiple indicators (Beebe et al. 2003a, 2013, 2016; Batista Napotnik et al. 2012).

Blebbing is commonly associated with apoptosis in the formation of apoptotic bodies, as a morphological characteristic of apoptosis (Charras 2006, 2008; Zhang et al. 2018). These are generally rounded or circular, but can differ in shape. In apoptosis, they dissociate from the cell generating apoptotic bodies. In general, blebbing begins by separation of the plasma membrane from the actin cortex or by a localized actin rupture. Blebs expand rapidly and their extension is dependent on actin polymerization. When expansion discontinues, bleb retraction is driven by reassembly of the contractile cortex under the membrane. Ezrin, a cross-linker between actin and the plasma membrane, is recruited followed by actin, actin bundling protein and contractile proteins are recruited to the membrane. When the contractile cortex assembly is complete, filaments form a cage-like structure under the membrane. Blebbing associated with apoptosis is caspase-dependent; caspase-deficient cells do not undergo blebbing. Caspase-3 cleaves and activates ROCK (Rho-associated protein kinase) in a Rho-independent manner leading to phosphorylation of myosin light chain (MLC) (Coleman et al. 2001; Sebbagh et al. 2001). MLC regulates assembly of myosin filaments, cell contractility, actomyosin contraction and the separation of the plasma membrane from the cortical cytoskeleton forming plasma membrane blebs. During apoptosis as the cell shrinks, intracellular pressure transients are generated by myosin II contraction of the actin cortex to drive inflation. Apoptotic bleb are ~1 to 5 μm in size and are filled with cytoplasm as well as fragments of DNA, nucleus and organelles. These apoptotic bodies are eventually phagocytized and recycled. Overall, blebbing during apoptosis is mechanistically similar to that occurring during blebbing in response to plasma membrane injury.

As previously indicated, blebbing is induced by usEPs. A unique formation of blebs occurred in U937 cells in response to usEPs. These blebs were structurally different than apoptotic blebs and were referred to as pseudo-like blebs (PLBs) (Rassokhin and Pakhomov 2012, 2014). Instead of the small, rounded, apoptotic membrane blebs, PLB were cylindrical extensions that could extend beyond the diameter of the cell in minutes. They formed on the anodal-facing side of the cell in response to colloidal-osmotic water uptake as cells swelled in response to plasma membrane permeabilization. Like other blebbing mechanisms, PLB extension depended on actin polymerization. Formation of PLB as well as their shape and growth were dependent on actin polymerization. Polymerization inhibitors cytochalasin D and latrunculin prevented PLB formation, while inhibition of WASP (Wiskott-Aldrich syndrome protein) with wiskostatin prevented PLB growth and extension.

Fluorescently labeled actin was shown to enter the PLB interior during elongation, forming the bleb, cortex, and scaffold. In contrast, blebbistatin, a myosin inhibitor, reduce retraction and a Rho-ROCK pathway inhibitor, Y-27632, prevented retraction. Rho GTPases effectors are well known regulators of actin and microtubules regulations. It is possible that cdc42, a Rho GTPase family member that is up upstream of WASP, is involved in actin polymerization. Alternatively, or in addition, ROCK, which is down-stream of RhoA, may also be involved in actin polymerization in this usEP-induced blebbing. However, ROCK was not shown to be Rho-dependent, so it could be activated by proteases that were shown for caspase-mediated ROCK activation in apoptosis. In any event, while the usEP-induced blebbing in U937 cells was unique and quite extensive, basic, common blebbing mechanisms were present.

It is also noteworthy and significant that the permeabilization that drives bleb formation is sensed by the cells as a plasma membrane injury. The entry of extracellular fluids and especially calcium is well-known wounding alert, as indicated above (Jimenez and Perez 2017). As already indicated, permeabilization depolarizes the transmembrane potential, activating pumping mechanisms to re-establish cellular homeostasis. These are series of rapid events from a number of different signaling pathways that participate in membrane repair including signaling through pathways that exploit calcium, oxidation–reduction (redox) reactions, lipids, kinases, GTPases and cytoplasmic dynamics (Horn and Jaiswal 2018). Calcium-sensitive proteins are damage sensors and participate early and rapidly in membrane plugging. As indicated in the blebbing responses, Rho family of GTPases regulate cytoskeletal dynamics. Even brief membrane injury leads to activation of Rho and/or cdc42 that are involved in actin and myosin dynamics as indicated in usEP-induced blebbing. Phospholipases contribute to modifying lipid composition in membrane wounds. As already mentioned, injury induces phospholipase C (PLC), generating diacylglycerol (DAG) and inositol triphosphate (IP3), which activates downstream effects and signaling platforms for PKC and Rho GTPases. As already indicated, usEPs induces PIP2 hydrolysis and PLC activity and have been shown to be responsible for remodeling of the actin cytoskeleton and membrane blebbing (Tolstykh et al. 2016, 2017). Redox signaling is initiated and proteins and lipid become oxidized. There is also significant cross-talk between Ca^{2+} and redox signaling. Some kinases, phosphatases and GTPases are Ca^{2+}- and redox-sensitive, so these signaling pathways have multiple activation mechanisms. Some reactions can be friend or foe, such as the Ca^{2+} activated protease calpain and redox reactions depending on the level of activation. Excessive calpain activation and ROS generation can be damaging and can lead to RCD and must be spatially and temporally controlled for purposes of membrane repair (Hyun et al. 2006; Jimenez and Perez 2017; Horn and Jaiswal 2018; Nakamura et al. 2018).

2.9 Effects of usEPs on Plasma Membrane Redox Systems (PMRS)

The model of the PMRS in Fig. 2.4 shows three major component including NAD(P)H as the intracellular electron donor, antioxidants CoQ and ά-tocopherol and quinone reductases. CoQ is abundant not only on the plasma membrane but also in mitochondria serving as a good one-electron pro-oxidant oxygen reduction. CoQ with or without ά-tocopherol scavenges ROS forming ubisemiquinone radical (CoQ^-) and prevent the dissemination of oxidative stress. The CoQ^- is transferred back to ubiquinol by PM redox enzymes, which includes cytochrome b5 reductase and NADH-quinone oxidoreductase (NQO1) shown in Fig. 2.4, as well as NADH-ferricyanide reductase, CoQ oxidase, an iron component and diferric transferrin (Hyun et al. 2006 and references within).

One of the mechanisms that had not been considered as responders to usEPs was the plasma membrane redox system (PMRS). The PMRS maintains the redox state of proteins, reduction of lipid hydroperoxides, stimulates cell proliferation, and as indicated above is reported to play an essential role in reducing oxidative stress, which has a significant impact on the rate of aging, lifespan, and many pathological conditions associated with increased oxidative stress (Rizvi and Srivastava 2010, Rizvi et al. 2011). In general, the PMRS efficiently functions to control cellular maintenance and homeostasis in the presence ROS generation from the intracellular or extracellular environment. ROS generation is simply a necessary problem caused by the use of oxygen in respiration and redox signaling and nitrogen in these and other cellular processes. A point to be reinforced is that low levels of ROS are used as necessary cell signaling mechanisms for many functions involving redox-regulated proteins. At intermediate levels, ROS induces inflammation and at higher excessive

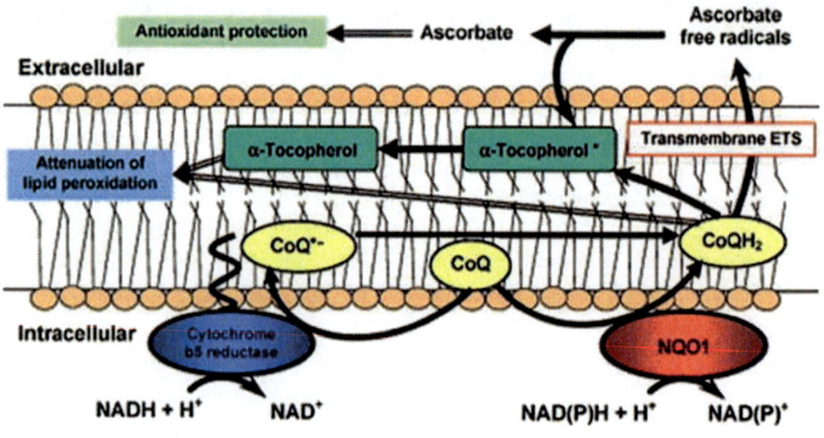

Fig. 2.4 The plasma membrane redox system (PMRS)—Abbreviations: ETS, electron transport system; and NQO1, NADH-quinone oxidoreductase. From Hyun et al. (2006) (Elsevier)

levels ROS induces RCD. It should be pointed out, that when a cell initiates and executes RCD mechanisms in response to a signal from a foe, the RCD mechanism is responding as a friend to the rest of the cells in the environment; RCD mechanisms are altruistic.

Oxidative phosphorylation (OXPHOS) is a system within mitochondria that functions to supply electrons from several sources to the electron transport chain (ETC) that is coupled to mechanisms that transport protons out of the mitochondrial matrix to establish a proton motive force that is ultimately used to produce ATP. During this process, electrons are ultimately transferred to oxygen, which is reduced to water or to generate H_2O_2 for a number of purposes, including cell signal transduction for coordinating cellular functions. When electrons are moved through the ETC to complex IV, they reduce oxygen to water. Depending on a number of factors, electrons can leak from the ETC, mostly from complexes I and III, which produce superoxide anions when oxygen is incompletely reduced. Superoxide dismutase converts the superoxides into H_2O_2. These systems will be discussed in Chap. 8.

When energy is scarce, such as with heavy muscular activities, caloric restriction or mitochondria dysfunction, glycolysis and fermentation in the cytosol supply alternative energy sources. An important part of this system is provided by the PMRS. During energy deprivation, the PMRS is activated to maintain a balanced NAD(P) + /NAD(P)H ratio that is critical for normal metabolism that protects cells against oxidative stress and apoptosis. Cells that lack mitochondria or mitochondrial function (ρ^o cells) have higher PMRS activity and produce less ROS, suggesting the PMRS has been upregulated to compensate for the absence of mitochondria (Scarlett et al. 2004). While the full physiological significance of the PMRS requires further investigation, it is known to be involved in regulation of cell physiology, hormonal signaling, and a primary defense system to protect against exogenous oxidative stress (Hyun et al. 2006; Horn and Jaiswal 2018).

The PMRS consists of three primary components including an intracellular electron donor, NAD(P)H; antioxidants, such as CoQ (Co-enzyme Q or ubiquinone) and alpha-tocopherol; and quinone reductase. CoQ and alpha-tocopherol scavenge superoxide anions and other ROS molecules preventing oxidative stress by forming ubisemiquinone radicals (CoQ^-). The CoQ radicals are converted back to ubiquinone by the plasma membrane redox enzymes. This protects membranes against lipid peroxidation and protein oxidation and resets the antioxidant for additional ROS protection. Plasma membrane redox enzymes include cell type-specific enzymes such as cytochrome b reductase and NAD(P)H quinone oxidoreductase, among others. A key characteristic of these endogenous antioxidants is their transient existence. This means that redox signaling can be reversible, specific, and localized for appropriate control in specific membrane damaged sites. Thus, when a specific plasma membrane site is breached, the response is localized to this vicinity.

In this context, it is interesting to consider how these redox systems function in the face of usEP-induced nanopore formation. This depends on the intensity of the usEPs and whether the PMRS antioxidant systems are sufficient to handle the ROS load induced by usEPs. When the PMRS are overloaded, reversible or irreversible oxidative damage occurs to lipids and proteins depending on whether damage is salvageable or not. Cells can attempt to repair such damages by inducing autophagy to sacrifice weaker cellular components for recycling salvage repair or can induce RCD if the damage is beyond repair. It is also possible and likely that usEPs can interfere with the function of the PMRS as it can in the mitochondrial electron transport chain. In order to determine effects of usEPs on the PMRS, we exposed cells to usEPs and analyzed redox behavior in locations in and near the plasma membrane. In other words, can usEPs affect the PMRS and how does the PMRS respond in different cell types.

A convenient way to analyze the PMRS is to use a redox sensing agent that does not enter cells and therefore assays redox systems in the plasma membrane. We used the CCK8 reagent (WST-8), which determines cell number-dependent redox reactions as a determinant of viability. The CCK-8 reagent [2-(2-methoxy-4-nitrophenyl)-3-(4-nitrophenyl)-5-(2,4-disulfophenyl)-2H-tetrazolium, monosodium salt], is highly water soluble and cell impermeable since it carries net negative charges. The WST8 reagent is reduced as it receives electrons from activity of NAD(P)H dehydrogenases at or near the plasma membrane. These electrons are then used to reduce a water-soluble orange formazan dye that is detected as a product of the PMRS. The equivalent cellular reaction would transfer electrons from NAD(P)H to CoQ by cytochrome b5 reductase among others, which would form the CoQ radical.

In order to use this redox system, we treated high concentrations of cells ($1 \times 10^6/100\ \mu l$) with usEPs so initial enzyme rates could be measured from reaction products that would reach sufficient measurable levels in times soon after treating cells. We used two different cell lines: H9c2 cardiac myoblasts, which are derived from non-cancerous cardiac tissues and exhibit good mitochondrial respiration rates and 4T1 mammary cancer cells, which we have used in cancer ablation studies (Guo et al. 2018; Beebe et al. 2018) and primarily use glycolysis as a major metabolic sustenance. H9c2 cardiac myoblast cells have robust mitochondrial OXPHOS functions and 4T1-luc mouse mammary cancer cells do not and depend primarily on glycolysis for energy production. In data not shown, H9c2 cells consumed oxygen about 4-times greater than 4T1-luc cells under basal conditions, in the presence of succinate and ADP and when uncoupled from OXPHOS.

Figure 2.5 show activities of the PMRS in H9c2 cells and 4T1 mammary cancer cells exhibit quite different sensitivities to usEPs. Figure 2.5 shows full time courses of the PMRS activity when cells were treated by different pulse numbers with 60 ns durations and electric fields of 40 kV/cm (1 pulse 2.4×10^{-4} Vs/cm times square root of pulse number). In both cell types lower pulse numbers increased PMRS activities. Figure 2.5 show a typical experiment for effects of usEPs on the redox activity in treated rat cardiac myoblasts (A) and mouse 4T1-luc mammary cancer cells (B). There are three phases in the kinetic of redox activity under these conditions. The first phase is linear with time, but relatively slow, which is likely due to reagent localization and early limited substrate availability. The second phase is linear with time and faster than the initial phase. This phase was used to determine PMRS rates shown in a typical experiment in Fig. 2.6. The third rate phase losses linearity with time due to substrate depletion and limitations of instrumentation sensitivity. Figure 2.7 shows experiments analyzing effects of usEPs on enzymatic rates of the

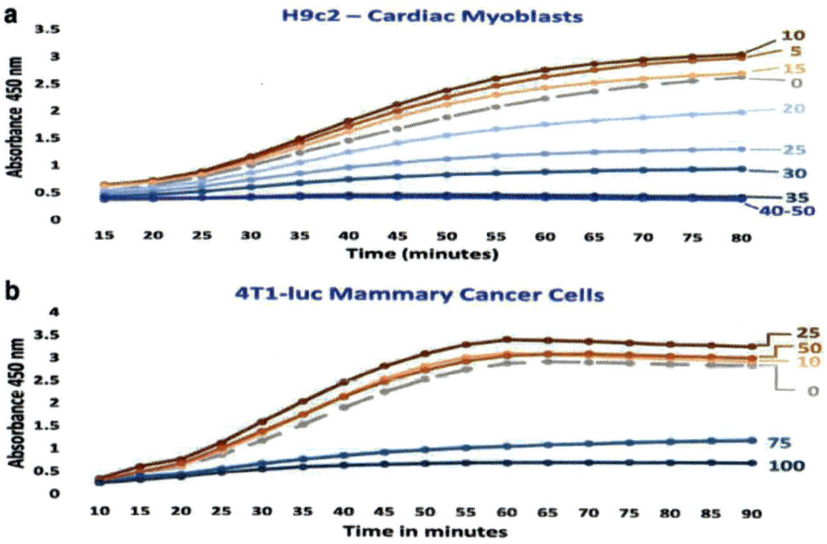

Fig. 2.5 usEPs attenuate the plasma membrane redox system—H9c2 cardiac myoblast (A) and 4T1-luc mammary cancer cells were treated with various pulse numbers as indicated with durations of 60 ns and electric fields of 40 kV/cm. Cells (1×10^6) were incubated with the WST-8 reagent for various times and enzyme activity was determine by absorbance of the dye at 450 nm. The numbers associated with each of the reaction phases indicate the number of pulses applied. The image represents a typical experiment

PMRS kinetics determined in the second kinetic phase of redox activity in Fig. 2.6. It includes a statistical comparison of the PMRS enzymatic rates compared to viability 24 h after usEP treatment. It is noteworthy that the H9c2 cells are a clone derived from embryonic BD1X rat heart tissue (Kimes and Brandt 1976) and have shown usefulness as an alternative to primary cardiomyocytes in vitro (Watkins et al. 2011). Because the heart cells are rich in mitochondria, they primarily use oxidative phosphorylation OXPHOS) for ATP production to produce energy needed for contractile function. In contrast, 4T1-luc mammary cancer cells are tumorigenic and metastatic and depend more on glycolysis instead of OXPHOS. It is known that cancer cells function under higher ROS levels, which promote proliferations, and exhibit higher thioredoxin antioxidant levels than normal cells (Schumacker 2006, 2015). Interestingly, regardless of primary metabolism, both cells have biphasic responses to usEPs showing increased redox rates at low pulse numbers and attenuated rates at higher pulse numbers. H9c2 cells have a slightly higher (28%) basal, unstimulated redox rate than 4T1-luc cells (0.419 vs. 0.327 absorbance units/min). However, the usEP-induced increase in redox rate in 4T1 was greater than in H9c2 cells (32% vs. 9%).

Figure 2.7 shows relationships between immediate usEP effects on the PMRS and the viability 24 h later given the same treatment. There are some unique differences between these two cellular PMRSs. In general, the 4T1 cell PMRS is more robust than

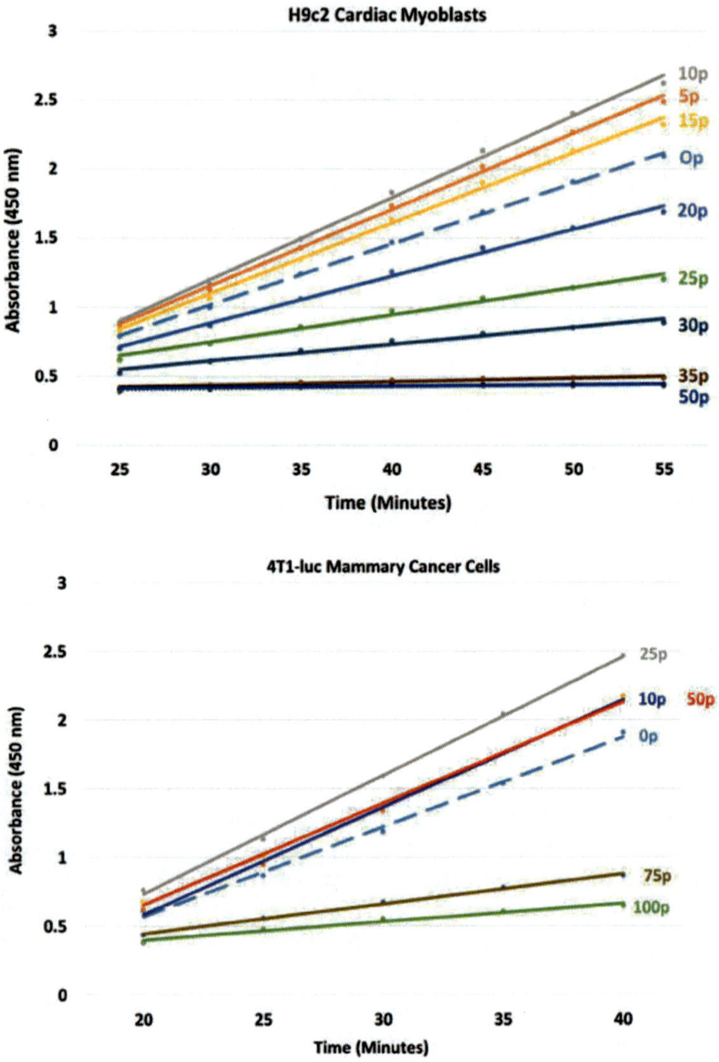

Fig. 2.6 Plasma membrane redox system activity rates in H9c2 cardiac myoblasts and 4T1-luc mammary cancer cells—A typical figure illustrates the linear rates of redox reactions in H9c2 heart cells and 4T1-luc cancer cells. Rates were determined in the second phase of reactions shown in Fig. 2.6. The number by each rate indicates the pulse number followed by the redox rates as absorbance units/min using the WST-8 reagent

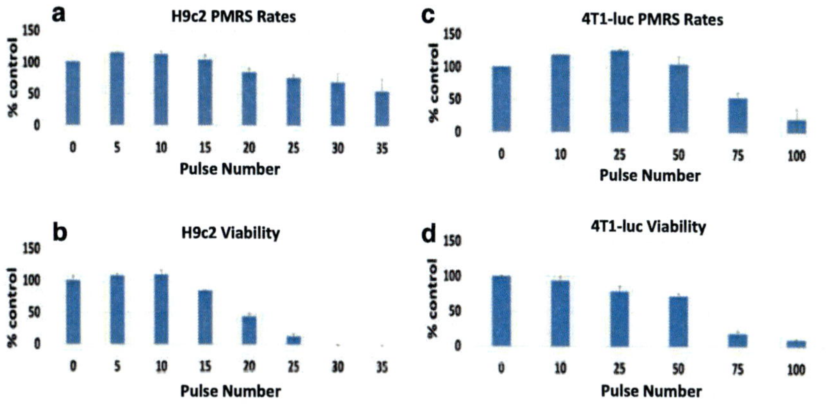

Fig. 2.7 PLRS rates and viabilities of H9c2 cardiac myoblasts and 4T1-luc mammary cancer cells—H9c2 and 4T1-luc cells (1×10^6) were treated with usEP conditions of 60 ns and 40 kV/cm with increasing pulse numbers as indicate. Redox reaction rates were determined immediately after pulse treatment as shown in Fig. 2.6 and redox rates were determined as shown in Fig. 2.7. Viability was determined using the WST-8 reagent 24 h after treatment. The results indicate the mean and stand deviations of 4 experiments

H9c2 cells. For example, 25 pulses increased 4t1-luc redox activity by about 25%, but inhibited H9c2 redox activity by about the same percentage. Pulsing condition that almost completely inhibited H9c2 redox activity had little or no effect on the 4T1-luc cell PMRS. For the same 50 pulse conditions, H9c2 cells have lost all viability while 4T1-luc viability has decreased only about 30%. In addition, 4T1-luc cells were more dependent on PMRS than H9c2 cells. When usEPs began to decrease 4T1-luc cell redox activity, cells began to loss viability; the decrease in 4T1-luc viability closely close mirrored decrease in PMRS rates. H9c2 cells still exhibited PMRS activity after treatment even when they would lose all cell viability 24 h later. Since heart cells have large numbers of mitochondria and depend on oxidative phosphorylation for their high energy demand, the PMRS provided only a limited energy backup compared to 4T1 cancer cells and is highly sensitive to usEPs. So, viability in H9c2 cells is less dependent on PMRS or the PMRS can only supply limited redox balance for cells that are primarily OXPHOS-dependent. In contrast, 4T1-luc cells, which have lower respiration through OXPHOS and depend mainly on glycolysis for ATP production, have robust PMRS activity that is relatively insensitive to usEPs until the pulse number are quite high. In data not shown here, the antioxidant vitamin C reduces inhibitory effects of usEP on PM redox reactions, but has no effects on usEP effects on plasma membrane permeability, suggesting that usEP effects on the PMRS are independent of plasma membrane permeabilization.

It is noteworthy that usEPs affect redox reactions, which are essentially electron transfer mechanisms. This indicates that usEP affect electron transfer in the plasma membranes of these cells. As seen in the Chap. 8 on mitochondria, usEPs also attenuate electron transfer in the electron transport chain (ETC) at complexes I and

IV as determined by decreases on oxygen flux or oxygen consumption in H9c2 cells. This suggests that usEPs can generally decrease electron flow in the PMRS and in the mitochondria ETC. UsEP may affect the enzymatic activity of redox enzymes and/or other mechanisms.

Other have analyzed effects of usEP on intracellular biochemical reaction using NADH autofluorescence microscopy (Awasthi et al. 2016). Changes in the life-time fluorescence of HADH can be used to differentiate between NADH bound or unbound to proteins. The life-time decay of NADH is much faster when it is not bound to protein. Therefore, an increase in lifetime indicates that NADH is more protein bound. In addition, an increase in protein bound NADH is more common in OXPHOS metabolism than in glycolysis. Using this approach, it was found that usEPs increases and then decreased the fluorescence lifetime intensity of NADH in HeLa cells in a stepwise manner regardless of the pulse width between 10 and 50 ns. However, the fluorescence intensity, which suggests increases in NADH, increased as pulse width increased and was distributed throughout the entire cell. These usEPs were applied at 45 kV/cm with a repetition rate of 1 kHz for 2 and 3 min to observe these changes. The authors found that these changes in fluorescent lifetime were very similar to those observed when these cells were treated with staurosporine, which is well known to induce apoptosis. The authors interpreted these results to indicate that usEPs induced apoptosis since both stimuli resulted in cell swelling and membrane blebbing. Especially at longer pulse duration the spread of fluorescence throughout the cell was suggested to result from NADH leaking into the cytoplasm as an indication that pores had opened in the mitochondria, especially the mitochondria permeability transition pore (mPTP), which is likely responsible for the induction of apoptosis. There are other discussions about effects of usEPs on mitochondria in the section of usEP effects on mitochondria (see Chap. 8) and in usEPs induce RCD mechanisms (see Chap. 8).

2.10 Summary

While the unique effects of usEPs occur inside the plasma membrane, the membrane itself is a sensor for such short pulses. However, usEP effects on the plasma membrane are distinctly different than conventional electroporation pulses with durations in the micro- and milli-second ranges. Specifically, usEP induce dense pore formation on all cell membranes with nanometer sizes, thus the name nanopores. These pores are unique because they have channel characteristics such as voltage-sensitivity, opening at high negative voltages, and inwardly rectifying pores. Formation of nanopores can result in cellular injury. Not surprisingly, cells are adept at repairing injury since this is the definition of the existence from the rest of the world. UsEP also have effects on channel activities that affect ion entry. This can occur in response to nanopore formation or perhaps directly on a channel itself. Finally, usEPs affect electron transport in the PMRS. As pointed out in the section of mitochondria, usEPs affect electron transport in the mitochondria. The mechanisms for this effect on electron transport remain to be determined.

References

Aglietti RA, Estevez A, Gupta A, Ramirez MG, Liu PS, Kayagaki N, Ciferri C, Dixit VM, Dueber EC (2016) GsdmD p30 elicited by caspase-11 during pyroptosis forms pores in membranes. Proc Natl Acad Sci USA 113:7858–7863

Alberts B, Johnson A, Lewis J, Raff M, Roberts K, Walter P (eds) (2014) Molecular biology of the cell, 6th edn. New York and Abingdon, UK. ISBN: 9780815344643

Armstrong CM, Hille B (1998) Voltage-gated ion channels and electrical excitability. Neuron Rev 20:371–380

Awasthi K, Nakabayashi T, Ohta N (2016) Effects of nanosecond pulsed electric fields on the intracellular function of HeLa cells as revealed by NADH autofluorescence microscopy. ACS Omega 1:396–406

Azarov JE, Semenov I, Casciola M, Pakhomov AG (2018) Excitation of murine cardiac myocytes by nanosecond pulsed electric field. J Cardiovasc Electrophysiol. https://doi.org/10.1111/jce.13834

Bailey RW, Nguyen T, Robertson L, Gibbons E, Nelson J, Christensen RE, Bell JP, Judd AM, Bell JD (2009) Sequence of physical changes to the cell membrane during glucocorticoid-induced apoptosis in S49 lymphoma cells. Biophys J 96:2709–2718

Batista Napotnik T, Wu YH, Gundersen MA, Miklavčič D, Vernier PT (2012) Nanosecond electric pulses cause mitochondrial membrane permeabilization in Jurkat cells. Bioelectromagnetics 33:257–264

Batista Napotnik T, Reberšek M, Vernier PT, Mali B, Miklavčič D (2016) Effects of high voltage nanosecond electric pulses on eukaryotic cells (in vitro): A systematic review. Bioelectrochemistry 110:1–12

Beebe SJ (2015) Considering effects of nanosecond pulsed electric fields on proteins. Bioelectrochemistry 103:52–59

Beebe, SJ (2016) Preclinical studies with nanosecond pulses. In: Miklavčič D (ed) Handbook of electroporation. Springer, Berlin, pp 1543–1562. ISBN: 978-3-319-26779-1 (Print) 978-3-319-26779-1

Beebe SJ, Fox PM, Rec LH, Buescher ES, Somers K, Schoenbach KH (2002) Nanosecond pulsed electric field (nsPEF) effects on cells and tissues: apoptosis induction and tumor growth inhibition. IEEE Trans Plasma Sci 30:286–292

Beebe SJ, White J, Blackmore PF, Deng Y, Somers K, Schoenbach KH (2003a) Diverse effects of nanosecond pulsed electric fields on cells and tissues. DNA Cell Biol 22:785–796

Beebe SJ, Fox PM, Rec LJ, Willis EL, Schoenbach KH (2003b) Nanosecond, high-intensity pulsed electric fields induce apoptosis in human cells. FASEB J 17:1493–2145

Beebe SJ, Chen YJ, Sain NM, Schoenbach KH, Xiao S (2012) Transient features in nanosecond pulsed electric fields differentially modulate mitochondria and viability. PLoS One 7(12):e51349

Beebe SJ, Sain NM, Ren W (2013) Induction of cell death mechanisms and apoptosis by nanosecond pulsed electric fields (nsPEFs). Cells 2:136–162

Beebe SJ, Lassiter BP, Guo S (2018) Nanopulse stimulation (NPS) induces tumor ablation and immunity in orthotopic 4T1 mouse breast cancer: a review. Cancers (Basel) 10(pii):E97

Blazek AD, Paleo BJ, Weisleder N (2015) Plasma membrane repair: a central process for maintaining cellular homeostasis. Physiol (Bethesda) 30:438–448

Botto L, Beretta E, Daffara R, Miserocchi G, Palestini P (2006) Biochemical and morphological changes in endothelial cells in response to hypoxic interstitial edema. Respir Res 7:7

Bowman AM, Nesin OM, Pakhomova ON, Pakhomov AG (2010) Analysis of plasma membrane integrity by fluorescent detection of Tl(+) uptake. J Membr Biol 236:15–26

Bubb KJ, Birgisdottir AB, Tang O, Hansen T, Figtree GA (2017) Redox modification of caveolar proteins in the cardiovascular system-role in cellular signalling and disease. Free Radic Biol Med 109:61–74

Buescher ES, Smith RR, Schoenbach KH (2004) Submicrosecond intense pulsed electric field effects on intracellular free calcium: mechanisms and effects. IEEE Trans Plasma Sci 32:1563–1572

Cai C, Masumiya H, Weisleder N, Matsuda N, Nishi M, Hwang M, Ko JK, Lin P, Thornton A, Zhao X, Pan Z, Komazaki S, Brotto M, Takeshima H, Ma J (2009) MG53 nucleates assembly of cell membrane repair machinery. Nat Cell Biol 11:56–64

Cai Z, Jitkaew S, Zhao J, Chiang HC, Choksi S, Liu J, Ward Y, Wu LG, Liu ZG (2014) Plasma membrane translocation of trimerized MLKL protein is required for TNF-induced necroptosis. Nat Cell Biol 16(1):55–65

Cantu JC, Tarango M, Beier HT, Ibey BL (2016) The biological response of cells to nanosecond pulsed electric fields is dependent on plasma membrane cholesterol. Biochim Biophys Acta 1858:2636–2646

Casciola M, Bonhenry D, Liberti M, Apollonio F, Tarek M (2014) A molecular dynamic study of cholesterol rich lipid membranes: comparison of electroporation protocols. Bioelectrochemistry 100:11–17

Chan FK, Luz NF, Moriwaki K (2015) Programmed necrosis in the cross talk of cell death and inflammation. Annu Rev Immunol 33:79–106

Charras GT (2008) A short history of blebbing. J Microsc 231:466–478

Charras GT, Hu CK, Coughlin M, Mitchison TJ (2006) Reassembly of contractile actin cortex in cell blebs. J Cell Biol 75:477–490

Charras GT, Hu CK, Coughlin M, Mitchison TJ (2006) Reassembly of contractile actin cortex in cell blebs. J Cell Biol 175:477–490

Chen, Schoenbach KH, Kolb JF, James SR, Garner AL, Yang J, Joshi RP, Beebe SJ (2004) Leukemic cell intracellular responses to nanosecond electric fields. Biochem Biophys Res Commun 317:421–427

Chen X, He WT, Hu L, Li J, Fang Y, Wang X, Xu X, Wang Z, Huang K, Han J (2016) Pyroptosis is driven by non-selective gasdermin-D pore and its morphology is different from MLKL channel-mediated necroptosis. Cell Res 26:1007–1020

Cheng DK, Tung L, Sobie EA (1999) Nonuniform responses of transmembrane potential during electric field stimulation of single cardiac cells. Am J Physiol 277:H351-362

Cohen P (2000) The regulation of protein function by multisite phosphorylation–a 25 year update. Trends Biochem Sci 25:596–601

Coleman ML, Sahai EA, Yeo M, Bosch M, Dewar A, Olson MF (2001) Membrane blebbing during apoptosis results from caspase-mediated activation of ROCK I. Nat Cell Biol 3:339–345

Craviso GL, Chatterjee P, Maalouf G, Cerjanic A, Yoon J, Chatterjee I, Vernier PT (2009) Nanosecond electric pulse-induced increase in intracellular calcium in adrenal chromaffin cells triggers calcium-dependent catecholamine release. IEEE Trans Dielect Electr Ins 16:1294–1301

Craviso GL, Choe S, Chatterjee P, Chatterjee I, Vernier PT (2010) Nanosecond electric pulses: a novel stimulus for triggering Ca2+ influx into chromaffin cells via voltage-gated Ca2+ channels. Cell Mol Neurobiol 30:1259–1265

Davies SS, Guo L (2014) Lipid peroxidation generates biologically active phospholipids including oxidatively N-modified phospholipids. Chem Phys Lipids 181:1–33

Demonbreun AR, McNally EM (2016) Plasma membrane repair in health and disease. Curr Top Membr 77:67–96

Denes A, Lopez-Castejon G, Brough D (2012) Caspase-1: is IL-1 just the tip of the ICEberg? Cell Death Dis 3:e338

Deng J, Schoenbach KH, Buescher ES, Hair PS, Fox PM, Beebe SJ (2003) The effects of intense submicrosecond electrical pulses on cells. Biophys J 84:2709–2714

Ding J, Wang K, Liu W, She Y, Sun Q, Shi J, Sun H, Wang DC, Shao F (2016) Pore-forming activity and structural autoinhibition of the gasdermin family. Nature 535:111–116

Donate A, Bulysheva A, Edelblute C, Jung D, Malik MA, Guo S, Burcus N, Schoenbach K, Heller R (2016) Thermal assisted in vivo gene electrotransfer. Curr Gene Ther 16:83–89
Edelblute CM, Hornef J, Burcus NI, Norman T, Beebe SJ, Schoenbach K, Heller R, Jiang C, Guo S (2017) Controllable moderate heating enhances the therapeutic efficacy of irreversible electroporation for pancreatic cancer. Sci Rep 7:11767
Edelblute CM, Guo S, Hornef J, Yang E, Jiang C, Schoenbach K, Heller R (2018) Moderate heat application enhances the efficacy of nanosecond pulse stimulation for the treatment of squamous cell carcinoma. Technol Cancer Res Treat 17:1533033818802305
Esser AT, Smith KC, Gowrishankar TR, Weaver JC (2009) Towards solid tumor treatment by nanosecond pulsed electric fields. Technol Cancer Res Treat 8:289–306
Esser AT, Smith KC, Gowrishankar TR, Vasilkoski Z, Weaver JC (2010) Mechanisms for the intracellular manipulation of organelles by conventional electroporation. Biophys J 98:2506–2514
Estlack LE, Roth CC, Thompson GL 3rd, Lambert WA 3rd, Ibey BL (2014) Nanosecond pulsed electric fields modulate the expression of Fas/CD95 death receptor pathway regulators in U937 and Jurkat Cells. Apoptosis 19:1755–1768
Fernández ML, Marshall B, Sagués F, Reigada R (2010) Structural and kinetic molecular dynamics study of electroporation in cholesterol-containing bilayers. J Phys Chem B 114:6855–6865
Fielding CJ, Fielding PE (2000) Cholesterol and caveolae: structural and functional relationships. Biochim Biophys Acta 1529(1–3):210–222
Foster KR (2000) Thermal and nonthermal mechanisms of interaction of radio-frequency energy with biological systems. IEEE Trans Plasma Sci 28:15–23
Frey W, White JA, Price RO, Blackmore PF, Joshi RP, Nuccitelli R, Beebe SJ, Schoenbach KH, Kolb JF (2006) Plasma membrane voltage changes during nanosecond pulsed electric field exposure. Biophys J 90:3608–3615
Gowrishankar TR, Esser AT, Vasilkoski Z, Smith KC, Weaver JC (2006). Microdosimetry for conventional and supra-electroporation in cells with organelles. Biochem Biophys Res Commun 341(4):1266–1276
Graves JD, Krebs EG (1999) Protein phosphorylation and signal transduction. Pharmacol Ther 82:111–121
Guo S, Jing Y, Burcus NI, Lassiter BP, Tanaz R, Heller R, Beebe SJ (2018) Nano-pulse stimulation induces potent immune responses, eradicating local breast cancer while reducing distant metastases. Int J Cancer 142:629–640
Hall EH, Schoenbach KH, Beebe SJ (2007) Nanosecond pulsed electric fields induce apoptosis in p53-wildtype and p53-null HCT116 colon carcinoma cells. Apoptosis 12(9):1721 1731
Hayes MJ, Merrifield CJ, Shao D, Ayala-Sanmartin J, Schorey CD, Levine TP, Proust J, Curran J, Bailly M, Moss SE (2004) Annexin 2 binding to phosphatidylinositol 4,5-bisphosphate on endocytic vesicles is regulated by the stress response pathway. J Biol Chem 279:14157–14164
Horn A, Jaiswal JK (2018) Cellular mechanisms and signals that coordinate plasma membrane repair. Cell Mol Life Sci 75:3751–3770
Hu Q, Joshi RP, Schoenbach KH (2005a) Simulations of nanopore formation and phosphatidylserine externalization in lipid membranes subjected to a high-intensity, ultrashort electric pulse. Phys Rev E Stat Nonlin Soft Matter Phys 72:031902
Hu Q, Viswanadham S, Joshi RP, Schoenbach KH, Beebe SJ, Blackmore PF (2005b) Simulations of transient membrane behavior in cells subjected to a high-intensity, ultra-short electric pulse. Phys Rev E 71:031914
Huang D, Zheng X, Wang ZA, Chen X, He WT, Zhang Y, Xu JG, Zhao H, Shi W, Wang X, Zhu Y, Han J (2017) The MLKL channel in necroptosis is an octamer formed by tetramers in a dyadic. Process Mol Cell Biol 37(pii):e00497-16
Hyun DH, Hernandez JO, Mattson MP, de Cabo R (2006) The plasma membrane redox system in aging. Ageing Res Rev 5(2):209–220
Ibey BL, Xiao S, Schoenbach KH, Murphy MR, Pakhomov AG (2009) Plasma membrane permeabilization by 60- and 600-ns electric pulses is determined by the absorbed dose. Bioelectromagnetics 30:92–99

Ibey BL, Mixon DG, Payne JA, Bowman A, Sickendick K, Wilmink GJ, Roach WP, Pakhomov AG (2010). Plasma membrane permeabilization by trains of ultrashort electric pulses. Bioelectrochemistry 79:114–121. [PubMed: 20171148]

Jimenez AJ, Perez F (2017) Plasma membrane repair: the adaptable cell life-insurance. Curr Opin Cell Biol 47:99–107

Joshi RP, Hu Q, Schoenbach KH, Hjalmarson HP (2002) Improved energy model for membrane electroporation in biological cells subjected to electrical pulses. Phys Rev E Stat Nonlin Soft Matter Phys 65:041920

Joshi RP, Schoenbach KH, Beebe SJ, Blackmore PF (2005) Simulations of transient membrane behavior in cells subjected to a high-intensity ultrashort electric pulse. Hu Q, Viswanadham S, Phys Rev E Stat Nonlin Soft Matter Phys 71(3):031914

Kimes BW, Brandt BL (1976) Properties of a clonal muscle cell line from rat heart. Exp Cell Res 98:367–381

Kotnik T, Miklavcic D (2006) Theoretical evaluation of voltage inducement on internal membranes of biological cells exposed to electric fields. Biophys J 90:480–491

Lassiter BP, Guo S, Beebe SJ (2018) Nano-pulse stimulation ablates orthotopic rat hepatocellular carcinoma and induces innate and adaptive memory immune mechanisms that prevent recurrence. Cancers (Basel) 10(3):E69

Law RH, Lukoyanova N, Voskoboinik I, Caradoc-Davies TT, Baran K, Dunstone MA, D'Angelo ME, Orlova EV, Coulibaly F, Verschoor S, Browne KA, Ciccone A, Kuiper MJ, Bird PI, Trapani JA, Saibil HR, Whisstock JC (2010) The structural basis for membrane binding and pore formation by lymphocyte perforin. Nature 468:447–451

Liu X, Zhang Z, Ruan J, Pan Y, Magupalli VG, Wu H, Lieberman J (2016) Inflammasome-activated gasdermin D causes pyroptosis by forming membrane pores. Nature 535:153–158

Marracino P, Apollonio F, Liberti M, d'Inzeo G, Amadei A (2013) Effect of high exogenous electric pulses on protein conformation: myoglobin as a case study. J Phys Chem B 117:2273–2279

Mihalache CC, Yousefi S, Conus S, Villiger PM, Schneider EM, Simon HU (2011) Inflammation associated autophagy-related programmed necrotic death of human neutrophils characterized by organelle fusion events. J Immunol 186:6532–6542

Nakamura M, Dominguez ANM, Decker JR, Hull AJ, Verboon JM, Parkhurst SM (2018) Into the breach: how cells cope with wounds. Open Biol 8:180135

Nesin V, Pakhomov AG (2012a) Inhibition of voltage-gated Na(+) current by nanosecond pulsed electric field (nsPEF) is not mediated by Na(+) influx or Ca(2+) signaling. Bioelectromagnetics 33:443–451

Nesin V, Bowman AM, Xiao S, Pakhomov AG (2012b) Cell permeabilization and inhibition of voltage-gated Ca(2+) and Na(+) channel currents by nanosecond pulsed electric field. Bioelectromagnetics 33:394–404

Neumann E, Schaefer-Ridder M, Wang Y, Hofschneider PH (1982) Gene transfer into mouse lymphoma cells by electroporation in high electric fields. EMBO J 1(7):841–845

Neumann E, Kakorin S, Toensing K (1999) Fundamentals of electroporative delivery of drugs and genes. Bioelectrochem Bioenerg 48:3–16

Nordzieke DE, Medraño-Fernandez I (2018) The plasma membrane: a platform for intra- and intercellular redox signaling. Antioxidants (Basel) 7:E168

Okada Y, Maeno E, Shimizu T, Dezaki K, Wang J, Morishima S (2001) Receptor-mediated control of regulatory volume decrease (RVD) and apoptotic volume decrease (AVD). J Physiol 532:3–16

Op den Kamp JA (1979) Lipid asymmetry in membranes. Annu Rev Biochem 48:47–71

Pakhomov AG, Pakhomova ON (2010) Nanopores: a distinct transmembrane passageway in electroporated cells. In: Pakhomov AG, Miklavcic D, Markov MS (eds) Advanced electroporation techniques in biology in medicine. CRC Press, pp 178–194

Pakhomov AG, Shevin R, White JA, Kolb JF, Pakhomova ON, Joshi RP, Schoenbach KH (2007a) Membrane permeabilization and cell damage by ultrashort electric field shocks. Arch Biochem Biophys 465:109–118

Pakhomov AG, Kolb JF, White JA, Joshi RP, Xiao S, Schoenbach KH (2007b) Long-lasting plasma membrane permeabilization in mammalian cells by nanosecond pulsed electric field (nsPEF). Bioelectromagnetics 28:655–663

Pakhomov AG, Bowman AM, Ibey BL, Andre FM, Pakhomova ON, Schoenbach KH (2009) Lipid nanopores can form a stable, ion channel-like conduction pathway in cell membrane. Biochem Biophys Res Commun 385:181–186

Pakhomov AG, Semenov I, Casciola M, Xiao S (2017) Neuronal excitation and permeabilization by 200-ns pulsed electric field: an optical membrane potential study with FluoVolt dye. Biochim Biophys Acta Biomembr 1859:1273–1281

Papackova Z, Cahova M (2015) Fatty acid signaling: the new function of intracellular lipases. Int J Mol Sci 16:3831–3855

Pawson T, Scott JD (2005) Protein phosphorylation in signaling–50 years and counting. Trends Biochem Sci 30:286–290

Qu X, Zou Z, Sun Q, Luby-Phelps K, Cheng P, Hogan RN et al (2007) Autophagy gene dependent clearance of apoptotic cells during embryonic development. Cell 128:931–946

Rassokhin MA, Pakhomov AG (2012) Electric field exposure triggers and guides formation of pseudopod-like blebs in U937 monocytes. J Membr Biol 245:521–529

Rassokhin MA, Pakhomov AG (2014) Cellular regulation of extension and retraction of pseudopod-like blebs produced by nanosecond pulsed electric field (nsPEF). Cell Biochem Biophys 69:555–566

Ray S, Kassan A, Busija AR, Rangamani P, Patel HH (2016) The plasma membrane as a capacitor for energy and metabolism. Am J Physiol Cell Physiol 310:C181-192

Reale R, English NJ, Garate JA, Marracino P, Liberti M, Apollonio F (2013) Human aquaporin 4 gating dynamics under and after nanosecond-scale static and alternating electric-field impulses: a molecular dynamics study of field effects and relaxation. J Chem Phys 139:205101

Ren W, Sain NM, Beebe SJ (2012) Nanosecond pulsed electric fields (nsPEFs) activate intrinsic caspase-dependent and caspase-independent cell death in Jurkat cells. Biochem Biophys Res Commun 421(4):808–812

Rizvi SI, Srivastava N (2010) Erythrocyte plasma membrane redox system in first degree relatives of type 2 diabetic patients. Intern J Diab Mell 2:119–121

Rizvi SI, Kumar D, Chakravarti S, Singh P (2011) Erythrocyte plasma membrane redox system may determine maximum life span. Med Hypotheses 76:547–549

Romeo S, Zeni L, Sarti M, Sannino A, Scarfì MR, Vernier PT, Zeni O (2011) DNA electrophoretic migration patterns change after exposure of Jurkat cells to a single intense nanosecond electric pulse. PLoS ONE 6:e28419

Roth CC, Tolstykh GP, Payne JA, Kuipers MA, Thompson GL, DeSilva MN, Ibey BL (2013) Nanosecond pulsed electric field thresholds for nanopore formation in neural cells. J Biomed Opt 18(3):035005

Sborgi L, Rühl S, Mulvihill E, Pipercevic J, Heilig R, Stahlberg H, Farady CJ, Müller DJ, Broz P, Hiller S (2016) GSDMD membrane pore formation constitutes the mechanism of pyroptotic cell death. EMBO J 35:1766–1778

Scarlett DJ, Herst P, Tan A, Prata C, Berridge M (2004) Mitochondrial gene-knockout (rho0) cells: a versatile model for exploring the secrets of trans-plasma membrane electron transport. Biofactors Rev. 20:199–206

Schoenbach KH, Peterkin FE, Alden RW, Beebe SJ (1997) The effect of pulsed electric fields on biological cells: experiments and applications. IEEE Trans Plasma Sci 25:284–292

Schoenbach KH, Joshi RP, Stark RH, Dobbs F, Beebe SJ (2000) Bacterial decontamination of liquids with pulsed electric fields, invited review paper. IEEE Trans Dielectr Electr Insul 7:637

Schoenbach KH, Beebe SJ, Buescher ES (2001) Intracellular effect of ultrashort electrical pulses. Bioelectromagnetics 22:440–448

Schoenbach KH, Joshi RP, Kolb JF, Chen N, Stacey M, Blackmore PF, Buescher ES, Beebe SJ (2004) Ultrashort electrical pulses open a new gateway into biological cells. Proc IEEE 92:1122–1137

Schoenbach KH, Joshi RP, and Beebe SJ, Baum CE (2009) A scaling law for membrane permeabilization with nanopulses. IEEE Trans. Dielectrics and Electrical Insulation. 16:1224–1235
Schumacker PT (2006) Reactive oxygen species in cancer cells: live by the sword, die by the sword. Cancer Cell 10:175–176
Schwann HP (1985) Dielectric properties of cells and tissues. In: Interactions between Electromagnetic Fields and Cells. C. Chiabrera, C. Nicolini, and H. P. Schwan, editors. Pergamon Press, New York, and London 75–97
Schumacker PT (2015) Reactive oxygen species in cancer: a dance with the devil. Cancer Cell 27:156–157
Sebbagh M, Renvoizé C, Hamelin J, Riché N, Bertoglio J, Bréard J (2001) Caspase-3-mediated cleavage of ROCK I induces MLC phosphorylation and apoptotic membrane blebbing. Nat Cell Biol 3:346–352
Segawa K, Kurata S, Yanagihashi Y, Brummelkamp TR, Matsuda F, Nagata S (2014) Caspase-mediated cleavage of phospholipid flippase for apoptotic phosphatidylserine exposure. Science 344:1164–1168
Semenov I, Xiao S, Pakhomov AG (2013a) Primary pathways of intracellular Ca(2+) mobilization by nanosecond pulsed electric field. Biochim Biophys Acta 1828:981–989
Semenov I, Xiao S, Pakhomova ON, Pakhomov AG (2013b) Recruitment of the intracellular Ca2+ by ultrashort electric stimuli: the impact of pulse duration. Cell Calcium 54:145–150
Semenov I, Xiao S, Kang D, Schoenbach KH, Pakhomov AG (2015a) Cell stimulation and calcium mobilization by picosecond electric pulses. Bioelectrochemistry 105:65–71
Semenov I, Zemlin C, Pakhomova ON, Xiao S, Pakhomov AG (2015b) Diffuse, non-polar electropermeabilization and reduced propidium uptake distinguish the effect of nanosecond electric pulses. Biochim Biophys Acta 1848:2118–21125
Semenov I, Xiao S, Pakhomov AG (2016) Electroporation by subnanosecond pulses. Biochem Biophys Rep 6:253–259
Sharma V, Tung L (2002) Spatial heterogeneity of transmembrane potential responses of single guinea-pig cardiac cells during electric field stimulation. J Physiol 542:477–492
Shi J, Gao W, Shao F (2017) Pyroptosis: gasdermin-mediated programmed necrotic cell death. Trends Biochem Sci 42:245–254
Simons K, Toomre D (2000) Lipid rafts and signal transduction. Nat Rev Mol Cell Biol 1:31–39
Smyth KC, Gowrishankar TR, Esser AT, Stewart DA, Weaver JC (2006) Spatially distributed, dynamic transmembrane voltages of organelle and cell membranes due to 10 ns pulses: predictions of meshed and unmeshed transpoort network models. IEEE Trans Plasma Sci 34:1394–1404
Sridhara V, Joshi RP (2014) Numerical study of lipid translocation driven by nanoporation due to multiple high-intensity, ultrashort electrical pulses. Biochim Biophys Acta 1838(3):902–909
Stewart DA, Gowrishankar TR, Weaver JC (2004) Transport lattice approach to describing cell electroporation: use of a local asymptotic model. IEEE Trans Plasma Sci 32:1696–1708
Sun J, Nguyen T, Aponte AM, Menazza S, Kohr MJ, Roth DM, Patel HH, Murphy E, Steenbergen C (2015) Ischaemic preconditioning preferentially increases protein S-nitrosylation in subsarcolemmal mitochondria. Cardiovasc Res 106:227–236
Tarek M (2005) Membrane electroporation: a molecular dynamics simulation. Biophys J 88:4045–4053
Taylor SS, Keshwani MM, Steichen JM, Kornev AP (2012) Evolution of the eukaryotic protein kinases as dynamic molecular switches. Philos Trans R Soc Lond B Biol Sci 367(1602):2517–2528
Tekle E, Oubrahim H, Dzekunov SM, Kolb JF, Schoenbach KH, Chock PB (2005) Selective field effects on intracellular vacuoles and vesicle membranes with nanosecond electric pulses. Biophys J 89:274–284
Tekle E, Wolfe MD, Oubrahim H, Chock PB (2008) Phagocytic clearance of electric field induced 'apoptosis-mimetic' cells. Biochem Biophys Res Commun 376:256–260

Thompson GL, Roth C, Tolstykh G, Kuipers M, Ibey BL (2014) Disruption of the actin cortex contributes to susceptibility of mammalian cells to nanosecond pulsed electric fields. Bioelectromagnetics 35:262–272

Tieleman DP (2004) The molecular basis of electroporation. BMC Biochem 5:10

Togo T (2004) Long-term potentiation of wound-induced exocytosis and plasma membrane repair is dependent on cAMP-response element-mediated transcription via a protein kinase C- and p38 MAPK-dependent pathway. J Biol Chem 279:44996–445003

Tolstykh GP, Thompson GL, Beier HT, Steelman ZA, Ibey BL (2016) nsPEF-induced PIP2 depletion, PLC activity and actin cytoskeletal cortex remodeling are responsible for post-exposure cellular swelling and blebbing. Biochem Biophys Rep 9:36–41

Tolstykh GP, Tarango M, Roth CC, Ibey BL (2017) Nanosecond pulsed electric field induced dose dependent phosphatidylinositol-4,5-bisphosphate signaling and intracellular electrosensitization. Biochim Biophys Acta Biomembr 1859:438–445

Verhoven B, Schlegel, Williamson P (1995) Mechanisms of phosphatidylserine exposure, a phagocyte recognition signal, on apoptotic T lymphocytes. J Exp Med 182:1597–1601

Vernier PT, Sun Y, Marcu L, Salemi S, Craft CM, Gundersen MA (2003) Calcium bursts induced by nanosecond electric pulses. Biochem Biophys Res Commun 310:286–295

Vernier PT, Sun Y, Marcu L, Craft CM, Gundersen MA (2004a) Nanoelectropulse-induced phosphatidylserine translocation. Biophys J 86:4040–4048

Vernier PT, Sun Y, Marcu L, Craft CM, Gundersen MA (2004b) Nanosecond pulsed electric fields perturb membrane phospholipids in T lymphoblasts. FEBS Lett 572:103–108

Vernier PT, Sun Y, Gundersen MA (2006a) Nanoelectropulse-driven membrane perturbation and small molecule permeabilization. BMC Cell Biol 7:37

Vernier PT, Ziegler MJ, Sun Y, Gundersen MA, Tieleman DP (2006b) Nanopore-facilitated, voltage-driven phosphatidylserine translocation in lipid bilayers–in cells and in silico. Phys Biol 3:233–247

Vernier PT, Sun Y, Chen MT, Gundersen MA, Craviso GL (2008) Nanosecond electric pulse-induced calcium entry into chromaffin cells. Bioelectrochemistry 73:1–4

Wang S, Chen J, Chen MT, Vernier PT, Gundersen MA, Valderrábano M (2009a) Cardiac myocyte excitation by ultrashort high-field pulses. Biophys J 96:1640–1648

Wang Y, Dawson VL, Dawson TM (2009b) Poly(ADP-ribose) signals to mitochondrial AIF: a key event in parthanatos. Exp Neurol 218:193–202

Watkins SJ, Borthwick GM, Arthur HM (2011) The H9C2 cell line and primary neonatal cardiomyocyte cells show similar hypertrophic responses in vitro. Vitro Cell Dev Biol Anim 47:125–131

Weaver JC, Chizmadzhev YA (1996) Theory of electroporation: a review. Bioelectrochem Bioenerg 4:135–160

White JA, Blackmore PF, Schoenbach KH, Beebe SJ (2004) Stimulation of capacitative calcium entry in HL-60 cells by nanosecond pulsed electric fields. J Biol Chem 279:22964–22972

Xia B, Fang S, Chen X, Hu H, Chen P, Wang H, Gao Z (2016) MLKL forms cation channels. Cell Res 26(5):517–528

Zhang J, Blackmore PF, Hargrave BY, Xiao S, Beebe SJ, Schoenbach KH (2008) Nanosecond pulsed electric field (nanopulse): a novel non-ligand agonist for platelet activation. Arch Biochem Biophys 471(2):240–248

Zhang Y, Chen X, Gueydan C, Han J (2018) Plasma membrane changes during programmed cell deaths. Cell Res 28:9–21

Chapter 3
Simulations of Membrane Effects of Cells After Exposure to Ultrashort Pulses

Ravi Joshi

Abstract Externally applied nanosecond electric pulses are useful to trigger and tailor bioeffects in cells and tissues. However, the parameter space is large given the different types of cells, and the wide range of potential electrical parameters (such as pulse durations and waveforms, field intensities, and number of pulses) that are available for use. To maximize benefits, tailor a desired response, and devise an efficient system, it becomes necessary to first understand and quantify the biological behavior driven by the electrical input. Model development based on the inherent biophysical processes is an elegant and cost-effective option to help advance this technology. Given the merits and need for modeling then, this chapter focuses on the various schemes for analysis and simulations of the electrically driven bioeffects. Schemes ranging from molecular level descriptions to an averaged continuum analyses are presented and discussed in this chapter. Relevant examples and illustrative result are also given in this context of cellular bioelectrics.

3.1 Introduction

Under natural conditions, without the application of any external stimulus, endogenous electric fields play an important role in the maintenance of cells, wound healing, embryonic development and patterning, and even tissue regeneration (Puller 2011; McCaig et al. 2005; Robinson and Messerli 2003; Levin 2014). It thus becomes a natural progression to attempt to gain the control of the behavior and response of cells with external electric fields. For example, transmembrane potential patterns, which produce electric field profiles, have been shown to play a regulatory role in development and regeneration (Levin et al. 2017). Even in tissue engineering, cell proliferation on scaffolds can be controlled by the application of such fields (Meng et al. 2013). At the tissue level, electric fields are used for the measurement of body composition (Hart 2009), and the promotion of wound healing (Puller and Isseroff 2005).

Biological sensory systems can detect very weak steady fields. For example, variations as small as a few percent of the earth's magnetic field (5×10^{-5} T) can be detected by sea turtles (Lohmann et al. 1996), while sharks and rays sense extremely

S. J. Beebe et al., *Ultrashort Electric Pulse Effects in Biology and Medicine*,
Series in BioEngineering, https://doi.org/10.1007/978-981-10-5113-5_3

weak electric fields in sea water. However, the effects of weak fields are difficult to understand and any possible role in triggering or coordinating coupled biophysical mechanisms in organized cell systems is challenging to analyze. Thus, molecular changes or signaling induced by low fields can be difficult to extract and discern from other sources of change such as the natural biochemical pathways. By contrast, the application of high-intensity electric fields does not present such difficulties. Short exposures of cells and tissues to strong fields represent an important subset of this class of external excitation. The fields involved are orders of magnitude larger, and so can readily pass the conceptual tests based on signal-to-noise (SNR) criteria, in spite of the short exposure times.

Scientific interest in the interaction of electric and magnetic fields with biological systems has continued (Blank 1995; Polk and Postow 1996). Applications and mechanistic understanding of electric field bio-applications have grown with regards to therapies, health risks and hazards, bio-sensing and medical delivery technologies. For example, externally applied fields have also been used for cell fusion (Zimmermann 1982; Jordan et al. 2013), electrorotation (Arnold and Zimmermann 1984; Fuhr et al. 1986), dielectrophoresis (Pohl and Crane 1971; Sauer 1983) and separation of cancer cells (Becker et al. 1995; Gascoyne and Vykoukal 2002). Applications of electric pulses can also be used to create pores in biological cell membranes can lead to orders-of-magnitude increase in plasma membrane permeability and has many interesting biomedical applications (Neumann and Rosenheck 1972; Stampfli 1958; Tsong 1991; Weaver and Chizmadzhev 1996; Abidor et al. 1979; Meglic and Kotnik 2016; Joshi and Schoenbach 2010; Rols and Teissie 1998). Early on, Neumann and colleagues (Neumann et al. 1982) used pulsed electric fields to temporarily permeabilize cell and coined the term "electroporation". Since then rapid advances have been made, with more recent applications including gene electrotransfer (Gothelf and Gehl 2010; Pavlin and Kandušer M 2015), delivery of plasmid DNA and other exogenous molecules into cells (Aihara and Miyazaki 1998; Daud et al. 2008; Titomirov et al. 1991; André et al. 2008; Rosazza et al. 2016; Heller and Heller 2006), electrochemotherapy (Mir et al. 1991; Miklavčič et al. 2014), drug delivery (Prausnitz et al. 1993; Yarmush et al. 2014) and controlled immuno-therapy (Nuccitelli et al. 2015; Calvet and Mir 2016; Sersa et al. 2015). As a non-viral method, electrotransfection has the benefits of low cost, ease and safety, and independence of cell surface receptors. It is also capable of delivering a wide spectrum of genes with different sizes. Of the above application, an especially important application is towards cancer treatment (Low et al. 2009; Calvet et al. 2014). When the principle of electroporation is combined with certain chemotherapeutic drugs, the cytotoxicity of these drugs is increased ten-fold (or higher), leading to improved and dramatic responses in the target tumors. This is called electrochemotherapy (Marty et al. 2006; Miklavčič et al. 2014; Schmidt et al. 2014; Pucihar et al. 2001), defined as the local potentiation by means of electric pulses.

Recent developments involve the use of high intensity (~50–100 kV/cm), nanosecond duration pulsed electric fields (Joshi and Schoenbach 2011; Napotnik et al. 2016). Use of nanosecond duration pulses leads to the creation of nanopores and can produce orders-of-magnitude increase in plasma membrane permeability as first shown for millisecond pulses about fifty years ago (Neumann and Rosenheck 1972). Potential applications include electrically triggered intra-cellular calcium release (Beebe et al. 2004; Vernier et al. 2003), shrinkage of tumors (Nuccitelli et al. 2006, 2014), temporary blockage of action potential in nerves (Joshi et al. 2008), and activation of platelets for accelerated wound healing (Schoenbach et al. 2007). Moderate intensity electric pulses in the microsecond regime is another emerging modality for relatively safe, effective, and minimally invasive ablation (Jourabchi et al. 2014). The non-thermal nature of this excitation (unlike the heating caused by microwave or radiofrequency ablation) allows for treatment even in close proximity to critical structures and/or large vessels and mitigates muscle contractions (Ahmed et al. 2011; Sano et al. 2015). Eliminating contractions helps improve the procedural safety of patients, since the need for neuroparalytic drugs to inhibit muscle contraction is then virtually eliminated. The short duration also eliminates localized thermal heating. Significant progress in this field, based on irreversible electroporation and bipolar high-frequency pulses, has been demonstrated by the Davalos group (Latouche et al. 2018; Edd et al. 2006).

3.2 Computational Aspects

Evaluations of cellular effects can be based on either microscopic methods that fold in the many-body atomistic interactions, or macroscopic continuum approaches. The former are more rigorous, include inherent many-body physical processes, require far fewer fitting parameters, and can account for changes and heterogeneous details over the nanoscale. However, these simulation techniques are computationally very demanding since details of every atom are considered. Hence, the physical system under analysis is restricted to small nanoscale patches, with time variations simulations to less than ~100 ns. Consequently, while the MD techniques provide an elegant and accurate pathway to probe the inherent physics and mechanistic details of biophysical processes such as electroporation, the time and length scales are very small to be of great practical application. The MD schemes are nonetheless discussed first for completeness. Later in Sect. 3.2.2 and beyond, the macroscopic techniques for the analysis of electroporation are described.

3.2.1 Molecular Dynamics Methods

The main lipid constituents of natural membranes are phospholipids that arrange themselves into a two-layered sheet (a bilayer). Application of an external electric field onto cells induces rearrangements of the membrane components that ultimately lead to the formation of aqueous hydrophilic pores. The membrane components are water and lipids, which are essentially dipolar molecules. Hence, the membrane molecular units which are polarized, can experience torques and forces as any elementary dipole would upon being subjected to a nonuniform external electric field.

In erythrocyte membranes, large pores could be observed using electron microscopy (Chang 1992); but in general, direct observation of the formation of nano-sized pores is not possible with conventional techniques. Furthermore, due to the complexity and heterogeneity of cell membranes, it is difficult to describe and characterize their electroporation in terms of atomically resolved processes. In this realm, atomistic and molecular dynamics (MD) simulations, have proven to be effective in providing insights into both the structure and the dynamics of model lipid membrane systems in general (Tieleman et al. 1997; Tobias et al. 1997; Forrest and Sansom 2000; Feller 2008; Mashl et al. 2001; Saiz and Klein 2001; Anezo et al. 2003; Berkowitz et al. 2006; Lindahl and Sansom 2008; Bockmann et al. 2008; Edholm 2008; Marrink et al. 2009). In particular, recent studies have shown that the method is suitable for investigating electroporation phenomena. Several MD simulations have recently been conducted to model the effect of electric field on membranes (Hu et al. 2005a, b; Tieleman 2004; Tarek 2005; Bockmann et al. 2008; Ziegler and Vernier 2008), which have provided physics-based insights and details of the electroporation process of lipid bilayers.

Molecular dynamics refers to computational methods aimed at simulating macroscopic behavior through the numerical integration of the classical equations of motion of a microscopic, many-body molecular system. Macroscopic properties are expressed as functions of particle coordinates and/or momenta, which are computed along a phase space trajectory generated by classical dynamics (Allen and Tildesley 1987; Leach 2001) MD simulations can thus be used to perform "computer experiments". Simulations are usually performed on a small number of molecules (usually up to a few hundred thousand atoms) due to the computational complexity which limits the size of the system being simulated. Typically, for membrane electroporation, a small patch of the membrane is used, with length scales on the order of 500 Å. In order to replicate the physical system better, periodic boundary conditions (PBCs) are typically used (Lindahl and Edholm 2000; Marrink and Mark 2001; Berkowitz 2009).

Traditionally, phospholipids have served as models for investigating the dynamical properties and response of membranes to external stimuli in such in silico experiments. Zwitterionic phosphatidylcholine (PC) lipid bilayers constitute the best-characterized membrane systems (Chiu et al. 1999; Feller et al. 2002). Actual membranes are very complicated structures with membrane proteins, ion channels, cholesterol etc. Despite the simple membrane structures built from phosphatidylcholine lipid molecules, these representations have proved to be remarkable test systems that provide very useful insights into the inherent physics, mechanisms, and to probe the physical properties of membranes. While most membrane models are consisted of fully hydrated, pure phospholipid bilayers, more complicated organization and account of external ions (e.g., the effect of salt concentrations) have emerged

(Hofsäß et al. 2003; Gurtovenko et al. 2005). For such systems, the average structure of the lipid–water interface at the atomic scale may be provided by the density distributions of different atom types along the bilayer normal.

The usual MD implementation involves the application of a constant electric field (coinciding with the externally applied value) perpendicular to the membrane plane. This involves adding a force, $\mathbf{F} = \mathbf{q_i E}$, to all the atoms bearing a charge, $\mathbf{q_i}$ (Tieleman et al. 2001; Crozier et al. 2001; Hu et al. 2005a, b; Tieleman 2004; Tarek 2005). The consequence resulting from high electric field application to the system stems from the properties of the membrane and from the simulations setup conditions: pure lipid membranes exhibit a heterogeneous atomic distribution across the bilayer, to which are associated charges and molecular dipole distributions. Besides, the atoms are in constant thermal motion and also subjected to the resultant force from all the many-body contributions. The permanent dipoles of the water molecules that surround a membrane also experience forces that can drive them towards the membrane. In addition, the presence of charged molecular groups within the membrane give rise to an electrostatic profile that plays a role in the dynamic evolution.

The MD simulations for electroporation calculations are carried out by selecting a segment of the lipid bilayer membrane and constructing an initial geometric arrangement of all the atoms and their bonding angles. Regions of water are then defined on either side of the membrane to form the total simulation space. The force fields for membrane molecular motion were taken from the literature (for example, van Gunsteren and Berendsen 1987; Siu et al. 2008). Simulations were usually carried out at constant particle number and temperature (e.g., 300 K), using a Berendsen thermostat (Berendsen et al. 1984) with a specified time constant, such as 0.1 ps. Long-range electrostatic interactions are computed with a particle mesh Ewald method (Darden et al. 1993) with a user-specified cutoff (e.g., 1 nm), and periodic boundary conditions. Similar cut-offs are specified for other potentials that may be used for the particle-particle interactions. The linear constraint solver (LINCS) algorithm (Hess et al. 1997) is typically used to constrain all the bond lengths within the lipids and on the water geometry.

MD simulations have shown that within a very short time scale (~1–2 ps), the external field induces an overall transmembrane potential (TMP). Simulations have shown the initiation and formation of water nanowires in the membrane (Tieleman 2004; Hu et al. 2013). Ultimately, water fingers forming at both sides of the membrane join up to form water channels that span the membrane. Within nanoseconds, a few lipid headgroups start to migrate from the membrane–water interface to the interior of the bilayer, stabilizing hydrophilic pores of ~1–3 nm diameter.

Electroporation is thus shown to be a twofold process. The first step involves a molecular rearrangement of the lipid within the membrane bilayer. This is facilitated by electrostatic forces and the Maxwell stress tensor arising from the applied electric field. However, the random opening of gaps in the membrane structure alone is insufficient for continued cellular flows and molecular/drug transport. Second, water entry is essential for continued transport and is facilitated by electrowetting, which depends on field-driven reorientation of the water dipoles. This process can function as an electric-driven toggling of a hydrophobic switch that changes the water permeability of the membrane gaps and can remain open for a long duration (Hu et al. 2013).

3.2.2 *Some Molecular Dynamics Examples in the Nanosecond Pulsing Context*

As an example of MD simulations related to cell membrane patches, the role of water on the lipid bilayer configuration and stability is discussed below. For this case there was no external electric field. The water-membrane system contained 6323 water molecules and 128 DPPC lipid molecules in a 6.9 nm × 7.4 nm × 7.0 nm simulation box. In addition, 847 water molecules were deliberately placed in between the two layers to determine their dynamical evolution. This membrane system was charge-neutral and represents a homogeneous section of a simple membrane as a test case of the lipidic system. Figure 3.1 show results of water expulsion from within the lipid bilayer after the initial insertion of the water molecules inside the membrane without any external electric field. The red and white spheres represent atoms. The gray "wires" denote the phospholipid tails. The 250 ps snapshot of Fig. 3.1b shows water molecules being expelled by the hydrophobic tails. In addition to water expulsion, molecular re-arrangement of the bilayer is predicted to be ongoing, as is easily seen from the 500 ps snapshot. Figure 3.1d reveals almost all water molecules to have been pushed out, with near recovery of the bilayer membrane structure. These MD results thus seem to suggest that water molecules in and of themselves (without the aid of an external driving electric field) would not have the capacity to initiate molecular displacements toward pore creation. Instead, the hydrophobic interaction would be sufficient to expel water away from the membrane.

We next turn to the issues of whether water should lead or follow pore formation in the dynamic sequence, and whether electric field assisted electrowetting is an important and inherent component of the overall process. To probe the importance of electrowetting, the MD simulation was run with an intact membrane, but with the application of a very short electric pulse so that initial field-driven lipid structural rearrangement could occur. If the electric pulse were then to be quickly switched off, the electrowetting process would immediately cease. So if electrowetting were important for water penetration, an ultrashort electric pulse would not be of sufficient duration to enable the complete penetrate of water in providing a contiguous path through the membrane. To probe this, a short 4.25 ns duration electric pulse with an intensity of 0.25 V/nm was used. The results obtained are shown in Fig. 3.2a–c. The state just after electric field termination shows a few water molecules having penetrated the membrane. However, snapshots later in time (2.5 and 90 ps after pulse

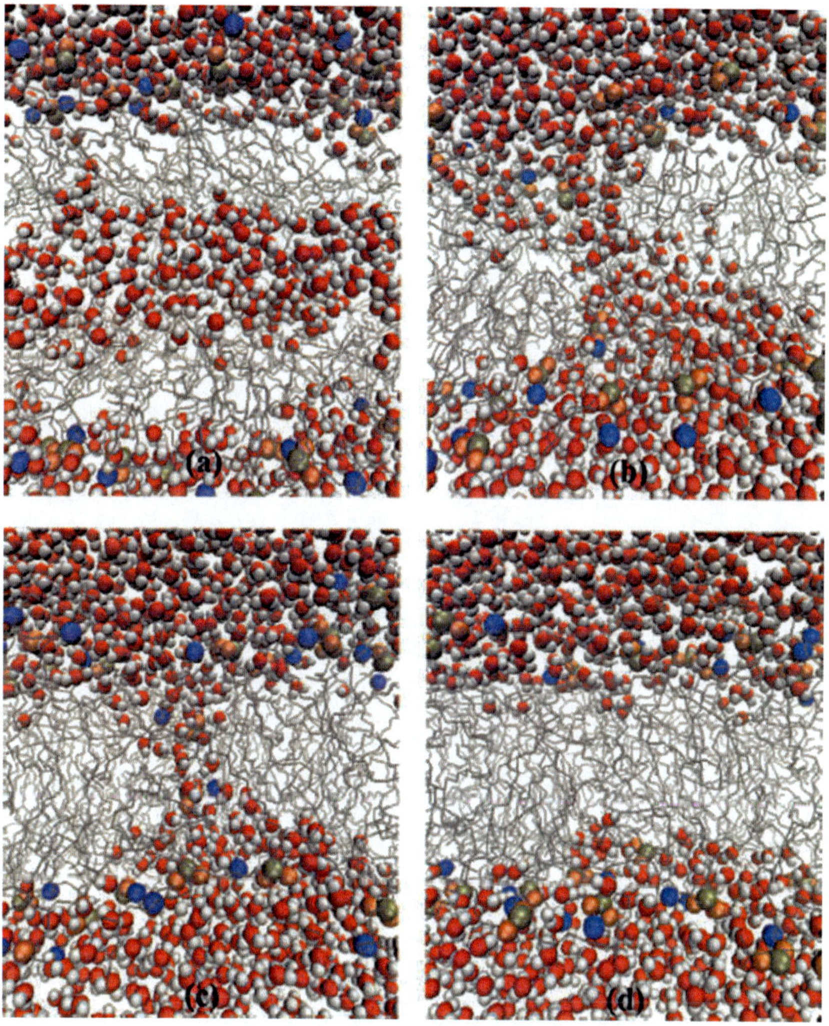

Fig. 3.1 MD results showing water being expelled from within the membrane. The red and white spheres represent oxygen and hydrogen atoms within water, gold and blue spheres denote phosphorus and nitrogen atoms within the lipid head groups, respectively, and the large orange spheres represent phospholipid oxygen atoms. Gray wires denote the phospholipid tails. **a** Initial bilayer membrane with 847 water molecules inserted in between. **b** Snapshot after 250 ps showing water molecules being expelled by the hydrophobic tails. **c** The t = 500 ps picture with molecular rearrangement in the form of oxygen and hydrogen atoms of the water, gold and blue spheres denote phosphorus and nitrogen atoms within the lipid head groups, respectively, and the large orange spheres represent phospholipid oxygen a hydrophilic pore, and **d** the situation after 1400 ps, with almost all water molecules pushed out and recovery of the bilayer membrane structure (After Hu et al. 2013)

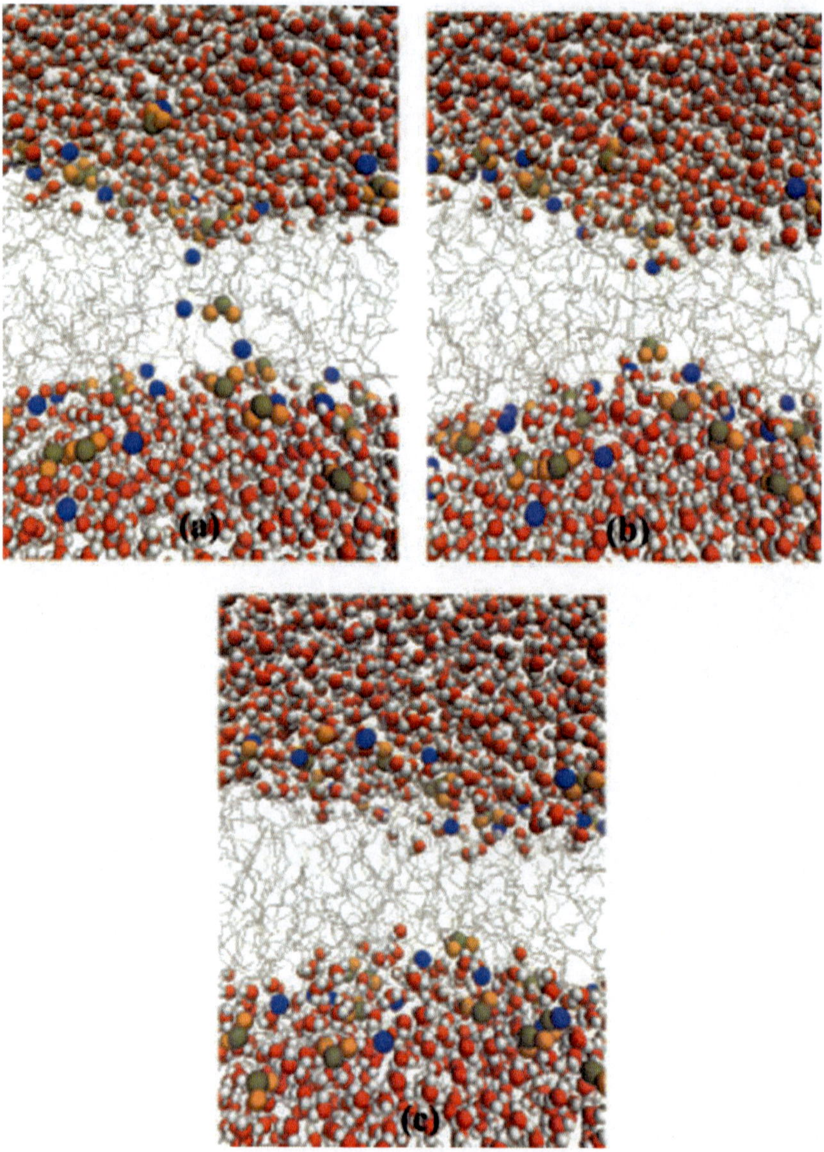

Fig. 3.2 The same lipid and color scheme as used in Fig. 3.1 for a short 0.25 V/nm, 4.25 ns duration electric pulse. **a** State just after electric field termination, **b** snapshot later in time at 2.5 ps, and **c** snapshot at t = 90 ps after pulse termination (After Hu et al. 2013)

termination) reveal water expulsion. This result underscores the role of electrowetting and the inability of water to continue membrane penetration in the absence of a driving electric field.

Two important aspects thus emerge from the above simulation results. First, the presence of a gap due to structural rearrangement of the lipidic bilayer molecules in itself would be insufficient to attract water molecules into the membrane region. Given the small radii of nanopores formed upon the application of a nanosecond, the high-intensity electric pulse (Esser et al. 2010), and the fact that the hydrophobic interaction preventing water throughput is stronger for smaller pores, mere membrane reorganization might not support cellular flows. However, if an electric field was applied long enough to facilitate electrowetting and lower the local surface tension, then water molecules could penetrate through the lipid membrane. Based on classical electrochemistry (Dujardin et al. 1994), it is well known that a local electric field can lower the interfacial tension γ according to $\gamma = \gamma_1 - C\ V^2/2$, where γ_1 is the reference "wettability" in the absence of any electric field, C is the interfacial capacitance, and V is the potential difference setup between the liquid and the lipid surface. So, regardless of the sign of the potential difference, the interfacial tension always decreases as an electric potential $|V|$ is applied. The process of electrowetting can alternatively be viewed as an outcome of a Maxwell stress at the boundary of water entering a nanopore.

Next, the proximity effects under high-intensity pulsing are discussed, since it has been speculated that high intensity nanosecond pulse application can create a high density of pores, termed supraporation (Stewart et al. 2004) due to the strong driving force. Since electric fields provide the primary driver, both for molecular rearrangement and electrowetting, it is perhaps instructive to assess the electric field characteristics within pores. Figure 3.3 shows the electric field results obtained using COMSOL MULTIPHYSICS (version 4.2b) within a membrane containing two partially filled pores in close proximity. The membrane thickness was taken to be 5 nm, with 1 nm diameter pores separated by a distance of 3 nm. Figure 3.3a shows the computed electric fields in the region for the case of symmetric water penetration at the two adjacent pores. As might be expected, the electric field distribution is symmetric, with the largest values at the "triple" water-membrane-pore boundary. As a result, the shape of the water will tend to flatten and reduce the contact angle, leading to larger electrowetting. However, in the case of asymmetry in water penetration between the two pores, the electric fields obtained from the simulations, as shown in Fig. 3.3b, have a different characteristic. The field at the upper pore with a larger water penetration remains similar, and is indicative of water wetting as before. However, at the bottom pore, the electric field is not as high at the "triple" water-membrane-pore boundary, but instead exhibits its highest value at the center. This suggests weaker surface wetting of the membrane and a tendency toward elongation at the pore center. However, if the top pore were to be completely solvated as assumed in Fig. 3.4 (possibly later in time with the electric field still applied), then the field distribution for the lower pore would go back to a pattern similar to Fig. 3.3a. These electric field results thus suggest that multiple pore formation is likely to be a sequential process.

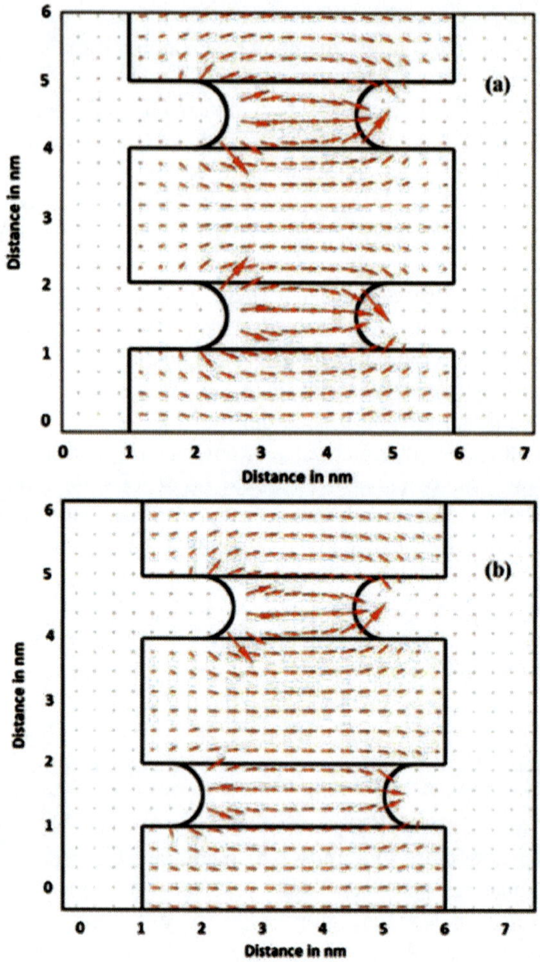

Fig. 3.3 Situation at a membrane containing two partially filled pores in close proximity. **a** Computed electric fields in the region for the case of symmetric water penetration. **b** Electric fields obtained from simulations with asymmetry in water penetration between the two pores (After Hu et al. 2013)

Another interesting result obtained from Molecular Dynamics simulations in the context of ultrashort (~ns) high intensity (~100 kV/cm) electric pulsing is field driven molecular flipping and/or translational motion. For example, externalization of phosphatidylserine (PS) was shown to occur in response to a nanosecond pulse for average electric fields above 2 MV/m (Vernier et al. 2004). A distinct polarity effect has also been observed, with the externalization predominantly occurring at the anode end. PS is an acidic phospholipid that is normally located on the inner leaflet of the lipid

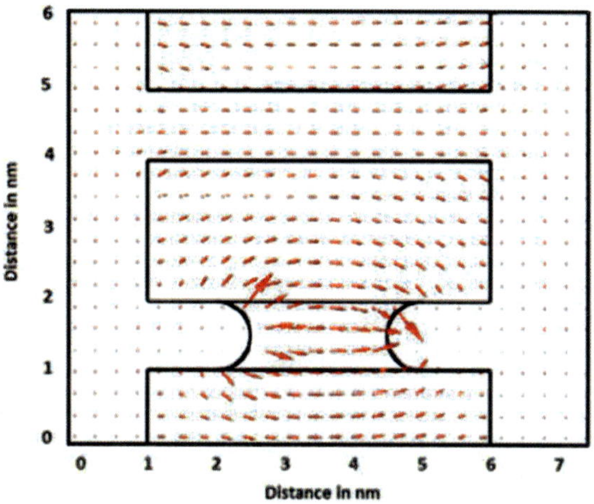

Fig. 3.4 Computed electric fields at a membrane containing pores in close proximity. Water is assumed to completely penetrate the upper pore and only partially penetrate the neighboring pore below (After Hu et al. 2013)

bilayer. The PS translocation event essentially marks cells for macrophage scavenging and ultimate death. The dynamical process details can be probed by MD simulations as discussed below.

As previously shown (Hu et al. 2005a, b), the initial structural rearrangement and molecular dipole re-orientation at the membrane is a critical step in the electroporation process. Once such an initial breakthrough is achieved, the poration process proceeds relatively quickly. Finally, the poration process has polarity dependence, and begins on the anodic side of a membrane. The related physics can easily be understood by considering the configuration within the DPPC membrane. Figure 3.5 shows a simple schematic of the membrane lipids with their dipoles located at the head groups. For each DPPC chain, the head group contains a dipole with positive charge on choline and negative charge centered on the phosphate group. Initially, with no electric field present, the dipoles are in random thermal motion with the positive charges residing on the outermost portions of the lipid. Thus Fig. 3.5a shows randomly distributed dipoles at the head groups on either side of the membrane as the initial configuration. Electrical field induced defects are initiated by the movement of dipoles on the surface of the membrane. Defects start to form on the anode side of the membrane because positively charged molecules (e.g., choline) on this side are forced to swing around (i.e., reorient in the presence of a strong external electric field) and enter the membrane. This same electrical field, however, when acting on the dipoles located at the cathodic surface, merely works to stretch the dipoles without any molecular movement into the membrane volume. Figure 3.5b shows the alignment of dipolar head groups on the anodic side gradually deviating from the equilibrium orientation, and a defect starting to form.

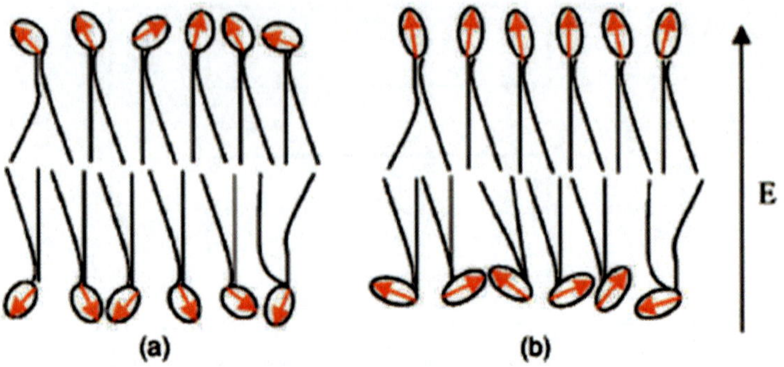

Fig. 3.5 Schematic of the pore initiation process at the DPPC membrane. The arrows present dipoles on the head groups of lipid molecules. **a** Initial configuration without an externally applied field, and **b** the configuration and expected movement of the dipole head groups upon the application of an external electric field (After Hu et al. 2005a, b)

Thus, pore formation is predicted to initiate at the membrane pole facing the anode. If PS externalization is pore-driven event, then this process should also start at the anode side. Experimental observations do indicate just such an anode side preference. The electrostatics provides an additional rationale for an anode-side event. Since the PS molecules residing on the inner leaflet are negatively charged, the externally applied electric field will tend to push out PS at the anode side, while pulling it inwards on the cathodic side. Coupling this with an anode-side pore formation event, leads to the collective effect of preferential PS externalization at the anode end. Figure 3.6 is a simple schematic demonstrating the dynamics. Initially (Fig. 3.6a), the dipoles at the DPPC lipids (shown as arrows) are randomly located, with a negative PS on the inner leaflet. With application of the electric field, the dipoles on the outer layer (anode-side) reorient leading to defect initiation at the outer membrane surface. Eventually a pore forms (Fig. 3.6c), and the negatively charged PS begins to drift and diffuse towards the exterior surface.

Results of MD simulations, shown in Fig. 3.6a–d demonstrate the above more clearly. An initial snapshot of the membrane system is shown in Fig. 3.7a, with the PS molecule located on one side of the membrane, opposite the anode. Figure 3.7b shows a pore starting to form with some translocation of the PS chain at 3.2 ns. Due to a large electrostatic force on the PS head group, the chain is dragged halfway to the anode side of the membrane along the wall of the nanopore at about 3.34 ns as shown in Fig. 3.7c. Finally, in Fig. 3.7d, the DPPS chain is on the other leaflet of the membrane at 3.61 ns. The MD simulation thus validates the pore-facilitated, field-assisted mechanism of PS externalization. An important difference between the DPPC-DPPS simulations of Fig. 3.7 and results for a pure DPPC membrane (not shown) is the shorter time duration for pore formation. Inclusion of a membrane defect in the form of a substitutional DPPS molecule, and the additional membrane force associated with the negative PS charge collectively contribute to the quicker poration.

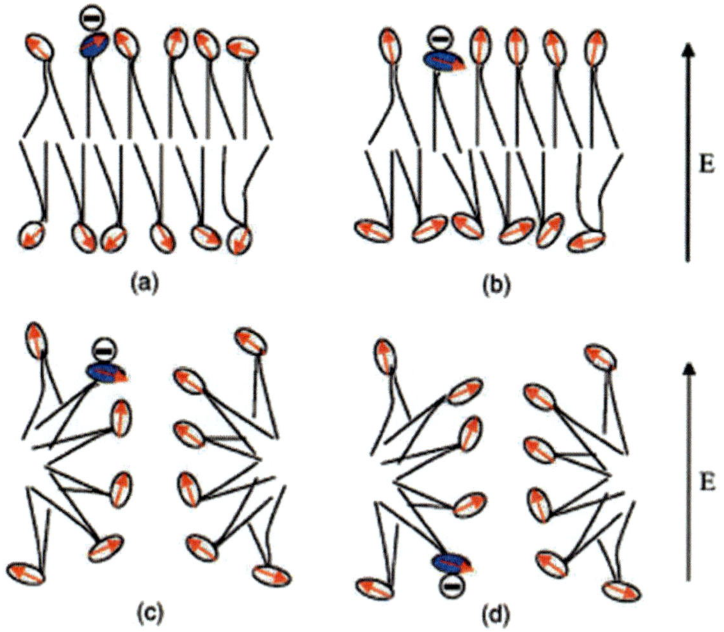

Fig. 3.6 Schematic of the pore initiation process within a DPPC-DPPS membrane. The arrows represent dipoles on the lipid head groups, and the shaded circle denotes the negative DPPS head group. **a** Initial configuration without an externally applied field; **b** expected configuration of the dipole head groups with an external field; **c** nanopore formation with a DPPS molecule close to the pore; and **d** DPPS molecular movement to the outer leaflet of the membrane (After Hu et al. 2005a, b)

This polarity effect and physics of the strong asymmetry in transmembrane potentials can be qualitatively understood by considering the pore-formation dynamics at each pole. The PS molecules are negatively charged, and prior to an electric-pulse event, occupy the inner membrane leaflet. This is shown schematically in Fig. 3.8. The anode and cathode are on the left and right sides of the overall figure. Upon the application of an external electric field, a pore begins to form at the two poles as the field strength is the strongest in this region. The charged PS molecules begin to move to the outer leaflet of the membrane on the anode side, first, through an ionic drift and then by diffusion Their movement at the anode pore leads to the formation of a strong "screening" layer close to the anodic nanopore region as shown in Fig. 3.8a. The screening layer consists of the positive ions from the inter-cellular aqueous fluid. The associated screening length (denoted by L_{SCR}) can be expected to be fairly large. Given the relative small pore diameter D (nanometer dimensions as shown by the present MD simulations) due to these ultrashort pulse durations, one can easily attain the $2\,L_{SCR} > D$ inequality. Hence, the screening layer, which is essentially a depletion region, effectively prevents the ions from going through the anodic pore. The transport is prevented (or is weak at best) until the pore diameter has increased (2 L_{SCR}

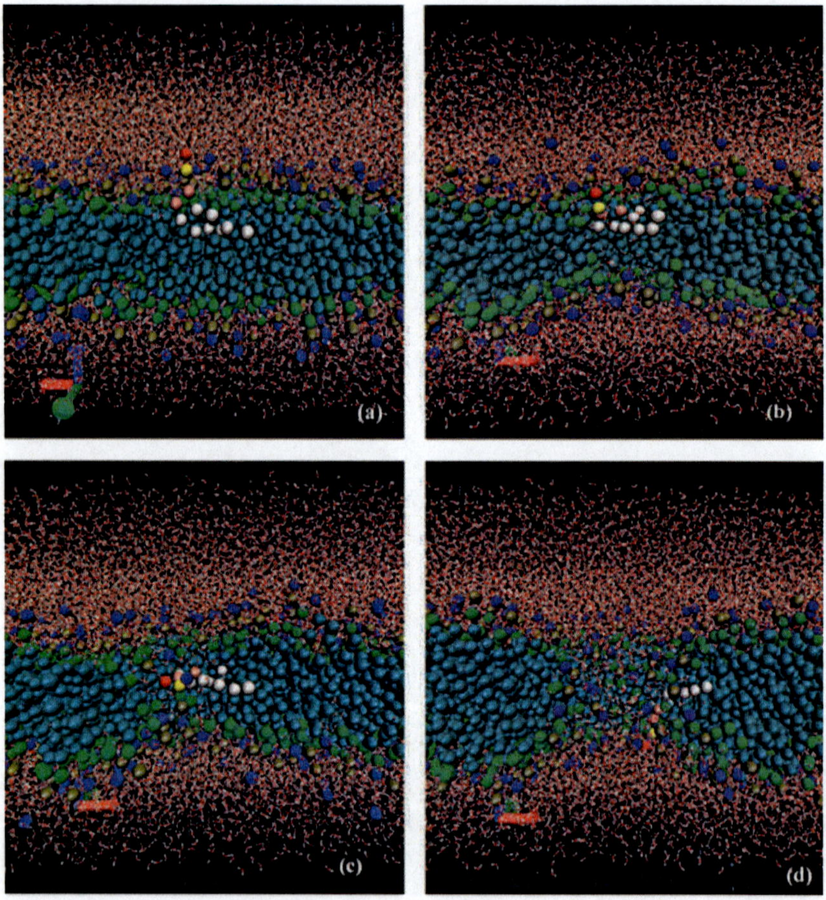

Fig. 3.7 Snapshots of the DPPS externalization process at a DPPC membrane. **a** Initial configuration at 0 ns; **b** a pore forms and DPPS starts to go along the pore wall at 3.2 ns; **c** DPPS halfway to externalization at 3.34 ns; and **d** DPPS on the other leaflet of the membrane at 3.61 ns. For DPPS, the positively polarized choline groups are colored red, and the negatively polarized phosphatidyl group yellow, the glycerol pink, the tails white. The oxygens in water molecules are red and hydrogens are white (After Hu et al. 2005a, b)

$< D$), or the PS charges have diffused sufficiently away on the outer leaflet from the pore location. The pore growth and increasing diameter scenario would be applicable to the traditional longer "electroporation" pulses. By contrast, on the cathodic side, the charged PS molecules are electrostatically driven away from the nanopore due to the directionality of the field and curvature of the membrane. The screening in the nanopore vicinity is therefore, not as strong, and the screening length L_{SCR} is relatively small. This is shown schematically in Fig. 3.8b, as $D > 2\ L_{SCR}$ The net effect is that higher conduction currents are predicted at the cathodic end, while the anodic

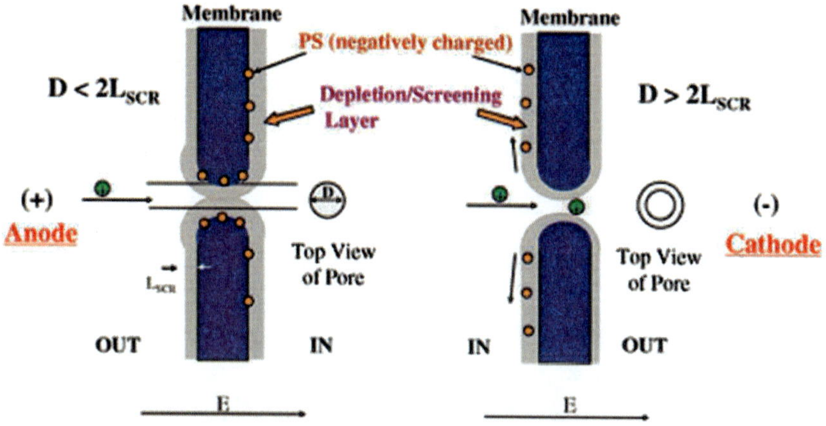

Fig. 3.8 Schematic diagram of the membrane polarity effect. "*D*" is the diameter of pores, and L_{SCR} is the length of the screening layers (After Hu et al. 2006)

nanopores though present, would not support large carrier transport. This qualitative physics underscores the notion that nanopores in themselves do not necessarily guarantee an ionic transport or strong conduction currents. Extrapolating on this concept then, the weaker conduction currents on the anodic side, would create a relatively larger displacement current since the total current must be conserved. This would lead to the larger electric-field buildup and transmembrane potential that has been seen experimentally (Frey et al. 2006).

Finally, the multiple-pore dynamics were studied on the basis of the MD simulations. At a constant but high electric field of 0.5 V/nm, more than one nanopore can be formed. However, the total number of molecules in a cell membrane is finite and fixed. Hence, any local nanopore formation is not a completely independent event. Instead, the pore formation affects the membrane surface tension (Joshi et al. 2001), leads to the local density and pressure changes, and changes the biosystem energy. For multiple pores, there is a mutual coupling as the growth of one pore begins to affect the dynamics of a neighboring nanopore. For example, an expanding pore would squeeze and move molecules in its vicinity that would lead to a "filling" effect on a neighboring pore. As a result of these collective dynamics, some pores can be expected to shrink (and completely disappear), while others would grow at their expense to some stable radius (Neu and Krassowska 2003) depending on the sizes and locations. Here, we probed this aspect through our MD simulations by applying the pressure coupling method. Figure 3.9a shows pores started to form at t = 4.2 ns. Figure 3.9b shows that two pores form at t = 4.6 ns. One of these is distinctly larger, and obviously has a faster growth dynamics. Figure 3.9c shows that a slightly longer time (time = 4.9 ns), the pore on the right continues to grow, while the pore on the left begins to shrink. Finally, the left pore is shown to shrink to a very small size, and the right nanopore becomes dominant at t = 5.6 ns following the electric pulse as shown in Fig. 3.9d. This clearly demonstrates that it is indeed possible for

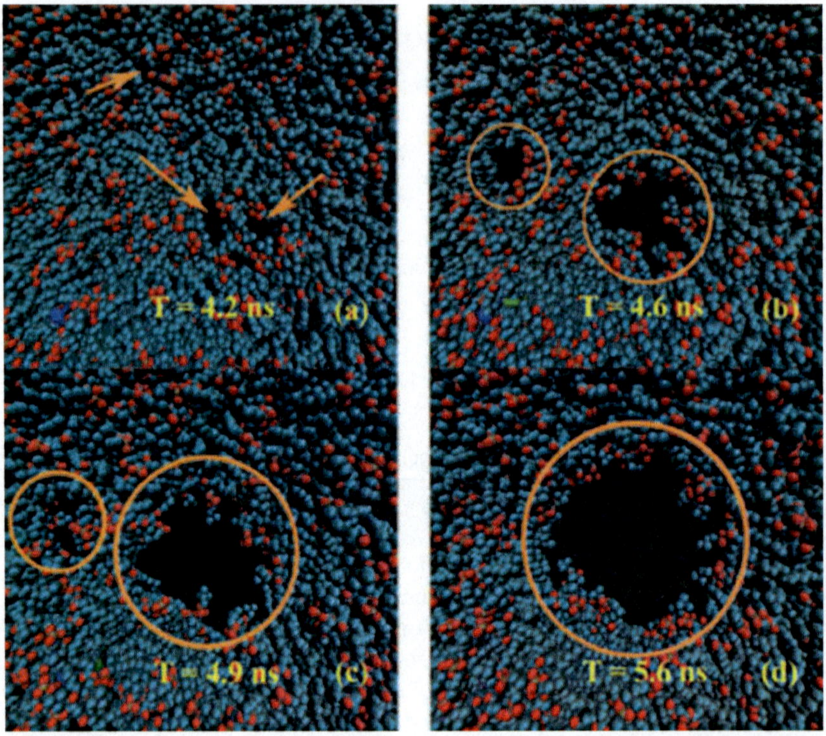

Fig. 3.9 MD snapshots of pore–pore interaction process in a DPPC membrane under the electric field of 0.5 V/nm. **a** Pores start to form at t = 4.2 ns. **b** Two pores form at t = 4.6 ns. **c** The pore on the right grows and the pore on the left shrinks at t = 4.9. **d** The left pore disappears and the right becomes dominant at t = 5.6 ns. Color method: Oxygen atoms are in deep gray (red in color graph) and carbon atoms in light gray (blue in color graph). Water molecules are not shown (After Hu et al. 2006)

the molecular system to potentially evolve from an initial multiple nanopore state to a system dominated by larger sized pores. This scenario would help explain the transport of molecules and fluorescent dyes at times well beyond the application of the ultrashort voltage pulse. Such delay uptake dyes, such as propidium iodide, have routinely been observed in the context of such nanosecond (ultrashort) electric pulsing experiments.

3.2.3 Analytic Methods for Transmembrane Potentials

Analysis of the cellular response to electric pulses requires as a first step, an evaluation of voltage and current distributions and their time-dependent evolution, within a cell. The problem of electric field coupling to biological cells and the transmembrane potential (TMP) induction was first studied by Schwan (Schwan 1983). Later contributions by Foster and Schwan (1995) and Foster (2000) extended these studies for exposures of spherical cells.

For cells with sufficiently regular shapes (spheres, spheroids, cylinders) that are sufficiently far apart (in dilute suspensions), the time dependence and spatial distribution of the induced transmembrane voltage can be derived analytically. In the case of irregular shapes, or cells close to each other (in dense suspensions, cell clusters, tissues), or both, the analytical approach fails. However, when modern computers and numerical methods are used, the transmembrane potential (TMP) induced on such cells can be evaluated sufficiently accurately. We now describe each of these approaches in more detail.

The derivation of the TMP is based on solving the partial differential equation:

$$e\nabla[(\sigma + \varepsilon\,\delta/\delta t)\nabla\Psi(x, y, z, t) = 0. \tag{3.1}$$

with σ denoting the electric conductivity, and ε denoting the dielectric permittivity of the point under consideration. In the steady state, the time derivatives are zero, and this equation simplifies into the Laplace equation $\nabla \cdot \nabla\Psi(x,y,z) = 0$. Solving this equation in a particular coordinate system and applying physically realistic boundary conditions (finiteness of Ψ, continuity of Ψ and its derivatives, asymptotic vanishing of the cell's effect on Ψ with increasing distance from the cell) yields the spatial distribution of Ψ. The induced transmembrane voltage is then calculated as the difference between the values of Ψ on both sides of the membrane.

For a single spherical cell with a nonconductive plasma membrane, the Laplace equation is solved in the spherical coordinate system, yielding the expression often referred to as the steady-state Schwan equation (Pauly and Schwan 1959):

$$\nabla\Psi_m = (3/2)ER\cos(\theta), \tag{3.2}$$

where R is the cell radius, E is the electric field, and θ is the angle measured from the center of the cell with respect to the direction of the field. Thus $\Delta\Psi_m$ is proportional to the applied electric field and the cell radius, and it varies as cos(θ), with extremal values at the points where the field is perpendicular to the membrane, i.e., at $\theta = 0°$ and $\theta = 180°$ (the "poles" of the cell).

The TMP as given by Eq. (3.2) is typically established within microseconds after the onset of the field. To describe the initial transient behavior, one uses the more general first-order Schwan equation (Kotnik et al. 1997):

$$\nabla\Psi_m = (3/2)ER\cos(\theta)\big[1 - \exp(-t/\tau_m)\big], \tag{3.3}$$

where τ_m is the time constant of membrane charging given by:

$$\tau_m = R\,\varepsilon_m/[2d\{(\sigma_i\sigma_e)/(\sigma_i + 2\sigma_e)\} + R\,\sigma_m]. \tag{3.4}$$

In the above, σ_i, σ_m, and σ_e denote the conductivities of the cytoplasm, cell membrane, and extra-cellular medium, respectively; while ε_m represents the dielectric permittivity of the membrane, and d is the membrane thickness. In certain in vitro experiments, where artificial extracellular media with conductivities far below typical physiological values are used, the factor 3/2 in Eqs. (3.2) and (3.3) can decrease (Kotnik et al. 1997). In general, Eq. (3.2) is applicable to exposures to all rectangular electric pulses longer than 1 μs, as well as all sine (AC) electric fields with frequencies up to 1 MHz (Kotnik et al. 2012). To determine $\Delta\Psi_m$ induced by shorter pulses or higher frequencies, the permittivities of the electrolytes surrounding the membrane also have to be taken into account, leading to a second-order model (Grosse and Schwan 1992; Kotnik and Miklavčič 2006).

Expressions similar to (3.2)–(3.4) can also be derived for nonspherical cells, provided they resemble a regular geometrical body such as a cylinder (e.g., muscle cells, axons of nerve cells), an oblate spheroid (e.g., erythrocytes), or a prolate spheroid (e.g., bacilli). To obtain the analogs of Schwan's equation for such cells, the Laplace equation is solved in a suitable coordinate system (Gimsa and Wachner 2001; Hu and Joshi 2009a, b; Joshi and Song 2010).

Almost all analyses and calculations of the TMP resulting from external pulsing have assumed spherical cells. In biology, examples of such spherical shapes are vesicles, protoplasts, murine myeloma cells (Gimsa and Wachner 1999), and some bacteria such as *Streptococcus* (Batzing 2002). However, most other cells are nonspherical in nature. As it turns out though, many cells deviating from the spherical shape do exhibit rotational symmetry, and can be represented either as prolate or oblate spheroids fairly accurately. For example, mammalian red blood cells are close to an oblate spheroidal shape, while retina photoreceptor cells (Radu et al. 2005), and many bacteria such as *E. coli*, *Pseudomonas* (Batzing 2002), and yeasts (Asencor et al. 1993) roughly have a prolate spheroidal geometry. This makes a compelling case, at least from the practical standpoint, to examine bioelectric pulsing effects in such spheroidal cell shapes for more realistic analyses and predictions. There have only been a few reports in the literature on analyses for irregularly shaped cells. A finite-element approach was also used (Pucihar et al. 2006) to solve for the TMP, though electroporation was not explicitly considered. The only other reports on spheroidal cells to our knowledge were those by Kotnik and Miklavčič (2000), and an analysis for ac voltage signals (Maswiwat et al. 2007). However, in neither case, electroporation was considered in a self-consistent manner. Hence, self-consistent calculations for spheroidal cell geometries that include time- and voltage-dependent membrane electroporation become necessary for arbitrary field orientations.

The schematic shown in Fig. 3.10 represents a double-shelled prolate spheroidal cell suspended in a medium. The geometric model is similar to that used previously (Jerry et al. 1996) for analyzing potentials induced by alternating fields. Since the spheroids have rotational symmetry in the x–y plane, the arbitrarily oriented, externally applied electric field $E_0(t)$ can be assumed to be in the x–z plane. In Fig. 3.10, α is the angle between $E_0(t)$ and the rotational z-axis. The cell shown is characterized

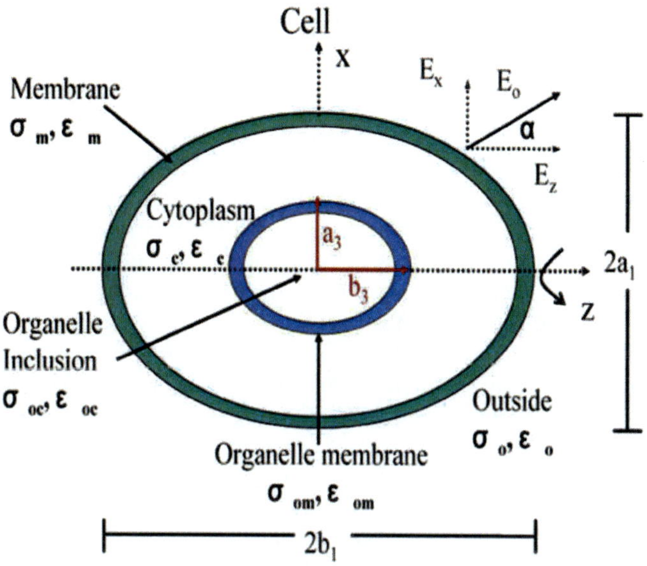

Fig. 3.10 Schematic of the prolate spheroidal cell model for the TMP calculation under arbitrary-oriented external nsPEF

by an outer shell (plasma membrane) and an inner shell (nuclear envelope). Each shell is represented here by two spheroids, and assumes that the four spheroids share the same foci. This geometry represents a simple first step, while more complicated "eccentric" positioning would require numerically based finite element (Corovic et al. 2008) or lattice models (Smith et al. 2006). The outer shell has an outer equatorial radius a_1, an inner equatorial radius a_2, an outer polar radius b_1, and an inner polar radius b_2. The focal distance c can then be found as: $c = b_{21} - a_{21} = b_{22} - a_{22}$ for prolate ($b_i > a_i$) spheroids of the outer shell. Thickness at the polar points is $h_o = b_i - b_2$. The inner shell has an outer equatorial radius a_3, an inner equatorial radius a_4, an outer polar radius b_3, and an inner polar radius b_4. Thickness at the polar points is $hi = b_3 - b_4$.The outer region has an assigned conductivity σ_0 and permittivity ε_0, while the corresponding parameters for the cell membrane are σ_m and ε_m. In Fig. 3.10, σ_c and ε_c are the conductivity and permittivity of the cytoplasm, σ_{om} and ε_{om} are the conductivity and permittivity of the organelle membrane, and σ_{oc} and ε_{oc} are the conductivity and permittivity of the organelle cytoplasm. Since the external electrical field $E_0(t)$ can be decomposed into its $E_x(t)$ and $E_z(t)$ components, the overall TMP $\Delta\Phi$ can be obtained by summing the separate contributions from $E_x(t)$ and $E_z(t)$. The Laplace equation can be expressed in prolate spheroids, and the electric fields solved (Hu and Joshi 2009a).

Invoking continuity in potential and current density at the membrane interfaces leads to appropriate boundary conditions, and a complete solution of the Laplace equation in prolate co-ordinates. Consequently, the electric field can be obtained as a derivative of this potential.

3.2.4 *Distributed Network Analysis for Transmembrane Potentials*

Analysis of the cellular response to electrical pulses requires, at the very least, an evaluation of the voltage and current distributions within the cell, their time-dependent evolution, and the effects of such fields on the membranes. From a simple qualitative standpoint, the presence of high electric fields at the membranes can be viewed as being the source of a strong Maxwell stress associated with the discontinuity in the dielectric permittivity between the membrane and its surrounding aqueous medium. This electric field driven stress is the source for pore opening, molecular transport and cellular responses. A possible approach to modeling that takes into account the different dielectric properties of the constituent components (e.g., outer plasma membrane, external medium, intracellular fluid, organelle membranes, endoplasmic reticulum etc.), is to represent the electrical characteristics of the cells by a distributed equivalent circuit. Node analysis can then be used to determine the voltages at a set of distributed discrete points as a function of time, in response to the temporal variations of the applied voltage. Using the node voltage values at the boundary end points, a system of N-equations involving N-grid points can be obtained. This system of linear equations can be represented via a sparse matrix, and easily solved using numerical techniques such as the Cholesky method (Dereniowski and Dariusz 2004). Volume and shape changes of cells can be ignored since the external applied pulse duration is short (~nanoseconds) for cells to deform much during this time interval. However, time-domain solvers such as SPICE are not suited to the present problem for the following reasons: (i) In SPICE, values of circuit parameters (e.g., resistors R, capacitors C, etc.) cannot be made time dependent; (ii) the stochastic nature of pore formation cannot be included; (iii) the geometric dependence of R and C make it inconvenient to run simulations with varying shapes and sizes; (iv) soft-thresholds inherent in the pore formation process are difficult to implement, and (v) it is difficult to use circuit models to simulate the gradual resealing of pores. Also, the resealing process takes a long time (~microseconds), and the long-lived, diffusion-driven currents cannot be modeled by circuit simulators.

Details of a time-domain nodal analysis involving a distributed equivalent circuit representation of a cell and its membrane structures have been discussed in the literature (Joshi and Schoenbach 2010; Joshi et al. 2004), Some salient features are discussed below for convenience. Essentially, the entire cell volume was broken up into finite segments, and each segment represented by a parallel resistive-capacitive (*RC*) combination to account for the current flow and charging effects. The computational region was a sphere that included the cell with its outer membrane, the aqueous intra-cellular medium, and surrounding suspension, discretized along the r, θ, and φ directions. For simplicity, the cell membrane itself was taken as an integral unit, i.e., this sub-region was not further discretized. For interior nodes, the current continuity equation is of the form:

$$\sum_{k=1}^{6}\left(\sigma E+\varepsilon\frac{\partial E}{\partial t}\right)_k\times A_k=\sum_{k=1}^{6}I_k=0, \tag{3.5}$$

where I_k are currents along the six faces of an elemental cube with surface areas A_k, the local electric field is E, while ε and σ are the average permittivity and conductivity at the site of each cube. A schematic of the discretized geometry is shown in Fig. 3.11a. Details of the equations for each elemental volume shown in Fig. 3.11a, are given in the next paragraph. Combining the current continuity then leads to N equations for the N unknown node voltages which can be solved using matrix decomposition techniques. Potentials on each node were updated at every time step based on the boundary conditions imposed by the externally applied field, and the dynamic membrane conductivity which changes over time due to electroporation. Nodes on the periphery of the simulation region of radius R, were assigned potential values as boundary conditions in keeping with the external electric fields. For example, for an external field of magnitude E_0 directed along the z-direction, the boundary nodes were set to values of—$E_0R\cos(\theta)$, with being the angular location of the node relative to the z-axis. For fields applied along two or more principal, appropriate superposition of voltages were applied as the potential at the boundary nodes.

The current continuity equation applied at each elemental volume shown in Fig. 3.11 leads to the following equation, which takes account of both conduction and displacement currents. The algebraic sum of the currents across each of the six surfaces on the elemental cube are taken into account.

$$\begin{aligned}
&\sigma_1\frac{V^t_{i-1,j,k}-V^t_{i,j,k}}{\Delta r}A_1+\sigma_2\frac{V^t_{i+1,j}-V^t_{i,j}}{\Delta r}A_2+\sigma_3\frac{V^t_{i,j-1,k}-V^t_{i,j,k}}{r\Delta\theta}A_3+\sigma_4\frac{V^t_{i,j+1,k}-V^t_{i,j,k}}{r\Delta\theta}A_4\\
&+\sigma_5\frac{V^t_{i,j,k-1}-V^t_{i,j,k}}{r\sin\theta\Delta\varphi}A_5+\sigma_6\frac{V^t_{i,j,k+1}-V^t_{i,j,k}}{r\sin\theta\Delta\varphi}A_6+\frac{\varepsilon_1}{\Delta t}\Big[\frac{V^{t+1}_{i-1,j,k}-V^{t+1}_{i,j,k}}{\Delta r}-\frac{V^t_{i-1,j,k}-V^t_{i,j,k}}{\Delta r}\Big]A_1\\
&+\frac{\varepsilon_2}{\Delta t}\Big[\frac{V^{t+1}_{i+1,j,k}-V^{t+1}_{i,j,k}}{\Delta r}-\frac{V^t_{i+1,j,k}-V^t_{i,j,k}}{\Delta r}\Big]A_2+\frac{\varepsilon_3}{\Delta t}\Big[\frac{V^{t+1}_{i,j-1,k}-V^{t+1}_{i,j,k}}{r\Delta\theta}-\frac{V^t_{i,j-1,k}-V^t_{i,j,k}}{r\Delta\theta}\Big]A_3\\
&+\frac{\varepsilon_4}{\Delta t}\Big[\frac{V^{t+1}_{i,j+1,k}-V^{t+1}_{i,j,k}}{r\Delta\theta}-\frac{V^t_{i,j+1,k}-V^t_{i,j,k}}{r\Delta\theta}\Big]A_4\\
&+\frac{\varepsilon_5}{\Delta t}\Big[\frac{V^{t+1}_{i,j,k-1}-V^{t+1}_{i,j,k}}{r\sin\theta\Delta\varphi}-\frac{V^t_{i,j,k-1}-V^t_{i,j,k}}{r\sin\theta\Delta\varphi}\Big]A_5+\frac{\varepsilon_6}{\Delta t}\Big[\frac{V^{t+1}_{i,j,k+1}-V^{t+1}_{i,j,k}}{r\sin\theta\Delta\varphi}-\frac{V^t_{i,j,k+1}-V^t_{i,j,k}}{r\sin\theta\Delta\varphi}\Big]A_6=0
\end{aligned}\quad . \tag{3.6}$$

In the above, the areas A_i are given as:

$$\begin{aligned}
A_1&=\left(r+\tfrac{\Delta r}{2}\right)^2\left[\cos\left(\theta-\tfrac{\Delta\theta}{2}\right)-\cos\left(\theta+\tfrac{\Delta\theta}{2}\right)\right]\Delta\varphi,\\
A_2&=\left(r-\tfrac{\Delta r}{2}\right)^2\left[\cos\left(\theta-\tfrac{\Delta\theta}{2}\right)-\cos\left(\theta+\tfrac{\Delta\theta}{2}\right)\right]\Delta\varphi,\\
A_3&=r\Delta r\sin\left(\theta-\tfrac{\Delta\theta}{2}\right)\Delta\varphi,\\
A_4&=r\Delta r\sin\left(\theta+\tfrac{\Delta\theta}{2}\right)\Delta\varphi.\\
A_5&=A_6=r\Delta r\Delta\theta,\\
A_6&=r\Delta r\Delta\theta.
\end{aligned} \tag{3.7}$$

Nodes with $j = 0$ and $j = m$ have to be treated carefully since these grid points are independent of k. Only I_1, I_2 and I_4 are nonzero in the equation, and the sum of I_5 and I_6 would be zero. For $j = 0$ and $1 < i < N_i$, Eq. (3.6) effectively becomes:

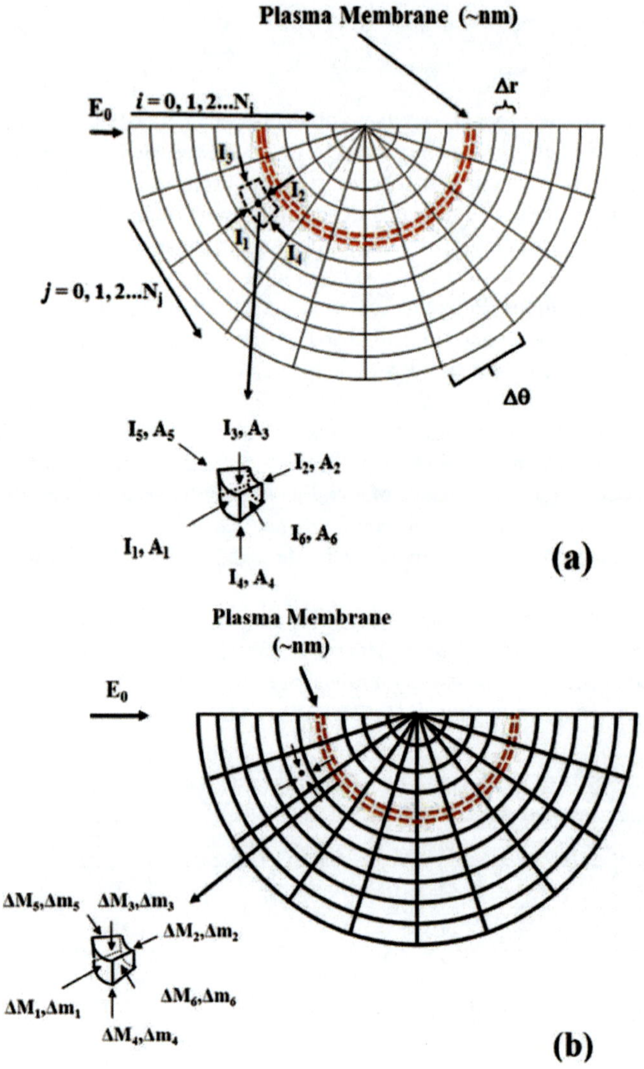

Fig. 3.11 Sketch of the discretization used for an elemental volume encompassing each grid point with six surfaces in: **a** transmembrane potential calculations, and **b** ion concentration calculations (After Hu et al. 2005a, b)

$$N_k\sigma_1\frac{V_{i-1,j}^t - V_{i,j}^t}{\Delta r}A_1 + n\sigma_2\frac{V_{i+1,j}^t - V_{i,j}^t}{\Delta r}A_2 + \sum_{k=0}^{N_k}\sigma_4\frac{V_{i,j+1,k}^t - V_{i,j}^t}{r\Delta\theta}A_4$$
$$+\frac{\varepsilon_1}{\Delta t}\left[\frac{V_{i-1,j}^{t+1} - V_{i,j}^{t+1}}{\Delta r} - \frac{V_{i-1,j}^t - V_{i,j}^t}{\Delta r}\right]A_1$$

$$+\frac{\varepsilon_2}{\Delta t}\left[\frac{V_{i+1,j}^{t+1}-V_{i,j}^{t+1}}{\Delta r}-\frac{V_{i+1,j}^{t}-V_{i,j}^{t}}{\Delta r}\right]A_2.$$

$$+\sum_{k=0}^{N_k}\frac{\varepsilon_4}{\Delta t}\left[\frac{V_{i,j+1,k}^{t+1}-V_{i,j}^{t+1}}{r\Delta\theta}-\frac{V_{i,j+1.k}^{t}-V_{i,j}^{t}}{r\Delta\theta}\right]A_4=0 \tag{3.8}$$

And for $j = N_j$ and $1 < i < N_i$, Eq. (3.6) effectively becomes:

$$N_k\sigma_1\frac{V_{i-1,j}^{t}-V_{i,j}^{t}}{\Delta r}A_1+n\sigma_2\frac{V_{i+1,j}^{t}-V_{i,j}^{t}}{\Delta r}A_2+\sum_{k=0}^{N_k}\sigma_4\frac{V_{i,j-1,k}^{t}-V_{i,j}^{t}}{r\Delta\theta}A_4$$

$$+\frac{\varepsilon_1}{\Delta t}\left[\frac{V_{i-1,j}^{t+1}-V_{i,j}^{t+1}}{\Delta r}-\frac{V_{i-1,j}^{t}-V_{i,j}^{t}}{\Delta r}\right]A_1$$

$$+\frac{\varepsilon_2}{\Delta t}\left[\frac{V_{i+1,j}^{t+1}-V_{i,j}^{t+1}}{\Delta r}-\frac{V_{i+1,j}^{t}-V_{i,j}^{t}}{\Delta r}\right]A_2.$$

$$+\sum_{k=0}^{N_k}\frac{\varepsilon_4}{\Delta t}\left[\frac{V_{i,j-1,k}^{t+1}-V_{i,j}^{t+1}}{r\Delta\theta}-\frac{V_{i,j-1.k}^{t}-V_{i,j}^{t}}{r\Delta\theta}\right]A_4=0 \tag{3.9}$$

The advantage of using short duration pulses in affecting membranes of inner organelles is perhaps not immediately obvious, and merits a quick discussion. Figure 3.12 shows a sketch of a spherical shell with a concentric inner organelle. We assume for simplicity that the conductivities of both membranes are zero, and that current continuity applies across line ABCDE shown in Fig. 3.12. For long-duration, slow-rising pulses, a near quasi-steady state prevails. Choosing the cell center as the reference voltage, the node potentials are roughly: $V_D \sim 0$ and $V_B \sim V_C$. Also, as there is no current flow through the membrane, under the quasi-steady state: $V_{CD} \sim V_{FG} \sim 0$. Thus, the potential across the inner membrane can be expected to be very modest, at best. This implies that membrane poration and other electrical effects would not be strong across cellular substructures and inner organelles. Fast rising, ultrashort pulses, on the other hand, would force a large nonequilibrium transient, and create substantially large values of V_{CD} and V_{FG} across the inner membranes.

For a quantitative demonstration of the above idea, a simple double-shelled cell model was used with the parameters given in Table 3.1 that are typical of biological cells. Most values were chosen from a report by Ermolina et al. (2001). The response of a low-intensity 280 ns trapezoidal pulse width, and 40 ns rise and fall times, was compared to that of a 11-ns high-intensity trapezoidal pulse width, and 1.75 ns rise and fall times. Figure 3.13 shows the transmembrane potentials across both the inner and outer membranes for the longer 280-ns pulse. The inner membrane potential is seen to go down to zero before it can reach the traditional electroporation threshold of ~ 1 V. This is due to inner membrane discharging and the evolution toward a low inner membrane voltage as qualitatively discussed in the context of Fig. 3.13. However, Fig. 3.14, for the much shorter 11 ns pulse, shows the inner membrane transmembrane

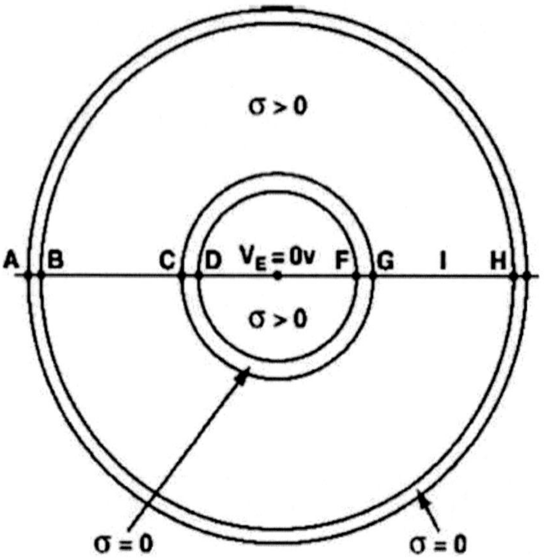

Fig. 3.12 Rough voltage analysis in the quasi-steady state for a model double-shelled cell in response to a slow rising, long pulse

Table 3.1 Parameters for simulation

Conductivities	
Environment	0.6 S/m
Cell membrane	0.0 S/m
Cytoplasm	0.6 S/m
Organelle membrane	0.0 S/m
Organelle inclusion	0.6 S/m
Relative permittivity	
Environment	80
Cell membrane	8
Cytoplasm	80
Organelle membrane	4
Organelle inclusion	80
Geometric parameters	
Radius of simulation region	10 μm
Radius of cell	5 μm
Thickness of cell membrane	5 nm
Radius of organelle	1 μm
Thickness of organelle membrane	10 nm

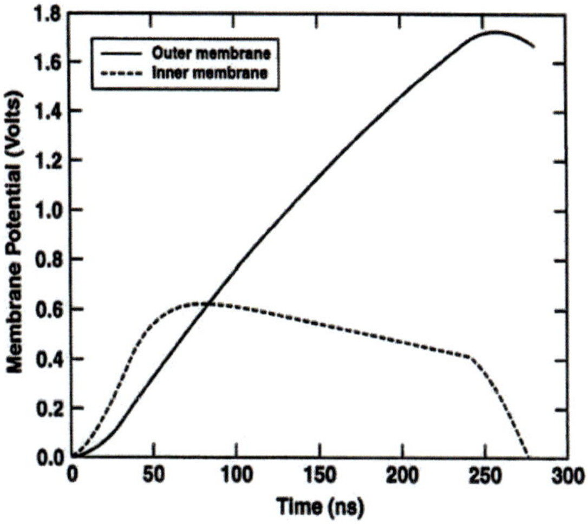

Fig. 3.13 Transmembrane potential of outer membrane and inner membranes for a 5 kV/cm, 280 ns duration trapezoidal pulse for the cell model of Fig. 3.2 (After Joshi and Schoenbach 2010)

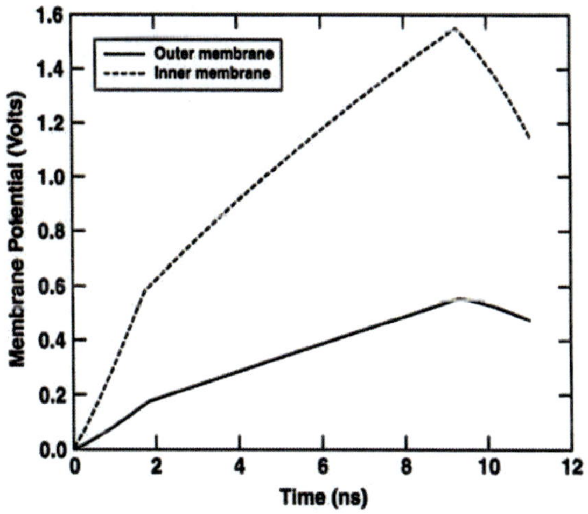

Fig. 3.14 Transmembrane potential of outer membrane and inner membranes for a shorter 25 kV/cm, 11-ns trapezoidal pulse

potential could easily reach the threshold. Its value is predicted to remain larger than that of the outer cell membrane. This result quantitatively demonstrates the utility and role of ultrashort pulses in bringing about electrical effects that permeate down to cellular substructures, while potentially leaving the outer cell membranes intact.

More applications and example results based on the distributed circuit approach discussed in this section will be presented in subsequent chapters.

3.3 Flows Through Membrane Nanopores

Pore generation, growth and size-evolution are required to characterize time-dependent membrane conductivity. These can be obtained based on the Smoluchowski continuum theory (Neu and Krassowska 1999; Barnett and Weaver 1991; Pastushenko and Chhizmadzhev 1982; Freeman et al. 1994; Glaser et al. 1988). The most accurate method for membrane pore transport is to solve the full Smoluchowski model, rather than using the asymptotic approximation (Neu and Krassowska 1999). This enabled calculations of the time-dependent pore density $N(t)$ at different locations on the membrane surface (Joshi et al. 2001). Application of the external electric field drives the pore generation and growth, through which ionic flow can occur. The current density $J_{EP}(t)$ through a membrane pore is given as: $J_{EP}(t) = i_{EP}(t)N(t)$, where $i_{EP}(t)$ is the current through a pore, and $N(t)$ is the pore density. The pore areas on the membrane at different mesh locations and times, as well as the electric field E can be obtained, which enables computations of ion flow across the cell membrane. Diffusive flow of ions through membrane pores can be modeled as (Schoenbach et al. 2015):

$$\mathrm{J_{diff}} = -\mathrm{D} * \mathrm{H} * \mathrm{K} * \frac{\partial \mathrm{C_{i,j}}}{\partial \mathrm{x}}. \tag{3.10}$$

In the above, J_{diff} is the flow rate per area, H denotes the Hindrance factor which is related to average pore sizes with $H \sim 1$ for pore sizes larger than ions, K is partition factor (= 0.83 for Ca^{+2}), D is the diffusion coefficient, and C_{ij} the concentration of the ith ion at grid point j. The number of ions going through membrane in unit time was calculated by: $\frac{\Delta m_1}{\Delta t} = J_{diff} * A_{pore}$, with being the cumulative sum of the pore areas. For other layers not adjacent to cell membrane, $J_{diff} = -D\frac{\partial C}{\partial x}$ was used. The drift pore currents $i_{EP}(t)$ and changes in ionic concentration ΔC were taken as:

$$\Delta \mathrm{C_{i,j}} = \sum_{i=1}^{6} \Delta \mathrm{M_i}/\mathrm{V_s}. \tag{3.11}$$

In the above expression:

$$\frac{\Delta \mathrm{M_1}}{\Delta \mathrm{t}} = \begin{cases} \mathrm{C_{i-1,j,k} V_s E_1 \mu A_{pores}}, & \mathrm{E_1} > 0 \\ \mathrm{C_{i,j,k} V_s E_1 \mu A_{pores}}, & \mathrm{E_1} \leq 0 \end{cases}; \tag{3.12}$$

where $C_{i,j,k}$ is the molarity of the unit volume inside cell, V_s is the space volume of the unit; μ is the ion mobility which equals $\frac{DqZ}{K_bT}$, with D being the ion diffusion coefficient (~0.8 × 10^{-5} cm^2/s), k_B the Boltzmann constant, T the temperature in Kelvin, q the electronic charge, z the valence charge of the ions (e.g., $z = 2$ for calcium ions), and E_1 the average electrical field across surface A_1 as shown in Fig. 3.12b. From Eq. (3.12), one can obtain the moles of ions (e.g., calcium) going into the cell through the pores for $E_1 > 0$, or getting out of the cell pores for $E_1 \leq 0$. Thus, the pore current $i_{EP}(t)$ in a time interval Δt is given as: $= \Delta M_1 \; \Delta t \; z \; N_A$, with N_A being the Avogadro number. The model and technique are general, and can be extended for use towards the analyses of multi-ion membrane flows. To maintain charge neutrality, an equal concentration of immovable anions can be assumed in the simulation space outside the cell.

References

Abidor IG, Arakelyan VB, Chemomordik Y, Chizmadzhev A, Pastushenko VF, Tarasevich MR (1979) The electrical breakdown of cell and lipid membranes: the similarity of phenomenologies. Bioelectrochem Bioenergy 6:37–52

Ahmed M, Brace CL, Lee FT Jr, Goldberg SN (2011) Principles of and advances in percutaneous ablation. Radiology 258:351–369

Aihara HJ, Miyazaki J (1998) Gene transfer into muscle by electroporation in vivo. Nat Biotechnol 16:867–870

Allen MP, Tildesley DJ (1987) Computer simulation of liquids. Clarendon Press, Oxford

André FM, Gehl J, Sersa G, Préat V, Hojman P, Eriksen J, Golzio M, Cemazar M, Pavselj N, Rols MP, Miklavčič D, Neumann E, Teissié J, Mir LM (2008) Efficiency of high- and low-voltage pulse combinations for gene electrotransfer in muscle, liver, tumor, and skin. Hum Gene Ther 19:1261–1271

Anezo C, de Vries AH, Holtje HD, Ticleman DP, Marrink SJ (2003) Methodological issues in lipid bilayer simulations. J Phys Chem B 107, 9424–9433

Arnold WM, Zimmermann U (1984) Electric field-induced fusion and rotation of cells. In: Biological membranes, vol 5. Academic Press, London, pp 389–454

Asencor FJ, Santamaria C, Iglesias FJ, Dominguez A (1993) Dielectric energy of orientation in dead and living cells of *Schizosaccharomyces pombe*: fitting of experimental results to a theoretical model. Biophys J 64:1626–1631

Barnett A, Weaver JC (1991) Electroporation: a unified, quantitative theory of reversible electrical breakdown and mechanical rupture in artificial planar bilayer membranes. Bioelectrochem Bioenerg 25:163–182

Batzing BL (2002) Microbiology: an introduction. Thompson Learning, Pacific Grove, CA

Becker FF, Wang XB, Huang Y, Pethig R, Vykoukal J, Gascoyne P (1995) Proc Natl Acad Sci 92:860–864

Beebe SJ, Blackmore PF, White J, Joshi RP, Schoenbach KH (2004) Nanosecond pulsed electric fields modulate cell function through intracellular signal transduction mechanisms. Physiol Meas 25:1077–1093

Berendsen HJC, Postma JPM, van Gunsteren WF, DiNola A, Haak JR (1984) Molecular dynamics with coupling to an external bath. J Chem Phys 81:3684–3690

Berkowitz ML (2009) Detailed molecular dynamics simulations of model biological membranes containing cholesterol. Biochim Biophys Acta (BBA) Biomembr 1788:86–96

Berkowitz ML, Bostick DL, Pandit S (2006) Aqueous solutions next to phospholipid membrane surfaces: insights from simulations. Chem Rev 106:1527–1539

Blank M (ed) (1995) Electromagnetic fields: biological interactions and mechanisms. American Chemical Society, New York

Bockmann RA, de Groot BL, Kakorin S, Neumann E, Grubmuller H (2008) Kinetics, statistics, and energetics of lipid membrane electroporation studied by molecular dynamics simulations. Biophys J 95:1837–1850

Calvet CY, Mir LM (2016) The promising alliance of anti-cancer electrochemotherapy with immunotherapy. Cancer Metastasis Rev 35:165–177

Calvet CY, André FM, Mir LM (2014) Dual therapeutic benefit of electroporation-mediated DNA vaccination in vivo. Oncoimmunology 3:e28540

Chang DC (1992) Structure and dynamics of electric field-induced membrane pores as revealed by rapid-freezing electron microscopy. In: Guide to electroporation and electrofusion. Academic Press, Orlando, pp 9–27

Chiu SW, Clark M, Jakobsson E, Subramaniam S, Scott HL (1999) Optimization of hydrocarbon chain interaction parameters: application to the simulation of fluid phase lipid bilayers. J Phys Chem B 103:6323–6327

COMSOL MULTIPHYSICS 4.2b _ 1997e2011, COMSOL, Inc. https://www.comsol.com/products/acdc/

Corovic S, Zupanic A, Miklavčič D (2008) Numerical modeling and optimization of electric field distribution in subcutaneous tumor treated with electrochemotherapy using needle electrodes. IEEE Trans Plasma Sci 36:1665–1672

Crozier PS, Henderson D, Rowley RL, Busath DD (2001) Model channel ion currents in NaCl extended simple point charge water solution with applied-field molecular dynamics. Biophys J 81:3077–3089

Darden T, York D, Pedersen L (1993) Particle mesh Ewald: An $N \cdot \log(N)$ method for Ewald sums in large systems. J Chem Phys 98:10089–10092

Daud AI, DeConti RC, Andrews S, Urbas P, Riker AI, Sondak VK, Munster PN, Sullivan DM, Ugen KE, Messina JL, Heller R (2008) Phase I trial of interleukin-12 plasmid electroporation in patients with metastatic melanoma. J Clin Oncol 26:5896–6903

Dereniowski D, Dariusz M (2004) Cholesky factorization of matrices in parallel and ranking of graphs. In: 5th international conference on parallel processing and applied mathematics, Lecture notes on computer science, vol 3019. Springer-Verlag, New York, pp 985–992

Dujardin E, Ebbesen TW, Hiura H, Tanigaki K (1994) Capillarity and wetting of carbon nanotubes. Science 265:1850–1852

Edd JF, Horowitz L, Davalos RV, Mir LM, Rubinsky B (2006) Vivo results of a new focal tissue ablation technique: irreversible electroporation. IEEE Trans Biomed Eng 53:1409–1415

Edholm O (2008) Time and length scales in lipid bilayer simulations. In: Feller SE (ed) Computational modeling of membrane bilayers, vol 60. Current topics in membranes. Elsevier, London, pp 91–110

Ermolina I, Polevaya Y, Feldman Y, Ginzburg B, Schlesinger M (2001) Study of normal and malignant white blood cells by time domain di-electric spectroscopy. IEEE Trans Dielect Elect Insulation 8:253–261

Esser AT, Smith KC, Gowrishankar TR, Vasilkoski Z, Weaver JC (2010) Mechanisms for the intracellular manipulation of organelles by conventional electroporation. Biophys J 98:2506–2514

Feller SE, Gawrisch K, MacKerell AD (2002) Polyunsaturated fatty acids in lipid bilayers: intrinsic and environmental contributions to their unique physical properties. J Am Chem Soc 124:318–326

Feller SE (ed) (2008) Computational modeling of membrane bilayers. In: Current topics in membranes, vol 60. Elsevier, London

Forrest LR, Sansom MSP (2000) Membrane simulations: bigger and better. Curr Opin Struct Biol 10:174–181

Foster KR (2000) Thermal and nonthermal mechanisms of interaction of radio-frequency energy with biological systems. IEEE Trans Plasma Sci 28:15–23

Foster KR, Schwan HP (1995) Dielectric properties of tissues—a review. In: Polk C, Postow E (eds) Handbook of biological effects of electromagnetic radiation. CRC Press, Boca Raton, pp 26–102

Freeman SA, Wang MA, Weaver JC (1994) Theory of electroporation of planar bilayer membranes: predictions of the aqueous area, change in capacitance, and pore-pore separation. Biophys J 67:42–56

Frey W, White JA, Price RO, Joshi RP, Nuccitelli R, Beebe SJ, Schoenbach KH, Kolb JF (2006) Plasma membrane voltage changes during nanosecond pulsed electric field exposure. Biophys J 90:3608–3615

Fuhr G, Glaser R, Hagedorn R (1986) Rotation of dielectrics in a rotating electric high-frequency field. Model experiments and theoretical explanation of the rotation effect of living cells. Biophys J 49:395–402

Gascoyne PR, Vykoukal J (2002) Particle separation by dielectrophoresis. Electrophoresis 23:1973–1983

Gimsa J, Wachner D (1999) A polarization model overcoming the geometric restrictions of the Laplace solution for spheroidal cells: Obtaining new equations for field-induced forces and transmembrane potential. Biophys J 77:1316–1326

Gimsa J, Wachner D (2001) Analytical description of the transmembrane voltage induced on arbitrarily oriented ellipsoidal and cylindrical cells. Biophys J 81:1888–1896

Glaser RW, Leikin SL, Chernomordik LV, Pastushenko VF, Sokirko I (1988) Reversible electrical breakdown of lipid bilayers: formation and evolution of pores. Biochim Biophys Acta 940:275–287

Gothelf A, Gehl J (2010) Gene electrotransfer to skin; Review of existing literature and clinical perspectives. Curr Gene Ther 10:287–299

Grosse C, Schwan HP (1992) Cellular membrane potentials induced by alternating fields. Biophys J 63:1632–1642

Gurtovenko AA, Miettinen M, Karttunen M, Vattulainen I (2005) Effect of monovalent salt on cationic lipid membranes as revealed by molecular dynamics simulations. J Phys Chem B 109:21126–21134

Hart FX (2009) Bioimpedance in the clinic. Zdravstveni Vestnik 78:782–790

Heller L, Heller R (2006) In vivo electroporation for gene therapy. Hum Gene Ther 17:890–897

Hess B, Bekker H, Berendsen HJ (1997) LINCS: A linear constraint solver for molecular simulations. J Comput Chem 18:1463–1472

Hofsaß C, Lindahl E, Edholm O (2003) Molecular dynamics simulations of phospholipid bilayers with cholesterol. Biophys J 84:2192–2206

Hu Q, Joshi RP (2009a) Analysis of intense, subnanosecond electrical pulse-induced transmembrane voltage in spheroidal cells with arbitrary orientation. IEEE Trans Biomed Eng 56:1617–1626

Hu Q, Joshi RP (2009b) Transmembrane voltage analyses in spheroidal cells in response to an intense subnanosecond electrical pulse. Phys Rev E 79:011901/1–8.

Hu Q, Viswanadham S, Joshi RP, Schoenbach KH, Beebe SJ, Blackmore PF (2005a) Simulations of transient membrane behavior in cells subjected to a high-intensity ultrashort electric pulse. Phys Rev E 71:031914

Hu Q, Joshi RP, Schoenbach KH (2005b) Simulations of nanopore formation and phosphatidylserine externalization in lipid membranes subjected to a high-intensity, ultrashort electric pulse. Phys Rev E 72:031902/1-10

Hu Q, Sridhara V, Joshi RP, Kolb JF, Schoenbach KH (2006) Molecular dynamics analysis of high electric pulse effects on bilayer membranes containing DPPC and DPPS. IEEE Trans Plasma Sci 34(4):1405–1411

Hu Q, Zhang Z, Qiu H, Kong M, Joshi RP (2013) Physics of nanoporation and water entry driven by a high-intensity, ultrashort electrical pulse in the presence of cellular hydrophobic interactions. Phys Rev E 87:032704/1-9

Jerry RA, Popel AS, Brownell WE (1996) Potential distribution for a spheroidal cell having a conductive membrane in an electric field. IEEE Trans Biomed Eng 43:970–972

Jordan CA, Neumann E, Sowers AE (2013) Electroporation and electrofusion in cell biology. Springer Science and Business Media, New York

Joshi RP, Schoenbach KH (2010) Bioelectric effects of intense ultrashort pulses. Critical Rev Bio-Med Eng 38:255–304

Joshi RP, Schoenbach KH (2011) Electric fields in biological cell and membranes. In: Electromagnetic fields in biological systems. CRC Press, Boca Raton, FL

Joshi RP, Song J (2010) Model analysis of electric fields induced by high-voltage pulsing in cylindrical nerves. IEEE Trans Plasma Sci 38:2894–2900

Joshi RP, Hu Q, Aly R, Schoenbach KH, Hjalmarson HP (2001) Self-consistent simulations of electroporation dynamics in biological cells subjected to ultrafast electrical pulses. Phys Rev E 64:11913

Joshi RP, Hu Q, Schoenbach KH (2004) Modeling studies of cell response to ultrashort, high-intensity electric fields—implications for intracellular manipulation. IEEE Trans Plasma Sci 32:1677–1686

Joshi RP, Mishra A, Song J, Pakhomov A, Schoenbach KH (2008) Simulation studies of ultra-short, high-intensity electric pulse induced action potential block in whole-animal nerves. IEEE Trans Biomed Eng 55:1391–1398

Jourabchi N, Beroukhim K, Tafti BA, Kee ST, Lee EW (2014) Irreversible electroporation (NanoKnife) in cancer treatment. Gastrointestinal Intervention 3:8–18

Kotnik T, Miklavčič D (2000) Analytical description of transmembrane voltage induced by electric fields on spheroidal cells. Biophys J 79:670–679

Kotnik T, Miklavčič D (2006) Theoretical evaluation of voltage inducement on internal membranes of biological cells exposed to electric fields. Biophys J 90:480–491

Kotnik T, Bobanović F, Miklavčič D (1997) Sensitivity of transmembrane voltage induced by applied electric fields—a theoretical analysis. Bioelectrochem Bioenerg 43:285–291

Kotnik T, Kramar P, Pucihar G, Miklavčič D, Tarek M (2012) Cell membrane electroporation—Part 1: The phenomenon. IEEE Electr Insul Mag 28:14–23

Latouche EL, Arena CB, Ivey JW, Garcia PA, Pancotto TE, Pavlisko N, Verbridge SS, Davalos RV, Rossmeisl JH (2018) High-frequency irreversible electroporation for intracranial meningioma: a feasibility study in a spontaneous canine tumor model. Technol Cancer Res Treatment 17:1–10

Leach AR (2001) Molecular modelling: principles and applications, 2nd edn. Prentice Hall, Englewood

Levin M (2014) Molecular bioelectricity: how endogeneous voltage potentials control cell behavior and instruct pattern regulation *in vivo*. Mol Biol Cell 25:3835–3850

Levin M, Pezzulo G, Finkelstein JM (2017) Endogenous bioelectric signaling networks: exploiting voltage gradients for control of growth and form. Annu Rev Biomed Eng 19:353–387

Lindahl E, Edholm O (2000) Mesoscopic undulations and thickness fluctuations in lipid bilayers from molecular dynamics simula-tions. Biophys J 79:426–433

Lindahl E, Sansom MSP (2008) Membrane proteins: molecular dynamics simulations. Curr Opin Struct Biol 18:425–431

Lohmann KJ, Lohmann CMF (1996) Detection of magnetic field intensity by sea turtles. Nature 380:59–61

Low L, Mander A, McCann K, Dearnaley D, Tjelle T, Mathisen I, Stevenson F, Ottensmeier CH (2009) DNA vaccination with electroporation induces increased antibody responses in patients with prostate cancer. Human Gene Ther 20:1269–1278

Marrink SJ, Mark AE (2001) Effect of undulations on surface tensioncin simulated bilayers. J Phys Chem B 105:6122–6127

Marrink SJ, de Vries AH, Tieleman DP (2009) Lipids on the move: simulations of membrane pores, domains, stalks and curves. Biochim Biophys Acta Biomembr 1788:149–168

Marty M, Sersa G, Garbay JR, Gehl J, Collins CG, Snoj M, Billard V, Geerrtsen PF, Larkin JO, Miklavčič D, Pavlovic I, Paulin-Kosir SM, Cemezar M, Morsli N, Soden DM, Rudolf Z, Robert C, O'Sullivan GC, Mir LM (2006) Eur J Cancer Suppl 4:3–13

Mashl RJ, Scott HL, Subramaniam S, Jakobsson E (2001) Molecular simulation of dioleylphosphatidylcholine bilayers at differing levels of hydration. Biophys J 81:3005–3015

Maswiwat K, Wachner D, Warnke R, Gima J (2007) Simplified equations for the transmembrane potential induced in ellipsoidal cells of rotational symmetry. J Phys D 40:914–923

McCaig CD, Rajnicek AM, Song B, Zhao M (2005) Controlling cell behavior electrically: current views and future potential. Physiol Rev 85:943–978

Meglic SH, Kotnick T (2016) Electroporation-based applications in biotechnology. In: Miclavčič (ed) Handbook of electroporation. Springer International Publishing, Cham, Switzerland

Meng S, Rouabhia M, Zhang Z (2013) Electrical stimulation modulates osteoblast proliferation and bone protein production through heparin-bioactivated conductive scaffolds. Bioelectromagnetics 34:189–199

Miklavčič D, Mali B, Kos B, Heller R, Serša G (2014) Electrochemotherapy: from the drawing board into medical practice. BioMed Eng OnLine 13:29

Mir LM, Belehradek M, Domenge C, Orlowski S, Poddevin B, Belehradek J Jr, Schwaab G, Luboinski B, Paoletti C (1991) C R Acad Sci III(313):613–618

Napotnik TB, Reberšek M, Vernier PT, Mali B, Miklavčič D (2016) Effects of high voltage nanosecond electric pulses on eukaryotic cells (in vitro): a systematic review. Bioelectrochemistry 110:1–12

Neu JC, Krassowska W (1999) Asymptotic model of electroporation. Phys Rev E 59:3471–3482

Neu JC, Krassowska W (2003) Modeling postshock evolution of large electropores. Phys Rev E 67:021915/1–12

Neumann E, Rosenheck K (1972) Permeability changes induced by electric impulses in vesicular membranes. J Membr Biol 10:279–290

Neumann E, Schaefer-Ridder M, Wang Y, Hofschneider PH (1982) Gene transfer into mouse lyoma cells by electroporation in high electric fields. EMBO J 1:841–845

Nuccitelli R, Pliquett U, Chen X, Ford W, Swanson RJ, Beebe SJ, Kolb JF, Schoenbach KH (2006) Nanosecond pulsed electric fields cause melanomas to self-destruct. Biochem Biophys Res Commun 343:351–360

Nuccitelli R, Wood R, Kreis M, Athos B, Huynh J, Liu K, Nuccitelli P, Epstein EH (2014) First-in-human trial of nanoelectroablation therapy for basal cell carcinoma: proof of method. Exp Dermatol 23:135–137

Nuccitelli R, Berridge JC, Mallon Z, Kreis M, Athos B, Nuccitelli P (2015) Nanoelectroablation of murine tumors triggers a CD8-dependent inhibition of secondary tumor growth. PLoS ONE 10:e0134364/1-17

Pastushenko VF, Chhizmadzhev YA (1982) The theory of bilayer lipid membrane electric clamp splitting of the energetic barrier. Biofizika 27:475–479

Pauly H, Schwan HP (1959) Über die Impedanz einer Suspension von kugelformigen Teilchen mit einer Schale. Z Naturforsch B 14:125–131

Pavlin M, Kandušer M (2015) New insights into the mechanisms of gene electrotransfer–experimental and theoretical analysis. Scientific Reports 5:9132 pp 1–11

Pohl HA, Crane JS (1971) Dielectrophoresis of Cells. Biophys J 11:711–727

Polk C, Postow E (1996) Handbook of biological effects of electromagnetic fields, 2nd edn. CRC Press, Boca Raton

Prausnitz MR, Bose VG, Langer R, Weaver JC (1993) Electroporation of mammalian skin: a mechanism to enhance transdermal drug delivery. Proc Natl Acad Sci USA 90:10504–10508

Pucihar G, Kotnik T, Kandušer M, Miklavčič D (2001) The influence of medium conductivity on electropermeabilization and survival of cells in vivo. Bioelectrochemistry 54:107–115

Pucihar G, Kotnik T, Valic B, Miklavčič D (2006) Numerical determination of transmembrane voltage induced on irregularly shaped cell. Ann Biomed Eng 34:642–652

Puller CE, Isseroff RR (2005) Cyclic AMP mediates keratinocyte directional migration in an electric field. J Cell Sci 118:2023–2034

Puller CE (ed) (2011) The physiology of bioelectricity in development, tissue regeneration and cancer. CRC Press, Boca Raton, FL, p 342

Radu M, Ionescu M, Irimescu N, Iliescu K, Pologea-Moraru R, Kovacs E (2005) Orientation behavior of retinal photoreceptors in alternating electric fields. Biophys J 89:3548–3554

Robinson KR, Messerli MA (2003) The role of endogenous electric fields as directional signals in development, repair and invasion. BioEssays 25:759–766

Rols MP, Teissie J (1998) Electropermeabilization of mammalian cells to macromolecules: control by pulse duration. Biophys J 75:1415–1423

Rosazza C, Meglic SH, Zumbusch A, Rols MP, Miklavčič D (2016) Curr Gene Ther 16:98–129

Saiz L, Klein ML (2001) Structural properties of a highly polyunsaturated lipid bilayer from molecular dynamics simulations. Biophys J 81:204–216
Sano MB, Arena CB, Bittleman KR, DeWitt MR, Cho HJ, Szot CS, Saur D, Cissel JM, Robertson J, Lee YW, Davalos RV (2015) Bursts of bipolar microsecond pulses inhibit tumor growth. Sci Rep 5:14999
Sauer FA (1983) Forces on suspended particles in the electromagnetic field. In: Coherent excitations in biological systems. Springer, New York, pp 134–144
Schmidt G, Juhasz-Böss I, Solomayer EF, Herr D (2014) Electrochemotherapy in breast cancer: a review of references. Geburtshilfe Frauenheilkd 74:557–562
Schoenbach KH, Hargrave B, Joshi RP, Kolb JF, Nuccitelli R, Osgood C, Pakhomov A, Stacey M, Swanson RJ, White J, Xiao S, Zhang J, Beebe SJ, Blackmore PF, Buescher ES (2007) Bioelectric effects of intense nanosecond pulses. IEEE Trans Dielectr Electr Insul 14:1088–1109
Schoenbach KH, Pakhomov AG, Semenov I, Xiao S, Pakhomova ON, Ibey BL (2015) Ion transport into cells exposed to monopolar and bipolar nanosecond pulses. Bioelectrochemistry 103:44–51
Schwan HP (1983) Biophysics of the interaction of electromagnetic energy will cells and membranes. In: Grandolfo M, Michaelson SM, Rindl A (eds) Biological effects and dosimetry of nonionizing radiation. Plenum Press, New York, pp 213–231
Sersa G, Teissie J, Cemazar M, Signori E, Kamensek U, Marshal G, Miklavčič D (2015) Electrochemotherapy of tumors as in situ vaccination boosted by immunogene electrotransfer. Cancer Immunol Immunother 64:1315–1327
Siu SWI, R. Vacha R, Jungwirth P, Bockmann RA (2008) Biomolecular simulations of membranes: physical properties from different force fields. J Chem Phys 128:125103
Smith KC, Gowrishankar TR, Esser AT, Stewart DA, Weaver JC (2006) The spatially distributed dynamic transmembrane voltage of cells and organelles due to 10 ns pulses: meshed transport networks. IEEE Trans Plasma Sci 34:1394–1404
Stampfli R (1958) Reversible electrical breakdown of the excitable membrane of a Ranvier node. An Acad Bras Cienc 30:57–63
Stewart DA, Gowrishankar TR, Weaver JC (2004) Transport lattice approach to describing cell electroporation: use of a local asymptotic model. IEEE Trans Plasma Sci 32:1696–1708
Tarek M (2005) Membrane electroporation: a molecular dynamics simulation. Biophys J 88:4045–4053
Tieleman DP (2004) The molecular basis of electroporation. BMC Biochem 5:10
Tieleman DP, Marrink SJ, Berendsen HJC (1997) A computer perspective of membranes: molecular dynamics studies of lipid bilayer systems. Biochim Biophys Acta 1331:235–270
Tieleman DP, Berendsen JHC, Sansom MSP (2001) Voltage-dependent insertion of alamethicin at phospholipid/water and octane water interfaces. Biophys J 80:331–346
Titomirov AV, Sukharev S, Kistanova E (1991) In vivo electroporation and stable transformation of skin cells of newborn mice by plasmid DNA. Biochim Biophys Acta 1088:131–134
Tobias DJ, Tu K, Klein ML (1997) Atomic-scale molecular dynamics simulations of lipid membranes. Curr Opin Colloid Interface Sci 2:15–26
Tsong TY (1991) Electroporation of cell membranes. Biophys J 60:297–306
van Gunsteren WF, Berendsen HJC (1987) Gromos User Manual, BIOMOS Biomolecular Software. Laboratory of Physical Chemistry, University of Groningen, Groningen
Vernier PT, Sun Y, Marcu L, Salemi S, Craft CM, Gundersen MA (2003) Calcium bursts induced by nanosecond electric pulses. Biochem Biophys Res Commun 310:286–295
Vernier PT, Sun Y, Marcu L, Craft CM, Gundersen MA (2004) Nanoelectropulse-induced phosphatidylserine translocation. Biophys J 86:4040–4048
Weaver JC, Chizmadzhev YA (1996) Theory of electroporation: a review. Bioelectrochem Bioenerg 41:135–160
Yarmush ML, Golberg A, Sersa G, Kotnik T, Miklavčič D (2014) Electroporation-based technologies for medicine: principles, applications, and challenges. Annu Rev Biomed Eng 16:295–320
Ziegler MJ, Vernier PT (2008) Interface water dynamics and porating electric fields for phospholipid bilayers. J Phys Chem B 112:13588–13596
Zimmermann U (1982) Electric field-mediated fusion and related electrical phenomena. Biochimica et Biophysica Acta (BBA)-Reviews on Biomembranes 694:227–277

Chapter 4
Comparison Between Monopolar and Bipolar Pulses for Effective Nanoporation

Ravi Joshi

Abstract The utility of electric pulsing for attaining and driving a range of bio-effects have already been discussed. This chapter focuses and probes the role of pulse shape (e.g., monopolar-vs.-bipolar), multiple electrode scenarios, and serial-versus-simultaneous pulsing. A possible analysis based on a three-dimensional time-dependent continuum model is discussed. Our results indicate that monopolar pulsing always leads to higher and stronger cellular uptake. This prediction is in agreement with experimental reports and observations. It is also demonstrated that multipronged electrode configurations influence and increase the degree of cellular uptake.

4.1 Introduction

From the standpoint of drug/gene delivery, besides the biologically-related variables (Pucihar et al. 2001; Prasanna and Panda 1997; Schoenbach et al. 2001; Kotnik et al. 2015; Teissie et al. 1999), other important factors are the electrode shape and geometries, as well as the electrical pulsing parameters (Santra et al. 2013; Kinosita et al. 1988). Membrane electropermeabilization begins when the applied field exceeds a critical transmembrane potential of about 1 V (Hibino et al. 1993) and is followed by electrophoresis–based molecular uptake (Pucihar et al. 2008; Sadik et al. 2014; Schoenbach et al. 2015; Mali et al. 2013). Thus, both permeabilization (which opens the cell to transport processes) and subsequent delivery are controlled in large part by electrical parameters such as the field strength, pulse duration, number of pulses, and pulse shape (Joshi and Schoenbach 2010). Multi-electrode systems have also been used (e.g., two parallel plates or six electrode arrays) for enhancing the delivery (Gilbert et al. 1997; Hofmann et al. 1996).

The role of the electrical pulse shape on biological cell membrane permeabilization has been the topic of multiple studies (Tekle et al. 1991; Teissie and Rols 1993; Kotnik et al. 2001; Kotnik et al. 2003; Faurie et al. 2010), though most have focused on millisecond pulses. Most of these previous studies concluded that bipolar pulses appear to have a stronger effect on electropermeabilization as compared to monopolar pulses. The reason given was that bipolar pulses compensate for the asymmetry inherent in the permeabilized areas at the poles of the cell that is introduced by

S. J. Beebe et al., *Ultrashort Electric Pulse Effects in Biology and Medicine*, Series in BioEngineering, https://doi.org/10.1007/978-981-10-5113-5_4

the resting transmembrane voltage. However, more recent studies with *much shorter pulses (300 and 600 ns)*, applied to Chinese Hamster Ovary cells showed *bipolar pulses to be less effective at electropermeabilization* (Ibey et al. 2014; Pakhomov et al. 2014). They were also found not to be as efficient in increasing intracellular calcium concentration. Now, application of pulses in this nanosecond domain (Beebe et al. 2004) is certainly important for a variety of reasons. For example, the short duration pulses mitigate muscle contractions, and so are a useful modality for the ablative treatment of solid tumors (Daskalov et al. 1999; Sano et al. 2015). Eliminating contractions helps improve the procedural safety of patients, since the need for neuroparalytic drugs to inhibit muscle contraction is then virtually eliminated. The short duration also eliminates localized thermal heating (Song et al. 2017) and reduces the potential for side-effects such as hyperthermia or cell damage.

The use of electrical pulses also arises in electrochemotherapy (ECT), which is used in clinics to treat patients with various types of cancers (Campana et al. 2019; Cadossi et al. 2014; Miklavčič et al. 2014). A common procedure for electrochemotherapy includes intra-tumoral or intravenous delivery of chemotherapeutic drugs, followed by the application of high voltage electrical monopolar pulses to the tumor area (Mir et al. 2006; Gehl et al. 2018; Yarmush et al. 2014). Two chemotherapeutics currently used against cancer in hospitals and clinics are bleomycin (Mir et al. 1996) and cisplatin (Spreckelmeyer et al. 2014). The use of electric pulses increases the permeabilization of the tumor membranes and facilitates the influx of the chemotherapeutic drugs (Tozon et al. 2001; Jaroszeski et al. 2000). A practical issue and question that arises in this context is whether to use monopolar or bipolar pulses. Traditionally, cell permeabilization was accomplished by application of micro- and millisecond monopolar electric pulses (Teissie et al. 1999; Weaver and Chizmadzhev 1996). In the context of such monopolar electrical excitation, however, the application of 100 μs long monopolar pulses at a 1 Hz has been reported to cause pain, discomfort and muscle contractions (Arena and Davalos 2012; Jiang et al. 2015). This necessitates the use of muscle relaxants and anesthesia (Ball et al. 2010). These problems can be alleviated for example by applying pulses at higher frequency (Županič et al. 2007), by using special electrode designs (Golberg and Rubinsky 2012; Yao et al. 2017), or by delivering bursts of short high-frequency bipolar pulses (Latouche et al. 2018; Dong et al. 2018).

Symmetric bipolar (BP) pulses, have been investigated as a method to improve permeabilization of cell membrane by efficiently porating both sides of the cell. For example, it was shown early on that 400 μs BP pulses increased transfection efficiency while reducing cell death (Tekle et al. 1991). Related work (Kotnik et al. 2001) involving studies of the impact of 1000 μs mono- and bipolar pulses on the permeabilization of cells to bleomycin, their survival, and uptake of Lucifer yellow. They concluded that BP pulses offer the advantage of increased cell permeabilization without the downside of increased cellular mortality. Furthermore, the use of BP pulses for in vivo tissue ablation (Ahmed et al. 2011) based on irreversible electroporation, was shown to be advantageous for reducing muscle contractions despite requiring higher-amplitude exposure to achieve a similar lethal effect. Later, a theoretical study (Arena et al. 2011), predicted that bipolar pulses *in the nanosecond range* would be advantageous in limiting joule heating and penetrating epithelial layers, resulting in more efficient electroporation of underlying tissues. These studies seemed to suggest an advantage in using BP pulses for electropermeabilization.

Nanosecond pulsed electric fields (nsPEFs) have been shown to permeabilize the plasma membrane (PM), to create a large population of smaller nanopores (Schoenbach et al. 2007; Joshi et al. 2008). A few papers have examined the issue of whether exposing cells to a bipolar pulse nsPEF would offer the same advantages as those observed with longer (microsecond or larger) pulses. Using bursts of ~50 extremely short monopolar (MP) and bipolar nanosecond pulses of 15 ns duration and an electric field of 28 kV/cm, membrane reorganization was shown to occurred symmetrically with BP pulses and asymmetrically with MP pulses (Vernier et al. 2006). However, comparative assessments between nanosecond and long-duration (μs) pulses were not presented or discussed. Another report based on bleomycin uptake, showed that extremely short (total duration 1.6 ns at full width half maximum) BP pulses were less effective than MP excitation at permeabilizing cells (French et al. 2009). More recent reports based on 600 ns pulse durations have shown bipolar nanosecond pulses to be less efficient at electropermeabilization and killing cells than monopolar pulses (Ibey et al. 2014). Experiments from the Slovenian group indicate the possibility of using high frequency bipolar pulses for electrochemotherapy, though at the expense of higher electric fields than for the classical monopolar pulses (Scuderi et al. 2019). The BP do have the potential to alleviate muscle contractions and pain in line with previous reports (Sano et al. 2015; Latouche et al. 2018; Scheffer et al. 2014).

Simulations have shown that monopolar pulsing lead to higher and stronger cellular uptake due to cellular electroporation. This prediction is in agreement with previous experimental reports and observations (Gianulis et al. 2015; Roth et al. 2015; Sano et al. 2014). For instance, in experiments (Gianulis et al. 2015) it was shown that the electric field polarity reversal hinders the electroporative efficiency. The data also noted that the rate and amount of uptake for the marker dye YO-PRO-1 was consistently two-to-three fold higher for unipolar pulses as compared to the bipolar treatments. Similar conclusions were reached by other researchers (Roth et al. 2015; Sano et al. 2014). Modeling work has also indicated that multi-pronged electrode configurations would influence the degree of cellular uptake. Judicious pulse sequencing would then add to the overall benefits.

4.2 Simulation Results

Here, the distributed circuit model already discussed is applied to evaluate and predict transport and ion flows into cells through porated membranes in response to external electric field pulsing. A general waveform, including monopolar or bipolar excitation, can be chosen. Table 4.1 lists the various parameters used. These are standard values that have typically been reported in the literature. For concreteness, the transport of calcium ions was simulated with a diffusion coefficient of 0.8×10^{-5} cm^2/s. For the ion concentration, a density of 2 mM was initially set at the start in the aqueous

Table 4.1 Simulation parameters used

Pulse types	Monopolar and bipolar
Pulse rise time	1.5 ns
Pulse fall time	1.5 ns
Membrane permittivity	$4 \times 8.85 \times 10^{-12}$ F/m
Membrane conductivity	5.3×10^{-6} S/m
Membrane thickness	5 nm
Cell radius	0.25 μm
Ca ion diffusion coefficient	0.8×10^{-5} cm²/s
Grid points R-direction	31 grid points
Grid points Θ-direction (0 – π/2)	11 grid points
Grid points Φ-direction (0 – 2 π)	20 grid points

medium outside the cell, while the initial values within each elemental the cell volume were set to zero. A representative schematic of bipolar and monopolar pulses with a near-rectangular shape used in the present simulations is shown in Fig. 4.1. The separation time between the end of any pulse and the beginning of the next pulse is denotes by T_s. Each of the repetitive cycles (varied for different simulations) were chosen to have short rise- and fall-times of 1.5 ns. The externally applied electric field magnitudes were varied, though for concreteness, a value of 100 kV/cm has

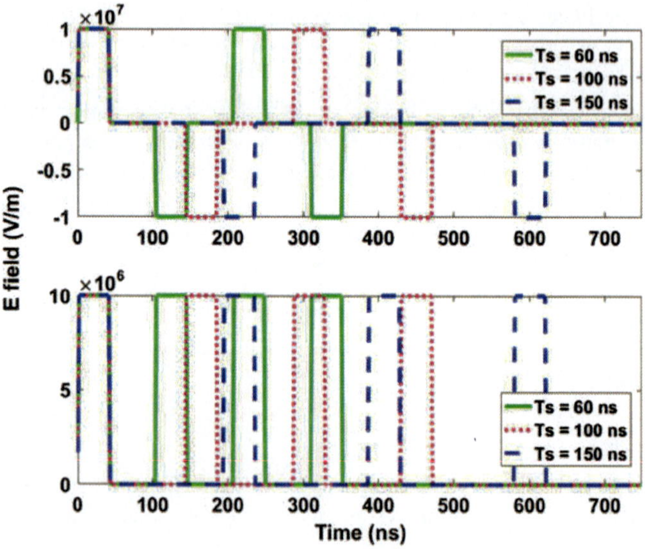

Fig. 4.1 A representative time-dependent bipolar and monopolar waveforms used for the simulations. The separation time between the end of any pulse and the beginning of the next pulse is denotes by T_s (After Hu and Joshi 2017)

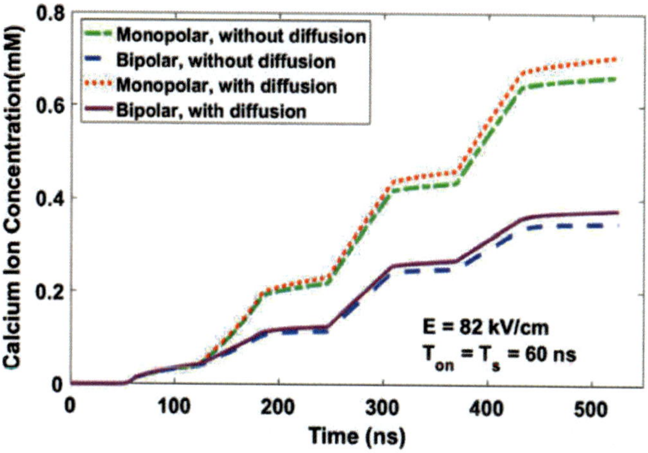

Fig. 4.2 Simulation results showing the temporal growth of the calcium concentration inside a cell subjected to bipolar and monopolar pulses, with and without diffusion taken into account (After Hu and Joshi 2017)

been shown in Fig. 4.1. Monopolar pulses are similar to the bipolar train, except that the negative fields are reversed to yield positive values at all times. For simplicity only calcium ions were assumed to be present in the extracellular medium though the approach is general and multi-ion situations can easily be treated.

Simulations of the time dependent calcium concentration within the cell in response to both bipolar and monopolar pulses of the type shown in Fig. 4.1 were carried out. These results are shown in Fig. 4.2, with a pulse "ON" time of 60 ns, a peak field of 82 kV/cm, and an "OFF"-time of 60 ns. The pulses were applied at the 9- and 3-o'clock positions, and hence, assumed to be along *only one of the three orthogonal axes*. The monopolar pulses had the same pulse parameters as the bipolar train except that all the pulses were positive in sign. For comparison, results with and without the inclusion of diffusive flows are also given. Figure 4.2 shows an initial delay for calcium ion entry since a finite time is required for membrane electroporation, which is then followed by ion inflows into the cell. The inflow due to drift is predicted to be rapid beyond the first pulse for the monopolar case, and so an upsurge is seen after about 121 ns. For the bipolar pulses, on the other hand, though pore formation is predicted to occur, the movement of ions keeps reversing in direction.

The difference between the two cases is perhaps best understood in terms of the transmembrane potentials (TMPs) over time, since the field values drive the flow. The TMP values are shown in Fig. 4.3a, b at the 9-o'clock position for the two polarities. With monopolar pulses, the TMP raises four times corresponding to the start of the four pulses. The highest TMP (and hence transmembrane electric field) value is reached for the first pulse. At subsequent times, due to pore formation, the membrane conductivity increases leading to lower voltage drops and reduced TMP values. In this monopolar pulse scenario, the positive external electric field direction is from left-to-right (i.e., from the 9- to 3-o'clock positions). Hence, positive ions go into the cell at the 9 o'clock position upon pore formation; but there is no ionic flow at the opposite end for the 3 o'clock position because the cell interior is devoid of any ions. The ionic throughput for this case continues from the left side and is most strong around the neighborhood of the 9-o'clock region. For bipolar pulses, the initial field direction is from left-to-right for the first pulse, leading to a positive TMP at the 9-o'clock position. However, this leads to a depletion of ions in the aqueous medium outside (but close to) the cell surface in the neighborhood of the 3-o'clock position. Since diffusion is a slow process, the recovery through ion flowback is slow, and hence the neighborhood around 3-o'clock remains relatively depleted. The second pulse, of reversed polarity, creates a negative TMP at the 9-o'clock location (as seen in Fig. 4.3b) with the field direction pointing from right-to-left. At this point in time, ion inflow into the cell is from the 3-o'clock neighborhood. Given the depletion of ions in the aqueous medium in this region though, the inflow is not as strong. Furthermore, reversal in the electric field leads to some outflow of ions from the pores previously formed at the 9-o'clock neighborhood. The net effect is that the ion content inside the cell is comparatively much lower than that for the monopolar case. So overall, the two main points to note from Fig. 4.2 are that: (i) with diffusion taken into account, the concentration in the cell is slightly higher, though the difference is not significant. (ii) The response in terms of ion inflow from monopolar pulsing is much stronger than that for bipolar pulses.

Next, simulations were carried out to gauge the amount of calcium inflows for two pulses having the same energy content but different electric field amplitudes and pulse durations. A simple calculation based on electrophoretic flow indicates that the flow volume F_{in} should roughly be dependent as: $v\,A\,dt$, where v is the drift velocity, dt the time duration, and A the effective pore area. The flow F_{in} would then scale as $F_{in} \sim E\,A\,dt$ at constant mobility. The energy G, on the other hand, scales as: $G \sim E^2 dt$. Hence, for two pulses having the same energy but different electric fields (E_1, E_2) and durations (dt_1, dt_2), one gets: $[F_{in1}/F_{in2}] \sim (A_1/A_2)\,(dt_1/dt_2)^{1/2}$. The simulation results for the time-dependent inflow are shown in Fig. 4.4 for pulse ON-times of 40 ns and 60 ns, having peak field amplitudes of 100 kV/cm and 82 kV/cm, respectively. The curves reveal that the shorter pulse (having the higher electric field of 100 kV/cm) drives more ions into the cell, as compared to a longer pulse of the same energy. This indicates that pore areal changes after the initial pulse are consequential and field-dependent. The outcome is a result of a higher density of pores at the higher fields, despite the shorter time duration. This is also indicative of a short pore formation time at these high fields, in keeping with experimental observations (Frey et al. 2006). Also, as with the previous result of Fig. 4.2, the monopolar pulsing is more effective at cellular loading.

The role of pulse separation (i.e., the pulse OFF time which represents the time interval separating the active pulses) on calcium ion inflows was probed next. Figure 4.5 shows simulation results for different pulse separation times T_s. A four-pulse monopolar sequence is shown in Fig. 4.5a. The ON-time remained at 40 ns, while the rise- and fall-times were set at 1.5 ns in all cases. The peak electric field

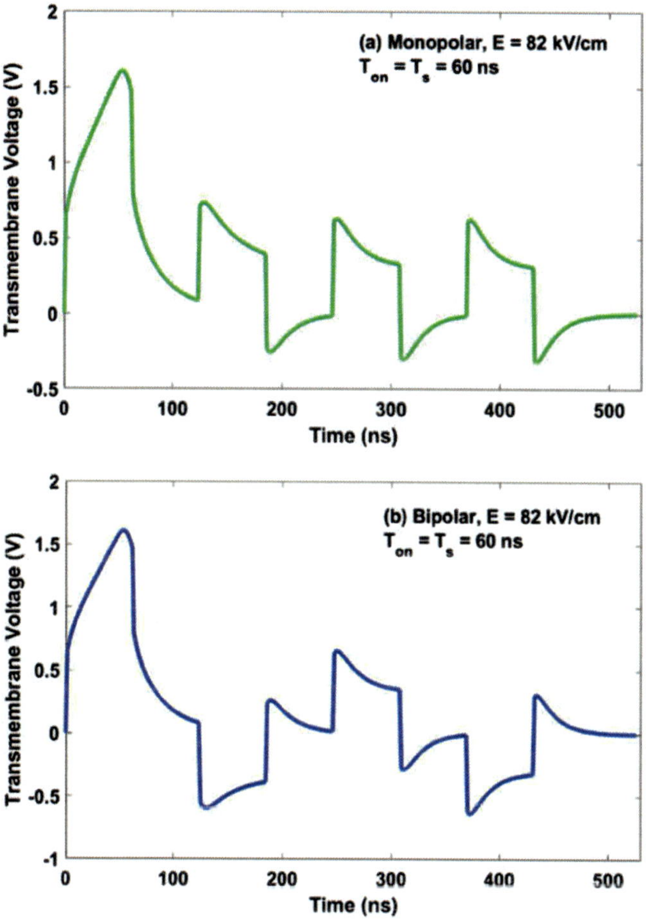

Fig. 4.3 Transmembrane potentials at the 9-o'clock position for the two pulsing cases. **a** Monopolar pulses, and **b** bipolar pulses (After Hu and Joshi 2017)

magnitude was taken to be 100 kV/cm. As might be expected, the calcium concentration in the cell decreases as the pulse delay is increased. This occurs due to two reasons. First, the calcium concentrations that might have been built up at the cell membrane boundary during the ON-time, diffuse and slowly recover back to a more uniform, equilibrium value. As a result, a lower density of calcium is available in the vicinity of the cell surface for re-entry. Second, the slow but gradual pore closing during the OFF state, reduces the total area of ion inflow through the membrane surface.

With the use of bipolar pulses, the scenario is somewhat different as shown in Fig. 4.5b for the same geometry and pulse parameters. Well after the pulse train is over, higher inflows are predicted to be from pulses with the *longer* delay. This result

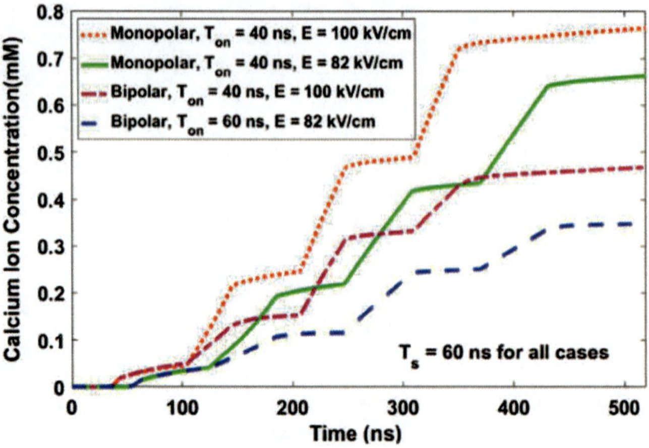

Fig. 4.4 Temporal growth of the calcium uptake for pulses of equal energy. Pulse ON-times of 40 ns and 60 ns were used, with corresponding field strengths of 100 kV/cm and 82 kV/cm, respectively. Results include both bipolar and monopolar pulsing scenarios (After Hu and Joshi 2017)

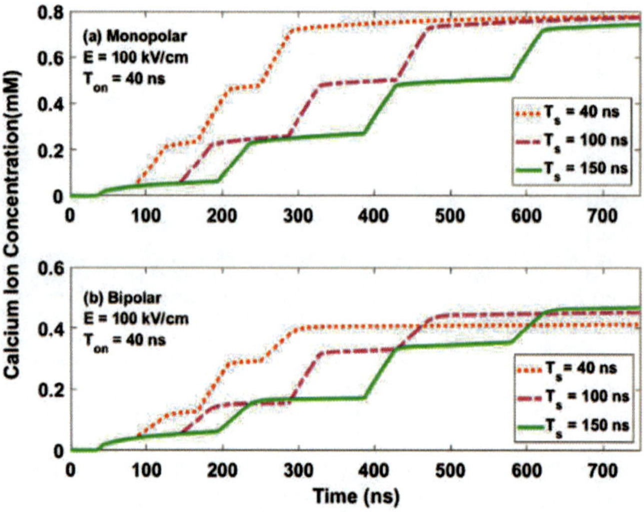

Fig. 4.5 Simulation results probing the effect of changes in the pulse separation times T_s. **a** Monopolar pulses, and **b** bipolar pulses (After Hu and Joshi 2017)

can qualitatively be understood as arising from the recovery of the ion concentrations outside, but in the immediate vicinity, of the cell surface. As already mentioned, one phase of the pulse drives the calcium away from the cell surface and deeper into the aqueous medium at either the 9- or 3-o'clock (or any pair of diagonal) positions. Hence, when the polarity is reversed for the succeeding stimulation, the amount

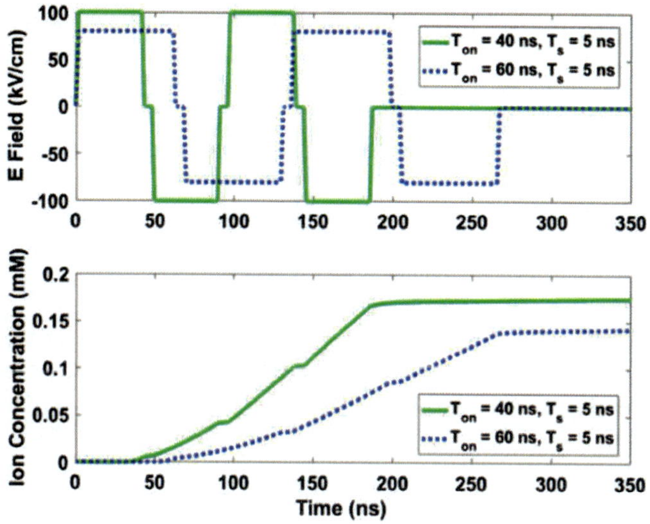

Fig. 4.6 Simulation results probing the effect of a small 5 ns pulse separation time T_s for bipolar pulses (After Hu and Joshi 2017)

of calcium available for inflow is lower due to the depletion that has taken place during the preceding pulse. However, as the pulse delay time T_s is increased, such depletions are quenched (and could nearly be restored) due to local diffusive flows. This increases the availability of calcium for cellular entry at the subsequent pulse. Expectedly though, the total magnitudes flowing into the cell are much lower for bipolar pulsing, as opposed to the previous monopolar case.

For completeness simulations probing the effect of a small 5 ns pulse separation time (T_s) for bipolar pulses were performed, with the results shown in Fig. 4.6 for two different ON-times of 40 ns and 60 ns. Based on equal energy, the electric field for the longer 60 ns ON time pulses was lower. In keeping with the trends of Fig. 4.4, the cellular inflow for the lower electric field (i.e., the 60 ns ON time case) was also corresponding lower. But more importantly, the uptake with the much shorter 5 ns pulse separation time, is predicted to be much less than the values with the longer separation. This is qualitatively in keeping with recent observations (Ibey et al. 2014; Pakhomov et al. 2014) who reported practically no calcium uptake for such bipolar pulses. Though a small amount of ion inflow is predicted in Fig. 4.6, the value is small enough that it might not have been observed experimentally. Besides, the temporal resolution available in the experimental measurements was not on such a fine nanosecond scale, but rather was probed on a time scale of seconds after the onset of the pulse. The long time scales could have meant a much greater internal diffusion, and hence dilution, of any meagre intracellular flows that might have occurred during the bipolar pulsing.

Based on the above results, the following conclusions seem to emerge with regards to the mechanisms for the lower cellular uptake by bipolar pulses. In this context, different mechanisms have been discussed (Gianulis et al. 2015) discussed three potential mechanisms previously proposed for the reduced uptake: (i) the assisted membrane discharge which shortens the time when membrane is above the critical breakdown potential, (ii) poration involves charge transfer and thus can be reverted by the electric field reversal, and (3) the possibility to drive charged ions out of the cell upon electric field reversal. Our calculations predict the uptake in the case of bipolar pulsing to be smaller because high transmembrane potentials (TMPs) are created over a shorter time duration. For example, a comparison between Fig. 4.3a, b reveals the absolute values of the TMP over the 200 ns -to- 400 ns time span remain relatively lower for the bipolar case. The primary reason is that with unidirectional pulsing, a subsequent pulse can start to raise the TMP to a higher level (as compared to a bipolar pulse) if the starting value was not at zero. For a bipolar pulse, on the other hand, if one is at a nonzero TMP value just prior to a subsequent pulse, the magnitude first goes through zero before being enhanced in the opposite direction.

Furthermore, based on the present results, the Ca-ions driven into the cell through the membrane pores do not flow out immediately from the entry sites upon polarity reversal. Instead, ions continue to flow in from the membrane region located diametrically on the opposite side. The inflow though is *less*, since the external ion concentration on the diametrically opposite side has fallen due to the prior movement from the first pulse. Thus, charge reversal is not seen to play a dominant role in the present simulations.

Next for completeness, the possibility of applying the voltage pulse in a sequential versus simultaneous manner was compared. The sequential (or series) application represented a situation wherein the electric pulse was first applied at the 9- and 3-o'clock positions (i.e. the field was taken to be along the horizontal direction). Then after a 60 ns delay, the same pulse was applied in an orthogonal direction to pass through the 12- and 6-o'clock positions. With simultaneous application, on the other hand, *both* the 9- and 3-o'clock positions, as well as the 12- and 6-o'clock locations, were exposed to the external voltage pulse simultaneously. Four sets of pulses were simulated. Thus, for the simultaneous application four distinct pairwise events occurred, while in the sequential case, eight separate events took place.

Results obtained for the cellular uptake are shown in Fig. 4.7 for a peak field amplitude of 80 kV/cm, a pulse ON-time of 40 ns with a OFF-time of 60 ns. Only the monopolar pulses were chosen since this modality has been shown to invoke a larger cell uptake, and hence is of higher interest. In Fig. 4.7, and the cellular uptake is predicted to be larger when both excitations are simultaneously applied at the 9–3 and 12–6 o'clock positions. Four distinct jumps in the ion concentration over time can be seen in the plot corresponding to the four distinct events for simultaneous pulsing. On the other hand, with serial pulses, eight distinct jumps in the ion concentration over time are evident in Fig. 4.7 as expected. Synergies for the monopolar case become apparent, for example, from the ion concentration around 400 ns. The predicted value of ~0.6 mM at this time for the simultaneous pulsing case is seen to be more than twice the concentration obtained for the serial case. Clearly, the two mutually

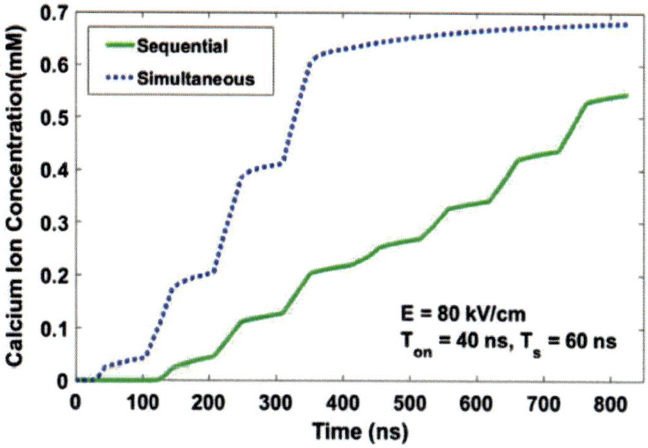

Fig. 4.7 Results of cellular uptake comparing sequential and simultaneous application of the external voltage pulse for monopolar pulses having a peak amplitude of 80 kV/cm. Four sets of two orthogonal simultaneous pulses were applied for the simultaneous case. For the sequential case, a total of 8 separate pulses were applied, first along one direction and then along the orthogonal direction with a 60 ns delay. This delayed pair was repeated four times (After Hu and Joshi 2017)

orthogonal pulses affect a larger fraction of the cell surface, and hence create a much larger effective pore area, to enhance the ion throughput for simultaneous excitations. This outcome seems reasonable since for this situation, pores get formed at both pairs of polar caps. Furthermore, poration is enhanced due to additive contributions of the electric fields even over the off-axis regions.

Finally, as a test of the possible synergies, simulations with multiple pulses were performed. Since monopolar pulses have consistently been shown to achieve higher cellular inflow, results from the bipolar pulse cases were omitted for brevity. Total ion concentration changes inside the cell were evaluated in response to monopolar pulses in the train having 40 ns duration, 1.5 ns rise-, 1.5 ns fall-times, and 60 ns off times. The electric field intensity was set at 80 kV/cm. In order to keep the overall energy delivered to the system equal, three cases are studied: (i) a scenario with six pulses applied along the x-direction, (ii) a second situation with three pulses applied along both the y- and z-directions simultaneously; and (iii) the simultaneous application of two pulses along the x-, y-, and z-directions for the greatest synergy. It may be emphasized, however, that there can be numerous ways of choosing values from a parameter space for comparisons. Here, we have opted to use one simple means, that of maintaining the input energy. Nonetheless, it must be clarified that electroporation is an especially complicated and sequential process, and there is no simple scaling formula that can factor in all the variables in a simple way.

In any event, results shown in Fig. 4.8 for the above three simulation cases reveal that the largest throughput occurs when all three pulses are applied simultaneously along the three mutually orthogonal directions. The weakest response is when pulses are only applied along one direction, even though their numbers are higher. All this

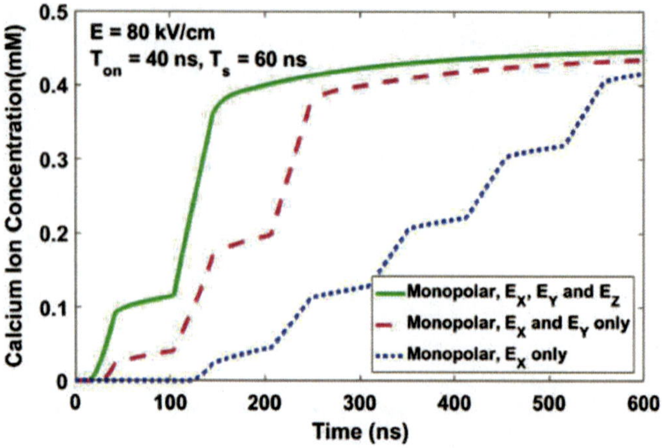

Fig. 4.8 Evolution of the ion concentration inside the cell in responses to pulses in a train with 40 ns durations, 1.5 ns rise- and fall-times, and 60 ns off times. The peak field amplitude was chosen to be 80 kV/cm. The three curves shown correspond to six pulses applied along the x-direction, three pulses applied along both the x- and y-directions simultaneously, and two pulses applied along the x-, y-, and z-directions simultaneously. The energy delivered to the system for all three cases was equal (After Hu and Joshi 2017)

collectively points to the inherent synergies, and suggests that judicious pulse manipulation, sequencing and multi-directional targeting can lead to enhanced cellular delivery. Though the three-dimensional simulations were carried out for spherical cells, it seems intuitive that such multi-directional pulsing would be even more advantageous for cells or tissues having irregular shape. By hitting the collective mass or various irregularly-shaped cells at multiple sites, a stronger and more prolific bio-response can be expected. This is quite likely in the case of tumor tissues. Ideas associated with phased array delivery could also be probed as the next logical step in this direction.

The strong performance of simultaneous pulsing in Fig. 4.8 is made clearer through the time-development of transmembrane potentials (TMPs) across various points on the membrane surface. Simulation results obtained are shown in Fig. 4.9 at the 30 ns time instant for the three cases. Since the pulses in the train had a 40 ns duration, with 1.5 ns rise-, 1.5 ns fall-times, and 60 ns off times, the field at 30 ns was 80 kV/cm. In Fig. 4.9a, the monopolar pulses were applied with the electric fields along the x-axis. For Fig. 4.9b, the external electric fields were along both the x- and y-directions; while Fig. 4.9c shows the results obtained when pulse trains were simultaneously applied along all three axes. In Fig. 4.9a, the green and bluish-green regions in the y–z plane have relatively low TMP values since this area is normal to the applied x–direction. The polar caps along the x-axis are the regions that exhibit high TMP values exceeding a 1 V magnitude (yellow color at the front end, dark blue towards the back). The result of Fig. 4.9b shows a much larger fraction of the surface area with higher TMP magnitudes, while Fig. 4.9c reveals the highest percentage of

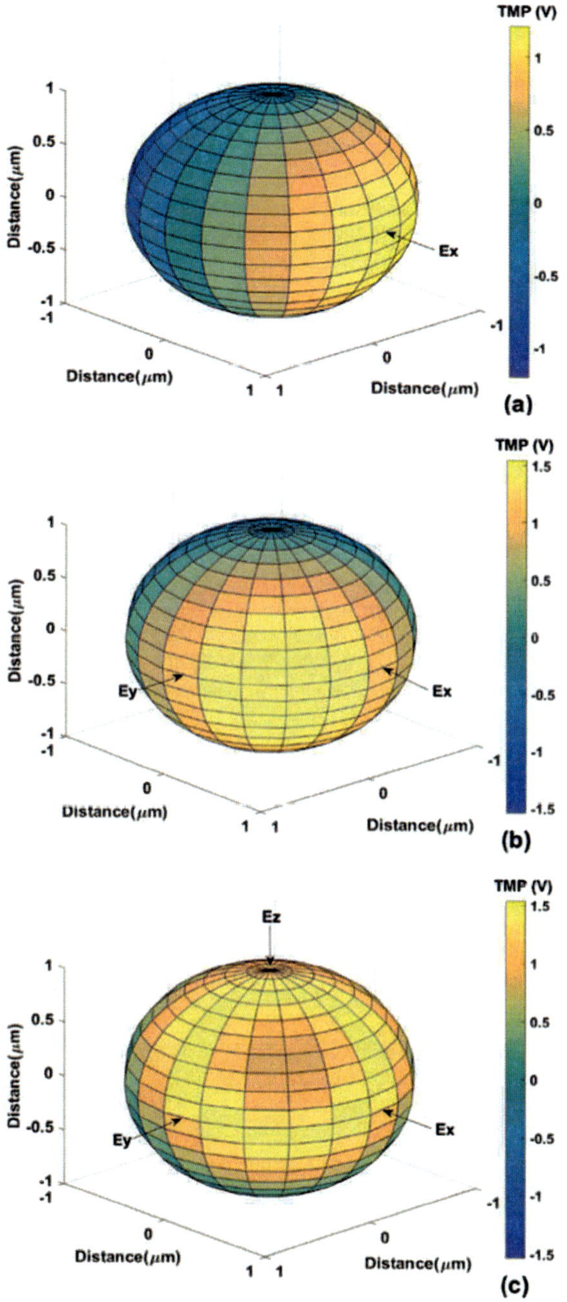

Fig. 4.9 Transmembrane potentials on the cell membrane surface for the pulse train of Fig. 4.8 at the 30 ns time instant. **a** Pulses applied only along the x-axis, **b** pulses applied along both the x- and y-directions, and **c** pulses applied along all three x-, y-, and z-directions simultaneously (After Hu and Joshi 2017)

the membrane having TMP magnitudes near, or in excess of, 1 V which is a convenient benchmark for electroporation. Since pore density scales nonlinearly with TMP magnitudes, a significantly high number of pores are predicted upon the simultaneous application of pulses in the three orthogonal directions. Though pore density distribution over the membrane surface could have been shown, the TMP profile was chosen here because of its simpler, narrow range of values. For completeness, the fractional values for the electroporated membrane area obtained from the simulations were 34, 36 and 63.95% for external fields applied along the x-, x and y-, and x-, y- and z-directions, respectively.

References

Ahmed M, Bruce CL, Lee FT Jr, Goldberg SN (2011) Principles of and advances in percutaneous ablation. Radiology 258:351–369

Arena CB, Davalos RV (2012) Advances in therapeutic electroporation to mitigate muscle contractions. J Membr Sci Technol 2:1–3

Arena CB, Sano MB, Rylander AN, Davalos RV (2011) Theoretical considerations of tissue electroporation with high frequency bipolar pulses. IEEE Trans Biomed Eng 58:1474–1482

Ball C, Thomson KR, Kavnoudias H (2010) Irreversible electroporation: a new challenge in "out of operating theater" anesthesia. Anesth Analg 110:1305–1309

Beebe SJ, Blackmore PF, White J, Joshi RP, Schoenbach KH (2004) Nanosecond pulsed electric fields modulate cell function through intracellular signal transduction mechanisms. Physiol Meas 25:1077–1093

Cadossi R, Ronchetti M, Cadossi M (2014) Locally enhanced chemotherapy by electroporation: clinical experiences and perspective of use of electrochemotherapy. Fut Oncol 10:877–890

Campana LG, Edhemović I, Soden D, Perrone AM, Scarpa M, Campanacci L, Cemazar M, Valpione S, Miklavčič D, Mocellin S, Sieni E, Serša G (2019) Electrochemotherapy—emerging applications technical advances, new indications, combined approaches, and multi-institutional collaboration. Eur J Surg Oncol 45:92–102

Daskalov I, Mudrov N, Peycheva E (1999) Exploring new instrumentation parameters for electrochemotherapy. Attacking tumors with bursts of biphasic pulses instead of single pulses. IEEE Eng Med Biol Mag 18:62–66

Dong S, Wang H, Zhao Y, Sun Y, Yao C (2018) First human trial of high-frequency irreversible electroporation therapy for prostate cancer. Technol Cancer Res Treat 17:1–9

Faurie C, Rebersek M, Golzio M, Kanduser M, Escoffre JM, Pavlin M, Teissie J, Miklavčič D, Rols MP (2010) Electro-mediated gene transfer and expression are controlled by the life-time of DNA/membrane complex formation. J Gene Med 12:117–125

Frey W, White JA, Price RO, Blackmore PF, Joshi RP, Nuccitelli R, Beebe SJ, Schoenbach KH, Kolb JF (2006) Plasma membrane voltage changes during nanosecond pulsed electric field exposure. Biophys J 90:3608–3615

French DM, Uhler MD, Gilgenbach RM, Lau YY (2009) Conductive versus capacitive coupling for cell electroporation with nanosecond pulses J Appl Phys 106:074701 (1–4)

Gehl J, Serša G, Matthiessen LW, Muir T, Soden D, Occhini A, Quaglino P, Curatolo P, Campana LG, Kunte C, Clover AJP, Bertino G, Farricha V, Odili J, Dahlstrom K, Benazzo M, Mir LM (2018) Updated standard operating procedures for electrochemotherapy of cutaneous tumours and skin metastases. Acta Oncol Stockh Swed 57:874–882

Gianulis EC, Lee J, Jiang C, Xiao S, Ibey BL, Pakhomov AG (2015) Electroporation of mammalian cells by nanosecond electric field oscillations and its inhibition by the electric field reversal. Sci Rep 5:13818

Gilbert RA, Jaroszeski MJ, Heller R (1997) Novel electrode designs for electrochemotherapy. Biochim Biophys 1334:9–14

Golberg A, Rubinsky B (2012) Towards electroporation based treatment planning considering electric field induced muscle contractions. Technol Cancer Res Treat 11:189–201

Hibino M, Itoh H, Kinosita K Jr (1993) Time courses of cell electroporation as revealed by submicrosecond imaging of transmembrane potential. Biophys J 64:1789–1800

Hofmann GA, Dev SB, Nanda GS (1996) Electrochemotherapy: transition from laboratory to the clinic. IEEE Eng Med Biol Mag 15:124–132

Hu Q, Joshi RP (2017) Comparative evaluation of transmembrane ion transport due to monopolar and bipolar nanosecond, high-intensity electroporation pulses based on full three-dimensional analyses. J Appl Phys 122:034701

Ibey BL, Ullery JC, Pakhomova ON, Roth CC, Semenov I, Beier HT, Tarango M, Xiao S, Schoenbach KS, Pakhomov AG (2014) Bipolar nanosecond electric pulses are less efficient at electropermeabilization and killing cells than monopolar pulses. Biochem Biophys Res Comm 443:568–573

Jaroszeski MJ, Dang V, Pottinger C, Hickey J, Gilbert R, Heller R (2000) Toxicity of anticancer agents mediated by electroporation in vitro. Anticancer Drugs 11:201–208

Jiang C, Davalos RV, Bischof JC (2015) A review of basic to clinical studies of irreversible electroporation therapy. IEEE Trans Biomed Eng 62:4–20

Joshi RP, Schoenbach KH (2010) Bioelectric effects of intense, ultrashort electric pulses. Crit Rev Bio-Med Eng 38:255–304

Joshi RP, Mishra A, Song J, Pakhomov A, Schoenbach KH (2008) Simulation studies of ultrashort, high-intensity electric pulse induced action potential block in whole-animal nerves. IEEE Trans Biomed Eng 55:1391–1398

Kinosita K Jr, Ashikawa I, Saita N, Yoshimura H, Itoh H, Nagayama K, Ikegami A (1988) Electroporation of cell membrane visualized under a pulsed-laser fluorescence microscope. Biophys J 53:1015–1019

Kotnik T, Mir LM, Flisar K, Puc M, Miklavčič D (2001) Cell membrane electropermeabilization by symmetrical bipolar rectangular pulses. Part I. Increased efficiency of permeabilization. Bioelectrochemistry 54: 83–90

Kotnik T, Pucihar G, Rebersek M, Miklavčič D, Mir LM (2003) Role of pulse shape in cell membrane electropermeabilization. Biochim Biophys Acta 1614:193–200

Kotnik T, Frey W, Sack M, Meglič SH, Peterka M, Miklavčič D (2015) Electroporation-based applications in biotechnology. Trends Biotechnol 33:480–488

Latouche EL, Arena CB, Ivey JW, Garcia PA, Pancotto TE, Pavlisko N, Verbridge SS, Davalos RV, Rossmeisl JH (2018) High-frequency irreversible electroporation for intracranial meningioma: a feasibility study in a spontaneous canine tumor model. Technol Cancer Res Treat 17:1–10

Mali B, Jarm T, Snoj M, Serša G, Miklavčič D (2013) Antitumor effectiveness of electrochemotherapy: a systematic review and meta-analysis. Eur J Surg Oncol 39:4–16

Miklavčič D, Mali B, Kos B, Heller R, Serša G (2014) Electrochemotherapy: from the drawing board into medical practice. Biomed Eng Online 13:29/1–20

Mir LM, Tounekti O, Orlowski S (1996) Bleomycin: revival of an old drug. Gen Pharmacol 27:745–748

Mir LM, Gehl J, Serša G, Collins CG, Garbay JR, Billard V, Geertsend PF, Rudolf Z, O'Sullivan GC, Marty M (2006) Standard operating procedures of the electrochemotherapy: instructions for the use of bleomycin or cisplatin administered either systemically or locally and electric pulses delivered by the CliniporatorTM by means of invasive or non-invasive electrodes. Eur J Cancer Suppl 57:874–882

Pakhomov AG, Semenov I, Xiao S, Pakhomova ON, Gregory B, Schoenbach KH, Ullery JC, Beier HT, Rajulapati SR, Ibey BL (2014) Cancellation of cellular responses to nanoelectroporation by reversing the stimulus polarity. Cell Mol Life Sci 71:4431–4441

Prasanna GL, Panda T (1997) Electroporation: basic principles, practical considerations and applications in molecular biology. Bioprocess Eng 16:261–264

Pucihar G, Kotnik T, Kandušer M, Miklavčič D (2001) The influence of medium conductivity on electropermeabilization and survival of cells in vitro. Bioelectrochemistry 54:107–115
Pucihar G, Kotnik T, Miklavčič D, Teissié J (2008) Kinetics of transmembrane transport of small molecules into electropermeabilized cells. Biophys J 95:2837–2848
Roth CC, Barnes RA Jr, Ibey BL, Beier HT, Mimun LC, Maswadi SM, Shadaram M, Glickman RD (2015) Characterization of pressure transients generated by nanosecond electrical pulse (nsEP) exposure. Sci Rep 5:15063
Sadik MM, Yu M, Zheng M, Zahn JD, Shan JW, Shreiber DI, Lin H (2014) Scaling relationship and optimization of double-pulse electroporation. Biophys J 106:801–812
Sano MB, Arena CB, DeWitt MR, Saur D, Davalos RV (2014) In-vitro bipolar nano- and microsecond electro-pulse bursts for irreversible electroporation therapies. Bioelectrochemistry 100:69–79
Sano MB, Arena CB, Bittleman KR, DeWitt MR, Cho HJ, Szot CS, Saur D, Cissel JM, Robertson J, Lee YW, Davalos RV (2015) Bursts of bipolar microsecond pulses inhibit tumor growth. Sci Rep 5:14999/1–13
Santra TS, Wang PC, Tseng FG (2013) Electroporation based drug delivery and its applications. In: Takahata K (ed) Advances in micro/nano electromechanical systems and fabrication technologies. InTech Publishers, London
Scheffer HJ, Nielsen K, de Jong MC, van Tilborg AA, Vieveen JM, Bouwman AR (2014) Irreversible electroporation for nonthermal tumor ablation in the clinical setting: a systematic review of safety and efficacy. J Vasc Interv Radiol 25:997–1011
Schoenbach KH, Beebe SJ, Buescher ES (2001) Intracellular effect of ultrashort electrical pulses. Bioelectromagnetics 22:440–448
Schoenbach KH, Hargrave BY, Joshi RP, Kolb JF, Osgood C, Nuccitelli R, Pakhomov AG, Swanson J, Stacey M, White JA, Xiao S, Zhang J (2007) Bioelectric effects of Intense nanosecond pulses. IEEE Trans Dielectr Electr Insul 14:1088–1109
Schoenbach KH, Pakhomov AG, Semenov I, Xiao S, Pakhomova ON, Ibey BL (2015) Ion transport into cells exposed to monopolar and bipolar nanosecond pulses. Bioelectrochemistry 103:44–51
Scuderi M, Rebersek M, Miklavčič D, Dermol-Cerne J (2019) The use of high-frequency short bipolar pulses in cisplatin electrochemotherapy in vitro. Radiol Oncol 53:194–205
Song J, Garner AL, Joshi RP (2017) Effect of thermal gradients created by electromagnetic fields on cell-membrane electroporation probed by molecular-dynamics simulations. Phys Rev Appl 7:024003
Spreckelmeyer S, Orvig C, Casini A (2014) Cellular transport mechanisms of cytotoxic metallodrugs: an overview beyond cisplatin. Molecules 19:15584–15610
Teissie J, Rols MP (1993) An experimental evaluation of the critical potential difference inducing cell membrane electropermeabilization. Biophys J 65:409–413
Teissie J, Eynard N, Gabriel B, Rols MP (1999) Electropermeabilization of cell membranes. Adv Drug Deliv Rev 35:3–19
Tekle E, Astumian RD, Chock PD (1991) Electroporation by using bipolar oscillating electric field: an improved method for DNA transfection of NIH 3T3 cells. Proc Natl Acad Sci USA 88:4230–4234
Tozon N, Serša G, Čemažar M (2001) Electrochemotherapy: potentiation of local antitumour effectiveness of cisplatin in dogs and cats. Anticancer Res 21:2483–2488
Vernier PT, Sun Y, Gundersen MA (2006) Nanoelectropulse-driven membrane perturbation and small molecule permeabilization. BMC Cell Biol 7:37/1–16
Weaver JC, Chizmadzhev YA (1996) Theory of electroporation: a review. Bioelectrochem Bioenerg 41:135–160
Yao C, Dong S, Zhao Y, Lv Y, Liu H, Gong L, Ma J, Wang H, Sun Y (2017) Bipolar microsecond pulses and insulated needle electrodes for reducing muscle contractions during irreversible electroporation. IEEE Trans Biomed Eng 64:2924–2937
Yarmush ML, Golberg A, Serša G, Kotnik T, Miklavčič D (2014) Electroporation-based technologies for medicine: principles, applications, and challenges. Annu Rev Biomed Eng 16:295–320

Župančič A, Ribarič S, Miklavčič D (2007) Increasing the repetition frequency of electric pulse delivery reduces unpleasant sensations that occur in electrochemotherapy. Neoplasma 54:246–250

Chapter 5
usEP Effects on the Endoplasmic Reticulum (ER)

Stephen J. Beebe

Abstract The endoplasmic reticulum (ER) is a tubular membranous labyrinth throughout the cytoplasm that is continuous with the nuclear membrane, exhibits contact sites with mitochondria and the plasma membrane, and displays domains for specific functions. It is smooth or rough with ribosomes for protein synthesis and is the home for folding proteins in their proper tertiary structures. The ER includes cellular stress response sensors that can lead to unfolded protein response (UPR), leading to regulated cell death (RCD). The ER is also a primary storage site for Ca^{2+} that can be used for scores of Ca^{2+} mediated signal transduction responses such as neurotransmitter release, muscle contraction, or contributors to RCD, among others. Since usEPs were unique for intracellular electric field effects, usEP-induced Ca^{2+} release was an excellent way to define these intracellular effects, albeit not without caveats. Because usEPs also induced plasma membrane permeabilization for Ca^{2+} influx, which occurred at lower charging conditions than ER-induced nanopore formation and Ca^{2+} release; because some cells expressed voltage-gated Ca^{2+} channels (VGCC), which could be directly activated or activated due to usEP-induced plasma membrane depolarization; because capacitative Ca^{2+} increases could increase intracellular Ca^{2+}; and Ca^{2+} increases from Ca^{2+}-induced Ca^{2+} release, significant care, and experimental manipulations were required to be sure that increases in intracellular Ca^{2+} were due to release from internal stores. Also, because there were other intracellular stores for Ca^{2+}, other approaches were needed to ensure that the source of intracellular Ca^{2+} release was from the ER. This chapter provides details from several studies using many different experimental techniques that lead to the conclusion that usEPs could induce Ca^{2+} release from the ER by forming ER nanopores. Notably, another series of experimental studies supported theoretical evidence that shorter pulse durations lead to more significant increases in intracellular Ca^{2+} than longer pulse durations.

S. J. Beebe et al., *Ultrashort Electric Pulse Effects in Biology and Medicine*,
Series in BioEngineering, https://doi.org/10.1007/978-981-10-5113-5_5

5.1 Introduction—ER Structure and Function-Interactions with the Nucleus and Mitochondria

The ER is an extensively elaborated tubular network that covers the entire cellular cytoplasm with intimate associations with the nucleus, mitochondria, and other cellular organelles. The ER, which occupies more than 10% of the total cellular volume, is the most single membranous structure in eukaryotic cells forming specific structures that include the nuclear envelop, flattened sacs called ER lumen or cisternal space, and tubular networks (English and Voeltz 2013). The ER membrane interfaces with the cell cytoplasm and regulates selective transport between the two compartments (Alberts et al. 2002) ER is present as rough ER (ER), containing ribosomes for protein synthesis, and smooth ER (SER) without ribosomes (Fig. 5.1). The major functions of the RER is protein synthesis, especially secretory and membrane proteins. The ribosomes are attached to the outer ER membrane with ribophorin I and II (Crimaudo et al. 1987) and are then transported into the ER lumen for proteolytic processing and for other post translation modifications, most commonly, N-glycosylation. Ribophorin I delivers the proteins as substrates too the oligosaccharyltransferase that glycosylates asparagine residues of newly synthesized proteins (Wilson et al. 2008) The RER also secretes proteins and some hormones from the ER lumen. The SER is primarily involved in producing lipids and proteins and detoxification of lipid drugs and endogenous toxic substances by the cytochrome P450 enzyme family (Neve and Ingelman-Sundberg 2010). The chapter of usEP effects on DNA and the nucleus (Chap. 7) discuss pulse-induced DNA damage, nuclear membrane perturbations, and on possible telomer damage as it connects to the inner mitochondrial membrane (Stacey et al. 2011). However, there are no studies that have revealed how usEP might effect communications between the ER and the nucleus.

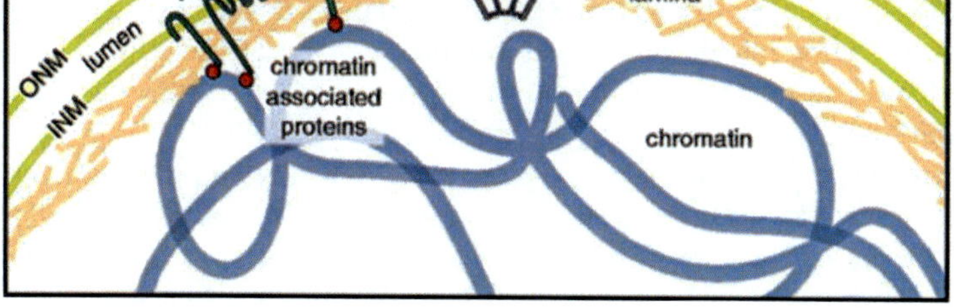

Fig. 5.1 The endoplasmic reticulum (ER) is continuous with the nuclear membrane and contains ribosomes as rough ER. Modified from Schooley et al. (2012)

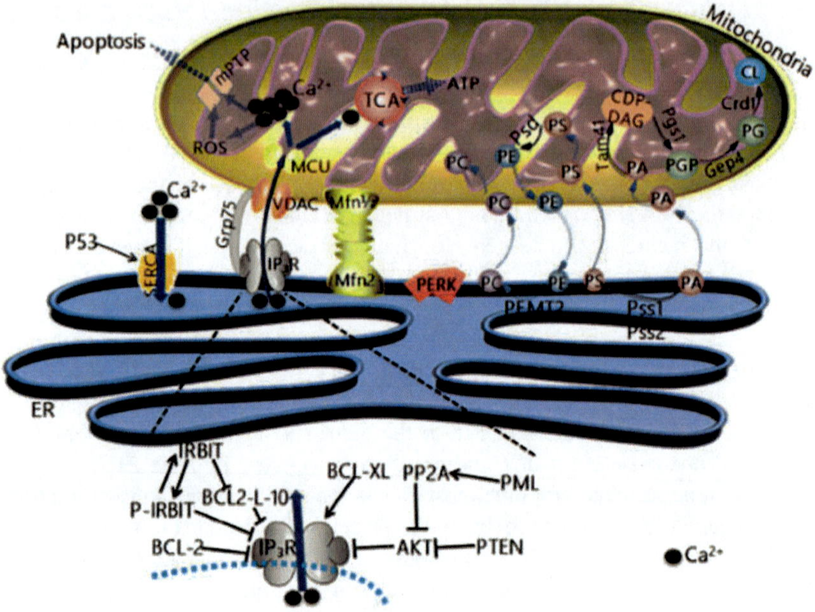

Fig. 5.2 The interaction between mitochondria and ER. The figure shows interactions between the ER and Mitochondria. from (Xia et al. 2019). See text for details. There are also interactions of phospholipids between the mitochondria and ER that are not discussed here

Nevertheless, there are no doubt usEP effects here, however, the reagents, imaging tools, and investigator intentions remain to be tapped.

The major sections of the ER are continuous with the nuclear envelop, which consists of two-membrane bilayers separated by a luminal space (Fig. 5.1). The ER is composed of distinct structural and functional domains with numerous contact sites with the plasma membrane, mitochondria (Fig. 5.2), and other organelles such as Golgi apparatus, endosomes, and peroxisomes (English and Voeltz 2013). These tubular connections allow inter-organellar communication that facilitate various homeostatic mechanisms. As DNA is transcribed in the nucleus, the resulting mRNAs are transported out of the nucleus through nuclear pores into the ribosomes that are attached to the ER called rough ER (RER) and the coded messages are translated into proteins and folded into the appropriate conformation by chaperone proteins within the ER. One of the important ER functions not yet mentioned and that is significantly important is Ca^{2+} mobilization and homeostasis.

Ca^{2+} is a major signal transduction protein and is involved in a myriad of cellular functions. The ER and mitochondria are major Ca^{2+} storage sites. The cellular concentrations of Ca^{2+} are kept low by storage and selective, transient release primarily from the ER, which is critical for Ca^{2+} signaling. Ca^{2+} is released from the ER into the cytoplasm by the inositol 1,4,5 trisphosphate receptor (IP3R) and the ryanodine receptor (RyR) (Berridge 2009; Lanner et al. 2010). When these receptors are activated and Ca^{2+} released, there is an additional Ca^{2+} release called Ca^{2+}-induced Ca^{2+} release (CICR), which causes rapid Ca^{2+} transients (Verkhratsky and Shmigol 1996; Berridge 1998). The hormone IP3 can induce Ca^{2+} release from theIP3R and calcium alone can induce Ca^{2+} release from the RyR (Seo et al. 2015). To ensure that ER Ca^{2+} levels are not depleted, these Ca^{2+} release processes activate a sarco/endoplasmic reticulum calcium ATPase (SERCA) in the ER membrane that pumps Ca^{2+} from the cytosol back into the ER (Vandecaetsbeek et al. 2011).

Cellular stress can lead to an unfolded protein response (UPR) in the ER (see Chap. 9 for cell stress response to usEPs in more detail). This stops protein translation, and the unfolded proteins are degraded. Cell signaling mechanism are activated and chaperones participate in protein folding to assist maturation of functional proteins, which requires ATP. In these situations, the contact sites between ER and mitochondria are tightened as mitofusion I (MfnI) or MfnII on the OMM interacts with and MfnII on the ER. Increased release of Ca^{2+} from the IP3R on ER is now immediately adjacent to the mitochondrial voltage-dependent anion channels (VDACs) and the IMM low-affinity, mitochondrial calcium uniporter (MCU) (Xia et al. 2019; Liu and Zhu 2017) (see Fig. 5.2). Ca^{2+} entry into the mitochondria activates dehydrogenases in the Krebs cycle (Wan et al. 1989; Traaseth et al. 2004) leading to increased ATP synthesis to meet the continued need for energy sources.

The ER UPR exhibits three stress response sensors including protein kinase R-like ER kinase (PERK) (Fig. 5.2), inositol-requiring enzyme 1 alpha (IRE1α), and activating transcription factor 6 (ATF6). As discussed in Chap. 19, usEPs induced activation of PERK in CT-26 colon carcinoma and EL-4 lymphoma cells. Correlated with PERK activation is the plasma membrane externalization of calreticulin (CRT) as it is transported from the ER. As a danger-associated molecular paterns (DAMP), CRT is associated with activation of caspases 3/7 as apoptosis is induced (Rossi et al. 2019). CRT, ATP, and HMGB-1 are DAMPs that have been shown to be released in response to usEPs (Nuccitelli et al. 2015; Guo et al. 2018; Rossi et al. 2019). In contrast, human dermal fibroblast exposed to usEPs one hundred (1 Hz) 10 ns pulses at 150 kV/cm (4.7×10^{-3} Vs/cm), unlike tunicamycin, a known ER stress inducer, did not exhibit UPR as determined by changes in ER stress genes or caspase-genes determined by RTqPCR (Martens et al. 2018).

5.2 usEPs Induce Ca^{2+} Release

Because usEPs were known to have intracellular effects, one of the earliest studied effects of usEPs was induction of Ca^{2+} mobilization, especially from the ER. By the early twenty-first century studies by Vernier et al. (2003) on the West Coast at USC and Beebe et al. (2003), Buescher and Schoenbach (2003), and White and colleagues (2004) on the East Coast at Old Dominion University demonstrated that usEP induced Ca^{2+} release from the ER. Vernier et al. (2003) used a MOSFET-based pulser to pulse Jurkat cells in small chambers under a microscope. They demonstrated intracellular Ca^{2+} bursts from the ER using Calcium Green as a fluorescent

Ca^{2+} sensor. By this time, it was becoming realized that usEPs induced pores in the plasma membrane that were smaller than the commonly used permeability marker propidium iodide (PI, ~1 nm), but large enough to allow Ca^{2+} to enter from the outside (nanopores). So, it was critical to show that the Ca^{2+} was not entering through nanopores in the plasma membrane Vernier et at showed that the intracellular Ca^{2+} fluorescence was uniform throughout the cell and occurred within seconds after ten 30 ns pulses at 25 kV/cm (2.4×10^{-3} Vs/cm). There was still increases in Calcium Green fluorescence when EGTA chelated the extracellular Ca^{2+}, so the Ca^{2+} source was intracellular. When thapsigargin was used to specifically deplete Ca^{2+} from the ER by inhibiting Ca^{2+} uptake into the ER (Thastrup et al. 1990), usEPs did not cause intracellular Calcium Green fluorescence. The use of EGTA to eliminate extracellular Ca^{2+} and thapsigargin to deplete ER calcium provided strong evidence that the usEP-induced mobilization of Ca^{2+} was from the ER and not through nanopores in the plasma membranes. The authors suggested that this Ca^{2+} release could be by one or more of at least three different mechanism. These included activation and release from ER voltage-activated calcium channels. The release of the second messenger IP3 or other signaling molecule was also suggested as a possible mechanism for ER Ca^{2+} release (see Fig. 5.2). UsEPs have been shown to induce increases in IP3 release in response to single 600 ns pulses at 16.2 kV/cm (9.6×10^{-3} Vs/cm), which are much more intense than those causing ER Ca^{2+} release. Electric field-induced cytoskeletal stress was also suggested by Vernier et al.as a cause for ER Ca^{2+} release. As indicated above, Rossi et al. (2019) demonstrated stress from the UPR with usEP conditions that were much greater than those used by Vernier and colleagues. usEPs were shown to induce cytoskeletal stress defined by disassembly of actin structure after cell swelling. However, these conditions were induced by four 600 ns pulses at 1.9 kV/cm or 2.2×10^{-3} Vs/cm, which was much higher than the charging effect in the ER Ca^{2+} mobilization studies (Pakhomov et al. 2014). Since Jurkat cells exhibit T-lymphocytes the authors also considered direct or indirect effects on the T-cell receptor associated with IP3 and tyrosine kinase activation. There are presently no studies that address this as a possible mechanism. Although threshold studies were not carried out, it appears that the usEP mobilization of internal Ca^{2+} occurs at lower usEP-initiated charging levels that those shown for mechanisms related to stress or IP3 signaling. From studies that followed the demonstration of ER Ca^{2+} release, it is most likely that this response occurs by nanoporation of the ER, as will be addressed by other studies that coincided with or followed those from Vernier et al. (2003).

Other early studies by Beebe et al. (2003) and White et al. (2004) addressed the concept of intracellular Ca^{2+} release from a different perspective using a Blumlein pulser. These author used real time microscopic analysis of cells with the fluorescent calcium indicator Fluo-3, and fluorometric analysis of Fura-2 as the Ca^{2+} sensor. The advantage of Fura-2 was that the Ca^{2+} concentration could be quantified. In the microscopic analysis a significant majority of HL-60 cells showed increases in intracellular calcium when exposed to pulses with 60 ns durations and 30 kV/cm (1.8×10^{-3} Vs/cm). Cells exhibited various fluorescence intensities with a mean of about a tenfold increase in fluorescence over control. Under these same conditions there were no increases in PI suggesting typical plasma membrane permeabilization did not occur. However, it was later shown that these usEP conditions induce nanoporation and Ca^{2+} entry from the outside. Nevertheless, the increase in Ca^{2+} was the same in the presence and absence of EGTA. However, these experiments were complicated by an HL-60 cell response to leak and ultimately drain their intracellular Ca^{2+} stores when extracellular Ca^{2+} was chelated with EGTA. HL-60 cells are unable to maintain their internal calcium stores for very long (~100 s) when placed in a calcium-deficient environment. Thus, like thapsigargin, EGTA depleted intracellular Ca^{2+} stores but by a different mechanism. The use of EGTA was important to eliminate influxes of Ca^{2+} from the culture media, so an EGTA dose response study indicated an EGTA level that minimized the intracellular Ca^{2+} leak in the presence of EGTA. The authors compared usEPs to the purinergic agonist UTP. This agonist increases intracellular Ca^{2+} by binding to purinergic P2Y receptors present in HL-60 cells that involve increases in the second messenger IP3, releasing Ca^{2+} from IP3-sensitive intracellular Ca^{2+} stores into the cytoplasm. This intracellular Ca^{2+} release stimulates capacitive Ca^{2+} entry by opening of store-operated channels in the plasma membrane permitting Ca^{2+} influx into the cell for replenishment of the internal stores (Alemany et al. 2000).

Experiments with UTP demonstrated that by conducting experiments with extracellular Ca^{2+} of 3 μM there was a Ca^{2+} gradient with the outside concentration higher. In this way the addition of EGTA cause a Ca^{2+} efflux from intracellular stores such that by 200 s after adding EGTA the intracellular Ca^{2+} was ~ 50% of that seen at zero time. The cells then were stimulated with 10 μM UTP at 200 s, and a modest yet transient response of about 50 nM was still seen, confirming that the increase in calcium was due to release from internal stores.

In order to measure the rapid response of intracellular calcium to UTP and usEPs in the absence of extracellular calcium, fiber optic light guides were used to monitor fluorescence in the cuvettes after stimulation. When EGTA was added to the medium, there was an immediate and steady decline in intracellular Ca^{2+}, which was similar to that seen with 10 μM UTP. When cells were pulsed for 60 ns at 15 kV/cm (9×10^{-4} Vs/cm) in the presence of 1.5 mM EGTA, there was an immediate increase in intracellular Ca^{2+} that began to decrease after ~10 s, and after ~60 s the intracellular Ca^{2+} was back to base line and continued to decline below base line. The difference in peak heights between the presence of calcium and when EGTA was present was not significant. These data showed that usEP treatment produces a rapid and transient mobilization of intracellular calcium similar to that observed with UTP.

To determine saturated Ca^{2+} responses, dose–response experiments were carried out with UPT and usEPs. 10 μM UTP induced a maximum Ca^{2+} mobilization, and a single 60 ns 15 kV/cm (9×10^{-4} Vs/cm) pulse gave a maximal response. To determine if the intracellular Ca^{2+} released in response to usEP and UTP, a series of studies were carried out with usEP, UTP, and thapsigargin. Experiments with usEP 15 kV/cm ($9 \text{x} 10^{-4}$ Vs/cm), 10 μM UTP, and 0.1 μM thapsigargin all gave similar responses with rapid and sustained increases in Ca^{2+} in the presence of Ca^{2+} and weaker but significant increases in Ca^{2+}-deficient media (3 μM Ca^{2+}). With all stimuli, after an initial intracellular Ca^{2+} release was induced with each agonist in Ca^{2+}-deficient media, the addition of 2 mM extracellular Ca^{2+} gave a rapid and sustained increases in Ca^{2+}. Such a response to UTP is known to occur after IP3-induced Ca^{2+} release from IP3-R in the ER by store-operated Ca^{2+} channels in the plasma membrane as discussed above (see Fig. 5.2). To further support the hypothesis that usEP are releasing Ca^{2+} from the ER and inducing capacitative Ca^{2+} entry through store-operated Ca^{2+} channels in the plasma membrane usEP, UTP, and thapsigargin were added sequentially in Ca^{2+} containing and Ca^{2+}-deficient media. So, when 4 or 15 kV/cm pulses (2.4 and 9×10^{-4} Vs/cm, respectively) were applied in either media, rapid and transient increases in Ca^{2+} occurred. Upon the post-stimulation, Ca^{2+} and UTP was added inducing another smaller but significant Ca^{2+} increase. The same results occurred when UTP induced the primary response and usEP induced the post-stimulation response. Given the similarity in response it was proposed the usEPs acted on the IP3-R to release Ca^{2+}, which activated the capacitative Ca^{2+} influx from store-operated Ca^{2+} (Harper et al. 2003). To provide further evident, thapsigargin, which results in Ca^{2+} release from the ER. Treatment of HL-60 cells with thapsigargin induced Ca^{2+} leaking of the sarco-ER calcium stores with subsequent stimulation of CCE. If usEP treatment depletes the sarco-ER of calcium, then it would be expected that usEP treatment after thapsigargin treatment would show a reduced release of calcium when compared with usEP treatment alone. One 15 kV/cm pulse to increase intracellular Ca^{2+} after thapsigargin treatment was reduced by ~63 to 20% S.E. ($n = 3$). This result supports the notion that usEP treatment and thapsigargin are depleting the same calcium pool, which is believed to be the sarco-ER.

Given the importance of Ca^{2+} in scores of cell functions, it is not surprising that biological cells have evolved multiple ways to respond to Ca^{2+} and to store intracellular Ca^{2+} for signal transduction and functional needs. Ca^{2+} can enter cells through voltage-gated Ca^{2+} channels (VGCC). UsEP with 4 ns durations and 80 kV/cm (3.2×10^{-3} Vs/cm) and 5 ns, 50–60 kV/cm pulses ($2.5–3.0 \times 10^{-3}$ Vs/cm) open VGCC in the absence of plasma membrane electroporation (determined by YO-PRO uptake) and no release from intracellular stores in the absence of extracellular Ca^{2+} (Vernier et al. 2008; Craviso et al. 2012). It was also possible for usEPs to induce nanopores that opened voltage gated Na^{+} and Ca^{2+} channels and Ca^{2+} release from intracellular stores (Wang et al. 2009). VGCC are opened by membrane depolarization of 30–50 mV, while electroporation occurs are a considerably higher magnitude (Kotnik and Miklavcic 2006). The usEP condition for plasma membrane nanoporation in CHO cells (5.4×10^{-4} Vs/cm) discussed below (Semenov et al. 2013b) is also above conditions for opening VGCC in chromaffin cells indicated above. Some of these findings leaves open the possibility that usEP can modulate receptors (Hristov et al. 2018) and possibly other proteins (Beebe 2015).

Two papers by Semenov and colleagues (2013a, b) provided new and valuable information concerning usEPs on Ca^{2+} mobilization. They used CHO cells, which lack voltage-gated Ca^{2+} channels, thereby eliminating one of the mechanisms for increasing intracellular Ca^{2+} so that effects of usEPs on intracellular Ca^{2+} could be attributed to electroporation. In the first paper (Semenov et al. 2013a) using one 60 ns pulse they observed two distinct thresholds for elevating intracellular Ca^{2+}. A plasma membrane threshold at 9 kV/cm (5.4×10^{-4} Vs/cm) and a threshold for release from the ER at 19 kV/cm (1.1×10^{-3} Vs/cm). After these thresholds were reached

Ca^{2+} levels increased linearly about 10 nM /for every kV increases. However, when the intracellular Ca^{2+} reached about 200–300 nM, there was as an usEP -induced Ca^{2+} elevation to near 3 μM caused by a mechanism known as Ca^{2+}-induced Ca^{2+} release (CICR). This CICR is important to amplify physiological mechanisms that require Ca^{2+}, especially skeletal and cardiac muscle for contraction and to ensure that intracellular Ca^{2+} stores are replenished. That this Ca^{2+} was coming from the ER was shown that the amplification response was prevented by pretreatment with thapsigargin, which depletes Ca^{2+} stores in the ER, and by preventing ER Ca^{2+} release by the ER inositol-1,4,5-trisphosphate receptors (IP3R) by 2-aminoethoxydiphenyl borate (2- APB). There was no response or change to the usEP-induced Ca^{2+} effects in the presence of caffeine or dantrolene, both inhibitor or ryanodine receptors (RyR). In fact, CHO cells lacked RyR, so the amplification of CICR was due to release from ER IP3 receptors alone. In contrast to CHO cells, CICR in cardiac muscle cells occurs through L-type Ca^{2+} channels in the plasma membrane that activates the RyR receptors. This coupling between influx and internal release is both structural, where the ensuring close proximity of Ca entry to sites of release, and functional, such that on a beat to beat basis Ca^{2+} influx equals Ca^{2+} efflux (Eisner et al. 2017).

In this study, the authors isolated and quantified response mechanisms and their sensitivities to usEP effects of intracellular calcium mobilization through nanopores in the plasma membrane, release of Ca^{2+} from the ER and when intracellular thresholds of 200–300 nM Ca^{2+}, CICR from the IP3 channels in the ER as an amplification mechanism to provide sufficient Ca^{2+} for a Ca^{2+}-mediated responses and to replenish Ca^{2+} stores for subsequent responses. The CICR was not a direct effect of usEPs, but a response to increases in intracellular Ca^{2+} through nanopores in the plasma membrane and ER. These conclusions were based on the use of Fura-2 with its ratiometric Ca^{2+} imaging to determine intracellular Ca^{2+} concentrations, the presence and absence of Ca^{2+} in the external media, a diverse range of electric fields with 60 ns usEPs, and inhibitors of critical receptors. Notably, for 60 ns pulses they determined that usEP threshold for ER Ca^{2+} release (19 kV/cm, 1.1×10^{-3} Vs/cm) occurs at about twofold higher than the threshold for Ca^{2+} influx through the plasma membrane (9 kV/cm, 5.4×10^{-4} Vs/cm).

In the second paper, Semenov and co-workers (2013b) also used CHO cells and ratiometic Fura-2 analyses. In 2 mM Ca^{2+} media, they observed Ca^{2+} influxes through the plasma membrane and in he absence of Ca^{2+} they only observed Ca^{2+} release from the ER. Using 10 ns, 60 ns and 300 ns pulses with short (fast) rises times (10%-90%) of 0.6 ns, 4.5 ns, and 4.5 ns, respectively, in Ca^{2+}-free media, Ca^{2+} was mobilized exclusively from the ER. In Ca^{2+}-containing media, usEP induced Ca^{2+} transients through the plasma membrane from the culture medium and release from intracellular stores. As observed in their previous paper, when the intracellular Ca^{2+} levels reached 200–300 ns Ca^{2+}, intracellular Ca^{2+} levels were amplified by CICR.

The most notable finding of the studies in this paper is the unencumbered finding that as the pulse duration is decreased there is a greater release Ca^{2+} from intracellular stores. Thus, usEPs with shorter durations have greater effects on intracellular structures, which in CHO cells is the ER. This is consistent with calculations and a simple cellular model, where the theory states that unlike conventional electroporation with pulse duration in the micro- and milli-second range, up to the electroporation threshold, usEPs can charge intracellular structures before the plasma membrane charges to protect them.

When 10 and 60 ns usEPs were adjusted to similar intracellular Ca^{2+} amplitudes, as expected, it was necessary for the 10 ns pulse to have a greater electric field amplitudes to match the 60 ns pulse. Generally, the electric field amplitudes were about tenfold greater for the 60 ns pulses over a range of electric fields. However, under all electric field amplitudes, all 10 ns pulses exhibited greater contributions from the ER.

Figure 5.3 provides a simple and clean presentation of these findings. It clearly and simply shows in red the Ca^{2+} transient in the presence of extracellular Ca^{2+} and in the blue those Ca^{2+} transients in the absence of extracellular Ca^{2+}. It can be observed that increasing the electric field increased the stimulus intensity in a non-linear manner because of the CICR response when the 200–300 nM Ca^{2+} threshold was reached as seen above from the previous paper (Semenov et al. 2013b). Regardless of the CICR, for all tested stimulus intensities, release of Ca^{2+} from the ER was greater for the 10 ns usEP, intermediate for the 60 ns pulse and least for the 300 ns pulse.

Ca^{2+} mobilization through IP3Rs plays in CHO cells was not specifically addressed, but it is well known that influxes of Ca^{2+} occur during many forms of regulated cell death including apoptosis (Beebe et al. 2003). It is also clear that ligand-gated channels such as the IP3R are an important component that modulates pro- and anti-apoptotic factors; however specific roles for Ca^{2+} in usEP-induced apoptosis has

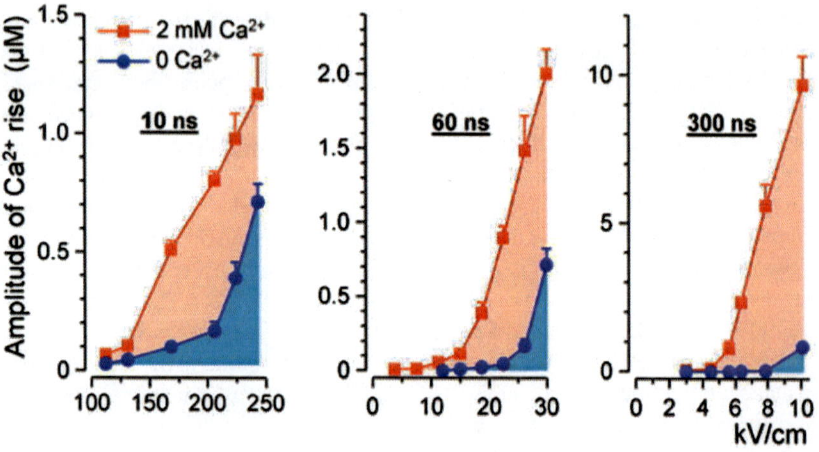

Fig. 5.3 Shorter usEPs induce greater release of Ca^{2+} from the ER. From Semenov et al. (2013b). Red areas indicates the increases in intracellular Ca^{2+} in Ca^{2+} containing media. The blue areas indicate the increases in intracellular Ca^{2+} from the ER in Ca^{2+}-free media

not been studied. Nevertheless, these Ca^{2+} mobilization studies and others (Zhang et al. 2008; Beebe 2013) indicate that usEPs can simulate Ca^{2+} signaling that can bypass the activation of membrane receptors and channels without invoking their specific participation.

While this study demonstrated the that shorter pulse duration induce greater effects from the ER, another study (Beebe et al. 2012, 2013) demonstrated that pulses with shorter (faster) rise-fall times had greater intracellular effect to dissipate the $\Delta\Psi m$ and induce cell death (see Fig. 8.12 in Chap. 8). Together, the two studies support the hypothesis that shorter duration usEPs with shorter rise-fall times have greater effects on intracellular structures and their function.

Although most of these Ca^{2+} mobilization studies focused on mechanisms of Ca^{2+} mobilization, only the study by Buescher et al. (2004) considered effects of usEP on cell functions. These studies showed the usEPs could interrupt cells functions. Although Ca^{2+} mobilization does play roles in inhibiting cell functions, it is best known for activating functions such as proliferation, apoptosis induction, muscle contraction and synaptic transmission. While usEPs were shown to induce IP3 signaling (Tolstykh et al. 2013), this is most likely due to a plasma membrane injury response (Demonbreun and McNally 2016). On study analyzed effects of usEP on Ca^{2+} mobilization and function in human platelets, here they were shown to induce Ca^{2+} influxes across the plasma membrane and release of Ca^{2+} from the ER (Zhang et al. 2008). Platelets are small discoid blood cells that are associated with diseases such as myocardial infarction and stroke when they form plugs in the vascular and block blood flow to the heart or brain. However, they are important for hemostasis and wound healing. Platelets harbor two type of granules that serve their functions. Dense granules contain ATP, ADP, glycogen, and Ca^{2+} and ά granules contain clotting and growth factors. When platelets are activated, they form gels that can be applied to wounds that promote healing. Thrombin from bovine has been used to activate platelets, but when used for wound healing they can cause problems such as bleeding, due to the development of cross-reactive anti-bovine antibodies that inhibit human coagulation factor V (Zehnder and Leung 1990). Possible contamination with bovine prions can cause Creutzfeldt-Jacob disease can resemble those of other dementia-like brain disorders, such as Alzheimer's, but develops more rapidly (Vacca 2016).

In these studies (Zhang et al. 2008), newly outdated platelets were received from the Red Cross and exposed to 300 ns pulses between 7 and 50 kV/cm (2.1×10^{-3} to 1.5×10^{-2} Vs/cm). Platelet were used washed or unwashed from sedimented platelet rich plasma (PRP). Platelet aggregation was measured as increases in light transmission as platelet coagulated in suspension. It was necessary for the suspension to be constantly stirred and as a distinct gel formed it would pass in and out of the light path causing a disruption of a constant light transmission. Overall, these studies demonstrated that usEPs represented a new modality that act as a non-ligand agonist for human platelet activation and can function as an alternative to thrombin and other agonists. In fact, usEP-induced platelet gel formation is superior to thrombin since it will not lead to thrombin side effects discussed above.

The usEP effects were quantified as electrical energy density (W) as Joules/cc according to the formula $W = \acute{\alpha}E^2\tau$ where $\acute{\alpha}$ is the conductivity of the media suspension and τ is the duration of the square wave pulse. This is a more traditional or conventional way to quantify usEP strength. As discussed earlier, this measurement is based on energy or thermal effects, yet usEP are low in energy and non-thermal. Nevertheless, usEPs with 300 ns durations and as low as 7 kV/cm (2.1×10^{-3} Vs/cm) with a maximum effect at 30 kV/cm (9×10^{-3} Vs/cm), platelets were activated to aggregate and form platelet gels. These could be used therapeutically instead of thrombin to enhance wound healing. This was the first clear example of usEP inducing a cellular activation process. Both 4 pulses (1.8×10^{-2} Vs/cm) and one pulse with one second intervals at 30 kV/cm, although the four pulse condition caused a more rapid aggregation. There was a sigmoidal increase in light transmission from about 5% at 0.1 J/cc to about 30–40% with about 1.1 J/cc. For a given pulse duration, this sigmodal relationship between percent light transmission and Joules/cc is also observed for the charging effect as Vs/cm.

In typical protocol, platelets were stimulated with usEPs in the absence of extracellular Ca^{2+} and increases in intracellular Ca^{2+} were measured, which represented intracellular Ca^{2+} release, then the increases in intracellular Ca^{2+} were measured after adding extracellular Ca^{2+}, which represents Ca^{2+} influx. UsEPs cause small increases in release of intracellular Ca^{2+} shown to be from the ER and a sigmoidal increases in Ca^{2+} influx across the plasma membrane and as a function of increasing energy density. This usEP-induced increase in intracellular Ca^{2+} was consistent with the well-known fact that aggregation of platelets by agonists, such as thrombin, requires an increase intracellular Ca^{2+}. The data for usEPs on Ca^{2+} release and Ca^{2+} influx was reminiscent of the well-known capacitative Ca^{2+} entry (CCE) or by channels referred to as store operated Ca^{2+} channels (SOCC) induced by hormones in many cell types (Taylor 2006; Smyth et al. 2006). In this process when intracellular Ca^{2+} is released from the ER by IP3, it is coupled with an activation of a Ca^{2+} influx process in the PM.

Since an increase in intracellular Ca^{2+} appeared to be associated with the usEP-induced activation of platelets, the authors determine the mechanism by which usEPs caused the intracellular Ca^{2+} release. Thus, usEPs were compared with two different agents that release intracellular Ca^{2+} by entirely different mechanisms, yet both activate SOCC. Relative to the small release of intracellular Ca^{2+} caused by usEPs and thapsigargin, thrombin caused a large increase intracellular Ca^{2+} release due to the influx of extracellular Ca^{2+}. Thapsigargin causes the increases in intracellular Ca^{2+} by inhibiting the Ca^{2+}-ATPase (SERCA) pump in the ER, thus causing Ca^{2+} to leak out of the ER through the pump into the cytoplasm (Berman 2000). This small increase intracellular Ca^{2+} induced by thapsigargin on Ca^{2+} release is explained by the fact that when Ca^{2+} slowly leaks out of the ER it is immediately pumped out of the cell by the Ca^{2+}-ATPase pump in the PM (Dean and Whiteheart 2004). Thrombin achieves this by increasing the second messenger IP3, which releases Ca^{2+} from the ER and subsequently stimulates SOCC in the PM. Thapsigargin and usEPs, compared to thrombin, produced smaller increases intracellular Ca^{2+} due to internal release, yet produced a much larger increase in Ca^{2+} due to Ca^{2+} influx.

The effects of usEPs on intracellular Ca^{2+} were more closely correlated with the effects of thapsigargin, thus it appears that usEPs are likely causing the formation of nanopores (Gowrishankar et al. 2006; Pakhomov et al. 2009) in the ER membrane and other intracellular membranes such as dense-granules. It is likely that if the leak process occurs in the ER, it will cause activation of SOCC in the PM. A pharmacological approach was used to inhibit SOCC with 2-aminoethoxydiphenyl borate (Dobrydneva and Blackmore 2001) and diethylstilbesterol (Zakharov et al. 2004). Neither SOCC inhibitor had an effect on usEP-induced Ca^{2+} influx while both inhibitors inhibited thapsigargin induced Ca^{2+} influx into platelets. Gd3 + , a non-selective Ca^{2+} channel blocker, was without effect on usEP-induced Ca^{2+} influx, except at concentration much higher than required to inhibit the thapsigargin response. These finding indicated influx of Ca^{2+} in response to usEP was not due to activation of SOCC but was due to formation of nanopores in the plasma membrane.

UsEPs were shown to not only release Ca^{2+} from dense granules but also release platelet-derived growth factor (PDGF) from ά-granules. There was a linear relationship between platelet aggregation as percent light transmittance when plotted against PDGF release. Thrombin also caused a release of PDGF consistent with platelet gel formation.

Overall, like other known platelets agonists, these studies demonstrated that usEPs induce Ca^{2+} mobilization that causes platelets to aggregate forming platelet gels and release the growth factor PDGF. The aggregation effects are electric field-dependent, yet multiple pulses produce less than additive effects. This is likely due to initial pulses desensitizing some step or steps in the aggregation process such as oxidizing membrane lipids or affecting ion channels or other signaling proteins. These studies also showed that the effects of usEPs and thrombin were additive, suggesting that usEPs were acting less specifically and in a more global manner to mobilize Ca^{2+} from a variety of intracellular stores through membrane nanopores. In contrast thrombin only acts on elements of the ER that contain IP3receptors. The release Ca^{2+} in response to IP3then leads to the activation of SOCC in the PM.

References

Alberts B, Johnson A, Lewis L, Raff M, Roberts K, Walter P (2002) Molecular biology of the cell. Garland Science, New York

Alemany R, Sichelschmidt B, Zu Heringdorf DM, Lass H, van Koppen CJ, Jakobs KH (2000) Stimulation of sphingosine-1-phosphate formation by the P2Y(2) receptor in HL-60 cells: Ca(2+) requirement and implication in receptor-mediated Ca(2+) mobilization, but not MAP kinase activation. Mol Pharmacol 58(3):491–497

Beebe SJ (2013) Cell responses without receptors and ligands, using nanosecond pulsed electric fields (nsPEFs). Int J Nanomed 8:3401–3404

Beebe SJ (2015) Considering effects of nanosecond pulsed electric fields on proteins. Bioelectrochemistry 103:52–59

Beebe SJ, White JA, Blackmore PF, Deng Y, Somers K, Schoenbach KH (2003) Diverse effects of nanosecond pulsed electric fields on cells and tissues. DNA Cell Biol 22:785–796

Beebe SJ, Chen YJ, Sain NM, Schoenbach KH, Xiao S (2012) Transient features in nanosecond pulsed electric fields differentially modulate mitochondria and viability. PLoS ONE 7(12):e51349

Beebe SJ, Sain NM, Ren W (2013) Induction of cell death mechanisms and apoptosis by nanosecond pulsed electric fields (nsPEFs). Cells 2(1):136–162

Berman MC (2000) Characterisation of thapsigargin-releasable Ca(2+) from the Ca(2+)-ATPase of sarcoplasmic reticulum at limiting (Ca(2+)). Biochim Biophys Acta 1509(1–2):42–54

Berridge MJ (1998) Neuronal calcium signaling. Neuron 21(1):13–26

Berridge MJ (2009) Inositol trisphosphate and calcium signalling mechanisms. Biochim Biophys Acta 1793(6):933–940

Buescher ES, and Schoenbach KH (2003) Submicrosecond intense pulsed electric field effects on intracellular free calcium: mechanisms and effects. IEEE Trans Dielectrics Electr Insul 10:788–794

Buescher ES, Smith RR, Schoenbach KH (2004) Submicrosecond intense pulsed electric field effects on intracellular free calcium: mechanisms and effects. IEEE Trans Plasma Sci 32:1563–1572

Craviso GL, Choe S, Chatterjee I, Vernier PT (2012) Modulation of intracellular Ca2+ levels in chromaffin cells by nanoelectropulses. Bioelectrochemistry 87:244–252

Crimaudo C, Hortsch M, Gausepohl H, Meyer DI (1987) Human ribophorins I and II: the primary structure and membrane topology of two highly conserved rough endoplasmic reticulum-specific glycoproteins. EMBO J 6(1):75–82

Dean WL, Whiteheart SW (2004) Plasma membrane Ca(2+)-ATPase (PMCA) translocates to filopodia during platelet activation. Thromb Haemost 91(2):325–333

Demonbreun AR, McNally EM (2016) Plasma membrane repair in health and disease. Curr Top Membr 77:67–96

Dobrydneva Y, Blackmore P (2001) 2-Aminoethoxydiphenyl borate directly inhibits store-operated calcium entry channels in human platelets. Mol Pharmacol 60:541–552

Eisner DA, Caldwell JL, Kistamás K, Trafford AW (2017) Calcium and excitation-contraction coupling in the heart. Circ Res 121(2):181–195

English AR, Voeltz GK (2013) Endoplasmic reticulum structure and interconnections with other organelles. Cold Spring Harb Perspect Biol 5(4):a013227

Gowrishankar TR, Esser AT, Vasilkoski Z, Smith KC, Weaver JC (2006) Microdosimetry for conventional and supra-electroporation in cells with organelles. Biochem Biophys Res Commun 341(4):1266–1276

Guo S, Burcus NI, Hornef J, Jing Y, Jiang C, Heller R, Beebe SJ (2018) Nano-pulse stimulation for the treatment of pancreatic cancer and the changes in immune profile. Cancers (Basel) 10(7):217

Harper JL, Camerini-Otero CS, Li AH, Kim SA, Jacobson KA, Daly JW (2003) Dihydropyridines as inhibitors of capacitative calcium entry in leukemic HL-60 cells. Biochem Pharmacol 65(3):329–338

Hristov K, Mangalanathan U, Casciola M, Pakhomova ON, Pakhomov AG (2018) Expression of voltage-gated calcium channels augments cell susceptibility to membrane disruption by nanosecond pulsed electric field. Biochim Biophys Acta Biomembr 1860(11):2175–2183

Kotnik T, Miklavcic D (2006) Theoretical evaluation of voltage inducement on internal membranes of biological cells exposed to electric fields. Biophys J 90:480–491

Lanner JT, Georgiou DK, Joshi AD, Hamilton SL (2010) Review Ryanodine receptors: structure, expression, molecular details, and function in calcium release. Cold Spring Harb Perspect Biol 2(11):a003996

Liu Y, Zhu X (2017) Endoplasmic reticulum-mitochondria tethering in neurodegenerative diseases. Transl Neurodegener 6:21

Martens SL, Roth CC, Ibey BL (2018) Nanosecond pulsed electric field exposure does not induce the unfolded protein response in adult human dermal fibroblasts. Bioelectromagnetics 39(6):491–499

Neve EPA, Ingelman-Sundberg M (2010) Cytochrome P450 proteins: retention and distribution from the endoplasmic reticulum. Curr Opin Drug Discov Devel 13(1):78–85

Nuccitelli R, Berridge JC, Mallon Z, Kreis M, Athos B, Nuccitelli P (2015) Nanoelectroablation of murine tumors triggers a CD8-dependent inhibition of secondary tumor growth. PLoS ONE 10(7):e0134364
Pakhomov AG, Bowman AM, Ibey BL, Andre FM, Pakhomova ON, Schoenbach KH (2009) Lipid nanopores can form a stable, ion channel-like conduction pathway in cell membrane. Biochem Biophys Res Commun 385(2):181–186
Pakhomov AG, Xiao S, Pakhomova ON, Semenov I, Kuipers MA, Ibey BL (2014) Disassembly of actin structures by nanosecond pulsed electric field is a downstream effect of cell swelling. Bioelectrochemistry 100:88–95
Rossi A, Pakhomova ON, Mollica PA, Casciola M, Mangalanathan U, Pakhomov AG, Muratori C (2019) Nanosecond pulsed electric fields induce endoplasmic reticulum stress accompanied by immunogenic cell death in murine models of lymphoma and colorectal cancer. Cancers (Basel) 11(12):2034
Schooley A, Vollmer B, Antonin W (2012) Building a nuclear envelope at the end of mitosis: coordinating membrane reorganization, nuclear pore complex assembly, and chromatin de-condensation. Chromosoma 121(6):539–554
Semenov I, Xiao S, Pakhomova ON, Pakhomov AG (2013) Recruitment of the intracellular Ca2+ by ultrashort electric stimuli: the impact of pulse duration. Cell Calcium 54(3):145–150
Semenov I, Xiao S, Pakhomov AG (2013) Primary pathways of intracellular Ca(2+) mobilization by nanosecond pulsed electric field. Biochim Biophys Acta 1828(3):981–989
Seo MD, Enomoto M, Ishiyama N, Stathopulos PB, Ikura M (2015) Structural insights into endoplasmic reticulum stored calcium regulation by inositol 1,4,5-trisphosphate and ryanodine receptors. Biochim Biophys Acta 1853(9):1980–1991
Smyth JT, Dehaven WI, Jones BF, Mercer JC, Trebak M, Vazquez G, Putney JW Jr (2006) Emerging perspectives in store-operated Ca2+ entry: roles of Orai, Stim and TRP. Biochim Biophys Acta 1763(11):1147–1160
Stacey M, Fox P, Buescher S, Kolb J (2011) Nanosecond pulsed electric field induced cytoskeleton, nuclear membrane and telomere damage adversely impact cell survival. Bioelectrochemistry 82(2):131–134
Taylor CW (2006) Store-operated Ca2+ entry: a stimulating store. Trends Biochem Sci 31:597–601
Thastrup O, Cullen PJ, Drøbak BK, Hanley MR, Dawson AP (1990) Thapsigargin, a tumor promoter, discharges intracellular Ca2+ stores by specific inhibition of the endoplasmic reticulum Ca2(+)-ATPase. Proc Natl Acad Sci USA 87(7):2466–2470
Tolstykh GP, Beier HT, Roth CC, Thompson GL, Payne JA, Kuipers MA, Ibey BL (2013) Activation of intracellular phosphoinositide signaling after a single 600 nanosecond electric pulse. Bioelectrochemistry 94:23–29
Traaseth N, Elfering S, Solien J, Haynes V, Giulivi C (2004) Role of calcium signaling in the activation of mitochondrial nitric oxide synthase and citric acid cycle. Biochim Biophys Acta 1658(1–2):64–71
Vacca VM Jr (2016) CJD: Understanding Creutzfeldt-Jakob disease. Nursing 46(3):36–42
Vandecaetsbeek I, Vangheluwe P, Raeymaekers L, Wuytack F, Vanoevelen J (2011) The Ca2+ pumps of the endoplasmic reticulum and Golgi apparatus. Cold Spring Harb Perspect Biol 3:a004184
Verkhratsky A, Shmigol A (1996) Calcium-induced calcium release in neurones. Cell Calcium 19(1):1–14
Vernier PT, Sun Y, Marcu L, Salemi S, Craft CM, Gundersen MA (2003) Calcium bursts induced by nanosecond electric pulses. Biochem Biophys Res Commun 310(2):286–295
Vernier PT, Sun Y, Chen MT, Gundersen MA, Craviso GL (2008) Nanosecond electric pulse-induced calcium entry into chromaffin cells. Bioelectrochemistry 73(1):1–4
Wan B, LaNoue KF, Cheung JY, Scaduto RC Jr (1989) Regulation of citric acid cycle by calcium. J Biol Chem 264(23):13430–13439
Wang S, Chen J, Chen MT, Vernier PT, Gundersen MA, Valderrábano M (2009) Cardiac myocyte excitation by ultrashort high-field pulses. Biophys J 96(4):1640–1648

White JA, Blackmore PF, Schoenbach KH, Beebe SJ (2004) Stimulation of capacitative calcium entry in HL-60 cells by nanosecond pulsed electric fields. J Biol Chem 279(22):22964–22972

Wilson CM, Roebuck Q, High S (2008) Ribophorin I regulates substrate delivery to the oligosaccharyltransferase core. Proc Natl Acad Sci USA 105(28):9534–9539

Xia M, Zhang Y, Jin K, Lu Z, Zeng Z, Xiong W (2019) Communication between mitochondria and other organelles: a brand-new perspective on mitochondria in cancer. Cell Biosci 9:27

Zakharov SI, Smani T, Dobrydneva Y, Monje F, Fichandler C, Blackmore PF, Bolotina VM (2004) Diethylstilbestrol is a potent inhibitor of store-operated channels and capacitative Ca(2+) influx. Mol Pharmacol 66(3):702–707

Zehnder JL, Leung LLK (1990) Blood 76:2011–2016

Zhang J, Blackmore PF, Hargrave BY, Xiao S, Beebe SJ, Schoenbach KH (2008) Nanosecond pulse electric field (nanopulse): a novel non-ligand agonist for platelet activation. Arch Biochem Biophys 471(2):240–248

Chapter 6
Intra-cellular Calcium Release Dynamics Due to Nanosecond Electric Pulsing

Ravi Joshi

Abstract Permeabilization of cell membranous structures by nanosecond electric field pulses triggers a transient rise of cytosolic calcium with multifarious downstream effects. Electroporation of intracellular membranes (such as those of the Endoplasmic Reticulum) are likely responsible for the calcium release. This is an important application of pulsed electric fields, since calcium is known as a ubiquitous second messenger molecule that regulates several responses in cell signaling, including enzyme activation, gene transcription, neurotransmitter release, secretion, muscle contraction etc. In this chapter, a model based analysis of the dynamical calcium release in response to an external electric pulse is discussed. The results obtained are shown to match experimental data fairly well.

6.1 Introduction

The use of very high electric fields (~50 kV/cm or higher) with pulse durations in the nanosecond range can lead to calcium release from the endoplasmic reticulum (Zimmermann and Neil 1996; Deng et al. 2003; Vernier et al. 2003; Beebe et al. 2004; White et al. 2004), with *many of the experiments being conducted in the absence of extra-cellular calcium.* External voltage triggering of Ca^{2+} release could be a vital element in artificially induced cell signaling. This aspect of electric field pulsing, especially nanosecond stimulation, has been analyzed (Semenov et al. 2013). Since intra-cellular Ca^{2+} release leads to some uptake by the mitochondria, such voltage-triggered Ca-release could potentially have secondary effects on the mitochondria. Alterations in the mitochondrial membrane potential due to cationic entry is a possibility. Such trans-membrane voltage changes are known to trigger opening of the permeability transition pore (Bernardi and Petronilli 1996; Ichas and Mazat 1998), leading to *cytochrome c* release and cellular apoptosis.

The resting calcium concentration in most cell types is around 100 nM, and its increases can lead to the activation of various Ca^{2+}-dependent cascades. Calcium is involved in effects as diverse as cell differentiation or division, cytoskeleton rearrangements, endo-and exocytosis, synthesis and release of neuromediators, activation of immune cells, apoptotic or necrotic cell death (Catterall 2000; Partridge

S. J. Beebe et al., *Ultrashort Electric Pulse Effects in Biology and Medicine*, Series in BioEngineering, https://doi.org/10.1007/978-981-10-5113-5_6

et al. 1994; Roberts-Crowley et al. 2009; Bootman et al. 2002). Elevation of Ca^{2+} concentration by nsPEF (by permeabilization of either plasma membrane or the endoplasmic reticulum) is a promising approach for non-chemical triggering of Ca^{2+}-signaling in various cell types. The role of nanosecond electric fields on intra-cellular calcium release was made more definitive by exposing CHO cells which lack voltage-gated Ca^{2+} channels., so that any changes in calcium density could only arise from membrane electroporation. Experiments (Semenov et al. 2013) showed that a single 60-ns pulse caused fast Ca^{2+} increase by influx from the outside and calcium efflux from the endoplasmic reticulum, with the E-field thresholds of about 9 and 19 kV/cm, respectively.

Calcium modeling calculations require determination of the time-dependent, transmembrane electric fields and the poration process that facilitates Ca^{2+} ion flux from the endoplasmic reticulum (ER). It is also critical to determine whether electroporation might occur at the ER, and the time required for such a potential event. An approach to calculating the dynamic electric fields and transmembrane potentials is through a time-domain nodal analysis based on a distributed equivalent circuit representation of a cell and its membrane structures (Hu et al. 2005). Essentially, the entire volume can be broken up into finite segments, and each segment represented by a parallel resistor–capacitor (RC) combination to account for the electrical current flow and charging effects. The ER center was taken to be the $r = 0$ point, while the intersection of the radial ray with the plasma membrane was the other boundary. The externally applied electric field was taken to be along the axis of symmetry for this spherical problem. Azimuthal symmetry was invoked to eliminate the third dimension. The plasma membrane (PM) was denoted as the "outer membrane," while the "inner membrane" represents the ER boundary. The model also included cytosol between these two membranes, as well as the region within the ER. The outer circular segment defined the overall simulation region.

Simple nodal analysis of circuit theory can be applied at each grid point to obtain the time-dependent node voltages in response to the external electric pulse $E_0(t)$. Potentials on each node are easily updated at every time step, and these values can be used as the drivers for the pore dynamics based on the Smoluchowski equation (Joshi et al. 2004). This would yield the time-dependent pore density and distribution of pore area at every membrane grid point. In the process, a dynamic membrane conductance taking electroporation into account would be naturally built in, and yield drift and diffusive fluxes through the nano-pores. The latter would provide the leak flux function and the time-dependent calcium ion outflow.

Intracellular Ca^{2+} dynamics has the unique feature of facilitating the generation of global events (often of a periodic nature) from local, thermally-activated stochastic opening/closing of channels on the ER membrane (Berridge et al. 2003, 1998; Joshi et al. 2007). Such channels are typically closely packed into clusters, called focal sites (Mak et al. 2000; Marchant and Parker 2001) with a random spatial distribution and an average spacing in the 2–6 μm range. A typical channel present in the ER of many cells that facilitates calcium movement, is the inositol 1,4,5-triphosphate receptor (IP3R) channel. Ryanodine receptors (RyR) though present, are more important in muscle cells (Sneyd and Falcke 2005). The IP3R channel has an activating binding site for the messenger molecule IP3 (m-gate), an activating site for Ca^{2+} (n-gate), and an inhibiting Ca^{2+} binding site (h-gate). Experimental findings suggest that the channel is open if both Ca^{2+} and IP3 are bound to the activating sites, and at the same time Ca^{2+} is not bound to the inhibiting site. Binding of Ca^{2+} to the inhibiting site of one of these subunits, closes the channel.

This "open channel" probability increases nonlinearly with the IP3 and calcium concentrations. Hence, any Ca^{2+} released by one channel increases the open probability of neighboring channels. This provides a self-amplifying, positive-feedback non-linear mechanism (A. Bugrim et al. 2003) referred to as "calcium-induced calcium release (CICR)". Very high Ca^{2+} concentrations inhibit the channels. The Ca^{2+} SERCA pumps remove Ca^{2+} from the intracellular space. This is necessary since elevated concentrations of Ca^{2+} are toxic for the cell.

Initial simulation efforts to quantify the intracellular Ca^{2+} dynamics primarily focused on deterministic continuum models (De Young and Keizer 1992; Li and Rinzel 1994; Jafri and Keizer 1995; Atri et al. 1993; Falcke et al. 1999). These reaction-diffusion models were able to explain the observed wave patterns, oscillatory, or bistable phenomena. Extensions to these models that allowed the channels to act as discrete Ca^{2+} sources facilitated the transition from localized to traveling structures (Bugrim et al. 1997). The stochastic behavior seen in spark and puff formation and the rather small number of channels creating a localized event, motivated the introduction of stochastic models (Falcke et al. 2000; Falcke 2003; Swillens et al. 1998; Shuai and Jung 2002, 2003; Keizer and Smith 1998; Keizer et al. 1998). Mesoscopic aspects of the calcium-release phenomena were also probed (Shuai and Jung 2002, 2003). They found that site clustering could effectively allow for collectively enhanced, coherent calcium responses to signals. Homogeneously distributed channels, on the other hand, would not be capable of producing the same large response.

Discrete models have successfully predicted oscillatory dynamical regimes and random, collective calcium enhancements. *However, such localized, discrete stochastic models are perhaps not necessary in the present context of field-assisted calcium release for a variety of reasons.* First, it has become apparent based on improved spatial imaging (Haak et al. 2001) that the overall calcium release varies in a continuous fashion despite the stochastic variations in the numbers of individual channels recruited for release and the durations of their openings. Thus, macroscopically, a continuum model remains relevant. Furthermore, the present focus is on very high (>15 kV/cm) electric fields. This external stimulus produces very strong electrostatic driving forces and gives rise to highly non-equilibrium conditions. Hence, it can safely be assumed that all of the channels are effectively driven into the calcium release (or "open") state by the strong external electric signal. Statistical variability can be expected to be minimal, and that all of the discrete sites would be collectively forced past the calcium-release threshold.

6.2 Intra-cellular Calcium Release Dynamics Due to High-Intensity, Short Electric Pulse

Calcium release from the ER predominantly occurs through IP3R sites (Shuai and Jung 2002). For example, the *Xenopus* oocyte primarily has IP3 channels. Ryanodine receptors are more important in muscle cells (Sneyd and Falcke 2005). It may be mentioned that some open issues regarding the RyR mathematical models remain. Hence for simplicity, here we have focused on the IP3R sites as the primary calcium release channel. For other applications, such as cardiac fibrillation that are thought to be coupled to calcium signaling in the ventricles, the RyR channels would need to be included. Our study utilizes the Li-Rinzel two-variable simplification (Li and Rinzel 1994) of the De Young–Keizer model (De Young and Keizer 1992) with appropriate modifications to account for electric field effects. The basic model has been used in the past (but without any external electric stimulation) for various analyses (Bugrim et al. 2003; Zwanzig 1961). According to this model, calcium flux between the ER and the intracellular space is driven by the three following processes: (i) Ca^{2+} outflow mediated by the IP3 channel, (ii) a small diffusion-driven Ca^{2+} leakage from the ER into the cytosol, and (iii) the SERCA ATPase pumps that drive Ca^{2+} back into the ER to maintain the resting (basal) calcium levels.

Here we include the additional mechanism of electric field driven Ca^{2+} outflow from the ER to the cytosol. Electroporation of the ER membrane, coupled with the electrical driving force on the calcium cations contributes to this outflow. In a sense, the ER becomes more "leaky." This effect has been included in our one-dimensional, time-dependent reaction–diffusion model through a time-dependent leak-flux term. The equations for $Ca^{2+}(z, t)$ density changes in the cytosol and ER are expressed, in general, by the continuity equation as:

$$\partial\{[Ca^{2+}]\}/\partial t = -\partial F(z, t)/\partial z + G(z, t) - R(z, t), \tag{6.1}$$

where F(z,t) is the passive calcium flux at any location "z" and time "t," while G(z, t) and R(z, t) are possible $[Ca^{2+}]$ generation and recombination/attachment rates that could include pumps, channels and pores. In our treatment, generation, recombination/attachment of Ca^{+2} ions, action of pumps, channels, pores etc. has been ignored within the bulk and only taken into account at the ER-cytosol boundary. The passive flow of $[Ca^{2+}]$ ions can be expressed in terms of drift and diffusive processes. Assuming that the drift velocity of the $[Ca^{2+}]$ ions in response to the highest local electric fields is much smaller than their thermal velocities, and that the role of scattering can be approximated by a relaxation time "τ", the flux F(z, t) can be cast into the following "drift–diffusion" form:

$$F(z, t) = [Ca^{2+}]\mu E(z, t) - D(z, t)\left(\partial[Ca^{2+}]/\partial z\right), \tag{6.2}$$

where μ is the ion mobility, E(z, t) is the space- and time-dependent local electric field, and D the diffusion coefficient. These parameters are simply related to the relaxation time "τ" as: $\mu = q < v^2 \tau > /(3 k_B T)$, and $D = (k_B T/q) \mu$, where q is the elementary charge, T the temperature, k_B the Boltzmann constant, v is the individual ionic velocity, while < > denotes an ensemble average over Ca^{2+} ions.

For completeness, it may be pointed out that there are some approximations inherent in our use of the above drift-diffusion scheme. For example, the diffusion coefficient and mobility are taken as fixed, invariant parameters. Strictly, this can only be done if the system is stationary, not far from equilibrium, and strong local inhomogeneities do not exist. For non-stationary, non-Markovian processes, the diffusion coefficient needs to be evaluated as an integral over a two-time velocity autocorrelation function (Zwanzig 1961). Due to the ultra-short electric pulse, we assumed near stationarity. Treatments of time-variations of transport parameters in an aqueous medium over time scales of nanoseconds or shorter, and in the presence of external electric fields, have been reported by our group elsewhere (Sridhara et al. 2006). Also, since the $[Ca^{2+}]$ release and other bio-chemical changes are relatively perturbative in nature with almost no impact or deviations in overall internal scattering rates, the μ and D transport parameters can essentially be assumed homogeneous. Thus, using Eq. (6.2) in Eq. (6.1) yields:

$$\partial\{[Ca^{2+}]\}/\partial t = D(z,t)(\partial^2[Ca^{2+}]/\partial z^2) \\ -\mu\partial\{[Ca^{2+}]\, E(z,t)\}/\partial z + G(z,t) - R(z,t). \qquad (6.3)$$

Since an analytical solution cannot be obtained for the above equation, a numerical approach was used based on a uniform discretization in space (spacing "dz") and time (interval "dt"). The details are briefly given next for completeness. The entire simulation region was divided into a set of "N" uniformly sized boxes of thickness "dz." In this discretized representation, index "i" denotes the ith time step (t = {i − 1} * dt), and "j" denotes the jth spatial step (z = {j − 1} * dz). Thus a total of (N + 1) grid points were chosen. Voltages were calculated at all of the (N + 1) grid points, while the Ca^{2+} concentration and electric field were defined at the center of each box (i.e. between adjacent grid points). The indices j = (J − 1) and j = J represent the two boxes on either side of the ER-cytosol boundary. The results of discretizing Eq. (6.3) are given below for various regions within the simulation zone.

The governing transport equation in the outer cytosol region can be expressed as:

$$\left\{Ca^{2+}_{i+1,j} - Ca^{2+}_{i,j}\right\}/(dt) = D_c\left\{Ca^{2+}_{i,j+1} - 2Ca^{2+}_{i,j} + Ca^{2+}_{i,j-1}\right\}/(dz)^2 \\ -\mu_c\left[\left\{Ca^{2+}_{i,j+1}E_{i,j+1}\right\} - \left\{Ca^{2+}_{i,j-1}E_{i,j-1}\right\}\right]/(2dz), \\ \text{for } J < j < N \qquad (6.4a)$$

where $Ca^{2+}_{i,j}$ denotes the calcium ion concentration $[Ca^{2+}]$ at the ith time step, and $E_{i,j}$ similarly denotes the value of the discretized electric field. In Eq. (6.4a), D_c and μ_c denote the diffusion coefficient and mobility values in the cytosol. At the last box next to the plasma membrane boundary (j = N), the governing transport equation is:

$$\left\{Ca^{2+}_{i+1,j}-Ca^{2+}_{i,j}\right\}/(dt) = -D_c\left\{Ca^{2+}_{i,j}- Ca^{2+}_{i,j-1}\right\}/(dz)^2 + \mu_c\left[\left\{Ca^{2+}_{i,j}E_{i,j}\right\} + \left\{Ca^{2+}_{i,j-1}E_{i,j-1}\right\}\right]/(2dz), \quad \text{for } j = N \tag{6.4b}$$

Similarly, the discretized continuity equation in the inner ER region can be expressed as:

$$\left\{Ca^{2+}_{i+1,j} - Ca^{2+}_{i,j}\right\}/(dt) = D_E\left\{Ca^{2+}_{i,j+1}- 2Ca^{2+}_{i,j} + Ca^{2+}_{i,j-1}\right\}/(dz)^2 - \mu_E\left[\left\{Ca^{2+}_{i,j+1}E_{i,j+1}\right\}-\left\{Ca^{2+}_{i,j-1}E_{i,j-1}\right\}\right]/(2dz), \quad \text{for } 1 < j < (J-1) \tag{6.4c}$$

where D_E and μ_E denote the diffusion coefficient and mobility values in the ER. At the $j = 1$ grid point (corresponding to the first box within the ER), the discretized equation is:

$$\left\{Ca^{2+}_{i+1,j} - Ca^{2+}_{i,j}\right\}/(dt) = D_E\left\{Ca^{2+}_{i,j+1}- Ca^{2+}_{i,j}\right\}/(dz)^2 - \mu_E\left[\left\{Ca^{2+}_{i,j}E_{i,j}\right\} + \left\{Ca^{2+}_{i,j+1}E_{i,j+1}\right\}\right]/(2dz), \quad \text{for } j = 1. \tag{6.4d}$$

For $j = J + 1$ (cytosol box just to the right of the ER boundary), the continuity equation becomes:

$$\left\{Ca^{2+}_{i+1,j}- Ca^{2+}_{i,j}\right\}/(dt) = D_c\left\{Ca^{2+}_{i,j+1}- Ca^{2+}_{i,j}\right\}/(dz)^2 + g(t) - \mu_c\left[\left\{Ca^{2+}_{i,j+1}E_{i,j+1}\right\} + \left\{Ca^{2+}_{i,j}E_{i,j}\right\}\right]/(2dz) + \left[J_{IP3R}-J_{SERCA} + J_{Leak}\right]/F_C, \tag{6.4e}$$

where g(t) denotes a poration dependent "leak flux" function. In Eq. (6.4e), J_{IP3R}, J_{SERCA} and J_{Leak} denote the usual flux terms at the ER boundary, while F_C is the volume fraction of the cytosol relative to the total cell volume. Similarly, for $j = (J - 1)$, one has:

$$\left\{Ca^{2+}_{i+1,j} - Ca^{2+}_{i,j}\right\}/(dt) = -D_E\left\{Ca^{2+}_{i,j}-Ca^{2+}_{i,j-1}\right\}/(dz)^2-g(t) + \mu_E\left[\left\{Ca^{2+}_{i,j}E_{i,j}\right\} + \left\{Ca^{2+}_{i,j-1}E_{i,j-1}\right\}\right]/(2dz) - \left[J_{IP3R}-J_{SERCA} + J_{Leak}\right]/F_E, \tag{6.4f}$$

where F_E is the volume fraction of the ER relative to the total cell volume (i.e. $F_E + F_C = 1$).

Initial starting concentrations were set on the basis of $E_{i,j} = 0$ and $[J_{IP3R} - J_{SERCA} + J_{Leak}] = 0$. A Ca^{2+} value of 0.1 μM was chosen inside the cytosol (i.e., $Ca^{2+}{}_{i=0,j} = 0.1$ μM for $J \le j \le N$), while the concentration inside the ER (say $Ca^{2+}{}_{ER0}$) was obtained from the initial steady-state condition of $[J_{IP3R} - J_{SERCA} + J_{Leak}] = 0$. Thus: $Ca^{2+}{}_{i=0,j} = Ca^{2+}{}_{ER0}$ for $1 \le j \le (J\text{-}1)$.

The discretized value of the electric field $E_{i,j}\{=E(z, t)\}$ is dictated by the duration of the externally applied voltage stimulation. The requisite internal electric fields were derived at each time instant from the distributed electric representation. Upon termination of the external voltage, the $E_{i,j}$ term quickly falls to zero at all grid points in the simulation region. Beyond such times, only the diffusion term is operative for Ca^{2+} transport. The "flux leak function" g(t) accounts for the ER membrane poration process, and includes the finite delay in its creation in response to the external voltage-pulse. It is obtained as an input parameter from calculations of membrane poration based on the Smoluchowski equation, discussed in the previous section. Thus, the function g(t) models time dependent "poration" and Ca^{2+} outflow from the ER at the ER-cytosol interface. In the scenario simulated here, the external voltage pulse is sufficiently large to porate the ER *before its termination*. Hence, during the later part of the voltage pulse (nanosecond regime), we have in effect a source term continually feeding Ca^{2+} into the cytosol, together with the drift of the Ca^{2+} ions away in the direction of the plasma membrane.

It may be mentioned that the above numerical implementation is based on an explicit scheme. An implicit scheme is easily realized by replacing terms such as $Ca^{2+}{}_{i,j}$ on the right side of Eqs. (6.4a)–(6.4f) by $[(Ca^{2+}{}_{i,j} + Ca^{2+}{}_{i+1,j})/2]$. This results in a set of coupled simultaneous equations that can be cast in a matrix form and solved to yield the relevant concentrations at the succeeding time steps. The implicit scheme has the advantage of greater numerical stability. In our case, the implicit scheme was implemented as well, and solution to the system of equations obtained through LU decomposition of the matrix. The difference between the two solutions was negligible, and there were no stability issues with the explicit scheme for the grid sizes chosen.

The mathematical expressions used for the various fluxes at the ER-boundary (for example indicated in Eqs. (6.4e) and (6.4f)) are given next. A similar discretized numerical implementation was used, and for simplicity, only the algebraic equations are given below. The constituent current densities J_{IP3R}, J_{SERCA}, and J_{Leak} can be expressed as (Wagner et al. 2004):

$$J_{IP3R} = V_{IP3R} m(t)^3 h(t)^3 \left([Ca^{2+}]_E - [Ca^{2+}]_C\right), \tag{6.5a}$$

$$J_{Leak} = V_{Leak}\left([Ca^{2+}]_E - [Ca^{2+}]_C\right), \tag{6.5b}$$

$$\text{and,}\quad J_{SERCA} = V_{SERCA}\left([Ca^{2+}]_C\right)^2 \Big/ \left[(K_{SERCA})^2 + \left([Ca^{2+}]_C\right)^2\right]. \tag{6.5c}$$

In the above formulation, V_{IP3R}, V_{Leak}, and V_{SERCA} are constants that have been reported in the literature. The variables m(t) and h(t) are time-dependent, and given by the following expressions:

$$m(t) = ([IP_3]/[IP_3 + K_{IP3}])([Ca^{2+}]_C/([Ca^{2+}]_C + K_{Act})), \qquad (6.6a)$$

$$\text{and,} \quad dh(t)/dt = [h_\infty - h(t)]/\tau, \qquad (6.6b)$$

$$\text{where } h_\infty = K_{Inh}/[(K_{Inh} + [Ca^{2+}]_C)]. \qquad (6.6c)$$

In Eq. (6.6a)–(6.6c), K_{Act} (and K_{Inh}) are the activating (and inhibiting) Ca^{2+} binding site dissociation constants, $[IP_3]$ is the time-dependent, spatially variable concentration of IP3 in the cytosol, while τ is the inactivation time constant. Values of the various constants and parameters are generally well established in the literature, and here many of those (e.g., Wagner et al. 2004) have been implemented. For completeness, the parameter set used in our simulations is given in Table 6.1.

Table 6.1 Parameters for the Ca model simulations

Model parameter	Value
K_{IP3}	0.15 μM
K_{Act}	0.8 μM
V_{IP3R}	8.5 s^{-1}
V_{Leak}	0.01 s^{-1}
V_{SERCA}	2.6 μM/s
K_{SERCA}	0.2 μM
V_{Prod}	0.075 μM/s
K_{Prod}	0.4 μM
V_1	0.001 μM/s
V_2	0.005 μM/s
V_3	0.02 μM/s
D_C	16 $\mu m^2/s$
D_E	16 $\mu m^2/s$
D_{IP3}	283 $\mu m^2/s$
K_{Inh}	1.9 μM
τ	2 s
λ	30.0
F_C	0.8
F_E	0.2

The complete dynamics requires inclusion of IP3 production and its degradation. Basically, Ca^{2+} -induced IP3 production at the ER and plasma membranes, is given by the following relation:

$$J_{IP3\_Prod} = V_{Prod}\left([Ca^{2+}]_C\right)^2 \Big/ \left(\left([Ca^{2+}]_C\right)^2 + K_{Prod}^2\right), \qquad (6.7a)$$

with appropriate constants V_{Prod} and K_{Prod}. The IP3 degradation model (e.g., Wagner et al. 2004) is based on IP3 hydrolysis by 5-phosphatase to inositol-1,4-bisphosphate (IP2) and the phosphorylation by 3-kinase to inositol-1,3,4,4-tetrakisphosphate (IP4). The corresponding rates, J_{Phos} and J_{Kinase} are given as:

$$J_{Phos} = V_3[IP3] \big/ ([IP3] + 30), \qquad (6.7b)$$

$$\text{and, } J_{Kinase} = (1-\theta)\, V_1[IP3] \big/ ([IP3] + 2.5) + \theta V_2[IP3] \big/ ([IP3] + 0.5), \qquad (6.7c)$$

where V_1, V_2, and V_3 are constants, and $\theta = [Ca^{2+}]_C/([Ca^{2+}]_C + 0.39)$. The concentrations [IP3], $[Ca^{2+}]_C$, and $[Ca^{2+}]_E$ are all in micro-molar units. The dynamical evolution of the IP3 concentration is then given in terms of the various creation-annihilation rates and the flux as:

$$\partial[IP3]/\partial t = D_{IP3}\partial^2[IP3]/\partial^2 x + [J_{IP3\_Prod} - \lambda\,(J_{Kinase} - J_{Phos})]\,/F_C, \qquad (6.7d)$$

with λ being the IP3 rate scaling factor.

It may be mentioned that theoretically, besides the ionic drift, diffusion through the electrically created nano-pores can also occur. However, such calcium diffusive fluxes can be expected to be relatively negligible for the following reasons. (i) Discontinuities in permittivity between the ER membrane and its surrounding aqueous medium impede ionic transport in the vicinity of a nano-pore. Such small-area nanopores (radii ~ 1 nm or less) within lipid membranes have been predicted based on continuum approaches (Vasilkoski et al. 2006) and also on the basis of molecular dynamics simulations (Hu et al. 2005). Their small diameters result from the ultrashort time scales during which the external driving fields are applied, thereby preventing sustained and large pore expansions. Experimentally, the nanopores have indirectly been detected by using various fluorescent dyes (Sun et al. 2006). For nanopores, both the electrostatic interactions and Born solvation energy (Cherepanov et al. 2003; Parsegian 1969) of ions approaching such apertures are strongly enhanced. Hence, in the absence of an electric field (upon termination of the ultra-short voltage pulse, for example), the ionic Ca^{2+} diffusion into the cytosol through the “nano-pipe” would be minimal. Nano-pore re-sealing can be expected to be fairly rapid after termination of the nano-second pulse given the small pore size. Very short resealing times (under 5 ns) have been reported in the literature (Tarek 2005) based on molecular dynamics simulations. Such resealing would further severely restrict any additional contributions to diffusive leak-out from the ER. For the above reasons, electrically induced

Ca^{2+} transfer from the ER was taken to be the operative via the drift mechanism only during external pulsing.

Some qualitative insights into the Ca^{2+} release dynamics are provided by the concentration-dependent rates. The J_{SERCA} function is sigmoidal and increases monotonically with the calcium concentration in the cytoplasm. A qualitative sketch is given in Fig. 6.1. The values were obtained from Eqs. (6.4–6.7) on applying the parameters listed in Table 6.1. The calcium release rate exhibits a bell-shaped curve as sketched in Fig. 6.1. The J_{IP3R} contribution to release depends on "m" and the differential $\{[Ca^{2+}]_E - [Ca^{2+}]_C\}$. The rate of calcium leakage is relatively very small, and does not have much bearing on the dynamics (Wagner et al. 1998). These rates, which do not take any account of electrically induced fluxes, underscore the possibility of attaining either bistable or oscillatory operating points. For example, in Fig. 6.1 the intersections of J_{SERCA} with $\{J_{IP3R} + J_{Leak}\}$ are represented by points "A", "B", and "C". Of these, "A" and "C" are stable operating points, while "B" denotes an unstable, oscillatory state. Thus, it would seem possible to drive the system from an initial state of low cytosolic Ca^{2+} levels towards relatively high calcium levels by initiating a triggered release.

The primary role of an external voltage is the additional release of Ca^{2+} due to nano-poration at the ER and the electrically induced drift. This outflow (to first order), would only depend on the calcium concentration within the ER, the voltage magnitude and pulse duration, and be independent of the Ca^{2+} concentration in the cell. Effectively, the electrical triggering would shift the calcium release curve vertically,

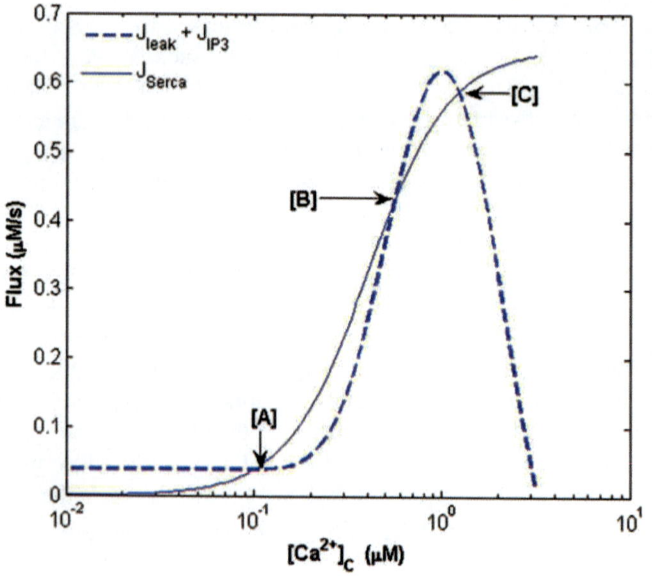

Fig. 6.1 Qualitative sketch of the calcium release and pumping rates as a function of the calcium concentration in the cytosol (After Joshi et al. 2007)

raising the outflow rates for a given $[Ca^{2+}]_C$ concentration. Two possible scenarios might then result. First, for long voltage pulses, the electrically driven outflow would be relatively long lived, and a high calcium concentration (with additional capacitive entry Putney 2005; Barritt 1999) from the extra-cellular medium) could be attained. A second outcome however, for short pulse durations, is that the electrically induced flow would start from an initial higher level (as compared to the equilibrium level), but merge with normal J_{IP3R} curve at the end of the voltage pulse. This could lead to two intersections with the sigmoidal J_{SERCA} curve—points "B" and "C".

For completeness, it may be pointed out that the above scenario changes slightly in the *absence of calcium within the extra-cellular medium.* Such conditions have been created experimentally through the use of chelating agents such as BAPTA, and the resulting calcium dynamics were reported (Beebe et al. 2003, 2004). If capacitive entry through store-operated calcium channels in the plasma membrane (Putney 2005; Barritt 1999)] was inhibited, then the high intra-cellular Ca^{2+} levels required for the bi-stable state "C" would be very difficult to attain.

Simulations for the time-dependent calcium concentrations within both the cytoplasm and endoplasmic reticulum, in response to a 60 ns, 15 kV/cm electric pulse were carried out. These electrical pulse parameters were chosen to match previous experimental reports (Beebe et al. 2004; White et al. 2004) on electric-field-induced calcium release. All calculations assumed an absence of extra-cellular calcium. The results for the averaged Ca^{2+} concentration in the cytosol are given in Fig. 6.2. The plot also includes experimental data points that have been reported in the literature (White et al. 2004). In Fig. 6.2, the Ca^{2+} concentration starts from an initial 0.1 μM value and increases in time to a maximum of about 0.23 μM. The data is

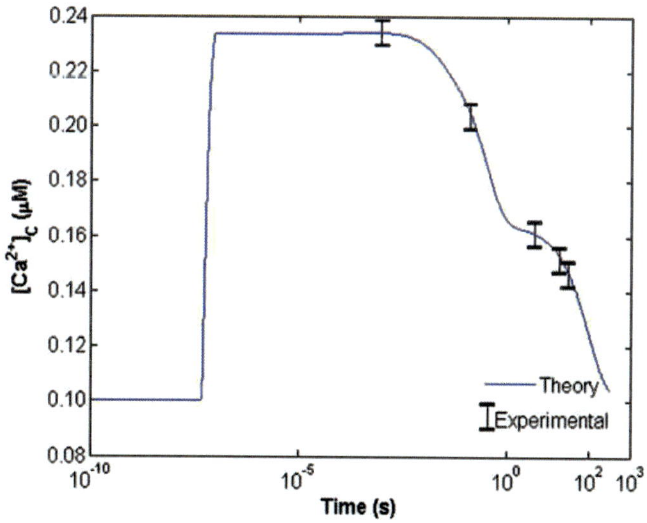

Fig. 6.2 Simulation results for average Ca-concentration in cytosol as a function of time. The experimental data points are from reports

for HL-60 cells. The sharp increase, seen during the initial phase, is due to electric field-induced poration within the ER that leads to a Ca^{2+} outflow. Drift rather than diffusion is the primary transport mechanism during this initial phase when the external voltage remains applied. Beyond this initial sharp Ca^{2+} release, IP3R-activated calcium transfer arising from the CICR mechanism begins to set in. A slow and more gradual recovery down to the initial concentrations is predicted, with final steady state being achieved after about 300 s. This ~4–5 min recovery time is in accord with reported measurements (Beebe et al. 2004; White et al. 2004) and involves a dynamical interplay between the CICR mechanism, action of the SERCA pumps and ER membrane leakage. The experimental data shown (White et al. 2004) shown in Fig. 6.2 matches the predictions well. The model predictions were within the experimental error deviations for all the points. Due to limitations in the temporal resolution of experimental techniques, values of the Ca^{2+} concentration at the earlier times could not be measured, and hence, such data points could not be shown. It is also interesting to note that our simulations demonstrate that very high densities, as needed to attain the bistable-state "C" of Fig. 6.1, cannot be achieved. Hence, a long-lived, Calcium-loaded state for cells is precluded in situations of extra-cellular free media. Even the oscillatory state is predicted to be difficult to attain at these relatively modest increases. The experimental data in Ca-free extra-cellular environments (Beebe et al. 2004; White et al. 2004) is in accord with this prediction, with no reports of oscillations.

The changes in the Ca^{2+} density with time within the ER are shown in Fig. 6.3. It exhibits an initial rapid decrease due to Ca^{2+} outflow following field-induced poration. The CICR mechanism continues to cause some Ca^{2+} outflow, and the density

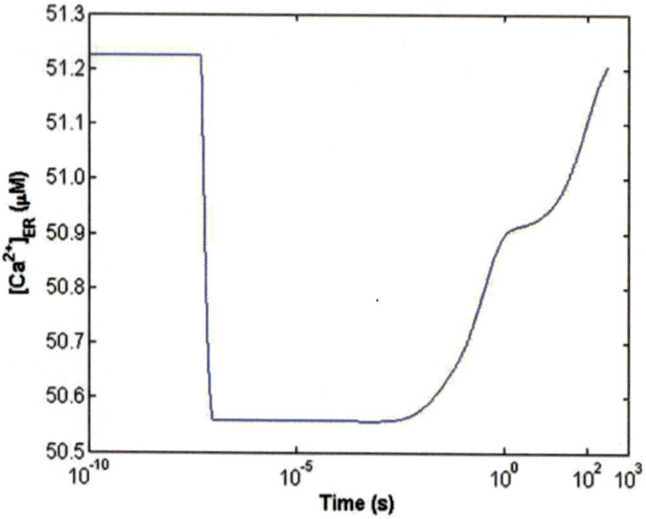

Fig. 6.3 Simulation results for average Ca-concentration in the ER as a function of time (After Joshi et al. 2007)

within the ER remains relatively low for a long duration. The overall plot of Fig. 6.3 roughly exhibits an inverted U-shape. Following initial depletion in the ER, the gradual restoration of the Ca^{2+} concentrations through SERCA pumping is predicted, after about 5 min.

References

Atri A, Amundson J, Clapham D, Sneyd J (1993) A single-pool model for intracellular calcium oscillations and waves in the Xenopus laevis oocyte. Biophys J 65:1727–1739

Barritt GJ (1999) Receptor-activated Ca^{2+} inflow in animal cells: a variety of pathways tailored to meet different intracellular Ca^{2+} signalling requirements. Biochem J 337:153–169

Beebe SJ, Fox PM, Rec LJ, Willis LK, Schoenbach KH (2003) Nanosecond, high intensity pulsed electric fields induce apoptosis in human cells. FASEB J 17:1493–1495

Beebe SJ, Blackmore PF, White J, Joshi RP, Schoenbach KH (2004) Nanosecond pulsed electric fields modulate cell function through intracellular signal transduction mechanisms. Physiol Meas 25:1077–1093

Bernardi P, Petronilli V (1996) The permeability transition pore as a mitochondrial calcium release channel: a critical appraisal. J Bioenerg Biomembr 28:131–138

Berridge M, Bootman MD, Lipp P (1998) Calcium—a life and death signal. Nature 395:645–648

Berridge MJ, Bootman MD, Roderick HL (2003) Calcium signalling: dynamics, homeostasis and remodeling. Nat Rev Mol Cell Biol 4:517–529

Bootman MD, Berridge MJ, Roderick HL (2002) Calcium signaling: more messengers, more channels, more complexity. Curr Biol 12:R563–R565

Bugrim AE, Zhabotinsky AM, Epstein IR (1997) Calcium waves in a model with a random spatially discrete distribution of Ca^{2+} release sites. Biophys J 73:2897–2906

Bugrim A, Fontanilla R, Eutenier BB, Keizer J, Nuccitelli R (2003) Sperm initiate a Ca^{2+} wave in frog eggs that is more similar to Ca^{2+} waves initiated by IP3 than by Ca^{2+}. Biophys J 84:1580–1590

Catterall WA (2000) Structure and regulation of voltage-gated Ca^{2+} channels. Annu Rev Cell Dev Biol 16:521–555

Cherepanov DA, Feniouk BA, Junge W, Mulkidjanian AY (2003) Low dielectric permittivity of water at the membrane interface: effect on the energy coupling mechanism in biological membranes. Biophys J 85:1307–1316

De Young GW, Keizer J (1992) A single-pool inositol 1,4,5-trisphosphate-receptor-based model for agonist-stimulated oscillations in Ca^{2+} concentration. Proc Natl Acad Sci U S A 8920:9895–9899

Deng J, Schoenbach KH, Buescher ES, Hair PS, Fox PM, Beebe SJ (2003) The effects of intense submicrosecond electrical pulses on cells. Biophysical J 84:2709–2714

Falcke M (2003) On the role of stochastic channel behavior in intracellular Ca^{2+} dynamics. Biophys J 84:42–56

Falcke MB, Lechleiter JD, Hudson JL (1999) Spiral breakup and defect dynamics in a model for intracellular Ca^{2+} dynamics. Physica D 129:236–252

Falcke M, Tsimring L, Levine H (2000) Stochastic spreading of intracellular Ca^{2+} release. Phys Rev E 62:2636–2643

Haak LL, Song L, Molinski TF, Pessah IN, Cheng H, Russell JT (2001) Sparks and puffs in oligodendrocyte progenitors: cross talk between ryanodine receptors and inositol trisphosphate receptors. J Neurosci 21:3860–3870

Hu Q, Viswanadham S, Joshi RP, Schoenbach KH, Beebe SJ, Blackmore PF (2005) Simulations of transient membrane behavior in cells subjected to a high-intensity ultrashort electric pulse. Phys Rev E.71:031914/1–9

Ichas F, Mazat JP (1998) From calcium signaling to cell death: two conformations for the mitochondrial permeability transition pore. Switching from low- to high-conductance state. Biochim Biophys Acta 1366:33–50

Jafri MS, Keizer J (1995) On the roles of Ca^{2+} diffusion, Ca^{2+} buffers, and the endoplasmic reticulum in IP3-induced Ca^{2+} waves. Biophys J 69:2139–2153

Joshi RP, Hu Q, Schoenbach KH (2004) Modeling studies of cell response to ultrashort, high-intensity electric fields—implications for intracellular manipulation. IEEE Trans Plasma Sci 32:1677–1686

Joshi RP, Nguyen A, Sridhara V, Hu Q, Nuccitelli R, Schoenbach KH (2007) Simulations of intracellular calcium release dynamics in response to a high-intensity, ultra-short electric pulse. Phys Rev E 75:041920/1–10

Keizer J, Smith GD (1998) Spark-to-wave transition: saltatory transmission of calcium waves in cardiac myocytes. Biophys Chem 72:87–100

Keizer J, Smith GD, Ponce-Dawson S, Pearson JE (1998) Saltatory propagation of Ca^{2+} waves by Ca^{2+} sparks. Biophys J 75:595–600

Li YX, Rinzel J (1994) Equations for InsP3 receptor-mediated [Ca2+]i oscillations derived from a detailed kinetic model: a Hodgkin-Huxley like formalism. J Theor Biol 166:461–473

Mak D, McBride S, Raghuram V, Yue Y, Joseph S, Fosken J (2000) Single-channel properties in endoplasmic reticulum membrane of recombinant type 3 inositol trisphosphate receptor. J Gen Physiol 115:241–256

Marchant JS, Parker I (2001) Role of elementary Ca(2+) puffs in generating repetitive Ca(2+) oscillations. EMBO J 20:65–76

Parsegian VA (1969) Energy of an ion crossing a low dielectric membrane: solutions to four relevant electrostatic problems. Nature 221:844–846

Partridge LD, Muller TH, Swandulla D (1994) Calcium-activated non-selective channels in the nervous system. Brain Res Rev 19:319–325

Putney JW (2005) Capacitative calcium entry: sensing the calcium stores. J Cell Biol 169:381–382

Roberts-Crowley ML, Mitra-Ganguli T, Liu L, Rittenhouse AR (2009) Regulation of voltage-gated Ca^{2+} channels by lipids. Cell Calcium 45:589–601

Semenov I, Xiao S, Pakomov AG (2013) Primary pathways of intracellular Ca^{2+} mobilization by nanosecond pulsed electric field. Biochim Biophys Acta 1828:981–989

Shuai JW, Jung P (2002) Optimal intracellular calcium signaling. Phys Rev Lett 88:68102/1–4

Shuai JW, Jung P (2003) Optimal ion channel clustering for intracellular calcium signaling. Proc Natl Acad Sci 100:506–510

Sneyd J, Falcke M (2005) Models of the inositol trisphosphate receptor. Progress Biophys Mol Biol 89:207–245

Sridhara V, Joshi RP, Schoenbach KH (2006) Microscopic calculations of local lipid membrane permittivities and diffusion coefficients for application to electroporation analyses. Biochem Biophys Res Comm 348:643–648

Sun Y, Vernier PT, Behrend M, Wang J, Thu MM, Gundersen MA, Marcu L (2006) Fluorescence microscopy imaging of electroperturbation in mammalian cells. J Biomed Opt 11:024010/1–10

Swillens S, Champeil P, Combettes L, Dupont G (1998) From calcium blips to calcium puffs: theoretical analysis of the requirements for interchannel communication. J Physiol (Lond) 509:67–80

Tarek M (2005) Membrane electroporation: a molecular dynamics simulation. Biophys J 88:4045–4053

Vasilkoski Z, Esser AT, Gowrishankar TR, Weaver JC (2006) Membrane electroporation: the absolute rate equation and nanosecond time scale pore creation. Phys Rev E 74:21904/1–12

Vernier PT, Sun Y, Marcu L, Salemi S, Craft CM, Gundersen MA (2003) Calcium bursts induced by nanosecond electric pulses. Biochem Biophys Res Commun 310:286–295

Wagner J, Li YX, Pearson J, Keizer J (1998) Simulation of the fertilization Ca^{2+} wave in Xenopus laevis eggs. Biophys J 75:2088–2097

Wagner J, Fall CP, Hong F, Sims CE, Allbritton NL, Fontanilla RA, Moraru II, Loew LM, Nuccitelli R (2004) A wave of IP3 production accompanies the fertilization Ca^{2+} wave in the egg of the frog, Xenopus laevis: theoretical and experimental support. Cell Calcium 35:433–447
White JA, Blackmore PF, Schoenbach KH, Beebe SJ (2004) Stimulation of capacitive calcium entry in HL-60 cells by nanosecond pulsed electric fields. J Biol Chem 279:22964–22972
Zimmermann U, Neil GA (1996) Electromanipulation of Cells. CRC Press, Boca Raton, FL
Zwanzig RW (1961) Lectures in theoretical physics. Interscience, New York

Chapter 7
Effects of usEPs on DNA, Nuclear, and Subnuclear Compartments

Stephen J. Beebe

Abstract As the largest intracellular structure in mammalian cells, the nucleus, its double phospholipid nuclear envelope, and its chromatin/DNA content were suspected targets for usEPs. Many different methods were used to determine DNA/ nuclear damage including analyses with the comet assay, DNA migration on agarose gels, mitotic indices, and chromatid structures, fluorescent in situ hybridization (FISH). In Jurkat cells exposed to (3.6×10^{-3} Vs/cm), the telomers were displaced from the nucleus, and nuclear membranes were sheared from the nucleus. SV40 fibroblasts did not show this apparent telomer and nuclear membrane damage, indicating cell-type differences., It was also shown as adherent cells were less susceptible to usEP-induced damage. Other studies showed that usEP induced significant physical damage to the nuclear membrane, cytoskeleton, and telomers, which form protein–protein or protein-DNA interactions with the nuclear envelope. Some usEP-induced nuclear/DNA damages were suspected due to effects similar to ionizing radiation caused by reactive oxygen species (ROS). Other studies using stably transfected cells with fluorescently labeled Histone-2b (H2B), which is tightly wound with DNA, and PCNA (proliferating cell nuclear antigen), which is loosely associated with DNA, indicated that H2B remained in the nucleus. In contrast, translocation of PCNA from the nucleus to the cytoplasm showed permeabilization of the nuclear membrane. So, usEPs had relatively severe and seemly rapid effects on nuclear structures. Yet, when phosphorylated Histone 2AX (γH2AX) was used as an early indicator of DNA damage in Jurkat cells, the damage appeared to be related to apoptosis's end stages since it was caspase-dependent. While all these methods are valid indicators of effects on DNA and/or the nucleus, the results between the comet assay and γH2AX under similar conditions with the same cell type are not readily reconcilable. One other study demonstrated that usEP also had effects on nuclear substructures called nuclear speckles, which are part of splicing factors and small nuclear ribonucleoproteins (snRNPs) that exhibit roles to provide splicing factors at transcription sites. So, there are apparent effects of usEPs on DNA, the nucleus, and subnuclear factors; the full extent of these effects requires additional experimentation.

S. J. Beebe et al., *Ultrashort Electric Pulse Effects in Biology and Medicine*,
Series in BioEngineering, https://doi.org/10.1007/978-981-10-5113-5_7

7.1 Introduction

Given that usEP affects intracellular structures and the nucleus is the biggest and most impressive structure in cells, thereby the first intracellular structure to be identified by microscopy, nuclei, are likely targets for usEPs in vitro and in vivo. In fact, what happened to the nucleus and DNA in response to usEPs was one of the earliest topics of interests for investigating pulse power effects on biological cells. Before discussing effects of usEPs on DNA, nuclear membrane, and nuclear sub-structures, some general background material will be presented to set the stage for how these structures respond to usEPs.

7.2 Structure of DNA and the Nucleus

Eukaryotic DNA is not a free-floating structure in the nucleus (Lodish et al. 2020). DNA is bound to proteins, especially histones to form chromatin. This structure includes intervals of 147 base-pairs of DNA wrapped around cores of histone proteins. This structure is the basis of 147 base pair DNA fragmentation cleaved during apoptosis. The N- and C-terminal of these histone extend from the nucleosome and can be modified reversibly. Modifications such as acetylation affects the condensation of the chromatin and thereby its accessibility to proteins that are required for gene transcription. The length of DNA in single cells is about 2 m and must be wound into a cell with a diameter of about 10–20 μm. Specialized proteins associate, fold, and organize nuclear DNA into structures of protein and DNA that appear as individual chromosomes during mitosis. Important functional units of DNA are the genes that code for all structures and function of living existence. A gene is defined as an entire nucleotide sequence that is needed for the synthesis of a functional gene product of RNA and a translated polypeptide. However, this so-called coding region cannot be realized for function alone. The coding region lies with other nucleotide sequences required for the synthesis of the gene product. These include transcription control regions called enhancers that lie around 50 kb away from the coding regions. Other DNA sequences needed for production of the gene product include sequences of the 5′ cap site and sequences that 3′ cleavage and polyadenylation called poly(A) sites, slicing sites needed for splicing of the primary RNA transcript. The coding region of the gene includes exons that are interspersed with other nucleotides call introns that are removed during transcription as the exons are spliced together. All DNA coding regions and many other nucleotide sequences lie on chromosomes that include a centromere in the middle, which binds to spindle microtubules during mitosis, meiosis, and telomeres, which are DNS-protein structures located at the ends of the chromosomes.

The non-coding DNA sequences within the genome make up the greatest numbers of sequences such that exons that are coded for gene products includes only about 1.5% of the total DNA in the genome. Only about 5% of human DNA codes for proteins and functional RNA and regulatory sequences that control their expression. A chromosome consists of a single sequence of DNA as long as ~280 Mb organized at various levels of condensation by the histone and non-histone proteins with which it is complexed (Lodish et al. 2004).

This genome is protected by a nuclear membrane that is a double layered lipid nuclear envelop. This membrane separates the nucleoplasm from the cytoplasm. While the cell membrane is a phospholipid bilayer, the nuclear membrane is a double bilayer—that is two phospholipid bilayers or four rows of lipids surrounding the nucleoplasm with its DNA. The nucleoplasm is essentially a suspension media within the nucleus including chromosomes, nucleolus, nucleotides involved in DNA replication and enzymes involved in nuclear function, among other functional proteins involved in transcription and transport between the nucleus and cytoplasm. The nuclear envelop is tethered to the cytoskeleton (Goldberg 2017). The outer nuclear membrane is uninterrupted from rough endoplasmic reticulum and both are covered with ribosomes where proteins are synthesized. The integral proteins in the inner nuclear membrane interact with nuclear lamin proteins to form the nuclear lamina that provides additional support to the inner nuclear membrane. In addition to protecting the DNA, the nuclear membrane determines, in part, what enters and exits the nucleus. It contains nuclear pores containing complexes of proteins that span the double layered membrane (see Fig. 5.1). The nuclear pore complex (NPC) is a mega structure of more than 120 MDa in humans containing more than 450 separate proteins of more than 30 different types. The proteins that form the NPC are called nucleoporins. While the number of NPC varies with cell type and developmental stage, there are an average of 1000 NPCs in a typical mammalian cell (Maul and Deaven 1977). The diameter of the NPC is about 5 nm, which is about the size of a 40 kDa protein. Macromolecules larger than this requires transport receptors (Kabachinski and Schwartz 2015). The two membrane bilayers of the nuclear envelop are joined at the points of contact to the NPC. Electron microscopy of this complex shows structures describes as "rings", "spoke" complexes and "plug-spoke" complexes as structures features arranged with octagonal symmetry. The NPC is the largest protein complex in the cell (Reichelt et al. 1990). Structurally the NPC is composed of a cytoplasmic ring, an inner pore ring and a nuclear ring as well as cytoplasmic filaments and a nuclear basket. The LINK complex (Linker of Nucleoskeleton and Cytoskeleton) is the best known structure anchoring the nucleus to cell architecture. This structure bridges the nuclear envelop and the nucleoskeleton to the cytoskeleton. In effect, the LINC complex is formed by SUN and KASH-domain proteins inserted in the outer nuclear membrane interacting with the nuclear laminin, which covers the inner nuclear membrane (Meinke and Schirmer 2015; Demircioglu et al. 2016; Goldberg 2017; Beck and Hurt 2017). These integral proteins are generally called NETs (nuclear envelop transmembrane proteins). This complex coordinates nuclear movement and anchorage and organizes chromatin. Forces that impact the cytoskeleton are transferred through the nuclear envelope to the nuclear lamina through this LINC complex. In this way communications between the cytoskeleton and the nucleus affected the entire cytoskeletal structure (Starr and Fridolfsson 2010; Lee and Burke 2018).

As an understanding of the NPC became clearer, the older notion that the nuclear envelop served to merely protect DNA became much too abridged. These macromolecular structures that cover the nuclear envelop not only serve diverse functions, but are also dynamic. The NPC is involved in nucleocytoplasmic transport, chromatin organization, and regulation of gene expression. The NPC allows transport into and out of the nucleus as bidirectional nucleocytoplasmic transport. Small molecules (<40 kDa, ~5 nm diameter) diffuse according to their concentration gradients. Active transport across the NPC dependent on receptor-mediated import and export routes occur for nuclear import of proteins, export of RNA and ribonucleoproteins (RNPs) and bi-directional shuttling of molecules including RNA, RNPs and proteins involved in signaling, such as SMAD, p53 and spliceosomal small nuclear RNP particles (snRNPs) (Beck and Hurt 2017). These active transport mechanisms are conducted by nuclear transport receptors (NTRs) called importins and exportins, which bind to the nuclear localization signal of cargo, facilitating its movement through the NPC, and with the functions of the small GTPase protein RAN, releases the cargo into the nucleoplasm and then export the NTR complex to the cytoplasm (Beck and Hurt 2017).

Some nucleoporins are also DNA binding proteins suggesting that they interact with chromatin affecting both chromatin structure and gene transcription. During interphase, there is a 3-dimensional architecture of non-random genome organization at these sites, providing structural and dynamic stability provided by the nuclear lamina at the inner nuclear membrane. NPCs, lamins, and NETs of the inner nuclear membrane can all interact with such nuclear components such as DNA, chromatin proteins like histone, epigenetic marks on chromatin and transcription factors, which can promote transcription (generally NPC) or sequester transcription factors (generally NETs) (Mattout-Drubezki and Gruenbaum 2003).

Telomeres at the ends of each chromosome are also transiently tethered to the nuclear envelop playing roles in nuclear organization and regulation of the cell cycle, gene expression, and telomere maintenance (Crabbe et al. 2012). During interphase, the telomeres are randomly distributed throughout the nucleus. At the end of cell division and while the nuclear membrane is re-assembling, telomeres are localized and bound to specific proteins at the nuclear rim, which was identified by co-localization with lamin A/C as structural components of the nuclear lamina associated with the nuclear envelop. Telomeres form protein–protein or protein-DNA interactions with the nuclear envelop or proteins associated with it during nuclear assembly. This appear to be a mechanism whereby telomeres bind to the nuclear envelop as a means to reorganize chromatin domains after cell division.

7.3 Effects of usEPs on the Nucleus and DNA

A number of studies have shown that usEP affects the nuclear membrane and subnuclear structures including telomeres and associated structures, splicing factors and/or nuclear DNA. The earliest study investigated 10 different cell lines that were either adherent or non-adherent including several cell lines from patients with genetic diseases that include hypersensitivity to genotoxic agents, defective DNA repair, and check point controls, which results in increased incidence of cancer (Stacey et al. 2003). These cell lines were treated with usEP conditions with 10, 60, and 300 ns durations and different electric fields, some of which were matched for the same energy density. Analyses included comet assays, DNA extractions immediately following cell lysis and agarose electrophoresis and cytogenetic analysis. As an overview, results indicated that cells grown in suspension were much more susceptible to DNA damage than adherent cells. It is possible that adherent cells have a more structured cytoskeleton, which by being bridged to the nucleus by the LINC complex, provides greater stability to the usEP charging effects. This is addressed by these and other authors as indicated below. Loss of cell viability was suspected to be due to immediate DNA damage based on comet assays or single cell gel electrophoresis assay. Measurement of the comet tails or DNA migration determine how much DNA damage is present in a given cell (Tice et al 2000; Kumaravel and Jha 2006). Another assay included longer DNA migration patterns on agarose gels after exposure to usEPs. When suspension cells were treated with one and five pulses, it was determined that essentially all damages occurred due to the first pulse since 5 additional pulses produced no greater loss in viability. It was also determined that there were no sub-populations of Jurkat cells that were resistant to usEPs, since cell surviving the first pulse regimen exhibited similar loss in viability as cells in the first treatment. Adherent cells were much less sensitive to usEPs showing survival higher than 60% survival after 5 pulses, except 3T3-L1 with ~6% survival after 1 pulse and SV-FA (SV-40 transformed fibroblasts with Fanconi's anemia with defective DNA repair) with 40% survival after 5 pulses.

The mitotic indices of these cells were analyzed, which indicate the percentage of cells that have entered mitosis. When DNA damage occurs, cells normally activate cell cycle checkpoints that prevent cells from dividing until DNA repair has taken place. After DNA damage the mitotic index is reduced until DNA repair is complete. AT-LCL cells, which exhibit mutations in the ATM gene that regulates cell division after DNA damage, and BS-LCL cells, which exhibit deficient DNA repair capability, did not recover for as long as 7 days after usEPs, with mitotic indexes remaining below the mitotic index of untreated cells. The mitotic index in Jurkat and HL-60 cells took 7 days (168 h) to recover while adherent cells that expressed higher survival levels recovered mitotic indices within 72–96 h after treatment. Compared to untreated cells, AT-LCL cells exposed to usEPs were shown to have increases in chromatid aberrations including chromatid gaps, chromatid breaks and fragments, which were consistent with these aberrations in cells treated with ionizing radiation.

In a subsequent paper, further support for these findings was reported suggesting usEP-induced physical damage to nuclear membranes, cytoskeleton, and chromosomal telomeres (Stacey et al. 2011). Again, Jurkat cells grown in suspension exhibited greater damage than HeLa and SV-40 normal fibroblast grown as adherent cells. When cells were exposed to one 60 ns pulse at 15 kV/cm (9.0×10^{-4} Vs/cm) Jurkat cells became rounded with intensely stained actin sites and did not survive. In contrast HeLa cells exposed to the same conditions, cells became more rounded and actin filaments became spotted as the cytoskeleton appeared to breakdown and cells no longer adhered to surfaces but recovered and remained mostly viable. When HeLa cells and SV-40 fibroblasts were treated with cytochalasin B to prevent cytoskeletal polymerization, they were more susceptible to usEPs indicating that a polymerized cytoskeleton provided support that enhance survival in response to usEPs.

This study also demonstrated disruption of the actin cytoskeleton and nuclear membrane and damage to telomeres that appeared to decrease Jurkat cell survival exposed to single 60 ns 60 kV/cm pulses (3.6×10^{-3} Vs/cm). Telomeres also appear to sustain damage to these conditions. The authors used fluorescent in situ hybridization (FISH), a molecular cytogenetic method that uses fluorescent probes that bind to specific complementary nucleic acid sequencies. Here the authors used a pan-telomere specific probe. In interphase Jurat cell nuclei, the telomers are adhered to the nuclear membrane. The results indicated to the authors that these telomer-nuclear membrane binding sites were potential targets for usEP nuclear damage. In untreated cells, the FISH-identified telomers were located within the DAPI stained nucleus. However, after a single 60 ns pulse at 60 kV/cm (3.6×10^{-3} Vs/cm), the telomers were displaced from the nucleus. The nuclear membrane appeared to be sheared from the nucleus. SV40 fibroblasts did not show this apparent telomer and nuclear membrane damage.

To quantify these results, the authors counted telomers in exposed and unexposed Jurkat cells and found that usEP-treated Jurkat cells exhibited only 11.5% of these structures compared to untreated cells. However, with software enhancement, these differences were not present. The author concluded that the probe was no longer binding efficiently in the usEP-treated cells. This suggested that usEPs induced conformational changes in the telomere-nuclear membrane structure interpreted as severe damage to these structures. The authors envisioned usEPs inducing electrochemical effects that reduced intracellular stability making the nuclear membrane more fragile, pulling it away from the nucleus. It was also possible that usEP caused depolarization of nuclear lamina increasing the weakness of the nuclear membrane. Such conformational changes could be responsible for the apparent lower affinity or efficiency of the telomer probe binding to telomer sequences associated with the nuclear membrane. The nearly absent effects in adherent cells are likely due to a more stable cytoskeletal structure such that the usEP effects are dispersed through the cell cytoskeletal structure rather than affecting the less stable nuclear membrane and associated telomers in cells growing in suspension.

Romeo et al. (2011) showed with Jurkat cells under "milder", non-lethal usEP conditions (single 60 ns, 10–25 kV/cm; 1.0×10^{-4} to 1.5×10^{-3} Vs/cm) that there was a transient electric field-dependent decrease in the comet assay DNA migration. The authors suggested that their results were due to pulse-induced transient conformational change in the living cell nucleoprotein. Others observed transiently disrupted cytoskeleton and transient increase in nuclear size in adherent HCT116 cells exposed to three 60 or 300 ns, 60 kV/cm pulses (3.6×10^{-3} and 1.8×10^{-2} Vs/cm, respectively) (Hall et al. 2005).

Data from both studies suggest that usEPs caused immediate and apparent direct DNA damage, especially to cells grown in suspension, presumably because they have a less well-developed cytoskeleton. The cytoskeleton, nuclear membrane, and telomeres were also damaged. In the first study (Stacey et al. 2003), the nature of the DNA damage appeared to be similar to that caused by ionizing radiation (IR). IR-mediated DNA damage is known to be caused by oxidative stress that results in oxidation of DNA as well as to oxidation/peroxidation of lipids and proteins. IR exposure to cellular water rapidly generates the reactive oxygen species (ROS), hydroxyl radical ($^{\bullet}OH$), and ionized water (H_2O^+). Within picoseconds (10^{-12} s), secondary ROS products such as superoxide ($O_2^{\bullet}$) and hydrogen peroxide (H_2O_2) materialized as IR as damaging agents. While this may be rapid it is not a direct effect or IR, but rather indirect effects (Thompson 2012; Reisz et al. 2014). In any event, there does not appear to be a sub-population of Jurkat cells that exhibit resistances to usEPs among the cells that are sensitive to usEPs. This indicates that usEPs have effects across an entire population regardless of cellular heterogeneity. In the second study (Stacey et al. 2011), using even lower electric field conditions, the damage to cytoskeleton, nuclear membrane, and telomeres suggests physical forces are involved in these effects and that they should be considered in contrast to or in addition to plasma membrane nanoporation as a cause of cell death.

Permeabilization of the nuclear membrane was also indicated in stably transfected CHO cells with fluorescently labeled Histone-2b (H2B) and PCNA (proliferating cell nuclear antigen (Thompson et al. 2016). H2B is tightly wound with DNA in structures called nucleosomes and not mobile within the nucleus. In contrast PCNA is loosely associated with DNA and is transiently associated during replication to secure polymerases. Both 10 ns and 600 ns usEPs, especially at high pulse numbers, permeabilize the nuclear membrane as indicated by translocation of PCNA from the nucleus to the cytoplasm, while H2B remained in the nucleus. These pulses also appear to cause disruption of the nuclear lamina, which are intermediate filaments that form a meshwork underlying the inner nuclear membrane and is associated with most the proteins in the inner nuclear membrane. They provide mechanical strength to the nuclear membrane (Goldberg et al. 2008; Chi et al. 2009).

In another study using acridine orange (AO), 10 ns 65 kV/cm (6.5×10^{-4} Vs/cm) or 60 ns 26 kV/cm (1.6×10^{-3} Vs/cm) pulses affected DNA in nuclei of HL-60 cells (Chen et al. 2004). AO binds to DNA in live cells. After usEP treatment, the intracellular AO fluorescence decreased. This was suspected to be due to one or more responses. It was possible, that AO alone or bound to nuclear DNA leaves the nucleus, the AO bound to DNA decreases, and/or the AO fluorescence is reduced by the usEP. This could mean that usEPs alter the characteristic of the nuclear membrane and plasma membrane or the pulses change the DNA molecules so that AO binding to DNA is decreased.

In the study shown in Fig. 7.1, effects of usEPs on DNA damage defined by repair mechanisms were shown to be related to caspase activation and apoptosis. Double-stranded DNA damage was defined by Histone 2AX (H2AX) phosphorylation on serine 139 (γH2AX) (Rogakou et al. 1998), one of the earliest and most sensitive indicators of DNA damage. In this study Jurkat cells were treated in the absence (middle panel) or presence (right panel) of z-VAD-fmk, a caspase inhibitor with ten 60 ns pulses at 60 kV/cm (1.1×10^{-2} Vs/cm), which was considerably more intense

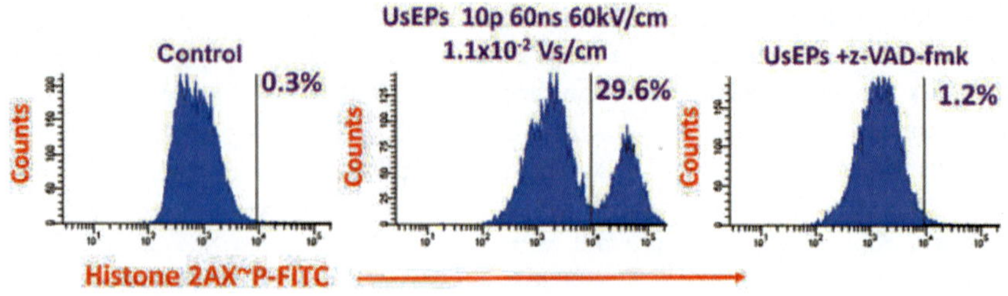

Fig. 7.1 UsEP induced DNA damage is caspase-dependent—Jurkat A3 clones were sham treated or treated with ten 60 ns pulses at 60 kV/cm (1.1×10^{-2} Vs/cm) in the presence and absence of the caspase inhibitor z-VAD-fmk, a caspase inhibitor. Six hours later, cells were permeabilized, labeled with an antibody to phosphorylated Histone 2AX (γH2AX) labeled with FITC and analyzed by flow cytometry

than the studies showing DNA or nuclear damage above. The binding of γH2AX appeared late (~4–6 h) and was attenuated by caspase inhibition, suggesting this DNA damage was related to late- stage apoptosis. When cells were analyzed 6 h after treatment for γH2AX, the presence of this DNA damage indicator increased from less than 1% in sham treated cells to nearly 30% in treated cells. However, in the presence of the caspase inhibitor, levels of γH2AX were near control levels. This demonstrated that the presence of DNA damage in usEP-treated Jurkat cells was caspase-dependent, indicating that the DNA damage under these conditions in Jurkat cells was related to apoptosis. This conclusion is also supported by the finding that DNA damage did not occur until late; it was not present at 2 h and very low at 4 h (data not shown). The timing of the presence of caspase-dependent DNA damage gives us some perspectives of the kinetics of caspase-mediated cell death in response to usEPs.

From a power perspective, it might be possible for usEPs to transiently disrupt single hydrogen bonds, but highly unlikely to disrupt millions of hydrogen bonds that stabilize DNA molecular structure. Nevertheless, a major factor proposed to determine usEP effects on cell membranes as well as intracellular structures and functions is immense charging effects on molecules, which could alter molecular structures, changing their conformation as suggested above. Based on the results of DNA damage reported by Stacey et al. (2003, 2011), Thompson et al. (2012, 2016) and Chen et al. (2004), and the results of Ren and Beebe above, provide contradictory evidence of usEPs on DNA damage. All of the damage assessment strategies are valid. The comet assay has been used in multiple laboratories as a method of risk assessment for characterization of biohazards (Møller 2006) and as measures of DNA damage and repair (Azqueta et al. 2019). The γH2AX biomarker has been suggested for use as a genotoxicity assay and as use for roles in genotoxicity risk assessment (Kopp et al. 2019). Thus, while it is clear that usEPs can damage DNA, it is difficult to determine the damage mechanism. Most experts suggest that the energy or power in usEPs is not sufficient to directly alter the double-stranded structure of DNA, so DNA damage is most likely due to indirect effects of usEPs on DNA caused by ROS or caspase-dependent mechanisms. It would be interesting if all of these approaches were applied in the same study so that conditions and cell types would be standardized.

The nucleus and DNA have been the focus of a number of studies. As reviewed above, many studies showed that usEPs induce DNA damaged determined by TUNEL, comet assays, fragmentation by flow cytometry, and γHistone-2AX. The earliest and perhaps the most thoroughly study analyzing effects of nsPEF on DNA demonstrated that nsPEF-induced genotoxicity was cell type-dependent, with non-adherent cells being more sensitive than adherent cells (Stacey et al. 2003). After a single 60 ns, 60 kV/cm pulse ($3.6x10^{-3}$), unlike all but one of seven adherent cell types, non-adherent cells exhibited 10% survival, induction of DNA damage and a decrease in the number of cells reaching mitosis. Comet assays immediately after usEP treatment of Jurkat cells increased comet tail lengths 1.3- and 2.6-fold with one and 10 pulses, respectively, with the same conditions indicated above, indicating that DNA damage was a direct effect of electric fields. HL-60 cells also showed similar effects, albeit less striking. When DNA was isolated immediately after pulsing and analyzed by electrophoresis on agarose gels, a smear of DNA occurred in cells exposed to five pulses with conditions above while DNA from unexposed cells was present as a tight band, again indicating direct usEP-induced DNA damage. Examining lymphoblastoid cell line (LCL) from a homozygous patient with ataxia telangiectasia (AT), who are unable to repair DNA damage in response to radiation, chromatid aberrations were identified in 78% of cells exposed to a single 600 ns, 60 kV/cm pulse ($3.6x10^{-2}$) Vs/cm), while only 22% were observed in control cells. In all, 41 types of damage were observed in exposed cells including more chromatin gaps, breaks and fragments than in control cells. However, the finding that DNA damage by γH2AX in Jurkat cells was late and caspase-dependent (Fig. 7.1) is inconsistent with results from the comet assay, which were analyzed immediately after pulsing. There are no simple explanations for these differences since both detect double strand breaks. The comet assay is less sensitive, more variable, and subject to other factors such as materials that generate variations and temperature, which is not a problem with a single pulse under the applied conditions. While these issues might raise questions, the comet assay is considered a sustainable method (Lynch et al. 2011). It would be suspected that such comet-detected and telomer-detected damages ultimately lead to signaling from the nucleus to the mitochondria ultimately leading to cytochrome c release and caspase-mediated apoptosis. However, these signals are reported to come from the nucleus via BH-3 only proteins Puma and Noxa in response to genotoxic stress such as a radiation, telomer erosion, and hypoxia, cytokine depletion, tumorigenic stress signals, among others (Villunger et al. 2003); however, Jurkat cells treated with the same conditions in Fig. 7.1, which were more intense than studies by Stacey et al. (2003, 2011) and Romeo (2011), did not exhibit either of these nuclear-mitochondrial signal molecules on immunoblots (Beebe et al. 2013), yet these same conditions induced cytochrome *c* release and caspase activation (Beebe et al. 2003, 2013; Ren et al. 2012). Taken together, additional studies are required to determine the mechanism of usEP-induced DNA damage.

7.4 Effects of usEPs on Nuclear Substructures

Intracellular usEP effects affect subnuclear structures. Figure 7.2 shows usEP effects on nuclear speckles (NSs) in Jurkat cells in both M-phase and G1-phase of the cell cycle (Chen et al. 2007). While NSs during mitosis are called mitotic interchromatin granules (MIGs) in the M-phase from metaphase to telophase, we refer to them in all cell cycle phases as NSs. Nuclear speckles, which are also called interchromatin granular complexes (IGCs) and SC35 domains and represent one of the subnuclear compartments or nuclear bodies along with nucleoli, paraspeckles, Cajol bodies and Gems (gemini of Cajol bodies), among other structures. An increasing number of other nuclear domains have been described including nuclear stress bodies, histone locus bodies, polycomb bodies, DNA damage foci, cleavage bodies, matrix-associated deacetylase bodies and clastosomes (Dundr 2012; Sleeman and Trinkle-Mulcahy 2014).

The earliest understanding of NSs, which included splicing factors and small nuclear ribonucleoproteins (snRNPs, U1, U2, U4,5, 6), was a role in splicing, primarily to provide splicing factors at transcription sites (Spector et al. 1983, 1991). Generally, the conventional view held that NSs function in assembly, modification, temporary storage, and recycling of splicing factors. It is now known that NSs are focal points or hubs that synchronize all nuclear gene expression regulation steps

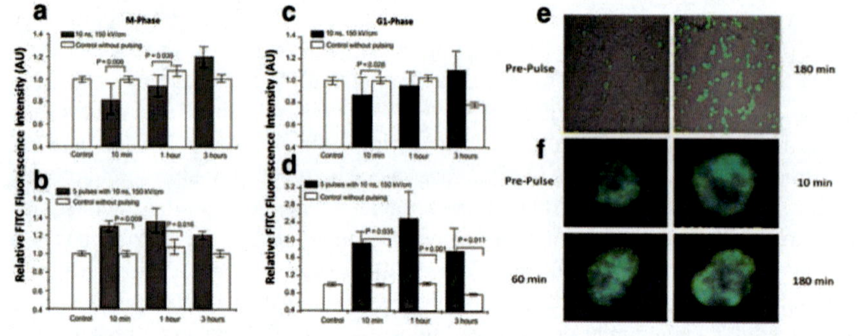

Fig. 7.2 Flow cytometric quantitative analysis in nuclear speckle changes in M-phase and G1-phase Jurkat cells at various times after usEP treatment with pulse(s) of 10 ns and 150 kV/cm. Human Jurkat cells were synchronized to the M-phase (95%) with nocodazole (**a** and **b**) or to the G1-phase (70%) with mimosine (**c** and **d**) and treated with 1 pulse ($1.5x10^{-3}$ Vs/cm)(**a** and **c**) or 5 pulses ($3.4x10^{-3}$ Vs/cm) at 1 Hz (**b** and **d**) usEPs with a duration of 10 ns and electric field intensity of 150 kV/cm. At indicated times (10, 60, and 180 min), nuclear speckles were detected by flow cytometry (relative FITC fluorescence intensity, Y axis) after intracellular immuno-labeling with Y12 antibody, a specific nuclear speckle marker. P value were determined by Student's test, two-tail of the log converted values. Error bars indicated SD. **e** Representative confocal microscopic images of Y12 antibody binding to speckles in M-phase Jurkat cells unexposed (left) and 3 h (right) after five consecutive pulses with 10 ns duration, 150 kV/cm electric field. **f** Typical confocal microscopic three-dimensional images of Y12 antibody binding to speckles in a Jurkat cell in M phase before, at 10 min, 1 h and 3 h after five consecutive 10 ns, 150 kV/cm pulses (400 × images) (Taken from Chen et al. 2007)

(Galganski et al. 2017). Importantly, NSs are dynamic structures with changing sizes, shapes and numbers depending on levels of transcription, metabolic activity and environmental signaling factors. Their constituents can constantly exchange within the nucleoplasm, other nuclear locations (Lamond and Spector 2003; Spector and Lamond 2011) and within the cytoplasm (Cáceres et al. 1998). NSs enhance the efficiency and integration of gene expression steps from transcription to mRNA export (Hall et al. 2006). Although transcriptions does not occur within most NSs (Cmarko et al. 1999) and DNA does not appear to be within NSs, but close to it (Hall et al. 2006), all of these steps are linked with RNA polymerase II transcription, which occurs in close association to and with NSs. Several reports have shown splicing activity occurs within NSs (Dias et al. 2010; Girard et al. 2012).

The NS-IGC proteome includes 146 proteins with 81% associated with IGC RNA metabolism and 54% as splicing factors. These NS-IGCs also include RNA associated proteins, cleavage and polyadenylation factors, RNA polymerase II subunits, transcription factors, mRNA export proteins, apoptosis proteins among others (Saitoh et al. 2004). Nuclear speckle proteins that are involved in chromosome localization, chromatin modification, transcription, splicing, 3 end processing, mRNA modification, mRNA coating with proteins and messenger ribonucleoprotein (mRNP) export are assembled in NSs (Biggiogera and Fakan 1998; Galganski et al. 2017).

Structurally, these proteins form NSs with SC35 domains of about 20–25 nm in diameter and assemble into 20–50 irregularly shaped "speckles" of 1–3 μM that are ultrastructures referred to as IGCs. These NS self-assemble. Although an underlying scaffold has not been clearly defined, but there are several structural proteins including lamin A and snRNP-associated actin have been identified in NSs (Lamond and Spector 2003). Actin and myosin Ic may be needed for transcriptional initiation and elongation (Galganski et al. 2017).

Hall and colleagues (2006) demonstrated several useful findings related to the structure and function of NSs. Nuclear speckles accumulate snRNPs and splicing factors, many of which contain RS motifs (Arg-Ser sequences) as part of their structures. One of these RS splicing factors is SC35, which is a common antigen for identifying NSs making them synonymous with SC35 domains. These SC35 domains are always enriched in poly(A) RNA. These poly (A) RNAs are present in transcripts of protein coding genes and SC35 domains can contain multiple specific pre-mRNAs from different genes that are often highly expressed in metabolically active cells. These pre-mRNAs accumulate within SC35 domains with their corresponding gene just on the periphery of the domain. Splicing occurs outside of the SC35 domain, yet spiced mRNA appears to pass through the domain. These splicing factor accumulations of NSs indicate locus-specific organizations of active genes with large domains of metabolic factors. There appears to be important links between SC35 domain accumulation for quality control and mRNA export from these domains. These SC35 domains (NSs, IGCs) function as nuclear foci that are spatially linked to specific pre-mRNAs transcription, splicing and export that rapidly recycle of RNA metabolic complexes (Hall et al. 2006).

An important characteristic of these NSs (SC35 domains, IGCs) is their dynamic nature. The shape and size of NS can be manipulated, and subsets of SR proteins have been shown to shuttle into and out of NS or remain within them depending, at least in part, to their phosphorylation state. These dynamic protein movements have been analyzed by inhibiting of RNA polymerase II with actinomycin D or DRB plus cycloheximide; by inhibiting phosphoprotein phosphatases PP2A with okadaic acid or cantharidin or inhibiting PP1 with tautomycin; or by overexpressing the SR protein kinase cdc-like-kinase (Clk)/STY (Cáceres et al. 1998; Sacco-Bubulya and Spector 2002; Hall et al. 2006). For example, inhibition of PP2A results in hyper-phosphorylated SC35 causing it to aggregate along with poly(A) RNA aggregation. In contrast, inhibition of PP1 or overexpression of Clk/Sty resulted in SC35 and poly(A) RNA disassembly. Inhibiting RNA polymerase II resulted in SR proteins SF2/ASF and SRp20 to shuttle into the cytoplasm while SC35 and SRp40 remained in the nucleus (Cáceres et al. 1998).

In Jurak cells, usEPs with 10 ns durations and electric fields of 150 kV/cm (1.5 × 10^{-4} Vs/cm), a relative low charging delivery, affected nuclear speckles (NSs), also known as small nuclear ribonucleoprotein particles (snRNPs) or interchromatin granule clusters (IGCs) in both the M-Phase and G1 phase of the cell cycle (Chen et al. 2007). Figure 7.2 shows NS in M-phase (Fig. 7.2a) and G1-phase (2C) that usEPs with a single 60 ns, 150 kV/cm pulse with transiently decreases NS formation 10 min after pulsing followed by time-dependent increases in NS formation 1 and 3 h post pulse. In contrast, 5 pulses at 1 Hz in M-phase (2B) and G1 phase (2D) with relative stability enhance NS formation up 10 3 h after pulses, especially in the G1-phase10 min and 1 h and 3 h. Increased NS formation was greater in the G1-phase than the M-phase. Figure 7.2e show M-phase microscopic images of increased cell numbers expressing NSs and Fig. 7.2f shows M-phase confocal images of time-dependent increases in NS formation after usEP treatment with 5 pulses. These studies indicate that NPS can also affect RNA–protein nuclear substructure complexes and/or mRNA processing. How might NPs have affect NSs?

It is known that the size (fluorescence intensity) (as in Fig. 7.2a–d) and shape of NSs reflect the steady state dynamics of protein constituents that arrive and leave the NS domains. For example, NS tend to round up (as in Fig. 7.2f) when mRNA Pol II transcription is inhibited or when pre-mRNA splicing is inhibited with siRNA (Lamond and Spector 2003; Spector and Lamond 2011). Decreased gene expression is associated with decrease in chromatin content and NS dispersion (as in Fig. Fig. 7.2a, C, 1 pulse). However, it is unclear if the usEP effects on NSs are direct effects on the complex structures themselves or effects on NS functions that affect RNA metabolism, SR protein phosphorylation status and/or inhibit mRNA export.

Given that there are a number of subnuclear structures, it is important to understand what the Y12 antibody and other antibodies specifically recognize. The Y12 antibody recognizes common core proteins of snRNPs involved in RNA processing (Lerner et al. 1981; Zieve and Sauterer 1990). More specifically, it recognizes snRNPs in SC35 domains that bind U-1, U-2, and U-4-6 small nuclear RNAs (snRNAs), so these are NS splicing factor proteins—RNA binding proteins. However, the Y12 antibody is now known to bind some coilin-rich Cajal bodies, so they may not be absolutely specific for NS, although they have been routinely used for decades (Hall et al. 2006).

In these studies antibodies could be used that identify splicing factor proteins that have been identified, although some transiently in dynamic NSs, including antibodies that identify RS (Arg-Ser motifs) domain containing proteins SC35, 9G8, S2/ASF, SRp40, SRp20 and SRm 300 (Fu and Maniatis 1990; Cavaloc et al. 1994; Cáceres et al. 1998; Hall et al. 2006). Among these, SC35, SRm 300 and SRp40 remain in NSs, while SF2/ASF, 9G8 and SRp20 shuttle in and out of NSs/nucleus (Cáceres et al. 1998; Hall et al. 2006). These antibodies specifically recognize their cognate antigens (proteins).

Koga and colleagues (2019) showed that usEPs induced the extracellular release of chromosomal DNA and histone citrullination in DMSO-differentiated HL-60 neutrophils. HL-60 cells are an acute myeloid leukemia cell line. The authors' intent was to show effects of usEPs on immune cells. They show that the cells were well-differentiated into neutrophil-like cells. The differentiated cells expressed neutrophil-like morphology with nuclear changes of lobulation with invaginated nuclear rim and expressed CD11b, which is typical of activated neutrophils and absent in undifferentiated HL-60 cells. These differentiated cells also exhibited down-regulation of the hTERT gene, a telomerase reverse transcriptase that is expressed in undifferentiated HL-60 cells.

The authors used SYTOX green, which stains DNA but does not enter intact cells. UsEPs with forty pulses with 80 ns durations at 20 kV/cm (1.6×10^{-3} Vs/cm) exposure to differentiated cells, there was a filamentous green staining in the extracellular space, indicating DNA extrusion from the cell, as well as intracellular nuclear staining. The results indicate that the usEPs permeabilized the plasma membrane allowing the cell impermeable CYTOX green to stain intracellular DNA and staining DNA that was released from the cell. As a positive control the authors used ionomycin, a Ca^{2+} ionophore, know to induce this response mimicked the effects of usEPs. The extrusion of chromosomal DNA in neutrophils is known as neutrophil extracellular traps (NETs) that function to trap pathogens as a mechanism of defense. This response leads to neutrophil cell death known as NETosis. These NETs, composed of decondensed chromosomal fiber aggregates, trap and kill pathogens. NETs also lead to phagocytosis of trapped pathogens by other neutrophils. Prior to extrusion, the NETs exhibit post-transcriptionally altered histones as arginine residues are hydrolyzed forming the un-conventional amino acid citrulline. This process is known as citrullination. This promotes the chromatin decondensation by reducing the histone positive charges (Wang et al. 2009). Kogo et al., also demonstrated that the extruded DNA was citrullinated in time-, electric field- and pulse number-dependent manners. The extrusion of DNA was enhanced in the presence of Ca^{2+} and citrullination as Ca^{2+}-dependent. Both responses required that the HL-60 cells be differentiated. These studies indicate that usEPs induce a non-infection or sterile activation of neutrophil-like cells. While the mechanism(s) responsible for usEP-induced NETosis, it is known to require ROS, which are increased by usEPs in a Ca^{2+}-enhanced manner (Nuccitelli et al. 2013; see Fig. 8.8). Pakhomov (2012) also demonstrated that usEPs increase ROS.

References

Azqueta A, Langie SAS, Boutet-Robinet E, Duthie S, Ladeira C, Møller P, Collins AR, Godschalk RWL (2019) Working Group 5 of the hCOMET project (CA15132). DNA repair as a human biomonitoring tool: comet assay approaches. Mutat Res 781:71–87

Beck M, Hurt E (2017) The nuclear pore complex: understanding its function through structural insight. Nat Rev Mol Cell Biol 18:73–89

Beebe SJ, Fox PM, Rec LJ, Willis EL, Schoenbach KH (2003) Nanosecond, high-intensity pulsed electric fields induce apoptosis in human cells. FASEB J 17:1493–1495

Beebe SJ, Sain NM, Ren W (2013) Induction of cell death mechanisms and apoptosis by nanosecond pulsed electric field (nsPEFs). Cells 2:136–162

Biggiogera M, Fakan S (1998) Fine structural specific visualization of RNA on ultrathin sections. J Histochem Cytochem 46:389–395

Cáceres JF, Screaton GR, Krainer AR (1998) A specific subset of SR proteins shuttles continuously between the nucleus and the cytoplasm. Genes Dev 12:55–66

Cavaloc Y, Popielarz M, Fuchs JP, Gattoni R, Stévenin J (1994) Characterization and cloning of the human splicing factor 9G8: a novel 35 kDa factor of the serine/arginine protein family. EMBO J 13:2639–2649

Chen N, Schoenbach KH, Kolb JF, James Swanson R, Garner AL, Yang J, Joshi RP, Beebe SJ (2004) Leukemic cell intracellular responses to nanosecond electric fields. Biochem Biophys Res Commun 317(2):421–427

Chen N, Garner AL, Chen G, Jing Y, Deng Y, Swanson RJ, Kolb JF, Beebe SJ, Joshi RP, Schoenbach KH (2007) Nanosecond electric pulses penetrate the nucleus and enhance speckle formation. Biochem Biophys Res Commun 364:220–225

Chi YH, Chen ZJ, Jeang KT (2009) The nuclear envelopathies and human diseases. J Biomed Sci 16:96

Cmarko D, Verschure PJ, Martin TE, Dahmus ME, Krause S, Fu XD, van Driel R, Fakan S (1999) Ultrastructural analysis of transcription and splicing in the cell nucleus after bromo-UTP microinjection. Mol Biol Cell 10:211–223

Crabbe L, Cesare AJ, Kasuboski JM, Fitzpatrick JA, Karlseder J (2012) Human telomeres are tethered to the nuclear envelope during postmitotic nuclear assembly. Cell Rep 2:1521–1529

Demircioglu FE, Cruz VE, Schwartz TU (2016) Chapter Four—Purification and structural analysis of SUN and KASH domain proteins. Meth Enzymol 569:63–78

Dias SM, Cerione RA, Wilson KF (2010) Unloading RNAs in the cytoplasm: an "importin" task. Nucleus 1:139–143

Dundr M (2012) Nuclear bodies: multifunctional companions of the genome. Curr Opin Cell Biol 24:415–422

Fu XD, Maniatis T (1990) Factor required for mammalian spliceosome assembly is localized to discrete regions in the nucleus. Nature 343:437–441

Galganski L, Urbanek MO, Krzyzosiak WJ (2017) Nuclear speckles: molecular organization, biological function and role in disease. Nucleic Acids Res 45:10350–10368

Girard C, Will CL, Peng J, Makarov EM, Kastner B, Lemm I, Urlaub H, Hartmuth K, Lührmann R (2012) Post-transcriptional spliceosomes are retained in nuclear speckles until splicing completion. Nat Commun 3:994

Goldberg MW, Fiserova J, Huttenlauch I, Stick R (2008) A new model for nuclear lamina organization. Biochem Soc Trans 36:1339–1343

Goldberg MW (2017) Nuclear pore complex tethers to the cytoskeleton. Semin Cell Dev Biol 68:52–58

Hall EH, Schoenbach KH, Beebe SJ (2005) Nanosecond pulsed electric fields (nsPEF) induce direct electric field effects and biological effects on human colon carcinoma cells. DNA Cell Biol 24:283–291

Hall LL, Smith KP, Byron M, Lawrence JB (2006) Molecular anatomy of a speckle. Anat Rec A Discov Mol Cell Evol Biol. 288:664–675

Kabachinski G, Schwartz TU (2015) The nuclear pore complex—structure and function at a glance. J Cell Sci 128:423–429

Koga T, Morotomi-Yano K, Sakugawa T, Saitoh H, Yano KI (2019) Nanosecond pulsed electric fields induce extracellular release of chromosomal DNA and histone citrullination in neutrophil-differentiated HL-60 cells. Sci Rep 9(1):8451

Kopp B, Khoury L, Audebert M (2019) Validation of the γH2AX biomarker for genotoxicity assessment: a review. Arch Toxicol 93(8):2103–2114

Kumaravel TS, Jha AN (2006) Reliable Comet assay measurements for detecting DNA damage induced by ionising radiation and chemicals. Mutat Res 605:7–16

Lamond AI, Spector DL (2003) Nuclear speckles: a model for nuclear organelles. Nat Rev Mol Cell Biol. 4:605–612

Lee YL, Burke B (2018) LINC complexes and nuclear positioning. Semin Cell Dev Biol 82:67–76

Lerner MR, Boyle JA, Hardin JA, Steitz JA (1981) Two novel classes of small ribonucleoproteins detected by antibodies associated with lupus erythematosus. Science 211:400–402

Lodish et al (2020) Molecular cell biology, 4th edn. H.H Freeman, New York

Lynch AM, Sasaki JC, Elespuru R, Jacobson-Kram D, Thybaud V, De Boeck M, Aardema MJ, Aubrecht J, Benz RD, Dertinger SD, Douglas GR, White PA, Escobar PA, Fornace A Jr, Honma M, Naven RT, Rusling JF, Schiestl RH, Walmsley RM, Yamamura E, van Benthem J, Kim JH (2011) New and emerging technologies for genetic toxicity testing. Environ Mol Mutagen 52(3):205–223

Mattout-Drubezki A, Gruenbaum Y (2003) Dynamic interactions of nuclear lamina proteins with chromatin and transcriptional machinery. Cell Mol Life Sci 60:2053–2063

Maul GG, Deaven L (1977) Quantitative determination of nuclear pore complexes in cycling cells with differing DNA content. J Cell Biol 73:748–760

Meinke P, Schirmer EC (2015) LINC'ing form and function at the nuclear envelope. FEBS Lett 589:2514–2521

Møller P (2013) The alkaline comet assay: towards validation in biomonitoring of DNA damaging exposures. Basic Clin Pharmacol Toxicol 98(4):336–345

Nuccitelli R, Lui K, Kreis M, Athos B, Nuccitelli P (2013) Nanosecond pulsed electric field stimulation of reactive oxygen species in human pancreatic cancer cells is Ca(2+)-dependent. Biochem Biophys Res Commun 435:580–585

Pakhomov AG (2012) Oxidative effects of nanosecond pulsed electric field exposure in cells and cell-free media. Arch Biochem Biophys 527:55–64

Reichelt R, Holzenburg A, Buhle EL Jr, Jarnik M, Engel A, Aebi U (1990) Correlation between structure and mass distribution of the nuclear pore complex and of distinct pore complex components. J Cell Biol 110:883–894

Reisz JA, Bansal N, Qian J, Zhao W, Furdui CM (2014) Effects of ionizing radiation on biological molecules–mechanisms of damage and emerging methods of detection. Antioxid Redox Signal 21:260–292

Ren W, Sain NM, Beebe SJ (2012) Nanosecond pulsed electric fields (nsPEFs) activate intrinsic caspase-dependent and caspase-independent cell death in Jurkat cells. Biochem Biophys Res Commun 421(4):808–812

Rogakou EP, Pilch DR, Orr AH, Ivanova VS, Bonner WM (1998) DNA double-stranded breaks induce histone H2AX phosphorylation on serine 139. J Biol Chem 273(10):5858–5868

Romeo S, Zeni L, Sarti M, Sannino A, Scarfì MR, Vernier PT, Zeni O (2011) DNA electrophoretic migration patterns change after exposure of Jurkat cells to a single intense nanosecond electric pulse. PLoS One 6(12):e28419

Sacco-Bubulya P, Spector DL (2002) Disassembly of interchromatin granule clusters alters the coordination of transcription and pre-mRNA splicing. J Cell Biol 156:425–436

Saitoh N, Spahr CS, Patterson SD, Bubulya P, Neuwald AF, Spector DL (2004) Proteomic analysis of interchromatin granule clusters. Mol Biol Cell 15:3876–3890

Sleeman JE, Trinkle-Mulcahy L (2014) Nuclear bodies: new insights into assembly/dynamics and disease relevance. Curr Opin Cell Biol 28:76–83

Spector DL, Schrier WH, Busch H (1983) Immunoelectron microscopic localization of snRNPs. Biol Cell 49:1–10
Spector DL, Fu XD, Maniatis T (1991) Associations between distinct pre-mRNA splicing components and the cell nucleus. EMBO J 10:3467–3481
Spector DL, Lamond AI (2011) Nuclear speckles. Cold Spring Harb Perspect Biol 3(2):a000646
Stacey M, Stickley J, Fox P, Statler V, Schoenbach K, Beebe SJ, Buescher ES (2003) Differential effects in cells exposed to ultra-short, high intensity electric fields: cell survival, DNA damage, and cell cycle analysis. Mutat Res 542:65–75
Stacey M, Fox P, Buescher S, Kolb J (2011) Nanosecond pulsed electric field induced cytoskeleton, nuclear membrane and telomere damage adversely impact cell survival. Bioelectrochemistry 82:131–134
Starr DA, Fridolfsson HN (2010) Interactions between nuclei and the cytoskeleton are mediated by SUN-KASH nuclear-envelope bridges. Annu Rev Cell Dev Biol. 26:421–444
Thompson LH (2012) Recognition, signaling, and repair of DNA double-strand breaks produced by ionizing radiation in mammalian cells: the molecular choreography. Mutat Res 751:158–246
Thompson GL, Roth CC, Kuipers MA, Tolstykh GP, Beier HT, Ibey BL (2016) Permeabilization of the nuclear envelope following nanosecond pulsed electric field exposure. Biochem Biophys Res Commun 470:35–40
Tice RR, Agurell E, Anderson D, Burlinson B, Hartmann A, Kobayashi H, Miyamae Y, Rojas E, Ryu JC, Sasaki YF (2000) Single cell gel/comet assay: guidelines for in vitro and in vivo genetic toxicology testing. Environ Mol Mutagen 35(3):206–221
Villunger A, Michalak EM, Coultas L, Müllauer F, Böck G, Ausserlechner MJ, Adams JM, Strasser A (2003) p53- and drug-induced apoptotic responses mediated by BH3-only proteins puma and noxa. Science 302(5647):1036–1038
Wang Y, Li M, Stadler S, Correll S, Li P, Wang D, Hayama R, Leonelli L, Han H, Grigoryev SA, Allis CD, Coonrod SA (2009) Histone hypercitrullination mediates chromatin decondensation and neutrophil extracellular trap formation. J Cell Biol 184(2):205–213
Zieve GW, Sauterer RA (1990) Cell biology of the snRNP particles. Crit Rev Biochem Mol Biol 25:1–46

Chapter 8
Mitochondria as usEP Sensors

Stephen J. Beebe

Abstract The endocytotic and symbiotic inclusion of a prokaryotc by an early eukaryote, its subsequent evolution as mitochondria, and its collaboration with the nucleus provided these new symbiotes with enough ATP to evolve a new world of extraordinarily diverse organisms. Mitochondria assumed roles for lives replete with energy from ATP and control over the death of cells when their usefulness was finished or when they malfunctioned or were injured beyond repair. The outer mitochondrial membrane (OMM) protects the electron transport chain (ETC) in the inner mitochondrial membrane and the mitochondria's DNA, which is used for some of the proteins in the ETC. The ETC is supplied with electrons from NADH, FADH2 produced by oxidative phosphorylation (OXPHOS) as Complexes I, III, and IV pump proton (H +) out of the matrix to generate a proton motive force and a mitochondrial membrane potential ($\Delta\Psi m$). H + reenter the matrix through ATP synthase for the production of ATP. All this complexity provides usEPs with multiple targets for effects on cell life and death. UsEP's role in cytochrome *c* release in apoptosis and other regulated cell death (RCD) mechanisms in cancer ablation has been a significant application with clinical medicine, which is still in developmental stages in clinical trials. UsEPs increase reactive oxygen species (ROS) and dissipate the $\Delta\Psi m$, which can occur without permeabilization of the IMM, especially in the presence of Ca^{2+} that enters cells through nanopores in the plasma membrane. This loss of $\Delta\Psi m$ is facilitated by usEP effects on the Ca^{2+}-dependent and redox-sensitive protein cyclophilin D (CypD). CypD regulates the mitochondrial permeability transition pore (mPTP) that dissipates the $\Delta\Psi m$, leading to regulated cell death and apoptosis if mitochondria release cytochrome *c* into the cytoplasm to activate caspases. We also discuss the possible identity of the mPTP as ATP synthase. Experiments continue to test this hypothesis. Experiments here also show that usEPs with a shorter (faster) rise-fall time are more effective to dissipate $\Delta\Psi m$ than usEPs with a longer (slower) rise-fall time. It also appears that over-expression of BCL-xl and BCL2 cannot protect the mitochondria from the effects of usEPs. Experiments measuring oxygen consumption in cells treated or not with usEPs indicate that the usEPs attenuate oxygen consumption in Complexes I and IV of the ETC. These results suggest that usEPs inhibit electron transport in the ETC. We also show that usEPs that ultimately lead to cell death in 4T1-luc mammary cancer cells up-regulates

S. J. Beebe et al., *Ultrashort Electric Pulse Effects in Biology and Medicine*,
Series in BioEngineering, https://doi.org/10.1007/978-981-10-5113-5_8

essential subunits in the ETC. Thus, usEPs target several mitochondrial components, including those that regulate $\Delta\Psi m$ and electron transport in the ETC.

8.1 Introduction

Given the dogma that usEPs affected intracellular structures and functions, the nucleus, mitochondria, and endoplasmic reticulum (ER) were primary suspected targets for major usEP effects. Given that the nucleus was considered the "executive suite" of the cell and was covered by a membrane with large complex pores, it was thought that for electric fields that "went through the cell", the nucleus seemed like an important target for intracellular electromanipulation. As early studies were being carried out on usEP effects on cell DNA damage, cell cycle analysis and viability (Stacey et al. 2003), usEPs were shown to induce apoptosis in human Jurkat cells (Beebe et al. 2002, 2003). Increases in caspase catalytic activity, late phosphatidylserine externalization and cytochrome *c* release were coincident in response to 10, 60, and 300 ns pulses. These results were time-, electric field-, and pulse duration-dependent, yet energy density-independent (Beebe et al. 2003; see *section usEPs induced RCD mechanisms, Chap. 9*). While some caspase mechanisms occur independently of mitochondria involvement, the importance of this organelle in apoptosis was well-known at that time and remains so. Other studies focused on the ER and Ca^{2+} mobilization (Vernier et al. 2003; Beebe et al. 2004; White et al. 2004; Buescher et al. 2004) as well as the most common usEP target, the plasma membrane (Beebe et al. 2002, 2003; Vernier et al. 2003, 2004a, b; Deng et al. 2003; Pakhomov et al. 2007a, b). However, given the high resting voltage in mitochondria, usEP effects on these membranes are expected to be principal intracellular targets (Kotnik and Miklavcic 2006). Other than the release of cytochrome *c*, it was not until 2010 and after that more specific usEP effects on these organelles were clearly shown. As indicated earlier, mitochondria are considered today to be a primary target for usEP effects on many aspects of cell physiology including oxygen consumption, reactive oxygen species (ROS) generation, cytochrome c release, regulated cell death (RCD), and others. Before additional studies on usEP effects of mitochondria are reviewed, it is of considerable interest to review the origins of mitochondria and consider how their roles in cell functions have affected animal and plant (chloroplasts) life on earth. It is also intriguing to contemplate whether the nucleus and/or mitochondria occupy the "executive suite" of the cell. Since mitochondria are provided by the mother, they could be considered the Queen of the cell; this would allow the nucleus to be the King of the cell. And while they occupy different "suites", they have excellent communications that resulted in quite a royal symbiosis for advancing life forms on planet earth.

8.2 From Prokaryotes to Eukaryotes—The Origin of Mitochondria

Prokaryotes, which include archaea and eubacteria (bacteria), do not have intracellular structural organization, such as organelles or membrane-bound intracellular structures (cyanobacteria are a notable exception). They do not have a nucleus but have a single circular chromosome of double stranded DNA in a cellular location called the nucleoid. Prokaryotes consist of a capsule-like cell wall and an inner cell membrane that surrounds the cytoplasm containing DNA, ribosomes for protein synthesis and some raw materials. Eukaryotes or multicellular organisms are more structurally organized and complex. They have a cytoplasm surrounded by a plasma membrane, but no cell wall. They also have a nucleus consisting of a double layered lipid nuclear membrane or nuclear envelop with a perinuclear space between the two lipid layers. The nuclear envelop surrounds and protects the chromosomal DNA genome. The double lipid bilayer contains many pores, which are complex structures containing more than 30 different proteins. These pores regulate the passage of large molecules such as RNA and ribosomal proteins out of the nucleus into the cytoplasm and signaling molecules and transcription factors from the cytoplasm into the nucleus (Maul and Deaven 1977; Lin et al. 2016). Eukaryotes also contain many mitochondria, which carry out oxidative phosphorylation and reduce oxygen to water and synthesize ATP for cellular energy. They also regulate cell death. Eukaryotes also have a cytoskeletal system and a system of membrane bound compartments called organelles. The cytoskeleton is a highly organized and intricate system of scaffolds built with tubulin-containing microtubules and actin-containing microfilaments. The dynamics of the eukaryotic cytoskeleton with molecular motors and self-organizing mechanisms can only best be described by computational models (Jekely 2014). The cytoplasm is filled with other membranous structures, Golgi apparatus, lysosomes, peroxisomes, and endoplasmic reticulum (ER) with (rough) and without (smooth) ribosomes. The ER is contiguous at sites with invaginations from plasma membrane and makes connections with mitochondria and is continuous with the nuclear membrane (see Chapter 5, Figs. 5.1 and 5.2). Communications between the ER and mitochondria maintain cellular homeostasis and determine cell fate, but these connections are also important for integrating nutritional and hormonal signals in normal metabolism as well as playing roles in metabolic diseases and other pathologies (Tubbs and Rieusset 2017). While the nucleus must communicate with the mitochondria, since most of the genes in the mitochondrial electron transport chain come from nuclear genes, the mitochondria also communicate with the nucleus in what is called retrograde signaling, which regulates gene transcription among other activities at the request/demand of the mitochondria (Cagin and Enriquez 2015).

Mitochondria are remarkable structures and have a most interesting origin in evolutionary history. They provide us with a rich source of ATP for life functions as well as intra-mitochondrial constituents for cell euthanasia when cells need to die.

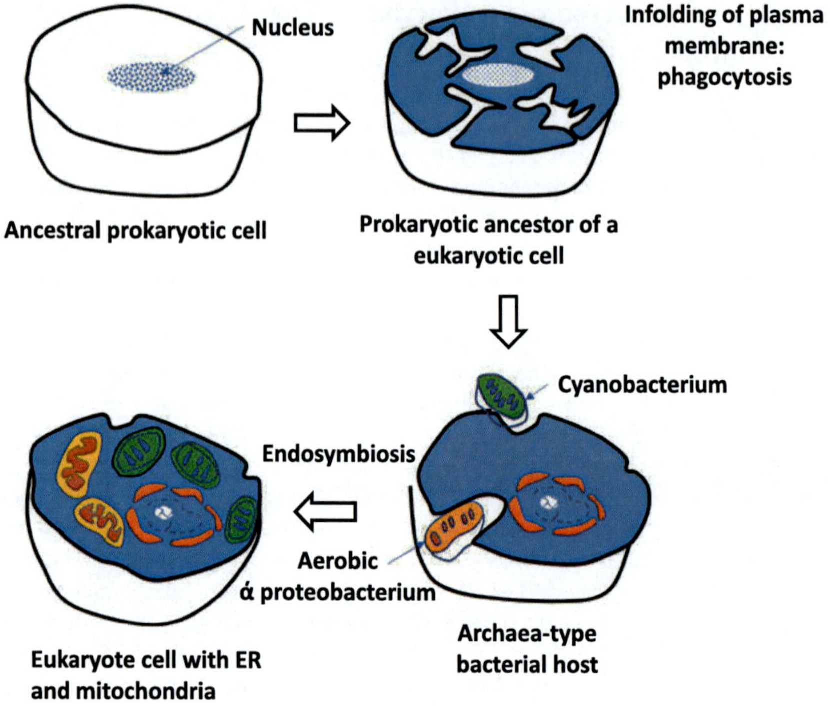

Fig. 8.1 The endosymbiotic origin of mitochondria

During evolution, eukaryotic cells are believed to have evolved from prokaryotic cells. These results had an extraordinary impact on bioenergetics and biosynthesis and consequently a historic and evolutionary influence on life on earth (Dyall et al. 2004) (Fig. 8.1).

Since these events occurred more than 1.5 billion years ago, there are many theories about how these incidents unfolded. All these ideas about the origin of mitochondria from bacteria are only theories and must remain so. These theories can never be scientifically proved because the evidence or the means to test hypotheses are lost in the past. Perhaps the most noteworthy hypothesis for the origin of mitochondria was the inclusion of an α-proto-mitochondria from a α-proteobacterium-like ancestor into an Archaea-type bacterial host (or into an already eukaryotic host) about 1.5 billion years ago. This event forever changed life on planet earth. This phenomenon is also fascinating because it was a once-occurring event in earth's evolutionary history (Dyall et al. 2004; Gray 2012). When cyanobacteria initiated the Great Oxygen Event or the Great Oxidation by producing enough oxygen by photosynthesis to saturate iron deposits on land and in sea, free oxygen escaped into the atmosphere. This epoch has also been called the Oxygen Catastrophe or Oxygen Crisis (even Oxygen Holocaust) because most of life on earth at that time could not utilize or tolerate oxygen and became extinct. About this time these α-proteobacterium-Archaea symbionts began to use oxygen and flourish with extraordinary and amazing diversity. Cyanobacteria used sunlight to oxidize water, thereby producing oxygen as a waste product. This Great Oxidation coupled with the utilization of oxygen by eukaryotes changed the direction of life on earth and commenced an astonishing evolution of plant and animal life (Sessions et al. 2009). Multicellular organisms independently evolved into at least 6 different eukaryotic groups (Rokas 2008). The utilization of oxygen by mitochondria through the electron transport chain (ETC) was a revolution of nature and provided the basis for ATP production by organisms in much greater quantities than was previously possible. The oxidation of water producing oxygen as waste by cyanobacteria and the reduction of oxygen producing water as waste by mitochondria were evolutionary genius and demonstrates an astounding conservation of nature's resources by completing an oxidation reduction cycle over a billion years—a kind of universal equilibrium.

Since this origin of mitochondria from bacteria is a hypothesis, there should be some evidence to support it. First, there are several overt features shared by bacteria and mitochondria. Bacteria and mitochondria have similar sizes and shapes, and both have double layered membranes. The outer and inner mitochondrial membrane (OMM, IMM) are suggestive of the bacterial cell wall and plasma membrane. Also, mitochondria divide independently of the cell, on their own, much like individual bacteria. Furthermore, like bacteria, the mitochondrial genome is circular, lacks introns and transcription is coupled to translation. However, the real evidence that α-proteobacteria evolved into mitochondria is based on comparative genomic, proteomic, and phylogenetic commonalities between mitochondria and α-proteobacteria such that they descended from a common evolutionary ancestor or group not shared by any other group—a monophyletic origin of bacteria from within α-proteobacteria (Gray 2012). Whichever hypothesis of origin is correct, pro-mitochondria arose only once in evolution. Thus, the origin of eukaryotes and humankind in their most complex form was a once-in-a-lifetime or planet-time phenomena! A eukaryote singularity.

In the symbiosis, the primitive eukaryote provided ample nutrients to the prokaryote, which then provided bountiful quantities of ATP for the primitive eukaryote. These new energy levels afforded new opportunities for additional developments. Since their proto-mitochondrial origin, mitochondria have advanced in drastically different directions, far and away from its α-proteobacterial origin (Gray 2012). In the evolution of the endosymbiont to an organelle, endosymbionts lost most of their genes and most of those that remain were transferred from the proto-mitochondria to the nucleus. It is estimated that 1000–3000 genes were lost in the bacterial endosymbiont to proto-organelle evolution. Interestingly, of the more than 800 human genes that carry α-proteobacteria signatures, only about 200 are found in the human mitochondrial genome, indicating that proto-mitochondrial contributions to eukaryote evolution extend beyond mitochondria to other cellular constituents (Gray 2012). Mitochondrial DNA code for only 13 subunits in the oxidative phosphorylation complexes that include the mitochondrial electron transport chain and

ATP synthase, while nuclear DNA codes for 77 subunits. Thus, most mitochondrial proteins are transcribed in the nucleus, synthesized in the cytoplasm, and must be transported through one or two mitochondrial membranes to find their final functional sites.

This enormously important incident of an α-proteobacterium-Archea symbiont was a prodigy of evolutionary nature. Yet during this vast evolutionary time and the incredible diversity giving rise to 8.7 million species of eukaryotes on earth today (Mora et al. 2011), none of the organisms, nor the unknown millions that are extinct, had ever been exposed to a usEPs until very recently. Thus, how different cells respond to usEPs and whether they are normal or distinctively cancerous can be expected to be diverse. This diversity will be amplified when we consider that low usEP conditions can activate cells and high or intense conditions can eliminate them. As will be discussed later, these differences can be observed within mitochondria themselves.

8.3 Up Close and Personal with Mitochondria

UsEP can affect mitochondria—There is enough evidence that usEPs can affect mitochondria, especially the mitochondrial membrane potential ($\Delta\Psi$m); however, there is evidence that other mitochondrial structures and functions are affected. Given known effects of electric fields on cell membranes, one obvious usEP-mediated mechanism is to permeabilize, or nanoporate, the IMM allowing ions to enter the matrix and dissipate the $\Delta\Psi$m (Vernier 2011; Batista Napotnik et al. 2012; Beebe 2012, 2013, 2015). While this does happen at relatively high electric fields, there is accumulating evidence that usEPs can induce non-permeabilizing effects on mitochondria affecting the $\Delta\Psi$m (Beebe et al. 2012, 2013, 2015) and affecting mitochondrial function in other ways including effects on the rate of oxygen consumption (Estlack et al. 2014a, b; Lassiter, Tanaz, Boni Hani, Lai, and Beebe unpublished; Fig. 8.16), electron transport chain complexes (Lane and Martin 2010; Figs. 8.4, 8.16 and 8.17), the mitochondria permeability transition pore (mPTP) (Beebe et al. 2012, 2015; Fig. 8.9) and perhaps ATP synthase (Figs. 8.14 and 8.15). To understand usEP effects on mitochondria, it will be helpful to review some basic principles of mitochondrial structures, generation, and function of $\Delta\Psi$m, structure and function of the electron transport chain (ETC) and their complexes as functional units of respiration, the mPTP, structure and function and ATP synthase, which is responsible for synthesizing ATP and may be a structural unit of the mPTP.

Mitochondria structure and function—The primary function of the mitochondria is to carry out cellular respiration in order to provide energy for cellular activities and to regulate metabolism. They do this by synthesizing ATP as a primary energy source. In so doing, mitochondria provide energy by oxidizing substrates from the citric acid or Krebs cycle and generate $\Delta\Psi$m. As with all biological systems, the function of mitochondrial components critically depends on their structure.

Figure 8.2 shows the basic structure of a mitochondrion. These rod-shaped mito-

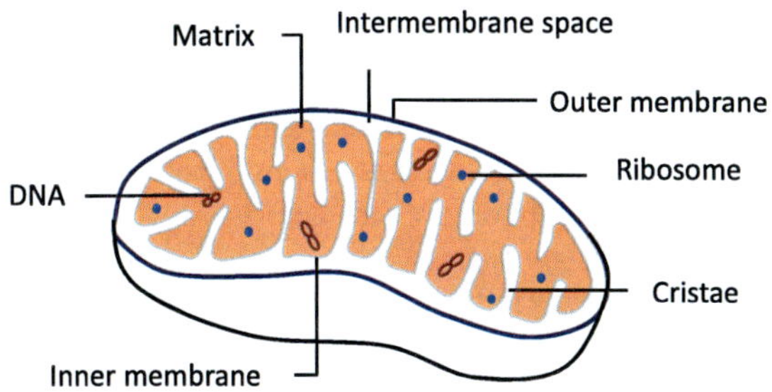

Fig. 8.2 A basic diagram of a mitochondrion. Mitochondria contain two membranes. The outer mitochondrial membrane (OMM). The OMM is permeable to small molecules. The inner mitochondrial membrane (IMM) is folded into cristae, which include the electron transport chain for ATP synthesis. The OMM and IMM are separated by an intermembrane space. The mitochondrial DNA, which contain 37 genes, exists in the matrix space within the IMM

chondria have two membranes—an outer mitochondrial membrane (OMM) and an inner mitochondrial membrane (IMM), creating an intermembrane space between them. Like the plasma membrane these are composed of phospholipid with imbedded proteins. The OMM is porous. It contains a significant number or proteins called porins. These structures allow exchange of materials between the intermembrane space and the cytoplasm, such as ions, nutrients, metabolites, ADP, and ATP among other relatively larger molecules. Because of the porous OMM, the intermembrane space has a similar composition to the cytoplasm. However, there are differences in the protein content of the inner membrane space compared to the cytoplasm. The IMM is a more complex, folded structure forming cristae providing a much larger surface area inside the organelle. This membrane is much more selective for molecular transport between the intermembrane space and the inside matrix. The IMM surrounds the matrix and contains an array of chemical reactions including the respiratory chain or electron transport chain and ATP synthase that work together to produce ATP and generate the $\Delta\Psi m$. The mitochondrial circular DNA is also contained within the mitochondrial matrix. The DNA is a small circular chromosome encodes only 37 genes (Anderson et al. 1981). It consists of heavy outer and a light inner strand encoding 13 proteins, including subunits of NADH dehydrogenases, cytochrome *c* oxidases, and ATP synthase; ribosomal RNAs and transfer RNAs. Note that the greatest number of genes encoding proteins in mitochondria are encoded in the nucleus, where they are transcribed, translated and the protein products imported into mitochondria. This requires mitochondrial proteins to transport these across the IMM. In humans and many other organisms, mitochondrial DNA is only inherited from the mother. This affords the basis for studying evolutionary history and relationships between and among individuals and groups of organisms (phylogenetics).

Mitochondrial DNA (mtDNA) has recently been shown to have a unique role distinct from producing mitochondrial transcripts and proteins. It is known that permeabilization of the OMM results in the release of pro-apoptotic factors for mitochondria, such as cytochrome *c*, to induce non-inflammatory apoptotic cell death. Under similar circumstances, mtDNA has been shown to be released into the cytoplasm as the inner mitochondrial membrane is ruptures (Riley et al. 2018). The mtDNA, which has a prokaryotic origin, then activates signaling pathways that are used by cells that are invaded by bacteria and viruses. These organisms also release DNA into the cytoplasm causing innate immune response and inflammasome activation. Such DNA fragments can activate cellular signaling systems sensed by Toll-like receptors (TLR) among other receptors as discussed in the section on usEPs induce RCD mechanisms. This can cause cytokine production such as IL-1β and/or type I IFN. It appears that caspase activation is important to prevent type I IFN production and inflammation during apoptosis (White et al. 2014; Rongvaux et al. 2014; Giampazolias et al. 2010).

The electron transport chain (ETC) and the mitochondrial membrane potential ($\Delta\Psi$m)—One of the most important structural and functional units in the IMM is the ETC comprising a series of four complexes referred to as complex I through IV. ATP synthase, which produces ATP, is complex V. During most efficient respiratory function, these complexes form a super-complex. With various conformations and stoichiometries, complexes I, III and IV form a stable supercomplex unit or respirasome (Guo et al. 2016; Melber and Winge 2016; Letts and Sazanov 2017), which "channels" electrons between and among active sites within and through the ETC. In these complexes, electrons are transferred from donor to acceptor molecules as oxidation (donating electrons) and reduction (accepting electrons) reactions or oxidation–reduction reactions, which is often referred to simply as redox reactions. These redox reactions are directly coupled with simultaneous transfer of proton across the IMM, which is from the matrix to the inner mitochondrial space (IMS). These same respiratory processes generate an electrochemical proton gradient or a proton motive force (PMF) that is used to synthesize ATP from ADP and inorganic phosphate (Pi) by ATP synthase or complex V. This chemiosmotic energy coupling generates the $\Delta\Psi$m, which is about -120 to -180 mV, negative inside. Under homeostatic conditions, the $\Delta\Psi$m is maintained by the balanced generation and utilization of ATP. This is a governing feature of mitochondria and is a sensitive indicator of cell function and fate.

Mitochondria have generally been considered to be individual "batteries" with a single unified mitochondrial membrane potential ($\Delta\Psi$m) along the cristae of the IMM of each mitochondria. This concept was supported by using membrane sensitive dyes and relative low resolution microscopy, showing that the $\Delta\Psi$m was continuously homogeneous within the organelle (Amchenkova et al. 1988; Skulachev 2001). The authors also showed that the $\Delta\Psi$m was lost simultaneously along the IMM with laser-induced injury to a ≤ 5 μm section. This appeared as if the filament was similar to a string of Christmas light, where loss of one light extinguished all the others. These studies and other used dyes and microscopy that lacked spatial resolution to visualize the fine structural filaments of along the IMM. Wolf and colleagues (Wolf et al.

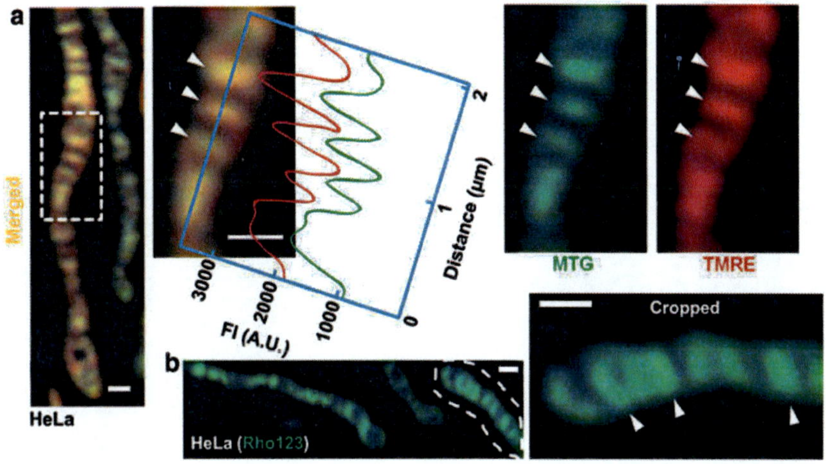

Fig. 8.3 HeLa cells are stained with mitoTracker green (MTG) and TMRE (red) **a** The dashed line indicates an expanded region, indicating that the signal intensities across the long axis (green and red lines) arrow heads indicate cristae membranes of the vary (scale bar 500 nm). **b** HeLa cell mitochondria stained with Rho123 to indicate,; dashed line is zoomed—in mitochondrion showing more intense signal at cristae (arrowheads). Scale bar 500 nm. (From Wolf 2019)

2019) resolved intramitochondrial structures in HeLa, L6 (rat myoblasts) and H1975 cells using LSM880 and Airyscan-based microscopic technology, which allowed separation of cristae (Fig. 8.3).

The authors used 0-N-nonyl acridine orange (NAO), a fluorescent probe that preferentially stains cardiolipin, and other stains to confirm they identified cristae in the IMM. Using TMRE and RHO123 to determine the $\Delta\Psi m$, they found there $\Delta\Psi m$ indicators were non-homogeneously associated with the IMM and the most robust TMRE signals were colocalized with NOA stained cristae. They observed that the $\Delta\Psi m$ appeared along segments of the cristae. Providing a new concept for the $\Delta\Psi m$, they found that different cristae within the mitochondria had different $\Delta\Psi m$. Over a length of about 3 μm along a mitochondrial filament the $\Delta\Psi m$ differed between −145 and −174 mV. So, the cristae act as individual energetic units such that each mitochondrion consist of hundreds of little "batteries" along the cristae that act independently of each other instead of one per mitochondria. It also prevents a dysfunction in a specific cristae ("small battery") from spreading the collapse to the rest of the unit. "The battery experts I had originally talked to were very excited to hear that they were right," Shirihai said. "It turns out that mitochondria and Teslas, with their many small batteries, are a case of convergent evolution" (Williams 2019) (Fig. 8.4).

Within the ETC, electrons enter complex I from NADH where they are oxidized, and electrons are transferred through co-factor FMN and iron-sulfur clusters to ubiquinone converting it to reduced ubiquinol. Electrons also enter complex II from succinate from the citric acid cycle to ubiquinone. Electrons from ubiquinol (CoQ)

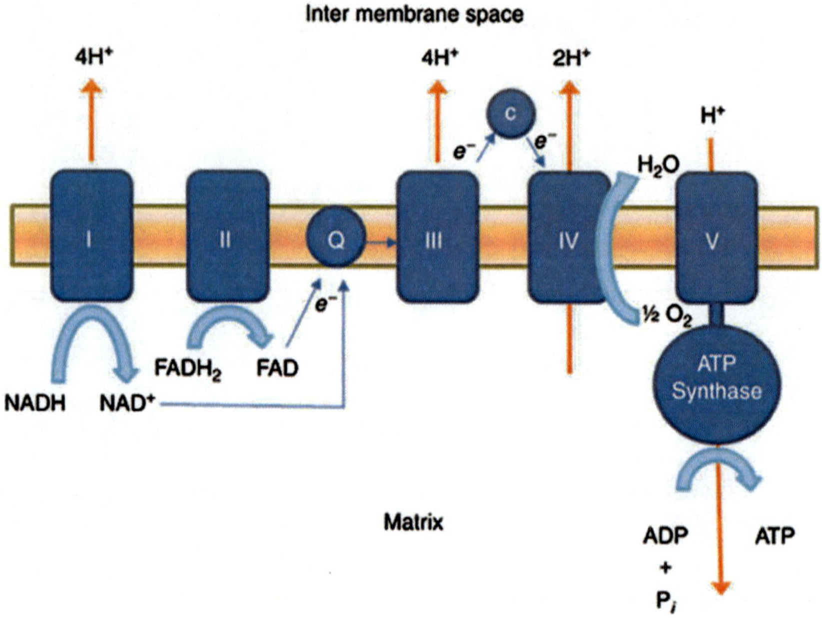

Fig. 8.4 The electron transport chain (Complexes I–IV) and ATP synthase (Complex V)

are transferred to complex III, then to cytochrome *c* and finally to complex IV where oxygen is reduced to water. When there electrons are transferring through the ETC, protons (H +) are pumped out of the mitochondrial matrix and into the intermembrane space by complexes I, III and IV. Together a proton (pH) gradient is produced and the mitochondrial membrane potential ($\Delta\Psi$m) of about −140 mV inside is generated. The energy in this electric field, referred to as $\Delta\Psi$m, is used to generate ATP from ADP and Pi as the protons pass through complex V or the ATP synthase (Zorova et al. 2018; Alpert et al. 2018).

The ECT functions as a supercomplex unit that transports electrons through the respiratory chain to molecular oxygen. Complex V is ATP synthase where ATP is synthesized. There will be more on ATP synthase later regarding the mitochondrial permeability transition pore (mPTP). Complexes I, III and IV are oxidoreductases that tightly couple ET to translocation of protons (H +) across the IMM to the intermembrane space; mutations, drugs or other factors that affect efficiency of electron transfer directly affect efficiency of proton translocation and therefore affect oxygen consumption and ATP synthesis.

Complex I, also known as NADH: ubiquinone oxidoreductase, NADH-CoQ reductase or NADH dehydrogenase, is L-shaped and is the largest complex, consisting of a total of 45–46 proteins, seven of which are coded by the mitochondrial genome, the rest are encoded by the nuclear genome (Guaras and Enríquez (2017)). Fourteen of these mammalian proteins are conserved from prokaryotes to humans.

This suggests that these proteins highly likely serve critical roles in redox reactions and proton translocation (Brandt et al. 2003; Yagi and Matsuno-Yagi 2003). The complex also consists of Fe-S complexes that mediate electron transfer. Electrons enter complex I from NADH to Coenzyme Q, then to Complex II, cytochrome *c* and finally to complex IV. Electron transport in complex I is coupled to proton transport from the inner to the outer side of the IMM, that is from the matrix to the IMS, thereby contributing to the proton motive force of the ETC, which also occurs in complexes III and IV. Along with complex III, complex I is a known source of reactive oxygen species (ROS) generation, which in small amounts functions as chemical messengers in signal transduction pathways in normal cellular function; however, intermediate amounts ROS can produce inflammation and in higher amounts serves as a source of oxidative damage to DNA and other cellular damage due to peroxidation of lipids and proteins. Complex I exhibits most mutations of respiratory complex proteins and is implicated in a number hereditary and several neurodegenerative and neuromuscular diseases of mitochondrial origin. Thus, understanding complex I functions has important implications for carcinogenesis, aging and mitochondrial diseases.

Complex II, also known as succinate dehydrogenase or succinate-CoQ reductase, is a more free-floating complex in the IMM because it functions as a component of the Krebs cycle as well as in the ETC (Cecchini 2003). It is the only complex that participates in the citric acid cycle and ETC. It consists of 4 protein subunits and received electrons from succinate that are used in the citric acid cycle to catalyze citrate to fumarate, contributing electrons to the ETC. Other electrons from fatty acids, for example, enter the ETC through complex II. From succinate, the electron transport sequence is Complex II to Coenzyme Q to Complex III to cytochrome *c* to Complex IV. Thus, complex II is a distinct electron transfer pathway starting at a site that is not a part of the NADH pathway. Unlike complex I, complex II is not a significant source of ROS. Also, unlike complex I, as well as complex II and IV, complex II does not couple electron transport with proton transport, so it does not contribute directly to the proton motive force and the $\Delta\Psi m$.

Complex III, also known as ubiquinol cytochrome c oxidoreductase or cytochrome bc_1 complex, is a multisubunit transmembrane protein that contains three transmembrane subunits that are conserved from bacteria to mammals contain all prosthetic groups involved in the redox reactions: the diheme cytochrome b encoded by the mitochondrial genome, cytochrome c1 encoded by the nuclear genome and a unique Fe-S protein cluster. Complex III has been associated with the generation of reactive oxygen species (ROS). It couples electron transport with proton pumping, thereby contributing to the $\Delta\Psi m$. Cytochrome c accepts the electrons and passes them to complex IV (Xia et al. 2013; Cooley 2013).

Complex IV or cytochrome c oxidase is the last step in the ETC catalyzing the transfer of electrons from ferrocytochrome c to molecular oxygen reducing it to water. Being a major regulatory site in oxidative phosphorylation makes it the most frequent cause of respiratory chain defects in humans. It is composed of 12 and 13 different subunits. The 3 largest subunits form the catalytic core and are encoded by mitochondrial DNA. The functions of the other polypeptides coded by nuclear DNA are less well established, but do not contribute to catalysis or protein pumping and likely serve roles in assembly and stability of the complex. Complex IV also couples electron transport with proton transport contributing to the ion gradient that establishes the $\Delta\Psi m$ (Barrientos et al. 2002; Li et al. 2006).

Complexes I, III and IV of the respiratory chain are all redox-coupled proton transporters involved in generating the PMF; however, each one uses a fundamentally unique mechanism to transport protons out of the mitochondrial matrix. Complex III moves protons across the IMM but is not a true pump. In contrast, complex I and complex IV are true proton pumps; however, differences in their structural organization indicate that their mechanisms for coupling electron and proton transport are different. Complex II is the entry point of electrons from the citric acid cycle and does not contribute proton transport or generation of the PMF.

As mentioned previously, the mitochondrial respiratory ETC complexes I, III and IV can assemble into a supermolecular structure called the respirasome. The respirasome contributes a significant kinetic advantage by facilitating catalysis through substrate channeling or the direct transfer of intermediates between active sites of two different catalytic mechanisms. For mobile subunits, efficiencies of ET and coupled proton translocation are facilitated by minimizing distances when ubiquinone and cytochrome c accept and then transfer electrons from complexes I / II and complex III, respectively. Other kinetic advantages may occur as the respirasome forms higher levels of organization consisting of respiratory "strings" or megacomplexes composed of repeating $I_2 + III_2$ and IV_2 units (Bultema et al. 2009; Guo et al. 2016; Melber and Winge 2016; Letts and Sazanov 2017). Thus, efficiency of ET is enhanced by the formation of these megacomplexes of the respirasome. So, the illustration of the ETC and complex V allows a simplistic representation of the mitochondrial ETC but allows an educational picture of electron entry and their coupled transport with proton transport.

The function of the ETC is to couple electron transport with proton transport producing a proton gradient across the IMM that the protons can flow back into the matrix along their electrochemical gradient through the ATP synthase to produce ATP. The processes of electron transport and the proton gradient it establishes are the basis for oxidative phosphorylation (OXPHOS) as the final stage of respiration. These transport mechanisms are fueled by electrons from NADH and succinate to generate and establish the proton motive force and maintain the $\Delta\Psi m$. Oxygen is the final electron acceptor as it is oxidized to water. This then removes electrons from the system so that the ETC can continue to transport protons, synthesize ATP, and maintain the $\Delta\Psi m$ and life's functions (Fig. 8.5).

Cytochrome *c* is an especially noteworthy component of the ETC. It is a heme containing protein that is loosely associated within IMM on the side of intermembrane space. As indicated above, one of its major functions is to transport electrons from complex III to complex IV and thereby it provides a vital function for ATP synthesis. The prosthetic heme of cytochrome *c* is covalently bound at positions Cys14 and Cys17.The prosthetic heme iron of cytochrome *c* is hexa-coordinated form with four covalent bonds from the nitrogen atom in the porphyrin ring and two axial ligand His18 and Met80. These ligands remain bound as it shifts from the ferric [Fe(III)] state to the ferrous state [Fe(II)] oxidation states. This provides for very efficient function

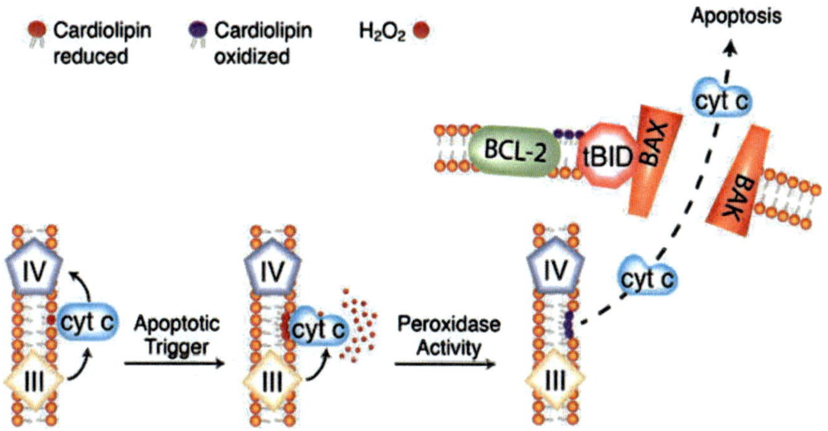

Fig. 8.5 Cardiolipin oxidation and cytochrome c release in the mitochondrial pathway to apoptosis. From Briehl (2015)

in electron transport and delivery from complex III to complexes IV (Bushnell et al. 1990). In another model, when cytochrome *c* binds complex III, Met80 moves away from the covalently bound iron to facilitate the reduction (Godoy et al. 2009).

These structural features become important when we more specifically consider electric field effects on mitochondrial structures and their functions and cytochrome *c* release during apoptosis. Cytochrome *c* loosely interacts by electrostatic interactions with a protein called cardiolipin (CL), which provides a new function to cytochrome *c*. CL is an important lipid in the IMM serving a structural role as well as a functional role for complexes III, IV and V in the ETC. This is the protein that Wolf et al. (2019) used to specifically identify the IMM with NAO binding [see above]. While cytochrome *c* has a life-supporting functions in electron transport activities in the ETC, when the cell receives an apoptotic signal, cytochrome *c* binds to cardiolipin, its conformation is changed, and it turns to life-terminating functions as a peroxidase, catalyzing the peroxidation of its substrate cardiolipin. In the presence of reactive oxygen species (ROS), cytochrome *c* is released through the OMM, by a mechanism that remains unclear (Mohammadyani et al. 2018). It has been suggested that cytochrome *c* has several functions that depend on levels of oxidative as well as nitrative stress serving as a redox sensor of levels of stress. In resting cells, it functions in electron transport while under low levels of stress a subpopulation may translocate to the nucleus and serve some preconditioning functions to further stresses. Under toxic stress, activated Bcl-2 leads to increased cytochrome c release and apoptotic cell death (Godoy et al. 2009).

One hypothesis is that the pro-apoptotic protein tBID is recruited to oxidized cardiolipin on the outer mitochondrial membrane, which enables cytochrome *c* release as well as other apoptotic proteins from the mitochondrial intermembrane space (Briehl 2015).The switch from redox activity to peroxidase activity initiates the kiss-of-death as cytochrome *c* leaves the mitochondria and binds to APAF 1 in the cytoplasm as the point-of-no-return to form the apoprotosome with APAF-1 and ATP [See usEP induce RCD mechanisms]. This complex recruits and activates initiator caspase-9 in the apoptosome, which then activates caspase-3 to begin apoptotic disassembly of the cell. Once cytochrome c is released, it is ten minutes to dead (Green 2005). More on in the RCD section.

Described from another perspective, cytochrome *c* is released from mitochondria in a two-step mechanism (Ott et al. 2002; Iverson and Orrenius 2004; Kagan et al. 2004). It must first dissociate from CL and then connect with pro-apoptotic pore formation. The peroxidation of CL is believed to function as a positive feedback mechanism leading to more CL peroxidation. This can lead to induction of apoptosis when CL peroxidation is sufficient. To differentiate between lethal and non-lethal stimuli, a threshold exists. Mitochondria respond to non-lethal injury by redistributing CL into the intermembrane space without peroxidation giving rise to mitophagy instead of apoptosis (Chu et al. 2013).

The cytochrome *c* heme protein plays a role that is carefully balanced between a life's respiration and death's apoptosis. In respiration it functions to shuttle electrons in the electron transport chain from complex III, cytochrome c reductase, to complex IV, cytochrome c oxidase. In apoptosis, it functions after its release from the IMM to form the apoptosome in the cytosol by binding to APAF-1, dATP and caspase-9, which activates the pro-caspase-9 and subsequently activates the executioner caspase-3 to induce apoptotic cell death.

The heme ligation in cytochrome *c* is in a flexible domain of the protein and dissociation of the heme Fe from Met80 transforms its respiratory role to its apoptosis role by activating its peroxidase activity (Bren and Raven 2017). As cytochrome *c* makes this respiratory—apoptotic transition, Molecular Dynamic (MD) simulations show that it undergoes a structural change from a B1 state to a B2 state, a major tertiary structural alteration. Unlike the B1 state, B2 cannot accept electrons from complex III, but functions as a peroxidase to catalyze the peroxidation of cardiolipin, which leads to its degradation and increased permeability of the IMM and the release of cytochrome *c*. MD simulations also show that electric fields of 2.5 mV/nm, which is one fourth of the upper value found at membrane interfaces, causes an increased mobility and structural distortion of cytochrome c, inducing a distortion of the sixth axial ligand Met80 from the heme. This five-coordinate species has a dramatically lower reduction potential and an increase in peroxidase activity. The positively charged region where negatively charged heme and Met80 reside and negatively charged regions on either side of this domain have been reported to be displace in opposite direction by an electric field. This structural movement results in a slight separation of Met80 from heme and the transition from B1 redox function to B2 peroxidase function (De Biase et al. 2009). Based on these simulations, it is possible that ultrashort electric fields with high enough amplitudes that effectively penetrate intracellular domains could cause an equilibrium shift from the B1 to B2 conformation of cytochrome *c*, which cannot accept electrons from complex III, but can increase the peroxidation of cardiolipin. This has been suggested to increase permeability of the IMM causing the release of cytochrome *c* into the cytoplasm and induce apoptosis. However, while these electric field effects on cardiolipin and cytochrome *c* are possible, it is not proven to be responsible for usEP-induced apoptosis or programmed cell death. There are other possible mechanisms for cytochrome *c* release that could occur in response to usEPs.

Like cytochrome *c*, cardiolipin (CL) plays important roles in promoting electron and proton transport, albeit indirectly through membrane dynamics, as well as inducing cell death. CL peroxidation by cytochrome c is essential for the release of cytochrome *c* as the point of no return from apoptosis. Cardiolipin is composed of four fatty acid acyl chains, two phosphates and three glycerol per molecule and is only found in mitochondria. It has a promiscuous affinity for proteins and facilitates protein associations, domain formation and clustering with robust organizing effects on the IMM. CL is an important component of the ADP/ATP carriers and respiratory complexes I, III and IV, functioning as a scaffold that facilitates formation of these complexes as well as ETC supercomplex formation. In addition, the small head group and long hydrophobic tails produces a conical shape that tends to impose negative curvature on lipid-water interfaces. The acyl groups can be remodeled and support mitochondrial dynamics in fission and fusion mechanisms as well as stabilize negative curvature in membrane invaginations and cristae tips and junctions. Generally, CL supports spatial organization of mitochondrial cristae and stimulates dynamic mitochondrial membrane behavior. Normally, CL is on the matrix side of the IMM while cytochrome *c* is on the other side facing the intermembrane space. When formation of ROS occurs, CL flips to the outer leaflet of the IMM where cytochrome *c* is. Upon binding CL, cytochrome *c* undergoes a structural conversion into a CL-specific peroxidase, which catalyzes the peroxidation of CL as indicated above. The transformation from electron carrier to a peroxidase involves specific and stoichiometric binding of cytochrome *c* to CL (Haines and Dencher 2002; Ren et al. 2014).

Cytochrome *c* exhibits two lipid binding sites. The so-called site A pool binds to ionic phospholipids, particularly CL, loosely by electrostatic interactions involving interactions of lysine 72 and 73 in cytochrome *c* with negative charged phosphate groups of CL, which can be disrupted with salt and does not cause any conformational changes. The other pool binds tightly to so-called site C with hydrogen bonding and hydrophobic interactions of nonpolar cytochrome *c* residues with one or more CL fatty acids. This binding causes conformational changes and interrupts the bond between Met80 and the heme iron. Like the possible electric field effect on cytochrome *c*, this tight association with the membrane significantly disrupts electron transport and transforms the protein into a peroxidase.

These two cytochrome c pools have different functions. About 85% is loosely bound while the remainder is tightly bound. The loosely bound participates in electron transport and superoxide anion scavenging thereby preventing oxidative stress. The tightly bound form does not participate in electron transport and functions as a peroxidase, eliminating hydrogen peroxide and preventing its diffusion into the cytosol. However, under apoptotic stimuli, mitochondrial electron transport is disrupted, large amounts of hydrogen peroxide and other reactive oxygen and nitrogen species are produced, the peroxidase activity of CL-bound cytochrome c is increased, and significant peroxidation of CL is induced (Kagan et al. 2004).

ATP synthase or complex V is a highly conserved enzyme complex that functions as two rotary reversible motors, catalyzes the synthesis of ATP from ADP and Pi with a clockwise rotation and hydrolyzes ATP as it rotates in a counterclockwise rotation. The ATP synthase is a rotary enzyme located in the IMM of mammalian cells and is the smallest molecular rotor motor known. The ATP synthase has evolved from that of the ancestral Archaea and has been amazingly conserved throughout evolution; their structures are essentially the same in bacteria and mitochondria of fungi, plants, and animals. However, there are some differences in how they are regulated depending on the needs of the organism. Through evolution, the enzyme evolved with ever increasing efficiency (Suzuki et al. 2014).

Structurally, the ATP synthase is a complex enzyme composed of a F_1 catalytic domain that faces the matrix and a hydrophilic F_o domain seated in the inner mitochondrial membrane. These two units are connected by a lateral and central stalk. The bacterial ATP synthase is a simpler structure (Fig. 8.6) (Yoshida et al. 2001).

The F_1 catalytic component consists of three α– and three β–subunits and a complex of γδε-subunits. The F_o domain consists of one α–, one β–subunits and 10–12 *c*-subunits. As the name implies, the vital function of ATP synthase is to synthesize ATP. To do this, the ATP synthase uses the electrochemical proton gradient generated

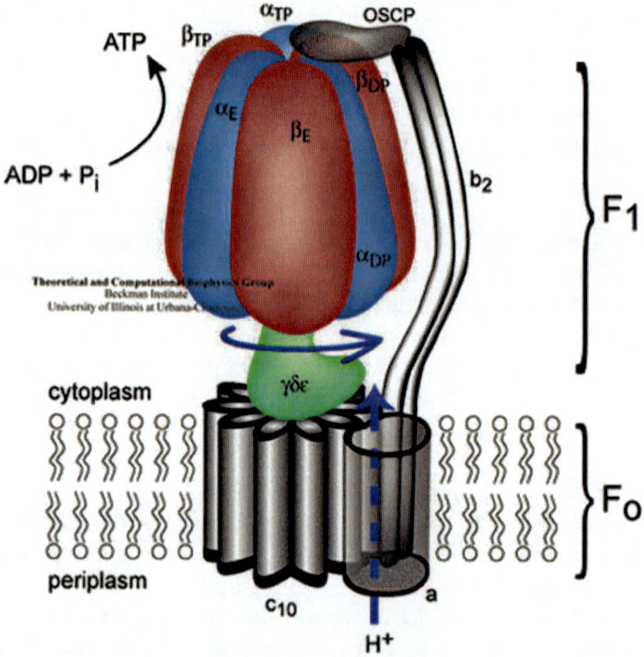

Fig. 8.6 ATP synthase—a marvellous rotary engine of the cell (from Klaus Schulten K, Stone J and Sene M. Theoretical and Computational Biophysics Group; Beckman Institute for Advanced Science and Technology. National Institutes of Health, National Science Foundation, Physics, Computer Science, and Biophysics. University of Illinois at Urbana-Champaign)

by the coupled electron-proton transport in the ETC so that the protons flow through the F_o subunit causing its rotation to induce a conformational change or rotate the F_1 subunit to drive ATP synthesis. In the reverse reaction, the F_0 motor rotates in the opposite direction and ATP is hydrolyzed. The synthesis of ATP is energetically unfavorable; it is more favorable to hydrolyze ATP.

The mammalian ATP synthase is more complex (Fig. 8.7) (Jonckheere et al. 2012). The F_1 catalytic components of mammalian ATP synthase are like the E-coli enzyme with the γδε-subunits forming the central stalk. The mammalian F_o domain contains rotary *c* subunits as well as subunits a, b, d, F6, OSCP and the accessory subunits e, f, g and A6L. The peripheral stalk is formed by subunits b, d, F6 and OSCP. Mechanically, the mammalian enzyme can be separated into "rotor" (c-ring, γ, δ, ε) and "stator" (α3β3, a, b, d, F6, OSCP) components (Devenish et al. 2000). The ATP synthase functional unit is thought to be a dimer. In the presence of Ca^{2+} and Pi, the regulatory protein CypD interacts with subunits in the lateral stalk to inhibit the enzyme. The fungal peptide, CsA reverts this inhibition by binding to and dissociating CyPD, thereby opening a Pi binding site that inhibits pore opening. CypD bonds to the extrinsic part of the lateral stock that faces the mitochondrial matrix. In cross-linking experiments, it binds the OSCP, b and d subunits. The ATP synthase is thought to exist in oligomer forms for optimal activity and binding CypD may enhance oligomeric formation of the enzyme or otherwise favor a more active conformation. CsA displacement of CypD may favor a less active conformation (Giorgio et al. 2010).

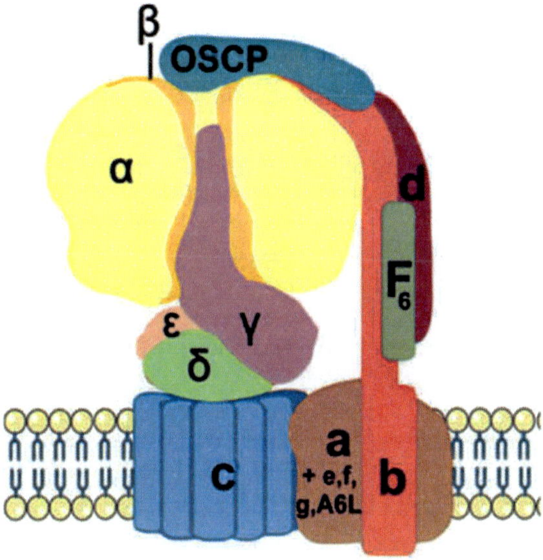

Fig. 8.7 Mitochondrial ATP synthase: architecture, function, and pathology. Jonckheere et al. (2012)

There are several remarkable distinctions concerning ATP synthase (Yoshida et al. 2001). Despite differences between the bacterial and mammalian enzymes, ATP synthases is one of the most conserved proteins during evolution with 60% amino acid homology among all catalytic β–subunits in nature. This is not necessarily surprising since the generation of ATP is required for essentially all energy requiring reactions in living systems. This enzyme has generated considerable interest over the years for several reasons. ATP synthase is the most abundant enzyme on the planet, and as might be expected, ATP synthesis is the most ubiquitous chemical reaction in nature. Furthermore, the molecular rotary mechanism of ATP synthesis is unique among all in nature; only the bacterial flagellum has a rotary motion. The scientific community expressed considerable skepticism concerning this rotary mechanism for more than a decade after it was proposed (Gresser et al. 1982) until an actin filament was observed to exhibit a rotary motion when attached to the γ-subunit of ATP synthase (Noji et al. 1997).

8.4 Effects of usEP on Mitochondria

Given this overview of the ETC and some of its associated functions, it is reasonable to consider that an applied usEP superimposed on the highly negative $\Delta\Psi$m generated by transport of negatively charged electrons through the ETC and positive charged protons across the IMM would have significant effects on mitochondrial functions. Given these lightening-like fluxes of charged ions in, around and through the IMM, it may also be reasonable to consider that there could be effects related and unrelated to permeabilization of the IMM. There are in fact effects on both the membrane itself and apparently on other membranes and structures that remain under investigation.

UsEPs induce cytochrome *c* release and apoptosis in a cell-type specific manner [see section on usEP induce RCD mechanisms, Fig. 9.5]—The first effects of usEP on mitochondria were shown to be cytochrome *c* release from mitochondria into the cytoplasm of human Jurkat cells exposed to three 10, 60 and 100 ns pulses at 1 Hz intervals. However, this in no way indicated that these were direct effects on mitochondria. The method for showing cytochrome c release included isolating cytoplasm and mitochondrial fractions and analyzing them on Western blots (immunoblots). Under these circumstances cytochrome *c* is released from the inner mitochondrial membrane space (IMS) through the OMM into the cytoplasm. Electric fields were adjusted for each pulse duration to provide near equal energy conditions of ~ 1.7 J/cc/pulse (Beebe et al. 2003). Importantly, the cytoplasm fractions were shown to be free of mitochondria as indicated by the absence of the integral mitochondrial protein cytochrome *c* oxidase (subunit IV), demonstrating that mitochondria were not present in the cytoplasm to account for the presence of cytochrome *c*. There was a pulse duration-dependent increase in cytochrome *c* release about 30–40 min after treatment as energy densities remained constant [see Fig. 9.5 in usEP induce RCD mechanisms]. This demonstrated that usEP-induced cytochrome *c* release was energy density-independent. These mitochondrial effects were used to demonstrate that usEP could induce apoptosis, since cytochrome *c* release was a well-characterized marker for apoptosis. Furthermore, caspase-3 activation was also demonstrated. Thus, these usEP-induced responses were down-stream biological effect of usEPs rather than a direct effect of usEPs on mitochondria. However, later data suggesting that usEP may have more direct effects on mitochondria will be presented; however, it is difficult to rule out indirect effects since usEP have numerous cellular targets. As indicated above, there are several different ways that usEPs may cause cytochrome *c* release from the mitochondria into the cytoplasm. As usual with biological systems, mitochondrial functions are complex.

Earlier it was described that cytochrome *c* is loosely bound to cardiolipin (CL) by electrostatic forces. It is not difficult to see how electric fields could interfere with such interactions. In fact, MD simulations have shown that a sufficiently large electric field could lower the reduction potential of cytochrome *c* and increase its peroxidase activity (De Biase et al. 2009). As pointed out earlier, in response to an apoptotic signal, the conformation of cytochrome *c* is altered to function as a peroxidase that oxidizes cardiolipin. Now in the presence of ROS, cytochrome *c* dissociates from cardiolipin and is released into the cytoplasm.

UsEPs generate reactive oxygen species—Reactive oxygen species (ROS) refers to several molecules or free radicals derived from oxygen. Unlike molecular oxygen, which has two un-paired, relatively non-reactive electrons with opposite spins in the outer orbit, superoxide anions have a single un-paired, highly reactive electron in the outer orbit and are powerful oxidants. Superoxide anions are the precursors of most ROS species and can react with superoxide dismutases to produce hydrogen peroxide (H_2O_2), which can be fully reduced to water or partially reduced to hydroxyl anions (OH−); both hydrogen peroxide and hydroxyl anions act as oxidants. Superoxide anions can also react with nitric oxide forming a reactive nitrogen species (RNS), peroxynitrite, another powerful oxidant.

Lower levels of ROS or RNS can act as signaling molecules that reversibly oxidize protein thiol groups that alter their structure and thereby their function. Several redox sensitive enzymes are affected by signaling ROS levels including kinases, phosphatases, GTPases and deacetylases among others. Roles for ROS signaling are important for many cell functions including cellular proliferation, differentiation tissue regeneration and prevention of aging (Finkel 2011). Redox signaling is based on cysteine (Cys) biochemistry in which reactive cysteines are reversibly oxidized to a sulfenic acid form. For example, ROS activates ERK and p38 pathways and inactivates protein tyrosine phosphatases (PTP). Many other kinases are activated by oxidation as are voltage-gated potassium channels and transcription factors such as hypoxia-induced factor, among others (Freund-Michel et al. 2013).

When ROS levels increase further, they can cause dysfunctions by damaging proteins, lipids, and DNA. This can cause activation of the innate immune system such as in response to damage-associated and pathogen-associated molecular patterns (DAMPs and PAMPs). This leads to inflammatory responses by activation of inflammasomes, such as through the Nod-like receptors (NLR), most commonly NLRP3

(Martinon et al. 2006; Anand et al. 2011). Common DAMP stimuli include extracellular ATP, urea crystals, bacteria, virus, and fungal infections. These lead to localization of NLRP3 to mitochondria, activation of caspase-1 and IL-1β production as a proinflammatory cytokine. Higher ROS levels produce oxidative stress that can target proteins, lipids and DNA induce inflammation, molecular malfunctions, and apoptosis (Turrens 2003; Circu and Aw 2010). As was discussed above, this can lead to oxidation of CL and cytochrome c release and as will be discussed below in more detail, oxidation of CypD, dissipation of the $\Delta\Psi m$ and initiation of regulated cell death pathways. Elevated ROS levels can also oxidize and active hypoxic-inducible factor (HIF-1) that enhances the survival and progression of tumors by upregulating cell metabolism, and angiogenesis (Gao et al. 2007).

The issue with low versus high ROS levels and cell function provides a conundrum for normal cell function, but an opportunity for successful cancer therapy. While considerable emphasis has been placed on uses for antioxidant to avert oxidative stress, it was intriguing to find that vitamin E and N-acetyl cysteine (NAC), acting as antioxidants, increased proliferation of Ras-mediated lung cancer models by decreasing ROS levels, DNA damage, and p53 expression. This suggested that some oxidant stress can promote cancer (Sayin et al. 2014). In general, some oxidants contribute to unstable genomes and tumor growth. However, the increased oxidant stress that is generated and used in newly forming tumors may make them more susceptible to excessive stress that can slow tumor progression and promote death. This is what Paul Schumacker aptly referred to ROS in cancer as "dancing with the devil" (Schumacker 2015).

It is generally considered that mitochondria are major sources of ROS; however, others have suggested that the endoplasmic reticulum and peroxisomes may produce even greater ROS levels, especially in the liver (Fridovich 2004). There are numbers of ROS producing centers in the mitochondria, cytosol, and plasma membrane. Superoxide are produced in the outer mitochondrial membrane, the matrix and on both sides of the inner mitochondrial membrane (Turrens 2003; Rimessi et al. 2016) and their ROS production depends upon activity of the respiratory chain. All complexes in the ETC can produce superoxides. These are produced by mass action, increasing when electron transport slows down and when oxygen tensions are high. Complex III is a major source in heart and lung mitochondria, while Complex I is a primary source in the brain and most often in pathological conditions (Turrens and Boveris 1980; Turrens 2003). Complex I superoxides primarily remain in the mitochondrial matrix and Complex III superoxides exit the mitochondria and enter the cytoplasm through the voltage-dependent anion channel (Han et al. 2003). Other sources of ROS in the cytosol often exit the mitochondria from cytochrome b5 reductase and monoamine oxidase as well as from cytosolic dehydrogenases and oxidoreductases (Turrens 2003; Rimessi et al. 2016). The plasma membrane has an active redox system that balances the NAD +/NADH ratio that is essential for energy metabolism. This plasma membrane redox system will be covered when usEP effects on the plasma membrane are presented (Chap. 2).

There is experimental evidence that usEPs in fact cause increases in intracellular oxidation products including ROS and H_2O_2; however, the role of oxidation itself in cell death did not seem likely (Pakhomov 2012; Pakhomova et al. 2012). In this well-designed and thorough study, usEPs were shown to not only cause intracellular oxidation (biological), but also extracellular oxidation (electrochemical), especially when using the non-specific ROS indicator (dihydro-dichlorodihydrofluorescein diacetate, H_2DCFDA). This commonly used dye is deacetylated intracellularly to H_2DCF, which is oxidized to highly fluorescent DCF. However, this dye is problematic because of self-oxidation, photo oxidation and leaking out of cells that were permeabilized by usEPs, which caused electrochemical oxidation in the extracellular media. Nevertheless, when this was considered, there were time-, electric field- and pulse number-dependent increases in intracellular oxidation products in response to usEPs. However, Jurkat cells were much more sensitive or more vulnerable for production of usEP-induced oxidation and cell death than U937 cells, so ROS generation in response to usEPs is cell type-dependent. Amplex Red was also used to show that H_2O_2 was also generated by usEPs. The authors concluded from experiments with Jurak cells that usEP-induced cell death was likely caused by some other mechanism(s) unrelated to H_2O_2 production. To rule out the possibility that dye leakage from electropermeabilized did not contribute directly and/or indirectly to the observed oxidation in Jurkat and U937 cells, to evaluate usEP effects of oxidation, the authors used highly adherent CHO cells, microscopy and dihydroethidium that forms DNA-binding ethidium when oxidized, which can much less readily leak out of permeabilized cells. It was also possible to exchange media with adherent cells with this approach. These experiments confirmed that usEPs induced oxidation of DHE. In all, these studies allowed valid conclusions that plasma membrane permeabilization, ROS formation and cellular anti-oxidant systems are factors that contribute to cell and tissue cytotoxicity in response to usEPs.

UsEPs were also shown to induce ROS production in human pancreatic cancer cells (Nuccitelli et al. 2013). Using pulses with durations of 100 ns and electric field strengths of 30 kV/cm and carboxy-H2DCFDA (carboxy-2′,70′-dichlorofluorescein diacetate) as a ROS-detection reagent, usEPs were shown to induce ROS production in a pulse number-dependent manner that was Ca^{2+}-dependent, Trolox-sensitive. Trolox C (6-hydroxy-2,2,5,7,8-pentamethylchroman) is an antioxidant that scavenges free radicals. The authors concluded that the requirement for ROS in apoptosis induction (Sato et al. 2004), suggested that usEP-induced increase in Ca^{2+} is an early step in the apoptosis pathway.

UsEP-induce generation of mitochondrial ROS is Ca^{2+}-dependent–While mitochondria are an obvious source of ROS, especially from Complex I and III, other organelles such as endoplasmic reticulum, nucleus and the plasma membranes can also produce superoxides. In addition, numbers of other catalytic reactions can produce ROS such as catabolism of purine nucleotides (xanthine oxidase), fatty acids metabolism including prostaglandins and leukotrienes synthesis from arachidonic acid (lipoxygenase reaction), and several biochemical reactions in peroxisomes (Freeman and Crapo 1982). Nevertheless, mitochondria are the predominant source of ROS produced in most apoptotic systems. While other studies measured effects of usEP effects of ROS in the cytoplasm, the most relevant source of reactive species that affect possible oxidation of CL and cytochrome *c* release as well as possible oxidation of CypD is superoxide anion production in the mitochondria as the major source of ROS in mitochondria.

Mitochondrial superoxide is generated as electrons leak primarily from Complexes I and III of the ETC and interact with molecular oxygen as byproducts of oxidative phosphorylation. However, there are other sources of ROS in mitochondria from pyruvate and α-ketoglutarate dehydrogenases, among others in the matrix and IMM (Remessi et al. 2016). In recent studies in the author's laboratory, usEPs were shown to increase mitochondrial superoxide anions in 4T1-Luc mammary cancer cells in a pulse number-dependent manner (Fig. 8.8), using MitoSOX™ Red, which is readily oxidized by superoxide, but not by other reactive oxygen or nitrogen species. Unstimulated cells exhibited low levels of superoxide anion and hydrogen peroxide, used as a positive control, significantly increased superoxides in mitochondria. As usEP numbers were increased, there were pulse number-dependent increases in cells expressing superoxide anions. These studies demonstrate that usEPs can generate superoxide anions as they leak from the ETC providing significant sources of ROS to induce regulated cell death. In theory this could be enough to cause opening of the mPTP, loss of $\Delta\Psi m$ and cytochrome c release; however, caspase-3 is not activated in 4T1-luc cells indicating that usEPs do not induce apoptosis in these cells, although cytochrome *c* release from these cells has not been determined (Beebe et al. 2018). It is not known if these ROS-mediated oxidation events have different thresholds

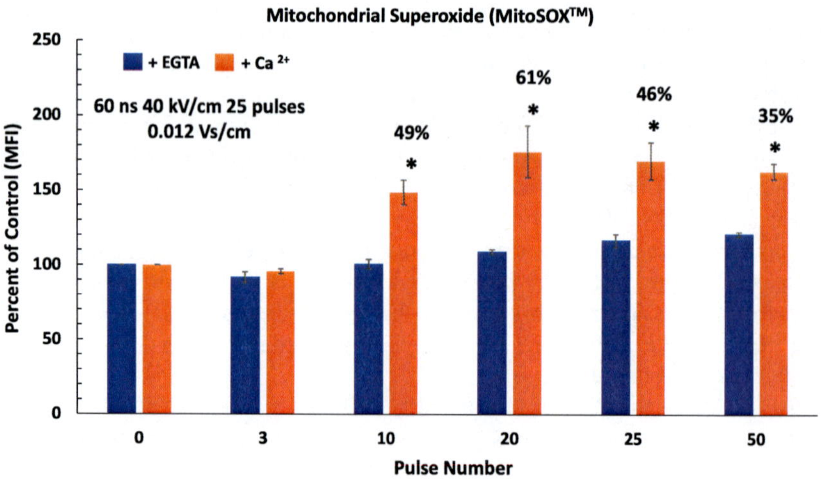

Fig. 8.8 UsEPs induced mitochondrial superoxide generation is Ca^{2+}-dependent. Mitochondrial superoxide was determined in H9c2 cells cardiac myocytes using MitoSox™ (dihydroethidium-triphenylphosphonium, 2 μM) fluorescence by flow cytometry 15–20 min after treatment. H9c2 cells were treated with different pulse numbers with 60 ns durations and 40 kV/cm Under conditions of 3, 10, 20, and 50 pulses, usEPs included 0.0042, 0.0071, 0.013, and 0.17 Vs/cm, respectively. N = 3, $p < 0.05$ (Ruedlinger, Potter, Bani Hani, Lassiter, and Beebe unpublished)

or if these increases in ROS are enough to cause these responses in these cells. It should be noted that cancer cells have enhanced antioxidant mechanisms because they express elevated levels of ROS compared to normal cells, which can enhance proliferation and adaptation to hypoxia as a survival strategy (Reczek and Chandel 2017).

There are many proteins that are regulated by Ca^{2+} and nearly all effects of Ca^{2+} are mediated by proteins. These are also many proteins that are redox sensitive and many of these are regulated by Ca^{2+}. So, it is possible the usEPs can affect proteins by elevating Ca^{2+} and/or elevating ROS. These data suggest that usEPs can have effect on proteins (Beebe 2015). It turns out that many of the Ca^{2+} regulated, redox sensitive proteins are in the mitochondria. One protein of primary interest in cyclphilin D, which is Ca^{2+}-dependent and redox sensitive. It is one of the best characterized regulators of the mPTP. Given that CypD is Ca^{2+}-dependent, can be redox-regulated, and usEPs induce significant levels of Ca^{2+} influx through the plasma membrane and increases in mitochondrial ROS, it is probable that usEP-induced cell death occurs as a result of opening the mPTP and loss of the mitochondrial membrane potential ($\Delta\Psi$m) as a result of Ca^{2+} and ROS-induced oxidization of CypD. In some instances, CL can be oxidized, and cytochrome *c* released to induce apoptosis. As indicated above, apoptosis does not appear to be induced by usEPs in all cells.

Before reporting studies of usEPs on these studies, a closer look at structure and function of CypD will be helpful because it is one of several suspected mitochondrial sensors of usEPs. Other likely sensors including complexes I and IV of the ETC and ATP synthase will be presented below. While the structural component of the mPTP is still in question, CypD is an undisputed regulator of that high conductance pore that forms in the IMM. CypD catalyzes the cis–trans isomerization of peptidyl propyl bonds (PPIase activity) and this activity is thought to be important in CypD regulation of the mPTP. CypD is also a well-known CsA binding protein, which inhibits CypD PPIase activity and thereby inhibits opening of the mPTP (Giorgio et al. 2010). In unrelated activities, CsA is also well-known immuno-suppressants inhibiting the phosphatase activity of calcineurin. The PPIase activity is required for CypD-induced activation of mPTP and CsA inhibition of this activity is required for inhibiting CypD effects on mPTP. The larger CypD family of proteins (16 isozymes) has been highly conserved during evolution with a characteristic 109 amino acid Cyp-like domain, suggesting important functions, likely associated with the PPIase activity as a chaperone for protein folding. However, data suggest that this activity is likely limited to select proteins in cell type-specific activities. CypD has been characterized as a mitochondrial redox sensor that participates in the thioredoxin system, which consists of NADPH, thioredoxin reductase and thioredoxin (Linard et al. 2009; Folda et al. 2016). Thioredoxins are critical for redox regulation of protein function and signaling via thiol redox control in several important biological processes (Arnér and Holmgren 2000). The role of CypD in this system appears to be important for controlling the oxidation state of mitochondrial membrane complexes. It has been shown that CypD activity is under redox control. CypD has four (4) Cys residues and the oxidation state of Cys 157 and Cys 203 affect its conformation, which has an impact on its catalytic activity and therefore its regulation of mPTP. Specifically, oxidation of CypD can lead to opening the mPTP, so it has greater activity in the reduced state. The thioredoxin system plays an important role in maintaining CypD in the reduced form since reducing and oxidizing conditions increase and decrease the PPIase activity of CypD, respectively, and thereby its control of mPTP (Linard et al. 2009; Folda et al. 2016). The role of these players in regulation of mPTP is emphasized since the mPTP can be inhibited by knocking out CypD, inhibiting CypD with CsA, nitrosylation of Cys 203 or mutation of Cys 203 to Ser in CypD. Alternatively, opening the mPTP can occur by inhibiting or knocking out thioredoxin, which keeps CypD in the oxidized state (Folda et al. 2016). Considering these data, we determined more fully whether CypD was a usEP sensor for dissipation of $\Delta\Psi$m by determining effects of Ca^{2+} and CsA on $\Delta\Psi$m and consider the finding that usEPs generate ROS within mitochondria, which could oxidize and inactivate CypD as a mechanism inactivate mPTP and dissipate $\Delta\Psi$m. One diagnostic indicator of CypD is inhibition or sensitivity to Cyclosporin A (CsA).

UsEP-induced dissipation of $\Delta\Psi$m is mediated by effects on CypD—Fig. 8.9 show results from a series of studies analyzing effects of Ca^{2+} and CsA on $\Delta\Psi$m in rat cardiomyocytes H9c2 myoblasts (Ruedlinger, Lassiter and Beebe 201 unpublished). These cells were chosen because they were derived from normal rat myocardium and contain a rich supply of mitochondria, and thereby sufficient levels of CypD, to provide adequate levels of ATP for cardiac function. In addition, these cells exhibit many normal skeletal muscle properties (Kimes and Brandt 1976). The strategy was to determine if CypD would inhibit usEP-induced dissipation of $\Delta\Psi$m in the absence

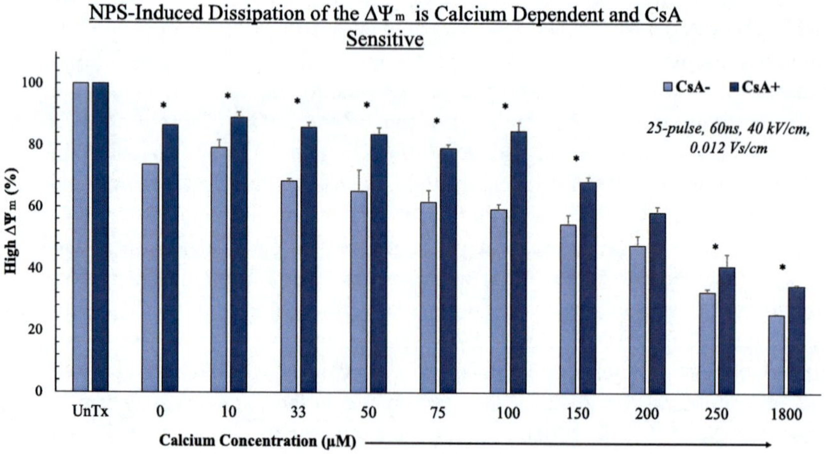

Fig. 8.9 UsEP effects on H9c2 cardiac myoblasts are Ca^{2+}-dependent and CsA-sensitive. H9c2 cardiac myoblasts were loaded with TMRE to determine $\Delta\Psi$m. Cells were treated with twenty-five (25) 60 ns (5 ns rise-fall time) pulses at 40 kV/cm or 0.012 Vs/cm in the presence of increasing concentrations of Ca^{2+} and in the presence and absence of CsA (5 μM). 15–30 min after pulsing, cells were analyzed by flow cytometry for TMRE expression. n = 3, mean ± SD. ± CsA—$p <$ *0.0001; # 0.001; 200 μM Ca^{2+}, $p = 0.069$ (Ruedlinger and Beebe unpublished)

and increasing presence of Ca^{2+}. To determine effects of Ca^{2+} and CsA on usEP-induced dissipation of ΔΨm, H9C2 cells were loaded with TMRE to determine effects on ΔΨm and treated with 25 pulses (1 Hz) with 60 ns durations (~ 5 ns rise time) and an electric field of 40 kV/cm in the presence of various concentrations of Ca^{2+} and the presence and absence of CsA (5 μM). This single usEP condition was chosen because it produced a significant effect on ΔΨm in the presence of Ca^{2+}. Unlike the usEP effects on ΔΨm in rat N1-S1 HCC and human Jurkat cells, which were essentially insensitive to usEP in the absence of Ca^{2+} until very intense conditions (Beebe et al. 2012, 2013), some H9c2 cells (20–25%) exhibited loss of ΔΨm in the absence of Ca^{2+}. Between 0 and 100 μM Ca^{2+}, there was a progressive loss of ΔΨm from 20% of cells to 40% of cells, respectively, that was attenuated by CsA. Over the range of Ca^{2+} concentrations (10–250 μM) the progressive Ca^{2+}-dependent loss of ΔΨm was inhibited by CsA by 15–30%. The Ca^{2+}-dependence and CsA-sensitivity is highly predictive for an effect of usEPs on mitochondrial CypD. Since CypD is a known regulator of the mPTP, is Ca^{2+}-dependent, redox sensitive, and CsA-sensitive, these studies demonstrate that usEP-induced dissipation of ΔΨm is highly likely due to effects on the mPTP, which dissipates the ΔΨm. Given that usEPs induce mitochondrial ROS levels (Fig. 8.8) and that oxidation of CypD leads to opening the mPTP, it is likely that the mechanism for usEP-induced opening the mPTP is by usEP-generated ROS that oxidizes CypD, especially in the presence of Ca^{2+} (Fig. 8.8).

We were interested to determine other characteristics of usEP effects of dissipating ΔΨm. In analyzing effects of EP effects on ΔΨm, we generally determined effects soon (10–15 min) after pulsing. To determine the determine time-dependent effects on ΔΨm after an usEPs treatment, we treated Jurkat A3 cells with ten 60 ns pulses at 60 kV/cm (1.9×10^{-3} Vs/cm) and analyzed population of cells with higher and lower ΔΨm by flow cytometry (Fig. 8.9). Only about 5% of untreated Jurkat cells exhibited low ΔΨm. There was a time-dependent increase in cells that lost their ΔΨm such that in the first 5–10 min about 40% of cells had a lowered ΔΨm, but by 1 h after treatment over 80% of cells had lost their ΔΨm. While this usEP treatment condition was generally lethal for a significant population of cells, some cells were less vulnerable, requiring longer times before dissipating their ΔΨm, indicating a rather heterogeneous population of cells based on sensitivity to usEP. Similar time-dependent decreases in ΔΨm were shown on N1-S1 cells between 1 and 30 min post pulse (Beebe et al. 2012) (Fig. 8.10).

When usEP conditions of electric field and/or pulse number are sufficiently intense and at least in some cases when Ca^{2+} is present, loss of ΔΨm happens relatively quickly, within 15–20 min, in populations of cells after usEP treatment. This is a seriously catastrophic cellular event if it is not corrected soon, because this would be disastrous for not only mitochondrial function, but also for cellular existence. Under these conditions, the basis for oxygen utilization and ATP production is lost. Given that mitochondria are the ATP-producing powerhouse of the cell as well as the executioner, mitochondria are valuable usEP targets when usEP conditions are designed to kill cancer.

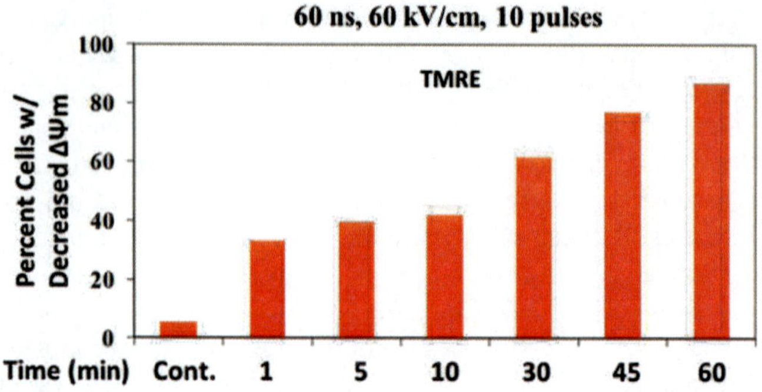

Fig. 8.10 UsEP-induced dissipation of ΔΨm are time-dependent. Jurkat A3 cells were loaded with TMRE as an indicator of the ΔΨm and then treated with ten 60 ns pulses (10 ns rise-fall time) and 60 kV/cm or 0.011 Vs/cm. At various times after treatment (1–60 min) cells were analyzed for TMRE fluorescence by flow cytometry. The data show the percentage cells with a decreased ΔΨm. (Sain and Beebe unpublished)

Loss of ΔΨm results in an influx of water and ions such as Ca^{2+} and other cations flow down their electrochemical gradients across the IMM, tending to equalize their concentrations between the mitochondrial matrix and the cytoplasm, thereby losing the proton gradient that is required for ATP production. Since electric field effects are best known for affecting lipid membranes, when even low usEP conditions were observed to dissipate the ΔΨm, it was reasonable to consider that this was likely due to nanoporation of the IMM. Yet permeabilization of the IMM could occur by opening the mPTP (Vernier 2011; Batista Napotnik et al. 2012). Jurkat cells were shown to lose their ΔΨm in a time-dependent manner over 5 h that was pulse number dependent in response to pulses with a pulse duration of 20 ns and an electric field of 20 kV/cm (4×10^{-4} Vs/vm /pulse) (Vernier 2011). Using three different ΔΨm indicators, 4 ns, 100 kV/cm pulses (4×10^{-4} Vs/cm/ pulse) were shown to induce a time-dependent and pulse number-dependent dissipation of ΔΨm 30 s to 3 min after pulsing. Under these same conditions their studies also showed a differential plasma membrane permeabilization of YO-PRO-1 over propidium iodide, indicating nanopore formation, especially at lower usEP conditions (Batista Napotnik et al. 2012).

Given that usEPs permeabilize plasma membranes, dissipate the ΔΨm and decrease cell viability, it was interesting to see the relationship among these characteristics. Figure 8.11 shows these relationships in Jurkat E6.1 cells 10–15 min after treatment with ten 60 ns pulses at various electric fields. There was an electric field-dependent increase in intracellular Ca^{2+} with electric fields as low as 10 kV/cm or (1.9×10^{-3} Vs/cm. As electric fields were increased there were electric field-dependent loss of ΔΨm and decreases in cell viability beginning at about 30 kV/cm or 5.7×10^{-3} Vs/cm. Decreases in ΔΨm and cell viability were closely coordinated. Thus, among these three cell indicators, influx of Ca^{2+} through the plasma membrane was the most sensitive cell response. Furthermore, intracellular Ca^{2+} were elevated

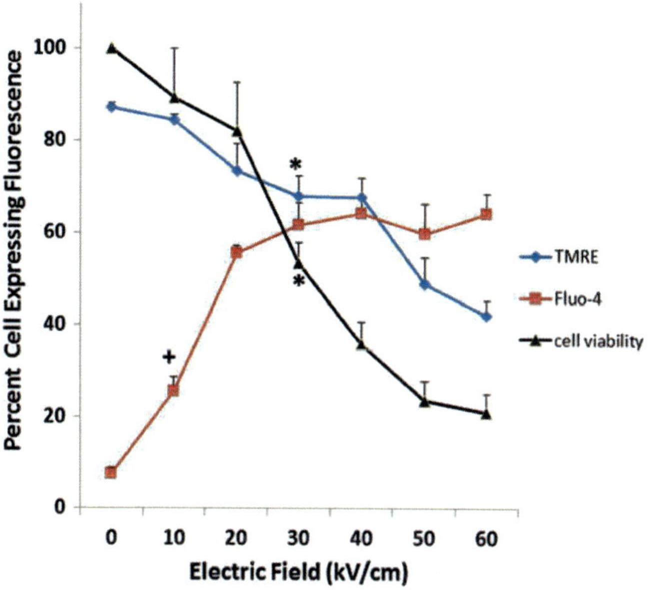

Fig. 8.11 UsEPs increase Ca^{2+} influx, dissipate $\Delta\Psi$m, and decrease cell viability. Jurkat E6.1 cells were loaded with Fluo-4 as a Ca^{2+} indicator and TMRE as an indicator of $\Delta\Psi$m. Cells were then treated with ten 60 ns pulses at various electric fields and analyzed by flow cytometry 10–15 min after pulsing. To determine cell viability cells were assayed 24 h later using the MTS assay. * n = 3, $p < 0.05$ vs control (Sain and Beebe unpublished)

at electric fields when significant decreases in $\Delta\Psi$m occurred. Similar results were observed in N1-S1 cells (Beebe et al. 2012).

UsEP waveforms with short rise-fall times have greater effects on $\Delta\Psi$m and cell death–An original hypothesis for usEP effects on cellular biology was that, unlike pulses in the micro- and milli-second range, shorter pulses with shorter rise-fall times and thereby higher frequency components, had greater probability for affecting intracellular structures and functions (Schoenbach et al. 2001). To address the question of possible differential effects of pulse rise-fall times, pulses with waveforms with short rise-fall times (15 ns) were compare with pulse waveforms with longer rise-fall times (150 ns) (Fig. 8.12). Relatively long pulse durations of 600 ns were tested, which were significantly longer than the membrane charging time constant of about 70 ns (Cole 1937; Schoenbach et al. 2001), so the tested rise-fall times were near and far, respectively, from the membrane charging time constant (Beebe et al. 2012).

Effects of usEPs on N1-S1 hepatocellular carcinoma (HCC) cells, a model for usEP liver cancer treatment, were used to determine effects on mitochondria ($\Delta\Psi$m with TMRE) and plasma membranes (Ca^{2+} with Fluo-4 and PI) using single 600 ns pulses with varying electric fields (0–80 kV/cm) and short (15 ns) vs. long (150 ns)

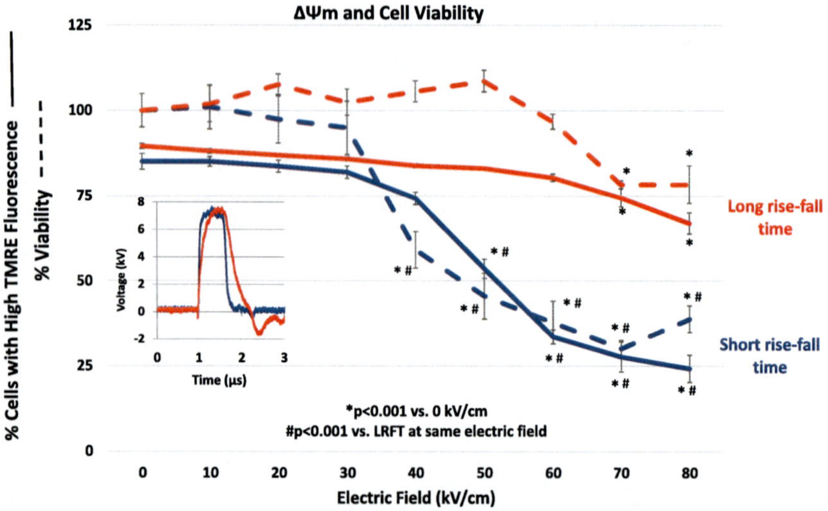

Fig. 8.12 UsEP short rise-fall time has great effects on ΔΨm ad viability. N1-S1 HCC cells were treated with durations of 600 ns with short (15 ns) and long (150 ns) rise-fall times at various electric fields delivering 0.006–0.048 Vs/cm. The ΔΨm was determine with TMRE 10–15 min post-pulse and viability was determined 24 h post-pulse by CellTiter-Glo Luminescent Cell Viability Assay Kit (Promega, Madison, WI) in Beebe et al. (2012). The inset shows the pulse waveforms for 15 (blue) and 150 ns (red) rise-all time pulses. The data represents a different experimental set from previous studies in Beebe et al. (2013)

rise and fall times. Figure 8.12 shows experiments of usEPs with 600 ns durations with short 15 ns and long 150 ns rise-fall times on the ΔΨm using TMRE as a mitochondrial indicator solid lines and cell viability (dotted lines). Cells exposed to pulses with a 15 ns rise fall time (solid blue trace) begin to lose their ΔΨm, as indicated by decreased TMRE fluorescence, at lower electric fields (40–50 kV/cm, 0.024 Vs/cm vs. 70–80 kV/cm, 0.042 Vs/cm) and with a greater slope than cells exposed to the 150 ns rise-fall time (solid red trace). When cell viability was determined 24 h later at the same usEP conditions, viability was closely correlated with loss of ΔΨm. There were no differences of rise-fall times on permeability of the plasma membrane, which occurred at about 10 kV/cm (0.006 Vs/cm) like that shown for the 15 ns rise-fall time in Fig. 8.11 above (Beebe et al. 2012). These results supported the concept that pulses with shorter rise-fall time have greater effects on ΔΨm, which appears to be a determinant of cell viability, than pulse with longer rise-fall times, This is consistent with the hypothesis that intracellular effect occur during the rise-fall times while effect on the plasma membrane occur at the pulse plateau (Schoenbach et al. 2001).

This study also demonstrated that under these pulsing conditions that Ca^{2+} permeabilization of the plasma membrane was more sensitive than dissipation of the ΔΨm (~10 kV/cm, 0.006 Vs/cm vs. ~ 40 kV/cm, 0.042 Vs/cm) (Beebe et al. 2012). This indicates that the elevated cytoplasmic Ca^{2+} is present when effects on ΔΨm are observed, consistent with Ca^{2+}-dependent effects on $\Delta\Psi m$ (Fig. 8.9). For these short rise fall time pulses, loss of cell viability for these cells occurs as charging effects exceed about 2–4 × 10^{-2} Vs/cm. In addition, since the shorter 15 ns rise-fall time pulse exhibits higher frequency components than longer 150 ns rise-fall time, it is possible that higher frequency components of usEPs are responsible for greater intracellular effects, especially on mitochondria and consequently on cell viability.

Since mitochondria are purveyors of cell death, Bcl-2 was shown to protect mitochondria (Thomenius and Distelhorst 2003; Shimuzi et al. 1998) and Bcl-xl was shown to inhibit redistribution of cytochrome c from the intermembrane space to the cytosol preceding loss of $\Delta\Psi m$ (Vander Heiden et al. 1997). To determine effects of Bcl-2 family members on usEP—induced loss of $\Delta\Psi m$, we treated Jurkat cells that over expressed Bcl-xl (Fig. 8.13) and Bcl-2 (not shown) for their effects on usEP-induced loss of $\Delta\Psi m$.

Bcl-xl overexpression does not protect usEP-induced dissipation of $\Delta\Psi m$—It is thought that loss of $\Delta\Psi m$ is an early step in apoptosis programs and that stabilization or prevention of this loss in $\Delta\Psi m$ can be prevented by anti-apoptotic proteins as a mechanism of resistance to chemotherapy (Decaudin et al. 1997). Bcl-2 and Bcl-xl bind to and prevent homo-oligomerization of pro-apoptotic family members, such as Bax and Bak, inhibiting their pro-apoptotic activation. Specifically, Bak adapts an amphipathic alpha-helix that interacts with Bcl-xl through hydrophobic and electrostatic interactions (Sattler et al. 1997). Another proposed mechanism indicates that Bcl-xl binds to the mitochondrial ATP synthase β subunit by stabilizing $\Delta\Psi m$ and thereby mitochondrial energetic capacity (Alavian et al. 2011; Chen et al. 2011). More recently it was proposed that another mechanism for Bcl-2 family member proteins to protect mitochondria against Ca^{2+}-mediated apoptosis is that the Bcl2/Bclxl BH4 domain binds and inhibit the IP_3 receptor from transferring excessive Ca^{2+} loads to mitochondria through the ER IP_3 receptor-mitochondria VDAC interaction site (Monaco et al. 2015).

To determine if Bcl-xl could protect Jurkat cells from usEP-induced dissipation of $\Delta\Psi m$, we treated Jurkat E6.1 cells that stably over-expressed Bcl-xl with usEPs and compared effects on $\Delta\Psi m$ and cell viability with controls cells containing an expression empty vector. The inset showing a western blot that Bcl-xl was over-expressed nearly sevenfold in the Jurkat clone. Jurkat cells were treated with ten 60 ns pulses with increasing electric fields and analyzed cells for effects on $\Delta\Psi m$ 10–15 min after treatment (Fig. 8.13a) and 24 h later for viability (Fig. 8.13b). There was an electric field-dependent decrease in $\Delta\Psi m$ and a corresponding decrease in cell viability with Jurkat cells that did and did not over-express Bcl-xl, with no detectable differences between them. In data not shown, similar results were observed in a Jurkat clone over-expressing Bcl-2. These studies indicate that over-expression of anti-apoptotic protein could not prevented the usEP-induced dissipation of $\Delta\Psi m$. Whether these anti-apoptotic proteins prevent homo-oligomerization of pro-apoptotic proteins at the mitochondria membrane or whether Bcl-xl binds to the β-subunit of ATP synthase to stabilize $\Delta\Psi m$ or whether Bcl2/Bclxl prevents

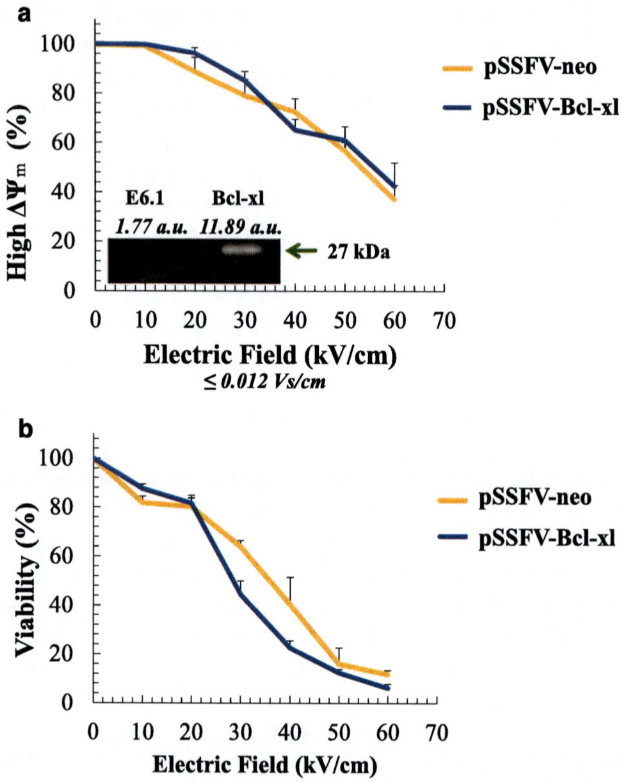

Fig. 8.13 Bcl-xl does not protect usEP-induced dissipation of ΔΨm. Jurkat clones that overexpressed Bcl-xl in psFFV neo or the empty vector were loaded with TMRE and treated with ten 60 ns pulses with electric fields 0–60 kV/cm or 0.002–0.011 Vs/cm. Ten-15 min post pulse cells were analyzed by flow cytometry for ΔΨm and 24 h later analyzed for viability as indicated in the legend to Fig. 8.12. The Jurkat clone were generous gifts from Dr. JD Robertson Department of Pharmacology, Toxicology and Therapeutics, University of Kansas Medical Center, Kansas City, Kansas. (Ruedlinger and Beebe et al. unpublished)

excessive Ca^{2+} transfer from the ER to the mitochondria, none of these mechanisms are effective to prevent or attenuate the usEP-induced loss of ΔΨm in Jurkat cells.

Mitochondrial permeability transition pore (mPTP)—is a megachannel connecting the cell cytoplasm to the mitochondrial matrix through the IMM and the OMM. In the late 1979s, Hawthorn and Hunter (1979) described the opening of a high conductance channel in the IMM induce by Ca^{2+} that resulted in swelling of mitochondria among other aspects of this megachannel in mitochondria (Hunter and Hawthorn 1979a, b). The pore is large enough to allow molecules as large as 1.5 kDa to cross the otherwise impermeable IMM. When this channel is fully open, the ΔΨm collapses. ATP is hydrolyzed instead of synthesized, oxidative phosphorylation is uncoupled, reactive oxygen species (ROS) are generated, matrix Ca^{2+} is released and respiratory control is lost. If the proton gradient across the IMM is not restored, mitochondria swell, the OMM is ruptured, mitochondrial proteins and DNA are released into the cytoplasm and cell death ensues. Depending on the context, cell death can be by apoptosis, if ATP levels are sufficient, or otherwise by regulated necrosis (Halestrap et al. 2006). However, as will be presented later regulated cell death is determined by factors other than ATP levels(See section of usEP induce RCD mechanisms, Chap. 9). The pore is modulated by several factors including activators such as Ca^{2+} with Pi and reactive oxygen species (ROS) and inhibitors such as adenine nucleotides, low pH, divalent cations like Mg^{2+} and cyclophilin inhibitors like cyclosporine A (CsA) (Crompton 1999). The mPTP is likely regulated by post-translation modifications such as phosphorylation (Halestrap et al. 2006). Unregulated or prolonged opening of the mPTP and cell death are associated with several diseases such as cerebral and myocardial ischemia/reperfusion injury, muscular dystrophy caused by defects in collagen IV and other degenerative disorders (Hurst et al. 2017; Rasola and Bernardi 2007; Pérez and Quintanilla 2017). Although regulation and dysfunction of the pore are understood, and it is agreed that mPTP is composed of proteins, the exact structural composition of the pore remains controversial.

Although the identity of the proteins that make up the mPTP is still debated, CypD is a well-characterized, genetically proven regulator of the mPTP (Giorgio et al. 2017). It has been suggested that in addition to the Na^+/Ca^{2+} and H^+/Ca^{2+} exchangers, flickering or transient opening of the mPTP may be a mechanism of Ca^{2+} efflux from the mitochondrial matrix to diminish sustained mitochondrial Ca^{2+} overload. Another possible physiological role for transient opening of the mPTP is in the production of brief "flashes" of ROS, which have been shown to be concurrent with decreases in $\Delta\Psi m$ and attenuated by CsA and CypD knockdown. This may be cell type-specific since it is not observed in skeletal muscle (Kwong and Molkentin 2015). However, under conditions of usEP-induced dissipation of $\Delta\Psi m$, this is likely to be non-reversible, since there is a time-dependent increase in cells with lower $\Delta\Psi m$ (Fig. 8.10) and this is closely correlated with numbers of cell that lose viability (Figs. 8.11 and 8.12) (Beebe et al. 2012, 2013, 2015). As might be suggested by the evolutionary conservations of cyclophilins (CyPs), they have been shown to be involved in a wide variety of pathophysiological conditions from inflammation, infections, vascular diseases, and innate immunity, in addition to its role in regulating the mPTP (Giorgio et al. 2017; Kwong and Molkentin 2015).

It is important to note that CypD is not a structural protein of the mPTP, but a regulator of the pore. The CypD bound form of the mPTP has a higher probability for the opening conformation. CsA binds to and removes CypD from its mPTP binding site, thereby revealing a Pi binding site, which upon binding decreases the probability of pore opening. Thus, CsA binding to CypD allows Pi to inhibit mPTP opening. Thus, CsA is really a desensitizing agent as the mPTP is more resistant to opening after the uptake of Ca^{2+}. In the presence of CsA, pore opening requires about twice the Ca^{2+}-Pi concentrations compared to when it is absent.

Figure 8.14a shows the original and new model of the mPTP. The earlier model of the mPTP included voltage-dependent anion channel (VDAC) in the OMM, adenine

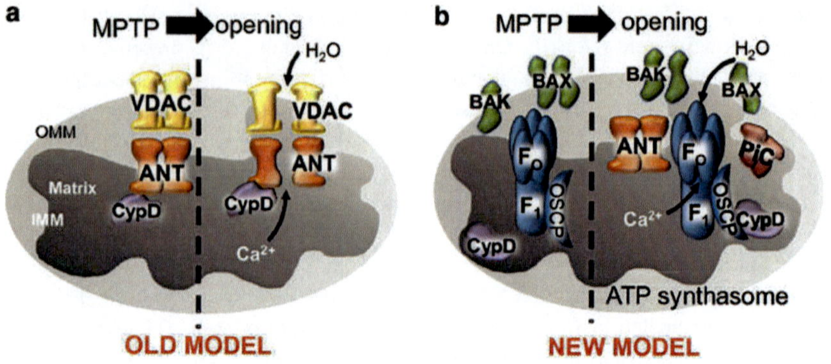

Fig. 8.14 Schemes of the original and new models of the mPTP. From Karch and Molkentin (2014)

nucleotide transporter (ANT) in the IMM and CypD in the matrix. These three components formed CsA-sensitive pore in lipid membranes. Further, VDAC exhibited similar electrophysiological characteristics to mPTP and it was a non-selective pore in the OMM. However, mitochondria lacking all three VDAC isoforms still exhibited permeability transition (PT). The ANT in the IMM was suggested because it exchanged ADP for ATP in the matrix and was an important ATP synthesis machinery. Furthermore, it directly interacted with CypD. However, deletion of ANT did not knockout Ca^{2+}-induced PT, ruling out ANT as a structural mPTP component. Figure 8.14b shows a newer model of the mPTP (Karch and Molkentin 2014).

Another candidate for the mPTP is the ATP synthase or complex V of the ETC (Fig. 8.15; also see Fig. 8.4 above and 8.17). The F_0 domain is in the IMM while the F_1 catalytic domain faces the matrix. These two components are connected by the central and peripheral stalk, which exhibits an OSCP (oligomycin sensitivity conferral protein) subunit that binds CypD. When ATP synthase dimerizes, which have been hypothesized to form the pore, were included in lipid membranes, typical mPTP

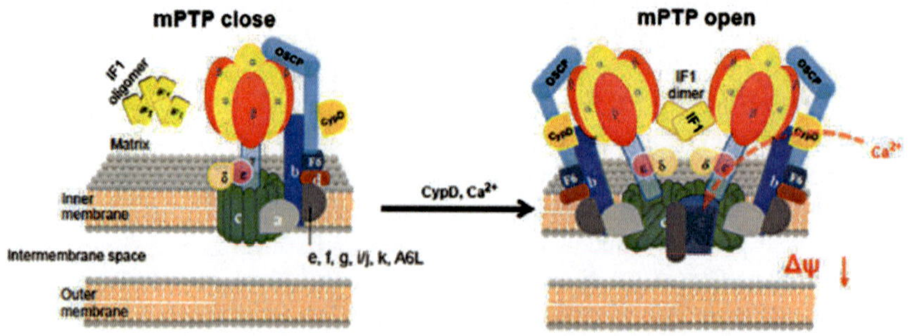

Fig. 8.15 The schematic models of ATP synthase serving as the mitochondrial permeability transition (mPTP). From Long et al. (2015)

characteristics were observed. When the purified *c*-subunit was included in liposomes, a voltage gated channel with high conductance was observed that was sensitive to adenine nucleotides, but not to CsA. However, when the F_0 and F_1 domains were included with the OSCP subunit from the peripheral stalk, a CsA-sensitive, CypD-regulated pore was observed. Finally, when shRNA was used to reduce the *c*-subunit, Ca^{2+}-mediated mPTP opening was prevented. These studies suggest that the *c*-subunit of the F_0 ATP synthase domains functions as the mPTP. However, when the *c*-subunit genes were disrupted, vestigial ATP synthase complexes were formed without c-subunits, mPT was still observed. These results suggest that *c*-subunit of ATP synthase is not an essential structural component of the mPTP. There are also requirements for hydrophilic surfaces when pores are open and hydrophobic surfaces when pores are closed, that do not appear to be met by ATP synthase dimers. Nevertheless, ATP synthase subunits e, f, and g have not been examined as structural components and could possibly fill this role. There are other possible candidates for the mPTP. For example, the pro-apoptotic proteins Bax and Bak have been connected as mPTP proteins. These proteins bind to the OMM, can cause increased permeability, mitochondrial swelling, OMM rupture, and cytochrome *c* release. Mitochondria that do not express Bax and Bak do not show mPT and are protective against cardiac ischemia–reperfusion injury, suggesting that Bax and Bak are essential components of the mPTP.

The earlier mPTP model with VDAC in the OMM and ANT as a CypD binding protein in the IMM is replaced with a developing model. This mPTP model, the ATP synthase associated with ANP and PiC, all CypD binding proteins, in the IMM and Bax and Bak in the OMM constitute a present working model for the mPTP. These proteins are also part of the ATP synthasome as an important IMM complex constituting an association among ANT that makes ADP available, and Pi that makes Pi available to the ATP synthase for the energy metabolism.

Effects of usEPs on mitochondrial respiration–While many studies with usEPs have been designed to use lethal effects for cancer ablation, not all effects of usEPs on cells are lethal. For example, when 600 ns pulses with long 150 ns rise-fall times were applied to N1-S1 cells with a mismatched load, Ca^{2+} influx occurred at low electric fields (~10 kV/cm), which was not different than when the load was matched with 600 ns durations and 150 ns or 15 ns rise-fall times, but there were no effects on $\Delta\Psi$m or cell death (Beebe et al. 2013). When ten 10 ns, 50 kV/cm pulses (0.005 Vs/cm) were applied to intact Jurkat and U937 cells, oxygen consumption was increased. However, when > 100 pulses were applied, oxygen consumption in Jurkat cells was decreased, but no effects were observed n U937 cells. When these similar pulses were applied to mitochondria isolated from U937 cells, no effects were observed (Estlack et al. 2014a, b). These studies indicated that usEPs could affect metabolic activity in cells.

In studies from the authors' laboratory, Fig. 8.16 shows effects of usEPs on metabolic activity determined by measuring oxygen utilization in Jurkat cells treated with ten 60 ns 20 kV/cm pulses, (0.0038 Vs/cm). It can be seen in Fig. 8.11 that under these same conditions (20 kV/cm) with this same Jurkat clone that only small effects were present on $\Delta\Psi$m and viability in the major population of cells compared to

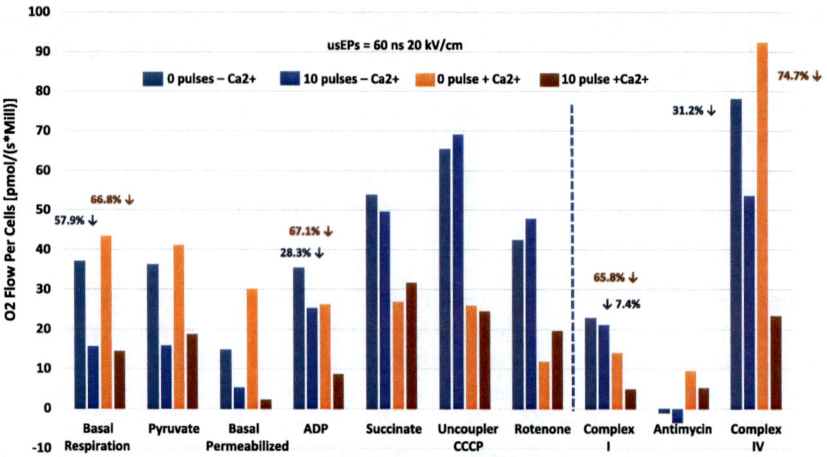

Fig. 8.16 High resolution respirometry of Jurkat cells–Jurkat E6.1 cells were treated with usEPs in Miro5 buffer with and without Ca^{2+} and analyzed for oxygen consumption with high resolution respirometry using the Clark electrode (Oroboros O2K). Cells were treated with ten 60 ns pulses at 20 kV/cm or 0.0038 Vs/cm. A typical experiment is illustrated (Lai and Beebe unpublished)

control. In order to measure mitochondrial respiration, it is necessary that the major population of cells exhibit a relatively high ΔΨm like those in the study shown here. Consequently, effects measured here on cell respiration were at electric fields that were below lethal levels and below conditions that caused (tense of the paragraph needs to be checked) significant loss of ΔΨm and opening of the mPTP. These studies were done under condition that allowed measurement of basal respiration levels as well as when cells are permeabilized and exposed to substrates such as succinate, ADP, uncoupling agents such as CCCP, which depolarizes mitochondria by abolishing the proton gradient across the inner mitochondrial membrane; rotenone, which inhibits complex I; and antimycin, which inhibits complex III, which is between cytochrome b and cytochrome c1.Using these reagents, it is possible to dissect the ETC to determine where usEPs affects respiration. These analyses are expected to measure the subtle effects of usEPs on mitochondrial metabolic function as determine by oxygen utilization.

When oxygen utilization is determined in intact cells, basal respiration is measured. It can be seen that usEPs attenuate respiration by around 60% in the presence or absence of Ca^{2+}. When the complex I substrate pyruvate was added, there was no effect. This demonstrated that the cells were intact because intact cells would not be permeable to pyruvate or malate, which was also added, but not shown. Then cells were permeabilized with a known amount of DMSO, which must be titrated for each cell type to ensure minimal mitochondrial damage. When ADP was added, respiration was stimulated except under non-pulsed conditions in the presence of Ca^{2+}. In the absence of Ca^{2+}, but not in the presence, effects of usEPs on basal respiration were less. When the complex II substrate succinate is added, there were no effects on usEPs regardless of the presence or absence of Ca^{2+}. Thus, the presence of complex II substrates resolved effects of usEPs on mitochondrial respiration. When electron transport was uncoupled for oxidative phosphorylation, no effects of usEPs were observed. In order to determine effects usEPs on complex I, rotenone, a complex I inhibitor was added. The difference between uncoupled respiration and that observed in the presence of rotenone, shows effects of usEPs on complex I. In the absence of Ca^{2+} there were only small effects of usEPs on complex I, suggesting that effects of usEPs on complex I were Ca^{2+}-dependent. This is reminiscent of the effects of Ca^{2+} on loss of $\Delta\Psi m$. In the presence of Ca^{2+}, usEPs inhibited complex I by more than 60%. When the complex III inhibitor antimycin was added, ascorbate was also added as an electron donor and TMPD facilitate electron transport from ascorbate to cytochrome c and then through Complex IV to determine its rate of electron transport. These observations indicate that usEPs have significant effects on complex IV in the absence and presence of Ca^{2+} by about 30% and 75% respectively. Thus, usEPs inhibited respiration at complex I and IV of the ETC.

Sites of superoxide anion and ROS production in the ETC—As shown in Fig. 8.16, high resolution respirometry of Jurkat cells shows that usEPs cause attenuation in respiration primarily from complex I and complex IV. As usEP induces attenuation of oxygen utilization, electron flow through the ETC is decreased and the probability for superoxide anion release is increased. It is reasonable that Complex I is considered the most vulnerable complex in the ETC (Papa et al. 2008)—it is a regulated entry site for electron transport, a primary site of ROS production only in the mitochondrial matrix (complex-III generated ROS also diffuses into the cytoplasm) and mitochondrial membranes and proteins are very near these ROS-generating mechanisms. Given these considerations, it is not surprising that usEPs affect complex I of the ETC. Antioxidant mechanisms are vital for protecting this essential respiratory machinery such that they do not become overloaded, which would give ROS free reign to induce oxidative damage and cause mitochondrial dysfunction. Much of these effects appear to be Ca^{2+}-dependent. Such effects on complexes I and IV are likely predictive of Ca^{2+}-dependent effects on $\Delta\Psi m$ and the mPTP when usEPs condition become more intense. In fact, complex I has been reported to regulate the mPTP (Li et al. 2012). Although mechanisms that affect oxygen consumption by attenuating electron transport at complex I and IV are not clear, it is likely that usEPs affect electron transport in the ETC. When oxygen flux is decreased, electron transport is decreased as is ATP production. UsEP also inhibit redox activity in the plasma membrane (see us effects on the plasma membrane, Figs. 2.5–2.7). UsEPs attenuate electron transport by some mechanism that remains to be determined.

Complexes I and III produce low levels of superoxide anions, which have a short half-life and are quickly converted to H_2O_2 by superoxide dismutase (SOD). H_2O_2 can be converted to highly toxic hydroxyl radical (HO^-) by the Fenton reaction when ferrous iron (Fe^{2+}) reacts with H_2O_2 to produce ferric iron (Fe^{3+}) hydroxide ion and hydroxyl radical. Damage due to hydroxyl radicals can be limited by converting H_2O_2 to water by mitochondrial antioxidants such as glutathione peroxidase (GPX) or thioredoxin-dependent peroxiredoxin (PRDX) (Kauppila et al. 2017).

Complex I contains 7–8 iron-sulfur complexes that transfers electrons in the forward direction (FET) between flavin mononucleotide (FNM, called I_F), where NADH is oxidized to NAD+, and Q-binding site (called I_Q) and where electrons are transferred to ubiquinone. I_F, I_Q, and iron sulfur complex N2 are the putative sites for superoxide anion production (Scialo et al. 2013, 2017). If usEPs inhibit electron transport through complex I like rotenone does, blocking electron transport through complex I at I_Q causes NADH to accumulate, over-reducing the I_F site where electrons leak to reduce oxygen to superoxide anions through FET (Scialo et al. 2017). When the pool of coenzyme Q (CoQ) is over-reduced, reverse electron transport (RET) occurs towards complex I. Thus, the redox state of the CoQ pool can determine electron leak and superoxide anion production as a mitochondrial signaling mechanism. This also occurs when succinate is supplied complex II as some of the electrons flow in the reverse direction. Interestingly, shifting oxidation of glucose to fatty acids alters the NADH:FADH2 ratio from 5:1 to 2:1 favoring oxidation through complex II and reshuffling complex I for ROS production by RET (Guaras et al. 2016). Under RET conditions, superoxide anion leak from the I_F or I_Q sites. ROS is produced by FET or RET at complex I, affecting oxidation of several proteins, suggesting site-specific ROS signaling that can activate specific signaling pathways (Bleier et al. 2015). For example, when macrophages respond to infection, they switch metabolism from oxidative phosphorylation to glycolysis, decrease levels of complex I and increase oxidation of succinate through complex II, which favors ROS production through RET. Inhibiting succinate oxidation blocks RET and ROS production and inhibits pro-inflammatory cytokine production implicating RET ROS signaling in metabolic immune cell programming (Scialo et al. 2016).

The model proposes that usEPs act on complex I to attenuate electron transport as determined by the decrease in oxygen consumption, which generates super oxide anions as ROS, which diffuse through the mitochondrial matrix. In the presence of Ca^{2+} ROS generation is enhanced. In the presence of superoxide anions (ROS) and Ca^{2+}, CypD is modified to open the permeability transition pore, which dissipates the $\Delta\Psi$m. The OSCP subunit of ATP synthase may also be affected by ROS and Ca^{2+} and participate in the mPTP response. Some superoxide anions may be provided from complex III, which diffuses into the matrix and cytoplasm. Complex I and III and their ROS products are in the matrix with complex V or ATP synthase/mPTP.

Possible mechanisms for usEP effects on oxygen consumption—How might usEPs affect oxygen consumption through the ETC? It is possible to understand how superimposition or "injection" of usEPs might be expected to affect movement of charged electrons and protons in the transport pathways like charged particle in among redox enzymes. Such effects would be expected to last if electric fields were applied, but not necessarily for long periods of time that are required for these measurements. It is possible that charging effects of usEP directly affect electron and proton movements, but it is unlikely that this is responsible for observed long lasting effects on oxygen consumption. Given that electron and proton transport are closely coupled, it is possible that usEPs could selectively affect either process directly and thereby affect overall transport of both ions and oxygen utilization. However, given the relationship between slowing electron transport and increases in superoxide anions, it is most likely that usEPs primarily attenuate electron transport. In any event, effects of usEPs to decrease the efficiencies of complex I and complex IV are long lasting, lasting more than an hour, which is required for making these measurements. These long-lasting effects suggest that more "stable" usEP effect have occurred. Long lasting effects could occur by direct effects of usEPs on proteins or more likely indirect effects on proteins and/or lipids by oxidation.

Water is very likely a usEP target; again, this is not likely to be long lasting unless it affects behaviors of other molecules due to stable hydration effects. Another possibility is that usEPs could affect charged mechanisms in water chains as reversible proton wires (Hummer and Wikstrom 2016). Because of differences in structural arrangements of ET and PT pathways in complex I and complex IV, the difference in transport mechanisms would appear to be due to how ET is coupled to PT. Movement of protons through hydrophobic membranes is not a simple task, so it is likely that mechanisms of PT in complex I and complex IV have some conserved elements. Molecular dynamic simulations in complex I indicate that water can penetrate the center of 3 antiporter-like subunits L, M and N in the membrane arm forming a half channel as part of a possible PT pathway (Hummer and Wikstrom 2016). A half channel on the positive side of the membrane is separated from another half channel on the negative side of the membrane. Such channels are filled with ionizable residues appropriate for water connections. Simulations clearly indicate water-mediated connections between half channels in the membrane arm of complex I. Since these channels are on different sides of the membrane, they are not always continuous, but appear to function as valves with wet and then dry periods that prevent proton leakage. Charged clusters of residues create a significant electric field within the membrane. Water is pulled into this inhomogeneous electric fields like an electroporation effect that wets the pore. Once the proton is pulled into the pore, the negative charge neutralizes and weakens the electric field and the water chain becomes unstable. It is certainly conceivable that usEPs could affect this electroporation-like effect on these proton conduits in membranes containing these proteins. However, such effects would be active only during the presence of the electric field, not for later times during oxygen consumption measurements. Once again, long lasting effects on oxygen consumption are likely due to other usEP mechanisms.

These effects on Complexes I and IV also appear to be rather stable since it takes more than an hour to determine respiration through this protocol and these usEP effects are still present. Like effects of usEPs on CypD, effects on Complexes I and IV are likely on proteins. Given the generous usEP-induced production of superoxides in mitochondria and the high affinity of superoxide anions for iron-sulfur cluster in proteins, it is possible that proteins given the potential vulnerability of complex I as a major superoxide anion producer, lipids and proteins in this ETC complexes are susceptible to effects of oxidation when oxidant exceed the antioxidant protective mechanisms. This can initiate the release of iron, which causes structural alterations of proteins, especially at Cys residues (Rimessi et al. 2016). Lesions in proteins can also occur by carbonylation, nitration or tyrosine derivatization as well as by formation of adducts with lipid peroxidation products, such as malondialdehyde (MDA) and 4-hydroxynonenal (HNE). Thus, it appears that complexes I and IV as well as CypD are targets that change functions due to usEP-induced oxidative effects. Since usEP-induced effects on oxygen flux or consumption do not result in significant cell death under these conditions, it is possible that usEP-induced decrease in electron transport through altered or possibly damage to complex I and IV may be like or related to aging (Kumaran et al. 2004, Preston et al. 2008; Tatarkova et al. 2011) or related to mitohormesis as an adaptation to stress. Mitohormesis is a synchronized mitochondrial response to mild stress that enhances subsequent responses to similar pressures (Yun and Finkel 2014).

There are other possible considerations for how usEPs could attenuate oxygen consumption. However, given that mutations in subunits most distant from the oxidoreductase and iron-sulfur clusters can affect both redox activity and proton pumping in complex I (Michel et al. 2011; Hummer and Wikstrom 2016), predicting usEP effects on specific subunits in complex I will be challenging. As suggested, it is likely that these usEP effects are due to relatively long-lasting effects on proteins and/or polypeptides in complex I and IV. The usEP effects are on genuine proton pumps that closely couple electron transport with (ET) proton translocation (PT). However, although the structural organization for electron channeling and PT pathways in these two complexes are different, there are likely similar structures that are affected by usEPs, suggesting that some aspects of their functional mechanisms are similar. Thus, it is possible that by identifying how usEPs affect both complexes, usEPs can serve as a tool to probe effects on complex I and IV and enhance our understanding of how these disparate structures function with such thermodynamic efficiency.

Recalled from effects of redox activity in the plasma membrane, [see usEP effects on the plasma membrane, Figs. 2.5–2.7] usEPs decreased redox activity, which is essentially electron transport, in the plasma membrane suggesting the usEPs attenuate electron transport in the mitochondria as determined by oxygen consumption and plasma membrane as determined by redox enzyme activity. So, usEPs have more generalized effect on electron transport. Thus, the redox enzymes that transport these electrons may be affected by usEPs, either directly or indirectly by effects of ROS-mediated modifications.

It is possible that usEPs could have direct effects on proteins as previously considered (Beebe 2015). Since proteins are essentially complex dipoles, it is possible they could be rearranged by modulating one or more of their local structures having effects on overall structure and function. This would be analogous to effects of phosphorylation, whereby negatively charged phosphate groups from ATP at specific phosphorylation sites alter conformation and functional properties of phospho-proteins. This is a highly conserved mechanism for reversibly regulating cell function by activities of kinases (adding phosphates) and phosphatases (removing phosphates) in cell function (Graves and Krebs 1999; Cohen 2000; Pawson and Scott 2005) and in mitochondria (Lim et al. 2016; Lucero et al. 2019).

Another way that usEPs could directly affect protein structures and functions is to disrupt their active structures. Proteins must assume a structure that supports their appropriate function. Many enzymes depend on chaperones to help them fold into functional conformations. Chaperones assist the covalent folding of proteins and prevent assembled subunits from aggregating and losing their functional structures. An example of possible denaturing effects of usEPs on a protein structure was shown by the inactivation of the catalytic activity of the catalytic (C) subunit of the cAMP-dependent protein kinase (PKA) in a pulse duration- and electric field-dependent manner (Beebe 2015). Given that PKA is the prototype protein kinase, this has broader implications for usEP effects on kinases, in particular and enzymes in general. This may also be important because PKA is known to be important for mitochondrial function (Acin-Perez et al. 2009; Monterisi and Zaccolo 2017).

UsEPs could also indirectly regulate proteins that are Ca^{2+}- and ROS-regulated as opposed to direct usEP actions. In fact, given the known mechanisms for regulation by many proteins by these signaling molecules, this indirect mechanism is hard to disprove. In fact, we hypothesize that usEPs affects CypD (Figs. 8.9 and 8.17) by increasing Ca^{2+} (Fig. 8.9) and generate ROS in a Ca^{2+}-dependent manner (Fig. 8.8). UsEPs attenuate oxygen consumption on Complex I (Fig. 8.16), which is likely to generate ROS.

Another possibility for usEP effects on redox proteins would to be affect the secondary or tertiary conformation of proteins in complex I and/or complex IV. It is also known that the ETC can establish supramolecular organizations forming supercomplexes with interactions among individual complexes forming a so-called respirasome, such as $I + III_2 + IV_{1-2}$ or "complex strings" with complex V or ATP synthase into oligomeric chains (Melber and Winge 2016; Guo et al. 2016). Such supercomplex organization exhibit greater efficiency in oxygen utilization and enhanced respiration and ATP production (Dudkina et al. 2010; Genova and Lenaz 2014). It is possible that usEPs attenuates respiration by affecting participation of

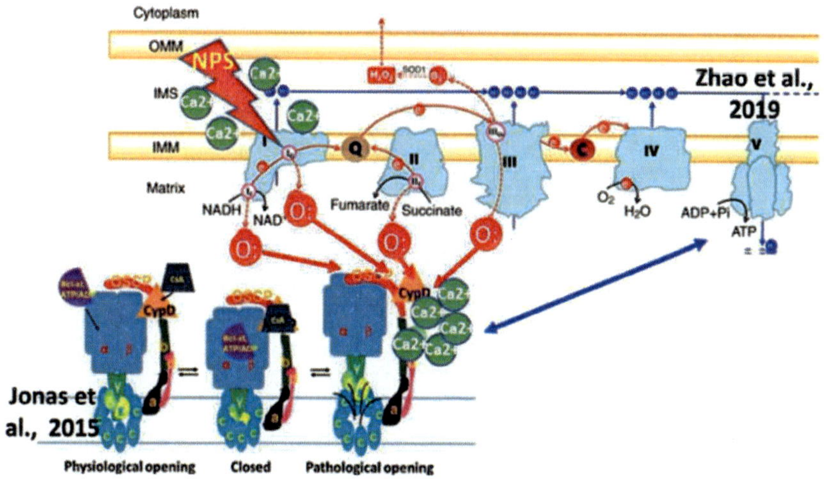

Fig. 8.17 The electron transport chain and ATP synthase (Complex V) as the mPTP— Adapted and modified from Zhou et al. (2019) and Jonas et al. (2015)

complex I and or IV in these supercomplexes. Since complexes III and IV are necessary for the assembly and stability of complex I (Blakely et al. 2005) and complex I is necessary for the full assembly of complex III (Ugalde et al. 2004), it is not difficult to see how decreased oxygen efficiency could be affected by usEP effects on complex I and/or IV in supercomplexes.

Yet these suggested mechanisms do not include consideration for the enhancement of Ca^{2+} on usEP-induced ROS formation. The intracellular Ca^{2+} is made available by the nanoporation of the plasma membrane and an influx of Ca^{2+} down its electrochemical gradient. One possibility for enhanced ROS production by complex I increased activity of the Krebs cycle with Ca^{2+} dependent dehydrogenases such as alpha-ketogluterate dehydrogenase and others, as the CoQ pool becomes over-reduced with excess electrons from complex II reducing capacity of forward electron transport. This would cause NAD + production from NADH with electrons from ubiquinol by reverse electron transfer (RET) generating ROS (Scialo et al. 2017; Adam-Vizia and Starkovb 2010; Bertero and Maack 2018).

Regardless of how usEPs slows respiration, it is most likely that usEPs attenuate electron transport through complexes I and IV as shown in Fig. 8.16. Furthermore, usEPs increase Ca^{2+} influx through the plasma membrane as shown in Fig. 8.11. Effects of usEPs on complex I and IV are exacerbated by increases in Ca^{2+} (Fig. 8.16). Since complex I is a primary site of superoxide anions production, the slowed electron transport is expected to increase production of ROS like that shown in Fig. 8.8. As the ROS levels increase and surpass antioxidant protection mechanisms in mitochondria, there is likely further damage to electron transport mechanisms, which further increases ROS generation. This is expected to slow electron transport and ATP production even more. As ROS levels accumulate, they begin to oxidize lipids and proteins including oxidation CypD which in the presence of elevated Ca^{2+} ultimately causes opening of the mPTP, loss of $\Delta\Psi m$ and regulated cell death.

8.5 usEPs Upregulate Genes in the Electron Transport Chain and ATP Synthase

In order to determine usEP effects on cells, we have used several cell types depending on the questions asked and the experimental design. In order to determine effects of usEPs on gene expression, we used a 4T1 mouse mammary cancer model, where we have demonstrated successful tumor ablation and an immune-mediated protective, vaccine effect after ablation (Guo et al. 2018; Beebe et al. 2018), which will be presented later. We carried out RNA sequencing on samples 2 and 4 h after treating cells with various usEP conditions, including ones that induced cell death by 25–90%. When these sequences were analyzed by Ingenuity Pathway Analysis (IPA), under condition of 80–90% cell death (24 h after treatment, 100 pulses, 60 ns 50 kV/cm), the two major changes in gene expression 2 h after treatment were analyzed under pathways classified as mitochondrial dysfunction and oxidative phosphorylation.

Figure 8.18 shows the electron transport chain complexes I-V and ATP synthase as complex V. Subunits from each complex that were up-regulated > *1.5-fold* are shown in red. There was increased gene expression in all complexes of the ETC and ATP synthase in response to these usEP treatment. There were 13 genes up-regulated in Complex I, 2 genes up-regulated in complexes II and III, 5 in complex IV and 3 in ATP synthase. The greatest number of up-regulated genes in complexes I and IV and ATP synthase agree with inhibitory effects on oxygen utilization observed in H9c2 cells (Complex I and IV) and for effects on CypD in ATP synthase as the mPTP. While these analyses are the results of a single RNA sequence analysis, these genes were up-regulated at the 2- and 4-h time points for 100 pulses, the 4 h time point for 50 pulses and the 2 h time point for 150 pulses. RNA at the 4-h time point for 150 pulses was degraded and not analyzed. Thus, for as long as 4 h at 50 and 100 pulses and for 2 h at 150 pulses, cells are up- and down-regulating (not yet analyzed) genes without significant degradation of RNA, demonstrating that they are not yet past a point of no-return, although 25–90% will be dead by 24 h after treatment.

Confirmation of RNA sequencing results by RTqPCR of 4T1 cell RNA—In order to confirm the RNA sequencing data, we concentrated on genes that were upregulated ≥ 1.5-fold in complex I based on single RNA sequencing samples. This is a reasonable usEP target because it is a regulatable and vernerable complex (Papa et al. 2008), is very sensitive to usEPs and is a common sourse of ROS. In addition, dysfunction of complex I arises from decreased catalytic activity caused by increased superoxide production (Hirst et al. 2013), which is induced by usEPs (Fig. 8.8). Primers were designed from 5 different genes in Complex I in the hydrophobic arm of complex I including mt-Nd1 and mt-Nd2, both expressed from mitochondrial

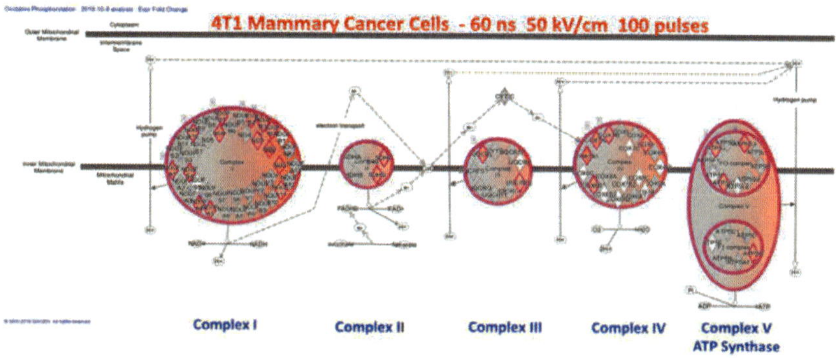

Fig. 8.18 UsEP treatment increase expression of genes in the electron transport chain and ATP synthase—4T1 mammary cancer cells were treated with 100 pulses, 60 ns 40 kV/cm or 0.024 Vs/cm. Two hours later, RNA was isolated and defined as intact by bioanalysis based on ratios of intact 28S and 18S ribosomal RNA. The mRNA is then reverse transcribed to cDNA and sequenced to analyze changes in gene expression in response to usEPs. Complexes in the ETC and ATP synthase are shown as indicated. The red colored images in each ETC complex indicates subunits of each complex that have been up-regulated. The results are based on a single mRNA sequencing analysis and are conformed by RTqPCR for selected genes (see below) (Bani-Hani and Beebe unpublished)

genes; and Ndufa1, Ndufa4, Ndufb4 and Ndufaf2, all genes encode in the nucleus. RNA sequencing changes as fold increase above control, non-pulsed 4T1 cells are shown for each complex I subunit for the indicated number of pulses. While RNA sequencing determined that the genome-encoded subunits Ndufa1, Ndufa4, Ndufb4 and Ndufaf2 were increased 1.5–1.7-fold, these small increases were not found to be statistically significent by RTqPCR. In contrast, the mitochondrial gene-encoded subunits Nd1 and Nd2 were significantly up-regulated two hours after 4T1 cells are exposed to 50 and 100 pulses at 60 ns and 40 kV/cm (Fig. 8.19).

Figure 8.19 show PTqPCR analyses on several genes that were p-regulated in complex I. The Nd1 subunit contains several important distinction regarding its role in the function and assembly of the most complex complex in the ETC (Sekiguchi et al. 2009; Gonzalez-Halphen et al. 2011; Zurita Rendón and Shoubridge 2012). Along with Nd2, Nd4 and Nd6, the Nd1 (Nqo8 in *T. thermophilus*) subunit it is a core protein of complex I and the most evolutionarily conserved subunits in the complex, suggesting it is a fundamental component important in complex I. It is at the interface of the hydrophobic membrane arm where proton transport occurs and the hydrophilic arm where electron transport occurs and also contributes to the quinone binding site or Q site xcx (Sekiguchi et al. 2009). Nd1 is included in the initial assembly of complex I. Assembly starts with two separate intermediates including the hydrophobic arm nuclear-coded subunits and the membrane-associated module

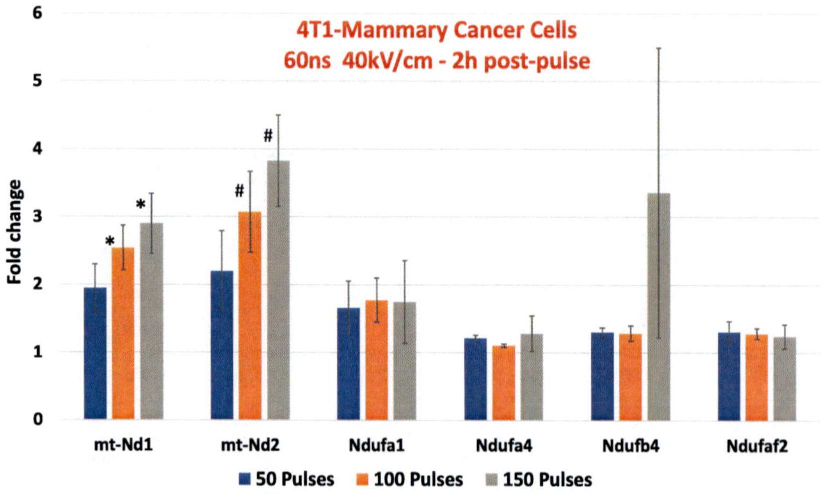

Fig. 8.19 Confirmation of RNA sequencing of ETC complex I subunits from 4T1 mammary cancer cells by RTqPCR—Specific primers for the indicated ETC complex I subunits were designed, and RNA isolated from 4T1 cells treated with 50, 100 and 150 pulses at 60 ns and 40 kV/cm. The RNA was isolated in 4 individual experiments (including that was RNA sequenced), reversed transcribed (RT), amplified by qPCR and indicated as the fold increased over control RNA from un-pulsed 4T1 cells. The Figure shows the mean and standard deviation of 4 separate RNA isolation experiments for RTqPCR and one from RNA sequencing, * $p < 0.01$; # $p < 0.05$

containing Nd1in a 400 kDa subcomplex. Nd1 is included in the module with 4 early assembly factors (NDUFAF3, NDUFAF4, C8orf38 and C20orf7), which dock and stabilize Nd1. If any one of this is knocked-out, Nd1 is not included and complex I assembly stalls and falls apart (Zurita Rendón and Shoubridge 2012). Nd1 appears to be part of a hot-spot in complex I for disease-causing mutations. Nd1 lies at the interface of the hydrophilic domain of the membrane domain of complex I, involving the large ubiquinone (Q) binding pocket, which is a structural component that is responsible for transmitting redox signals that induce conformatiomnal changes to proton pumping modules. Interestingly, subunit Nd2 was also significantly up-regulated by RNA sequencing and RTqPCR (Fig. 8.19). Along with Nd4 and Nd5, these subuniyts appear to be antiporter-like subunits due to their homology to the Na^+/H^+ antiporter family, so are believed to be sites of proton tranport and form the core of the proton transport mechanism for complex I (Hirst et al. 2013). The further importance of Nd1 is indicated by a commion site for disease-causing mutations. Several different diseases are caused by complex I dysfunctions that map to Nd1, Nd4, Nd5 and Nd6, which appear in a conserved hydrophobic loop that protrudes into the matrix. Nd1 is associated with several diseases including Leber's hereditary optic neuropathy (LHON) and severe,multi-systemic clinical diseases such as mitochondrial encephalomyopathy, lactic acidosis and stroke-like syndrome (MELAS) and Leigh syndrome (Gonzalez-Halphen et al. 2011). Based on using a software that predicts probability or possibility of mutation damaging effects on protein funtion Nd1, along with Nd4 and Nd5 and Nd6, are most affected by pathogenic mutations. This suggest these subunits form an operational nucleus in the bioenergetic machinery of complex I (Gonzalez-Halphen et al. 2011).

Yet another distinction of the Nd1 subunits is nature of effects of complex I inhibitors. In addition to being common sites for mutations, Nd1, Nd4 and Nd5 are binding sites for potent complx I inhibitors including rotenone, synthetic pesticides and plant acetogenins. Nd1 is at the interface between the membrane domain and the hydrophilic domain at the Q binding and reduction site. This is the site of the complex that produces large conformational changes that couple electron transport to proton pumping mechanically transmitted from Nd1along the membrane arm involving Nd2, Nd4 and Nd5. Complex I inhibitors such as fenperoxymate, asimicin, piericidin, rolliniastatins may displace Q and/or prevent the mechanical transmission of these redox-linked conformational changes. Some of these inhibitors more effectively inhibit proton pumping than blocking redox activity of mitochondrial NADH-Q recutase activity (Gonzalez-Halphen et al. 2011).

Regarding effects of chemical complex I inhibitors, it is interesting to considered usEPs as an complex I inhibitor of a different sort. While usEP-induced inhibition of complex I is generally considered to primarily affect electron transport, a biased and unsubstantiated opinion, like chemical inhibitors of complex I, usEP could affect proton pumping, which uncoupled from electron transport could produce ROS in the same way as effects on electron transport. In any event, the finding that usEPs induce upregulation of Nd1 and Nd2, suggest that profound inhibition of complex I induces these cells to upregulate at least two critical components of the largest ETC complex. It is easy to see why the Nd1 subunit of complex I is important. It is one of the most evolutionary conserved subunits in complex I, the site of disease-causing mutations and an apparent universal site of action for complex I inhibitors. It is at the crux of the the boundary between the membrane and hydrophilic arms of the complex where conformational changes occur that couple electron transport with proton pumping. Furthermore, its early and essential participation is required for the successful biogenesis of complex I. This further defnes complex I as a vulnerable ETC complex and a suseptable target for usEPs.

In summary,given that usEPs have effects on intracellular structures and consequently can modify their functions and affect their fate, it is not surprising that mitochondria are targets for these ultrashort pulses. This raises the question of what are the specific mitochondrial receptors or sensors for these usEPs. From the available data, it appears that there are several mitochondrial sensors for usEPs. These include components in the ETC, most likely iron-sulfur clusters in complex I and IV that transport electrons and pump protons; CypD as a regulator of the mPTP, which may be ATP synthase; and likely CL and cytochrome *c* in their roles in regulated cell death by apoptosis. However, evidence that CL and cytochrome *c* are actually usEP sensors is only circumstantial and may be cell-type specific occurring only when usEPs induce apoptosis. The usEP sensors in the ETC complexes I and IV are more senstive than CypD because they are affected at lower usEP conditions than loss of $\Delta\Psi m$ and opening of the mPTP. Based on available data from 2 laboratories with Jurkat cells, usEPs induce increases in oxygen utilization under low usEP conditions of 0.002 Vs/cm (Estlack et al. 2014a, b). Under similar conditions, we saw decreases in oxygen utilization with 0.004 Vs/cm (Fig. 8.16). At a higher usEP condition of 0.012 Vs/cm (Fig. 8.9), the mPTP is opened and $\Delta\Psi m$ was decreased. When usEP conditions are increased tenfold to 0.36 Vs/cm, loss of $\Delta\Psi m$ is Ca^{2+} independent and could be due to permeabilization of the IMM. Interestingly, permeabilization of Jurkat cell plasma membrane occurred at 0.006 Vs/cm. This is about 6-times lower than that required for dissipating $\Delta\Psi m$ by affecting CypD in a Ca^{2+}-dependent manner and 60-time lower than that required to permeabilize the IMM according considerations in Figs. 8.13 and 8.14.

Thus, there is a hierarchy of usEP sensors in the mitochondria and elsewhere. The following mechanistic scenerio is likely. As electric field charging conditions as Vs/cm increase, effects on mitochondria begin with increases and then decreases in oxygen consumption as usEPs affect electron and proton transport as they traverse the ETC. In actuality, usEPs are increasing and then decreasing oxidation—reduction (redox) reactions in the ETC. Such measurement could only be made with an intact mitochondrial membrane, so $\Delta\Psi m$ is intact. Since these effects on redox reactions are long lasting, they are most likely due to increasing effects of ROS by oxidizing redox-sensitive enzymes and other proteins.

Increase in ROS levels changes the redox balance in the ETC first and formost, which can shift electron transport from forward to reverse, generating more ROS in potentually feed-forward reactions that lead to progessive mitochondrial dysfunction until RCD intervenes. So, as oxygen utilization decreases in complex I and IV, there are greater increases in ROS production affecting their iron-sulfur clusters, setting in motion increases in oxydation state of an hierarchy of redox-regulated proteins. The oxidation-sensitivity of these proteins likely progresses from more sensitive (lower Km) ROS signaling second messenger sensors such as kinases and phosphtases, GTPases and deacetylases, among others (Finkel 2011; Freund-Michel et al. 2013), to inflammasome-mediated complexes that are sensitive to DAMPs and PAMPs in innate immune mechanisms (Jin and Flavell 2010; Rathinam et al. 2012). And finally CypD becomes oxidized and controls of the mPTP is lost and the permeability pore opens and $\Delta\Psi m$ is lost. About this time usEP conditions have lowered the reduction potential of cytochrome *c*, changing its conformation and its function to a peroxidase that oxidizes CL (De Biase et al. 2009). Cytochrome *c* dissociates from cardiolipin and is released into the cytoplasm to activate the apoptosome, caspase-9 and caspase-3. After cytochrome c release, the decision for and the point of life's no return, there is then "ten minutes to dead" (Green 2005).

References

Acin-Perez R, Salazar E, Kamenetsky M, Buck J, Levin LR, Manfredi G (2009) Cyclic AMP produced inside mitochondria regulates oxidative phosphorylation. Cell Metab. 9:265–276

Adam-Vizia V, Starkovb AA (2010) Calcium and mitochondrial reactive oxygen species generation: how to read the facts. J Alzheimers Dis 20(Suppl 2):S413–S426

Alavian KN, Li H, Collis L, Bonanni L, Zeng L, Sacchetti S, Lazrove E, Nabili P, Flaherty B, Graham M, Chen Y, Messerli SM, Mariggio MA, Rahner C, McNay E, Shore GC, Smith PJ, Hardwick JM, Jonas EA (2011) Bcl-xL regulates metabolic efficiency of neurons through interaction with the mitochondrial F1FO ATP synthase. Nat Cell Biol 13(10):1224–1233

Alpert NM, Guehl N, Ptaszek L, Pelletier-Galarneau M, Ruskin J, Mansour MC, Wooten D, Ma C, Takahashi K, Zhou Y, Shoup TM, Normandin MD, El Fakhri G. Quantitative in vivo mapping of myocardial mitochondrial membrane potential. PLoS One 13(1):e0190968

Amchenkova AA, Bakeeva LE, Chentsov YS, Skulachev VP, Zorov DB (1988) Coupling membranes as energy-transmitting cables. I. Filamentous mitochondria in fibroblasts and mitochondrial clusters in cardiomyocytes. J Cell Biol 107:481–495

Anand PK, Malireddi RK, Kanneganti TD (2011) Role of the nlrp3 inflammasome in microbial infection. Front Microbiol 2:12

Anderson S, Bankier AT, Barrell BG, de Bruijn MHL, Coulson AR, Drouin J, Eperon IC, Nierlich DP, Roe BA, Sanger F, Schreier PH, Smith AJH, Staden R, Young IG (1981) Sequence and organization of the human mitochondrial genome. Nature 1981(290):457–465

Arnér ES, Holmgren A (2000) Physiological functions of thioredoxin and thioredoxin reductase. Eur J Biochem 267:6102–6109

Decaudin D, Geley S, Hirsch T, Castedo M, Marchetti P, Macho A, Kofler R, Kroemer G (1997) Bcl-2 and Bcl-XL antagonize the mitochondrial dysfunction preceding nuclear apoptosis induced by chemotherapeutic agents. Cancer Res 57:62–67

Barrientos A, Barros MH, Valnot I, Rötig A, Rustin P, Tzagoloff A (2002) Cytochrome oxidase in health and disease. Gene 286:53–63

Batista Napotnik T, Wu YH, Gundersen MA, Miklavčič D, Vernier PT (2012) Nanosecond electric pulses cause mitochondrial membrane permeabilization in Jurkat cells. Bioelectromagnetics 33:257–264

Beebe SJ, Fox PM, Rec LH, Buescher ES, Somers K, Schoenbach KH (2002) Nanosecond pulsed electric field (nsPEF) effects on cells and tissues: apoptosis induction and tumor growth inhibition. IEEE Trans Plasma Sci 30:286–292

Beebe SJ, Fox PM, Rec LJ, Willis EL, Schoenbach KH (2003) Nanosecond, high-intensity pulsed electric fields induce apoptosis in human cells. FASEB J 17:1493–1495
Beebe SJ, Blackmore PF, White J, Joshi RP, Schoenbach KH (2004) Nanosecond pulsed electric fields modulate cell function through intracellular signal transduction mechanisms. Physiol Meas 25:1077–1093
Beebe SJ, Chen YJ, Sain NM, Schoenbach KH, Xiao S (2012) Transient features in nanosecond pulsed electric fields differentially modulate mitochondria and viability. PLoS ONE 7:e51349
Beebe SJ, Sain NM, Ren W (2013) Induction of cell death mechanisms and apoptosis by nanosecond pulsed electric fields (nsPEFs). Cells 2:136–162
Beebe SJ (2015) Considering effects of nanosecond pulsed electric fields on proteins. Bioelectrochemistry 103:52–59
Beebe SJ, Lassiter BP, Guo S (2018) Nanopulse stimulation (NPS) induces tumor ablation and immunity in orthotopic 4T1 mouse breast cancer: a review. Cancers (Basel) 10:pii:E97
Bertero E, Maack C (2018) Calcium signaling and reactive oxygen species in mitochondria. Circ Res 122:1460–1478
Buescher ES, Smith RR, Schoenbach KH (2004) Submicrosecond intense pulsed electric field effects on intracellular free calcium: mechanisms and effects. IEEE Trans Plasma Sci 32:1563–1572
Blakely EL, Mitchell AL, Fisher N, Meunier B, Nijtmans LG, Schaefer AM, Jackson MJ, Turnbull DM, Taylor RW (2005) A mitochondrial cytochrome b mutation causing severe respiratory chain enzyme deficiency in humans and yeast. FEBS J 272:3583–3592
Bleier L, Wittig I, Heide H, Steger M, Brandt U, Drose S-S (2015) Generator-specific targets of mitochondrial reactive oxygen species. Free Radic Biol Med 78:1–10
Brandt U, Kerscher S, Dröse S, Zwicker K, Zickermann V. Proton pumping by NADH: ubiquinone oxidoreductase. A redox driven conformational change mechanism? FEBS Lett 545:9–17 (Review)
Bren KL, Raven EL (2017) Locked and loaded for apoptosis. Science 356:1236
Briehl MM (2015) Oxygen in human health from life to death–an approach to teaching redox biology and signaling to graduate and medical students. Redox Biol 5:124–139
Bultema JB, Braun HP, Boekema EJ, Kouril R (2009) Megacomplex organization of the oxidative phosphorylation system by structural analysis of respiratory supercomplexes from potato. Biochim Biophys Acta 1787:60–67
Bushnell L, Bjorkman D, McGreevy J (1990) Ultrastructural changes in gastric epithelium caused by bile salt. J Surg Res 49(3):280–286
Cagin U, Enriquez JA (2015) The complex crosstalk between mitochondria and the nucleus: what goes in between? Int J Biochem Cell Biol 63:10–15
Cecchini G (2003) Function and structure of complex II of the respiratory chain. Annu Rev Biochem 72:77–109
Chu CT, Ji J, Dagda RK, Jiang JF, Tyurina YY, Kapralov AA, Tyurin VA, Yanamala N, Shrivastava IH, Mohammadyani D, Wang KZQ, Zhu J, Klein-Seetharaman J, Balasubramanian K, Amoscato AA, Borisenko G, Huang Z, Gusdon AM, Cheikhi A, Steer EK, Wang R, Baty C, Watkins S, Bahar I, Bayir H, Kagan VE (2013) Cardiolipin externalization to the outer mitochondrial membrane acts as an elimination signal for mitophagy in neuronal cells. Nat Cell Biol 15:1197–1205
Chen YB, Aon MA, Hsu YT, Soane L, Teng X, McCaffery JM, Cheng WC, Qi B, Li H, Alavian KN, Dayhoff-Brannigan M, Zou S, Pineda FJ, O'Rourke B, Ko YH, Pedersen PL, Kaczmarek LK, Jonas EA, Hardwick JM (2011) Bcl-xL regulates mitochondrial energetics by stabilizing the inner membrane potential. J Cell Biol 195:263–276
Circu ML, Aw TY (2010) Reactive oxygen species, cellular redox systems, and apoptosis. Free Radic Biol Med 48:749–762
Cohen P (2000) The regulation of protein function by multisite phosphorylation–a 25 year update. Trends Biochem Sci 25:596–601
Cole KS (1937) Electric impedance of marine egg membranes. Trans Faraday Soc 23:966

Cooley JW (2013) Protein conformational changes involved in the cytochrome bc1 complex catalytic cycle. Biochim Biophys Acta 1827:1340–1345
Crompton M (1999) The mitochondrial permeability transition pore and its role in cell death. Rev Biochem J 341:233–249
Deng J, Schoenbach KH, Buescher ES, Hair PS, Fox PM, Beebe SJ (2003) The effects of intense submicrosecond electrical pulses on cells. Biophys J 84:2709–2714
De Biase PM, Paggi DA, Doctorovich F, Hildebrandt P, Estrin DA, Murgida DH, Marti MA (2009) Molecular basis for the electric field modulation of cytochrome C structure and function. J Am Chem Soc 131:16248–16256
Devenish RJ, Prescott M, Boyle GM, Nagley P (2000) The oligomycin axis of mitochondrial ATP synthase: OSCP and the proton channel. J Bioenerg Biomembr 32:507–515
Dudkina NV, Kouril R, Peters K, Braun HP, Boekema EJ (2010) Structure and function of mitochondrial supercomplexes. Biochim Biophys Acta 1797:664–670
Dyall SD, Brown MT, Johnson PJ (2004) Ancient invasions: from endosymbionts to organelles. Science 304:253–257
Estlack LE, Roth CC, Cerna CZ, Wilmink GJ, Ibey BL (2014a) Investigation of a direct effect of nanosecond pulse electric fields on mitochondria. In: Proceeding of SPIE 8941, Optical interactions with tissue and cells XXV; and terahertz for biomedical applications 89411S (13 Mar 2014)
Estlack LE, Roth CC, Thompson GL 3rd, Lambert WA 3rd, Ibey BL (2014b) Nanosecond pulsed electric fields modulate the expression of Fas/CD95 death receptor pathway regulators in U937 and Jurkat cells. Apoptosis 19:1755–1768
Finkel T (2011) Signal transduction by reactive oxygen species. J Cell Biol 194:7–15
Folda A, Citta A, Scalcon V, Calì T, Zonta F, Scutari G, Bindoli A, Rigobello MP (2016) Mitochondrial thioredoxin system as a modulator of cyclophilin D redox state. Sci Rep 6:23071
Freeman BA, Crapo JD (1982) Biology of disease: free radicals and tissue injury. Lab Invest 47:412–426
Freund-Michel V, Guibert C, Dubois M, Courtois A, Marthan R, Savineau JP, Muller B (2013) Reactive oxygen species as therapeutic targets in pulmonary hypertension. Ther Adv Respir Dis 7:175–200
Fridovich I (2004) Mitochondria: are they the seat of senescence? Aging Cell 3(1):13–16 (Review)
Gao P, Zhang H, Dinavahi R, Li F, Xiang Y, Raman V, Bhujwalla ZM, Felsher DW, Cheng L, Pevsner J, Lee LA, Semenza GL, Dang CV (2007) HIF-dependent antitumorigenic effect of antioxidants in vivo. Cancer Cell 12:230–238
Godoy LC, Muñoz-Pinedo C, Castro L, Cardaci S, Schonhoff CM, King M, Tórtora V, Marín M, Miao Q, Jiang JF, Kapralov A, Jemmerson R, Silkstone GG, Patel JN, Evans JE, Wilson MT, Green DR, Kagan VE, Radi R, Mannick JB (2009) Disruption of the M80-Fe ligation stimulates the translocation of cytochrome c to the cytoplasm and nucleus in nonapoptotic cells. Proc Natl Acad Sci USA 106(8):2653–2658
Guarás A, Perales-Clemente E, Calvo E, Acín-Pérez R, Loureiro-Lopez M, Pujol C, Martínez-Carrascoso I, Nuñez E, García-Marqués F, Rodríguez-Hernández MA, Cortés A, Diaz F, Pérez-Martos A, Moraes CT, Fernández-Silva P, Trifunovic A, Navas P, Vazquez J, Enríquez JA (2016) The $CoQH_2$/CoQ ratio serves as a sensor of respiratory chain efficiency. Cell Rep 15:197–209
Giampazolias E, Zunino B, Dhayade S, Bock F, Cloix C, Cao K, Roca A, Lopez J, Ichim G, Proïcs E, Rubio-Patiño C, Fort L, Yatim N, Woodham E, Orozco S, Taraborrelli L, Peltzer N, Giorgio V, Soriano ME, Basso E, Bisetto E, Lippe G, Forte MA, Bernardi P (2010) Cyclophilin D in mitochondrial pathophysiology. Biochim Biophys Acta 1797(6–7):1113–11138
Giorgio V, Soriano ME, Basso E, Bisetto E, Lippe G, Forte MA, Bernardi P (2010) Cyclophilin D in mitochondrial pathophysiology. Biochim Biophys Acta 1797(6-7):1113–11138
Giorgio V, Burchell V, Schiavone M, Bassot C, Minervini G, Petronilli V, Argenton F, Forte M, Tosatto S, Lippe G, Bernardi P (2017) Ca^{2+} binding to F-ATP synthase β subunit triggers the mitochondrial permeability transition. EMBO Rep 18:1065–1076
Green DR (2005) Apoptotic pathways: ten minutes to dead. Cell 121:671–674

Genova ML, Lenaz G (2014) Functional role of mitochondrial respiratory supercomplexes. Biochim Biophys Acta 1837:427–443
Gonzalez-Halphen D, Ghelli A, Iommarini L, Carelli V, Esposti MD (2011) Mitochondrial complex I and cell death: a semi-automatic shotgun model. Cell Death Dis 2:e222
Graves JD, Krebs EG (1999) Protein phosphorylation and signal transduction. Pharmacol Ther 82:111–121
Gray MW (2012) Mitochondrial evolution. Cold Spring Harb Perspect Biol 4:a011403
Gresser MJ, Myers JA, Boyer PD (1982) Catalytic site cooperativity of beef heart mitochondrial F1 adenosine triphosphatase. Correlations of initial velocity, bound intermediate, and oxygen exchange measurements with an alternating three-site model. J Biol Chem 257:12,030–12,038
Guarás A, Perales-Clemente E, Calvo E, Acín-Pérez R, Loureiro-Lopez M, Pujol C, Martínez-Carrascoso I, Nuñez E, García-Marqués F, Rodríguez-Hernández MA, Cortés A, Diaz F, Pérez-Martos A, Moraes CT, Fernández-Silva P, Trifunovic A, Navas P, Vazquez J, Enríquez JA (2016) The CoQH2/CoQ ratio serves as a sensor of respiratory chain efficiency. Cell Rep 15(1):197–209
Guaras AM, Enríquez JA (2017) Building a beautiful beast: mammalian respiratory complex I. Cell Metab 25(1):4–5
Guo R, Gu J, Wu M, Yang M (2016) Amazing structure of respirasome: unveiling the secrets of cell respiration. Protein Cell 7:854–865
Guo S, Jing Y, Burcus NI, Lassiter BP, Tanaz R, Heller R, Beebe SJ (2018) Nano-pulse stimulation induces potent immune responses, eradicating local breast cancer while reducing distant metastases. Int J Cancer 142:629–640
Haines TH, Dencher NA (2002) Cardiolipin: a proton trap for oxidative phosphorylation. FEBS Lett 528:35–39
Halestrap AP, Clarke SJ Javadov SA (2006) Mitochondrial permeability transition pore opening during myocardial reperfusion—a target for cardioprotection. Cardiov Res 61:372–385
Han D, Antunes F, Canali R, Rettori D, Cadenas E (2003) Voltage-dependent anion channels control the release of the superoxide anion from mitochondria to cytosol. J Biol Chem 278:5557–5563
Haworth RA, Hunter DR (1979) The Ca^{2+}-induced membrane transition in mitochondria. II. Nature of the Ca^{2+} trigger site. Arch Biochem Biophys 195:460–467
Hirst SM, Karakoti A, Singh S, Self W, Tyler R, Seal S, Reilly CM (2013) Bio-distribution and in vivo antioxidant effects of cerium oxide nanoparticles in mice. Environ Toxicol 28:107–118
Hummer G, Wikström M (2016) Molecular simulation and modeling of complex I. Biochim Biophys Acta 1857:915–921
Hunter DR, Haworth RA (1979a) The Ca^{2+}-induced membrane transition in mitochondria. I. The protective mechanisms. Arch Biochem Biophys 195:453–459
Hunter DR, Haworth RA (1979b) The Ca^{2+}-induced membrane transition in mitochondria. III. Transitional Ca^{2+} release. Arch Biochem Biophys 195:468–477
Hurst S, Hoek J, Sheu SS (2017) Mitochondrial Ca^{2+} and regulation of the permeability transition pore. J Bioenerg Biomembr 49:27–47
Iverson SL, Orrenius S (2004) The cardiolipin-cytochrome c interaction and the mitochondrial regulation of apoptosis. Arch Biochem Biophys 423:37–46
Jékely G (2014) Origin and evolution of the self-organizing cytoskeleton in the network of eukaryotic organelles. Cold Spring Harb Perspect Biol 6:a016030
Jin C, Flavell RA (2010) Inflammasome activation. The missing link: how the inflammasome senses oxidative stress. Immunol Cell Biol 88(5):510–512
Jonas EA, Porter GA Jr, Beutner G, Mnatsakanyan N, Alavian KN (2015) Cell death disguised: the mitochondrial permeability transition pore as the c-subunit of the F(1)F(O) ATP synthase. Pharmacol Res 99:382–392
Jonckheere AI, Smeitink JA, Rodenburg RJ (2012) Mitochondrial ATP synthase: architecture, function, and pathology. J Inherit Metab Dis 35:211–225
Kagan VE, Borisenko GG, Tyurina YY, Tyurin VA, Jiang J, Potapovich AI, Kini V, Amoscato AA, Fujii Y (2004) Oxidative lipidomics of apoptosis: redox catalytic interactions of cytochrome c with cardiolipin and phosphatidylserine. Free Radic Biol Med 37:1963–1985

Karch J, Molkentin JD (2014) Identifying the components of the elusive mitochondrial permeability transition pore. Proc Natl Acad Sci U S A 111:10396–10407

Kauppila TES, Kauppila JHK, Larsson NG (2017) Mammalian mitochondria and aging: an update. Cell Metab 25:57–71

Kimes BW, Brandt BL (1976) Properties of a clonal muscle cell line from rat heart. Exp Cell Res 98:367–381

Kotnik T, Miklavcic D (2006) Theoretical evaluation of voltage inducement on internal membranes of biological cells exposed to electric fields. Biophys J 90:480–491

Kumaran S, Subathra M, Balu M, Panneerselvam C (2004) Age-associated decreased activities of mitochondrial electron transport chain complexes in heart and skeletal muscle: role of L-carnitine. Chem Biol Interact 148:11–18

Kwong JQ, Molkentin JD (2015) Physiological and pathological roles of the mitochondrial permeability transition pore in the heart. Cell Metab 21:206–214

Lane N, Martin W (2010) The energetics of genome complexity. Nature 467(7318):929–934

Letts JA, Sazanov LA (2017) Clarifying the supercomplex: the higher-order organization of the mitochondrial electron transport chain. Nat Struct Mol Biol 24:800–808

Li B, Chauvin C, De Paulis D, De Oliveira F, Gharib A, Vial G, Lablanche S, Leverve X, Bernardi P, Ovize M, Fontaine E (2012) Inhibition of complex I regulates the mitochondrial permeability transition through a phosphate-sensitive inhibitory site masked by cyclophilin D. Biochim Biophys Acta 1817:1628–1634

Li Y, Park JS, Deng JH, Bai Y (2006) Cytochrome c oxidase subunit IV is essential for assembly and respiratory function of the enzyme complex. J Bioenerg Biomembr 38:283–291

Lim S, Smith KR, Lim ST, Tian R, Lu J, Tan M (2016) Regulation of mitochondrial functions by protein phosphorylation and dephosphorylation. Cell Biosci 14(6):25

Lin DH, Stuwe T, Schilbach S, Rundlet EJ, Perriches T, Mobbs G, Fan Y, Thierbach K, Huber FM, Collins LN, Davenport AM, Jeon YE, Hoelz A (2016) Architecture of the symmetric core of the nuclear pore. Science 352(6283)

Linard D, Kandlbinder A, Degand H, Morsomme P, Dietz KJ, Knoops B (2009) Redox characterization of human cyclophilin D: identification of a new mammalian mitochondrial redox sensor? Arch Biochem Biophys 491:39–45

Long Q, Yang K, Yang Q (2015) Regulation of mitochondrial ATP synthase in cardiac pathophysiology. Am J Cardiovasc Dis 5:19–32

Lucero M, Suarez AE, Chambers JW (2019) Phosphoregulation on mitochondria: Integration of cell and organelle responses. CNS Neurosci Ther 25(7):837–858

Martinon F, Pétrilli V, Mayor A, Tardivel A, Tschopp J (2006) Gout-associated uric acid crystals activate the NALP3 inflammasome. Nature 440:237–241

Maul GG, Deaven L (1977) Quantitative determination of nuclear pore complexes in cycling cells with differing DNA content. J Cell Biol 73:748–760

Melber A, Winge DR (2016) Inner secrets of the respirasome. Cell 167:1450–1452

Michel J, DeLeon-Rangel J, Zhu S, Van Ree K, Vik SB (2011) Mutagenesis of the L, M, and N subunits of complex I from *Escherichia coli* indicates a common role in function. PLoS ONE 6:e17420

Mohammadyani D, Yanamala N, Samhan-Arias AK, Kapralov AA, Stepanov G, Nuar N, Planas-Iglesias J, Sanghera N, Kagan VE, Klein-Seetharaman J (2018) Structural characterization of cardiolipin-driven activation of cytochrome c into a peroxidase and membrane perturbation. Biochim Biophys Acta 1860:1057–1068

Monaco G, Decrock E, Arbel N, van Vliet AR, La Rovere RM, De Smedt H, Parys JB, Agostinis P, Leybaert L, Shoshan-Barmatz V, Bultynck G (2015) The BH4 domain of anti-apoptotic Bcl-XL, but not that of the related Bcl-2, limits the voltage-dependent anion channel 1 (VDAC1)-mediated transfer of pro-apoptotic Ca^{2+} signals to mitochondria. J Biol Chem 290:9150–9161

Monterisi S, Zaccolo M (2017) Components of the mitochondrial cAMP signalosome. Biochem Soc Trans 45:269–274

Mora C, Tittensor DP, Adl S, Simpson AG, Worm B (2011) How many species are there on Earth and in the ocean? PLoS Biol 9:e1001127
Noji H, Yasuda R, Yoshida M, Kinosita K Jr (1997) Direct observation of the rotation of F1-ATPase. Nature 386:299–302
Nuccitelli R, Lui K, Kreis M, Athos B, Nuccitelli P (2013) Nanosecond pulsed electric field stimulation of reactive oxygen species in human pancreatic cancer cells is $Ca^{(2+)}$-dependent. Biochem Biophys Res Commun 435:580–585
Ott M, Robertson JD, Gogvadze V, Zhivotovsky B, Orrenius S (2002) Cytochrome c release from mitochondria proceeds by a two-step process. Proc Natl Acad Sci U S A 99:1259–1263
Ren W, Beebe SJ (2011) An apoptosis targeted stimulus with nanosecond pulsed electric fields (nsPEFs) in E4 squamous cell carcinoma. Apoptosis 16:382–393
Pakhomov AG, Shevin R, White JA, Kolb JF, Pakhomova ON, Joshi RP, Schoenbach KH (2007) Membrane permeabilization and cell damage by ultrashort electric field shocks. Arch Biochem Biophys 465:109–118
Pakhomov AG, Kolb JF, White JA, Joshi RP, Xiao S, Schoenbach KH (2007) Long-lasting plasma membrane permeabilization in mammalian cells by nanosecond pulsed electric field (nsPEF). Bioelectromagnetics 28:655–663
Pakhomov AG (2012) Oxidative effects of nanosecond pulsed electric field exposure in cells and cell-free media. Arch Biochem Biophys 527:55–64
Pakhomova ON, Khorokhorina VA, Bowman AM, Rodaitė-Riševičienė R, Saulis G, Xiao S, Pakhomov AG (2012) Oxidative effects of nanosecond pulsed electric field exposure in cells and cell-free media. Arch Biochem Biophys 527:55–64
Papa S, De Rasmo D, Scacco S, Signorile A, Technikova-Dobrova Z, Palmisano G, Sardanelli AM, Papa F, Panelli D, Scaringi R, Santeramo A (2008) Mammalian complex I: a regulatable and vulnerable pacemaker in mitochondrial respiratory function. Biochim Biophys Acta 1777:719–728
Pawson T, Scott JD (2005) Protein phosphorylation in signaling–50 years and counting. Trends Biochem Sci 30:286–290
Preston CC, Oberlin AS, Holmuhamedov EL, Gupta A, Sagar S, Syed RH, Siddiqui SA, Raghavakaimal S, Terzic A, Jahangir A (2008) Aging-induced alterations in gene transcripts and functional activity of mitochondrial oxidative phosphorylation complexes in the heart. Mech Ageing Dev 129(6):304–12
Rasola A, Bernardi P (2007) The mitochondrial permeability transition pore and its involvement in cell death and in disease pathogenesis. Apoptosis 12:815–833
Rathinam VA, Vanaja SK, Fitzgerald KA (2012) Regulation of inflammasome signaling. Nat Immunol 13(4):333–342
Reczek CR, Chandel NS (2017) The two faces of reactive oxygen species in cancer. Ann Rev Cancer Biol 1:79–98
Ren M, Phoon CK, Schlame M (2014) Metabolism and function of mitochondrial cardiolipin. Prog Lipid Res 55:1–16
Riley JS, Quarato G, Cloix C, Lopez J, O'Prey J, Pearson M, Chapman J, Sesaki H, Carlin LM, Passos JF, Wheeler AP, Oberst A, Ryan KM, Tait SW (2018) Mitochondrial inner membrane permeabilisation enables mtDNA release during apoptosis. EMBO J 37:pii:e99238
Rimessi A, Previati M, Nigro F, Wieckowski MR, Pinton P (2016) Mitochondrial reactive oxygen species and inflammation: molecular mechanisms, diseases and promising therapies. Int J Biochem Cell Biol 81:281–293
Rokas A (2008) The origins of multicellularity and the early history of the genetic toolkit for animal development. Annu Rev Genet 42:235–251
Rongvaux A, Jackson R, Harman CC, Li T, West AP, de Zoete MR, Wu Y, Yordy B, Lakhani SA, Kuan CY, Taniguchi T, Shadel GS, Chen ZJ, Iwasaki A, Flavell RA (2014) Apoptotic caspases prevent the induction of type I interferons by mitochondrial DNA. Cell 159:1563–1577
Sato T, Machida T, Takahashi S, Iyama S, Sato Y, Kuribayashi K, Takada K, Oku T, Kawano Y, Okamoto T, Takimoto R, Matsunaga T, Takayama T, Takahashi M, Kato J, Niitsu Y (2004)

Fas-mediated apoptosome formation is dependent on reactive oxygen species derived from mitochondrial permeability transition in Jurkat cells. J Immunol 173:285–296
Sekiguchi K, Murai M, Miyoshi H (2009) Exploring the binding site of acetogenin in the ND1 subunit of bovine mitochondrial complex I. Biochim Biophys Acta 1787:1106–1111
Sessions AL, Doughty DM, Welander PV, Summons RE, Newman DK (2009) The continuing puzzle of the great oxidation event. Curr Biol 19:R567–574
Schoenbach KH, Beebe SJ, Buescher ES (2001) Intracellular effect of ultrashort electrical pulses. Bioelectromagnetics 22:440–448
Sattler M, Liang H, Nettesheim D, Meadows RP, Harlan JE, Eberstadt M, Yoon HS, Shuker SB, Chang BS, Minn AJ, Thompson CB, Fesik SW (1997) Structure of Bcl-xL-Bak peptide complex: recognition between regulators of apoptosis. Science 275:983–986
Sayin VI, Ibrahim MX, Larsson E, Nilsson JA, Lindahl P, Bergo MO (2014) Antioxidants accelerate lung cancer progression in mice. Sci Transl Med 6:221ra15
Schumacker PT (2015) Reactive oxygen species in cancer: a dance with the devil. Cancer Cell 27:156–157
Scialo F, Mallikarjun V, Stefanatos R, Sanz A (2013) Regulation of lifespan by the mitochondrial electron transport chain: reactive oxygen species dependent and reactive oxygen species-independent mechanisms. Antioxid Redox Signal 19:1953–1969
Scialo F, Sriram A, Fernandez-Ayala D, Gubina N, Lohmus M, Nelson G (2016) Mitochondrial ROS produced via reverse electron transport extend animal lifespan. Cell Metab 23:725–734
Scialò F, Fernández-Ayala DJ, Sanz A (2017) Role of mitochondrial reverse electron transport in ROS signaling: potential roles in health and disease. Front Physiol 8:428
Shimizu S, Eguchi Y, Kamiike W, Funahashi Y, Mignon A, Lacronique V, Matsuda H, Tsujimoto Y (1998) Bcl-2 prevents apoptotic mitochondrial dysfunction by regulating proton flux. Proc Natl Acad Sci USA 95(4):1455–1459
Skulachev VP (2001) (2001) Mitochondrial filaments and clusters as intracellular power-transmitting cables. Trends Biochem Sci 26:23–29
Stacey M, Stickley J, Fox P, Statler V, Schoenbach K, Beebe SJ, Buescher S (2003) Differential effects in cells exposed to ultra-short, high intensity electric fields: cell survival, DNA damage, and cell cycle analysis. Mutat Res 542:65–75
Suzuki T, Tanaka K, Wakabayashi C, Saita E, Yoshida M (2014) Chemomechanical coupling of human mitochondrial F1-ATPase motor. Nat Chem Biol 10:930–936
Tatarková Z, Kuka S, Račay P, Lehotský J, Dobrota D, Mištuna D, Kaplán P (2011) Effects of aging on activities of mitochondrial electron transport chain complexes and oxidative damage in rat heart. Physiol Res 60:281–289
Thomenius, Distelhorst (2003) Bcl-2 on the endoplasmic reticulum: protecting the mitochondria from a distance. J Cell Sci 116:4493–4499
Tubbs E, Rieusset J (2017) Metabolic signaling functions of ER-mitochondria contact sites: role in metabolic diseases. J Mol Endocrinol 58:R87–R106
Turrens JF, Boveris A (1980) Generation of superoxide anion by the NADH dehydrogenase of bovine heart mitochondria. Biochem J 191:421–427
Turrens JF (2003) Mitochondrial formation of reactive oxygen species. J Physiol 552(Pt 2):335–344
Ugalde C, Janssen RJ, van den Heuvel LP, Smeitink JA, Nijtmans LG (2004) Differences in assembly or stability of complex I and other mitochondrial OXPHOS complexes in inherited complex I deficiency. Hum Mol Genet 13:659–667
Vander Heiden MG, Chandel NS, Williamson EK, Schumacker PT, Thompson CB (1997) Bcl-xL regulates the membrane potential and volume homeostasis of mitochondria. Cell 91(5):627–637
Vernier PT, Sun Y, Marcu L, Salemi S, Craft CM, Gundersen MA (2003) Calcium bursts induced by nanosecond electric pulses. Biochem Biophys Res Commun 310:286–295
Vernier PT, Sun Y, Marcu L, Craft CM, Gundersen MA (2004) Nanoelectropulse-induced phosphatidylserine translocation. Biophys J 86:4040–4048
Vernier PT, Sun Y, Marcu L, Craft CM, Gundersen MA (2004) Nanosecond pulsed electric fields perturb membrane phospholipids in T lymphoblasts. FEBS Lett 572:103–108

Vernier PT (2011) Mitochondrial membrane permeabilization with nanosecond electric pulses. Conf Proc IEEE Eng Med Biol Soc 2011:743–745

White JA, Blackmore PF, Schoenbach KH, Beebe SJ (2004) Stimulation of capacitative calcium entry in HL-60 cells by nanosecond pulsed electric fields. J Biol Chem 279:22964–22972

White MJ, McArthur K, Metcalf D, Lane RM, Cambier JC, Herold MJ, van Delft MF, Bedoui S, Lessene G, Ritchie ME, Huang DC, Kile BT (2014) Apoptotic caspases suppress mtDNA-induced STING-mediated type I IFN production. Cell 159:1549–1562

Williams SCP (2019) Mitochondria and Tesla battery packs work pretty much the same way, study reports. UCLA Newsroom (14 Oct 2019)

Wolf DM, Segawa M, Kondadi AK, Anand R, Bailey ST, Reichert AS, van der Bliek AM, Shackelford DB, Liesa M, Shirihai OS (2019) Individual cristae within the same mitochondrion display different membrane potentials and are functionally independent. EMBO J 38(22):e101056

Xia D, Esser L, Tang WK, Zhou F, Zhou Y, Yu L, Yu CA (2013) Structural analysis of cytochrome bc1 complexes: implications to the mechanism of function. Biochim Biophys Acta 1827:1278–1294

Yagi T, Matsuno-Yagi A (2003) The proton-translocating NADH-quinone oxidoreductase in the respiratory chain: the secret unlocked. Biochemistry 42:2266–2274

Yun J, Finkel T (2014) Mitohormesis. Cell Metab 19:757–766

Zhao RZ, Jiang S, Zhang L, Yu ZB (2019) Mitochondrial electron transport chain, ROS generation and uncoupling (Review). Int J Mol Med 44(1):3–15

Zorova LD, Popkov VA, Plotnikov EY, Silachev DN, Pevzner IB, Jankauskas SS, Babenko VA, Zorov SD, Balakireva AV, Juhaszova M, Sollott SJ, Zorov DB (2018) Mitochondrial membrane potential. Anal Biochem 552:50–59

Zurita Rendón O, Shoubridge EA (2012) Early complex I assembly defects result in rapid turnover of the ND1 subunit. Hum Mol Genet 21:3815–3824

Chapter 9
usEP Induce Regulated Cell Death Mechanisms

Stephen J. Beebe

Abstract Cells employs many different mechanisms for their expiration. The best known and studied regulated cell death mechanism is apoptosis. Cell death subtypes, including regulated cell death (RCD), programmed cell death (PCD), and accidental cell death, are discussed. A discussion on and caution for the use of the term "necrosis" is included. Apoptosis was the earliest RCD mechanism shown in Jurkat cell responses to usEPs as determined by cytochrome c release and caspase activation. This apoptotic cell death was enhanced by usEP-induced supraelectroporation, as electric fields with nanosecond durations and short(fast) rise-fall times passed through the cell, while pulses with microsecond duration go around cells. However, using Jurkat clones that did and did not express APAF-1, which is an essential protein for apoptosome formation as a platform for caspase-9 and caspase-3 activation, it was also shown that usEPs induced caspase-dependent and caspase-independent cell death. A role for caspases depended on the usEP charging intensity with lower usEP impact causing caspase-dependent cell death while higher charging caused caspase-independent cell death. Although a full discussion of all RCD mechanisms is not included, evidence is presented that not all cell types responded to usEP by apoptotic cell death. The presence or absence of Ca^{2+} has an impact on the RCD mechanisms. Human triple-negative breast cancer cells expressed either or both necroptosis and parthanatos. Necroptosis is sometimes considered regulated necrosis because plasma membrane pores form from intracellular proteins. Finally, usEPs are also shown to induce cell responses downstream of toll-like receptors (TLRs). Considerations for immunogenic cell death (ICD) are also considered.

9.1 Introduction

The first considerations for regulated cell death (RCD) were with the discovery of apoptosis referred to as programmed cell death (PCD). Genetic studies of programmed cell death in C. elegans led to the identification of genes conserved from this worm to humans that identified homologous molecular mechanisms for how cells programmed their own death (Conradt et al. 2016). Many genes and conserved cell death pathways were identified that activated suicide or euthanasia programs that

S. J. Beebe et al., *Ultrashort Electric Pulse Effects in Biology and Medicine*,
Series in BioEngineering, https://doi.org/10.1007/978-981-10-5113-5_9

disassembled dying cells and removed them without inflammation and damage to adjacent cells. Since the discovery of apoptosis, many other PCD mechanisms have been defined. They all exhibit delayed death that is programmed or regulated by genetic mechanisms with specific signaling pathways that bring about cell demise (Tang et al. 2019; Stockwell et al. 2017; Linkerman et al. 2014; Galluzzi et al. 2018). RCDs are evolutionarily conserved mechanisms that are critical to cellular and tissues homeostasis. RCD is designed to eliminate unneeded cells or ones that are potentially dangerous. It can be silent immunologically or immunogenic. Cells can die passively by accidental cell death (ACD) due to immediate injury. In RCD, cells do not die immediately but attempt to respond to a stress and survive or to die actively by committing suicide or altruistic euthanasia to maintain the functions of surrounding cells. How quickly necrosis, or a final cell death, is reached, depends on the stimulus and its intensity. Viruses, bacteria, and cancers have scores of strategies that either induce or block specific RCD mechanisms, most notably recognized for apoptosis. Some bacteria inhibit apoptosis during infection by increasing expression of anti-apoptotic proteins, inhibiting cytochrome c release, scavenging ROS, and/or by activating cellular mechanisms that increase pro-survival proteins such as PI3K/Akt or Raf/MEK/EFK, which are also anti-apoptotic. PCD, RCD and ACD are discussed below (Fig. 9.1).

Successful exogenous and endogenous diseases, such as viruses and cancer, respectively, have evolved mechanisms that attempt to prevent the most common natural cell euthanasia or suicide mechanisms that affect cell survival. In fact, we have learned significantly about apoptosis and survival mechanisms based how they affected disease entities. This makes sense because the most successful viruses have evolved because they modulate the most effective RCD mechanisms in the host. The scientific community also used this understanding of disease mechanisms to target and stop them with therapeutic interventions. In any of these events, apoptosis appears to be one of the major RCD or PCD mechanisms that are affected by natural

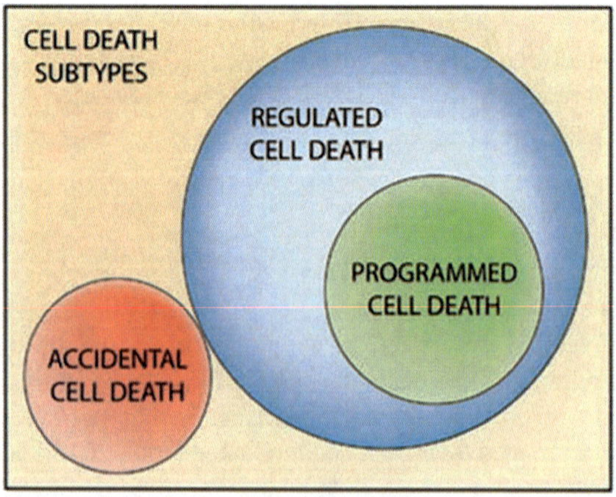

Fig. 9.1 Cell death subtypes: from Galluzzi et al. (2015)

invaders as well as by intrinsic mutations and other anti-cancer mechanisms that cause these diseases.

9.1.1 The Meaning of Necrosis

Before 1972, when apoptosis was realized (Kerr et al. 1972; Kerr and Searle 1972; Kerr 2002), the biological community considered one kind of cell death, which was called necrosis. There was not much to say about these cells and tissue other than defining the morphology of the dead cells. The membranes were ruptured, and cells were lysed leaking their intracellular contents into the surrounding environment.

Necrosis has other adjectives such as caseous, fat and fibroid necrosis (Kumar et al. 2010; D'Arcy 2019). Necrosis is specifically a descriptive characterization of "dead." The term necrosis comes from Greek nékrōsis, "death" or "state of death". And disappointedly, it does not tell us anything about molecular routes or passages that were taken to get to this deadness or necrotic end, because there were none; there is nothing programmed here. No signaling pathways of stress, hypoxia, anoxia, cell repair, DNA damage control or survival operations are defined. This is seen when cells or tissues are irreversibly injured with immediate membrane rupture. Necrosis was and is the result of death of tissues due to trauma, injury, which is now called accidental cell death (ACD). Necrosis is the end result of cell death. It is an uncontrolled cell death that has no specific pathways or mechanisms that regulate it. Necrosis is the observations of the histologists after the death process is completed. It should be noted that necrosis, membrane rupture, is the last event in any cell death mechanism, programmed or accidental.

However, there are several classifications of necrosis depending on other events that surround cell death. There is coagulative necrosis due to severe ischemia; gangrenous necrosis, a type of coagulative necrosis resembling mummification of tissues; and liquefactive necrosis, when dead cells are exposed to digestive enzymes due to inflammation or bacterial infection. The term "autolysis" is used to refer to death due to cell destruction by the action of its own enzymes. Necrosis is a cold and unimaginative term, a state of not being alive. It is like looking at a dead organism or piece of injured tissue, for example. After the death process is complete, all cells and tissues will be classified as necrotic. However, as a scientific term, "necrosis" is frequently misused, such as by saying "cell death by necrosis". This does not tell us anything about the death process. Nevertheless, there is an understandable conflict with the term necrosis. Before apoptosis and PCD was realized, historically cell death was referred to as necrosis. This is when cells have succumbed to death processes and lose their membrane integrity and become leaky. These cells are dead or are necrotic. Regardless of how a cell dies, it will eventually lose its membrane integrity and lyse, unless it is phagocytized. It is now said to be necrotic, no matter how it died. The morphological features of necrosis is cell swelling or oncosis and plasma membrane permeabilization. Again, this is best defined as accidental cell death (ACD).

9.1.2 *Pyroptosis and Necroptosis Are Called Regulated Necrosis*

As it turns out, there are at least two RCD mechanisms that end life by making pores in cell membranes causing oncosis and plasma membrane permeabilization; however, this occurs by regulated mechanisms that produce pores in cell membranes from the intracellular proteins. These have often been referred to as "regulated necrosis". However, these two mechanisms lead to a common result by different mechanisms and therefore have different names called necroptosis and pyroptosis. Both mechanisms establish plasma membrane channels and pores, respectively, by specific mechanisms that are part of inherent regulated pathways. These mechanisms will be discussed below.

9.2 Cell Death Can Be Programmed or Regulated

Circa 1972, a new concept concerning death was revealed. It actually took more than a decade to be accepted or noticed outside of a relatively small group of histologists and morphologists. However, when it did become known, it expanded and made a tremendous impact and immensely enhanced our understanding of events far beyond death itself, about causes of death, how cancers avoid cell death, and how mechanisms of cell death can be manipulated. When cell death is looked at closely, there is so much more to learn about the diverse and wonderful mechanisms of survival and the infamous and offensive yet shrewd and cunning mechanisms of cancer's survival. In coming to understand these complicated life and death processes, Nature has given scientists everlasting employment because there are so many processes and mechanisms that differ among cancers, which can change as the cancers are treated. It impossible to see the end of it all, until.... necrosis.

It was recognized that cell loss was necessary to compensate for new cell growth to maintain homeostatic control of cell function, form, and size. It was believed that cell turnover was due to cell death, but such processes had not been defined. The systematic physiological cell death was referred to as necrobiosis. In the late 1960s, textbooks used the term "coagulative necrosis" to define control of cell populations. For decades, dying cells in healthy tissue was observed, but sufficient notice of a uniquely characterized cell death mechanisms appeared in 1971–1972. Kerr (1971), Kerr et al. (1972), Kerr and Searle (1972) described a unique kind cell death mode referred to as shrinkage necrosis. The term "apoptosis" was taken from Greek apóptōsis, "falling off" as the falling off of petals of a flower or leaves from a tree. The verbal stress is on the penultimate syllable "ptosis" (silent "p") from the root word "to fall", which was already used to describe drooping of the eye (Kerr and Searle 1972). It is now known that apoptosis is responsible for the control of cell populations as a homeostatic complement to mitosis for maintaining normal population control. A series of studies demonstrated the presence of apoptosis in a wide range of cellular process (Kerr et al. 1972; Kerr and Searle 1972; Wyllie et al. 1972, 1973, 1980). These events and processes include the appearance of apoptotic bodies in normal organ development, embryonic mesenchyme, ACTH withdrawal in the adrenal cortex, germinal centers of lymphoid tissues, orchidectomy, involution of the corpus luteum and specific stages and positions in embryogenesis. Reading any of these early papers by Kerr, Wyllie, Currie, and colleagues provides a wealth of scientific perspective with breadth, depth and consideration of analyses and is especially instructive for applying the scientific method.

These earliest understandings of apoptotic characteristics came from histologists, electron microscopist and morphologists. Necrosis carried terms like pyknosis, where the nucleus shrinks and chromatin condenses; karyolysis, where the nucleus is destroyed by swelling and the chromatin losses its affinity for staining; and karyorrhexis, where the chromatin is distributed irregularly throughout the cytoplasm. In contrast, during apoptosis cells shrink, nuclei fragment, chromatin condenses and cell membranes bleb and expel apoptotic bodies. As time passed, it became clear that morphology was not the best way to define apoptosis because there are overlapping morphological characteristics among cell death mechanisms. For example, it was realized that phagocytosed apoptotic bodies were mistaken for autophagic vacuoles, which are vacuoles originating within cells that degrade their own dysfunctional cellular materials for recycling. Apoptotic bodies are phagocytized and recycled by other cells. Interestingly Kerr and Searle (1972) indicated that in mammary carcinoma induced by an anthracene (the simplest tricyclic aromatic hydrocarbon, a carcinogen), most of the apoptosis cells were phagocytized by the tumor cells. In addition, nothing in the morphology of apoptotic features foretold the biochemical, enzymatic, post-translational changes, and complex signaling events from the mitochondria, nucleus endoplasmic reticulum, and lysosomes that lead to these morphological characteristics. So, through the 1990s, there were hundreds of papers published describing this so-called programmed cell death. During these revelations, it became clear that these death processes occurred over time, that they appeared to be controlled so they were referred to as programmed cell death (PCD). As the CD mechanisms were being revealed, PCD became synonymous with apoptosis. Historically cell death based on morphology was classified with apoptosis as type I cell death, autophagy as type II cell death and) and necrosis as type III cell death. However, as the CD scene became more complicated, and as we looked more closely at how cells die through agents that kill cancer and ablate other tissue, and as we define these CD processes more closely, apoptosis does not necessarily have to be synonymous with programmed cell death, per se.

Again, cell fate is determined by a balance between anti-apoptotic survival mechanisms and pro-apoptotic CD mechanisms. Generally pro-apoptotic mechanisms are in inactive states but importantly can be rapidly activated (Green 2005); however, they need to be activated at appropriate times. These activation mechanisms demonstrate the importance of cells dying in a timely manner. When CD is eminent, default mechanisms are invoked, and RCD mechanisms overrule survival mechanisms. Under these conditions, cells commit suicide, a kind of cellular euthanasia.

These are programmed and therefore regulated events, meaning they function in pathways and ultimately target key components of CD machinery. Regulated cell death mechanisms are genetically encoded by and can be manipulated with pharmacologic and/or genetic intervention(s).

Cell death is generally classified as three subtypes including accidental cell death (ACD), which is independent of the other two subtypes, and programmed cell death (PCD), which is a subset of regulated cell death (RCD), the third subtype. There are many known RCD mechanisms that are the purveyors of life's end. Among these RCD mechanisms is PCD, which specifically occurs as a means of cell and tissue homeostasis, during embryonic development, and some immune responses and microenvironmental stresses (Galluzzi et al. 2015). These are natural mechanisms that remove unwanted or damaged cells without inflammation and without damage to surrounding clonal mates and adjacent tissues. Classical examples are apoptosis and elimination of cells that formed the frogs' tails during metamorphosis. Another example is sculpting our fingers and toes during embryogenesis, when cells between the digits die by apoptosis. There are other intrinsic mechanisms regulated by hormones as well as autocrine and paracrine factors that induce PCD to maintain homeostasis. For example, after ovulation the dominant follicle forms the corpus luteum, which produces progesterone to maintain the endometrium as preparation for pregnancy. If pregnancy does not occur, the cells of the corpus luteum die by apoptosis so new follicles establish during the next menstrual cycle. Another example is the turnover of millions, even billions of neutrophils in our bodies that are ready for fighting foreign invaders but are not needed because we stayed safe. Basically, PCD is nature's euthanasia (painless killing) mechanism. PCD is the counterpoint to mitosis and proliferation. Together they maintain appropriate cell numbers in tissues and organs that provide normal function. Set aside from RCD processes the term PCD are specific to natural mechanisms of development and the maintenance of homeostasis and does not lead to inflammation.

It is now clear that there are many different cell death mechanisms that are regulated, but they most often result in inflammation. These mechanism have been presented and reviewed in excellent detail (Tang et al. 2019) and will not be specifically presented here except in reference to induction of them by usEPs.

When studies with usEPs began in the mid-late 1990s, there were two defined mechanisms for cell death – apoptosis and "necrosis". Apoptosis was then called PCD. As now, necrosis was not programmed; it was CD by misfortune or murder or dead cells that could not be removed before inflammation occurred. The term necrosis is too commonly used but remains part of the medical terminology because the term has been used for centuries but has become jargon and often used when a real RCD mechanism is not known. Necrosis is not a cell death mechanism and does not tell us anything about how a died, only that it is dead. It is now equivalent to ACD However, Wyllie and Kerr (1980) used necrosis as the antithesis of apoptosis. Necrosis was indicted by irreversible changes in the nucleus (karyolysis, pyknosis, and karyorhexis) and in the cytoplasm (condensation and intense eosinophilia, loss of structure, and fragmentation). These are the characteristics of cells' cadavers. They do not tell us how the cells died—by toxin, ischemia, heat, cold, mechanical trauma, apoptosis, or other RCD mechanisms (Majno and Joris 1995). In early studies with usEPs it was noteworthy that cells did not die by sudden ACD, although not designated as ACD because the term was not in use until later (Beebe et al. 2002, 2003). However, when pathologists analyzed slices of tissues many hours after usEP treatment or in fixed samples of tissues, they called it necrosis and rightly so, because the cells in the tissue were dead. However, that terminology ignores how the cell really died. That is because the CD processes have reached their end, cells have lost their defining structures, and molecular mechanisms have terminated; regardless of how they died they show necrotic features. What has been lost is the defining characteristics of the RCD process; the occurrences of cell biology, enzymology, cell signaling, and gene expression that preceded this ultimate end. Many of the morphological features of cell death overlap. For these reasons, defining CD mechanisms by morphology is problematic (Galluzzi et al. 2015). It is not impossible to use morphology for RCD criteria, but it is dependent on when the observations are made. They need to caried out during the death processes; calling for time courses and carried out within a dose–response study during the kinetics of the RCD processes (see Parvathenani et al. 1998; Beebe et al. 2003). In larger tissue masses that do not allow rapid removal of the dead and dying cells by macrophages, such as in dying solid tumors, cells will appear necrotic at the end of their lives, regardless of how they died. In vitro, where there are no phagocytes, regardless of how a cell dies, the final end result when the RCD processes are complete will look like necrosis.

Since our overall behavior is defined by the responses from cells that make us, survival is paramount at the cellular and organismal level and is carried out by the same vital survival mechanisms. When an individual's survival is threatened, they will do all they can do to survive. So, when a cell receives a potentially lethal stimulus, like usEP signals, cells will try to survive, by whatever mechanisms available to them—and there are many mechanisms for survival, interestingly, like there are many mechanisms for RCD.

So, the question arises, when cells receive lethal usEPs, into which Zenn diagram sphere in Fig. 9.1 do they fit? Under usEP conditions that are used to treat cancer and to study cell death mechanisms in vitro, cell death is not in the ACD sphere. For essentially all cells and cancer types, when cell death is not ACD, the cell death mechanism is RCD. Since the PCD sphere fits inside the RCD domain, as discussed above, it is a specialized form of RCD that occurs when cells die naturally during development or during homeostatic control of cell number. There is nothing natural about usEPs.

So, philosophically, we could ask if cell death due to a sunburn, as death by heat (Fulda et al. 2010), or frost bite, as death by cold (Neutelings et al. 2013), would be considered RCD because the damage was not due to a physiological processes. However, these cell death stimuli are not natural. Cell death induced by chemotherapeutic agents or usEP, which induced one or another delayed cell death process would also cause damage to cells adjacent to cancer cells due to inflammation because dead and dying tumor cells would not be removed before they died and released their internal contents. These cells died by RCD or if death was immediate, they died by ACD; Nevertheless, they will all look necrotic in the end.

9.3 Cell Stress Responses to usEPs

usEPs do not exist naturally like heat, cold and radiation (Maier et al. 2016). So, cells have no evolutionary specifically established mechanisms to respond to usEPs. As might be expected, cells responses to usEPs can be defined as stress mechanisms that cells use to provide sufficient time to recover from the stress, to re-establish homeostasis and repair if necessary. Before discussing lethal stress and RCD in response to usEPs, it is was most interesting to discover how Morotomi-Yano et al. (2013) defined unique responses to usEPs compared to heat and radiation, which have well known responses to stress. Cell stress responses are initiated by phosphorylation from one or more four stress-related kinases. Among these include PERK, which is a double stranded-dependent RNA-dependent kinase localized in the endoplasmic reticulum (ER) and activated by unfolded protein response. Another stress-activated kinase in GCN2, which is known to be activated by amino acid deprivation and UV radiation. PKR (is it PERK?) is activated by double-stranded viral RNA to prevent viral protein synthesis as well as halting general host protein synthesis under viral stress. Finally, HRI is a heme-regulated inhibitory stress kinase. Downstream of all of these kinases is phosphorylation of the ά-subunit of eIF2, which is the rate-limiting enzyme in initiation of protein synthesis. This phosphorylation event inhibits protein synthesis to conserve energy. This is referred to as an integrated stress response under many stressful stimuli. Another regulatory mechanism is the phosphorylation of 4E binding protein (4E-BP), which blocks cap-dependent initiation of translation. This phosphorylation is independent of kinases that phosphorylate eIF2ά, but initiated by mTORc1(mammalian target of rapamycin complex 1), which regulates homeostasis and proliferation and is sensitive to energy, oxygen and nutrient levels and growth factors.

The authors used 80 ns pulses (Morotomi-Yano et al. (2012)). As electric fields and/or pulse numbers were increased usEPs rapidly phosphorylated eIF2ά, demonstrating a robust integrated stress response. A threshold for phosphorylation of eIF2ά as about 20 kV/cm with 20 pulses or 7.2×10^{-2} Vs/cm. Using cells that were devoid of PERK or GCN2, eIF2ά was still phosphorylated; however, when both kinases were absent phosphorylation was greatly reduced. The authors suggested that there was functional compensation between the two kinases as opposed to the presence of another kinase that could at least marginally phosphorylate eIF2ά. These results suggested that cells responded to usEPs like they did to UV radiation and unfolded proteins. However, additional studies demonstrated usEP elicited different characteristics than other ER stress responses when expression of other proteins were analyzed.

Overall, these studies showed that usEPs activated PERK/GCN2-mediated eIF2ά phosphorylation and the concurrent decrease in 4E-BP1 phosphorylation in a PERK/GCN2-independent manner. Both of these events inhibit translation. The authors considered the ER stress response because PERK, which is involved. PERK is located with its N-terminus inside the ER, which senses unfolded proteins, and the C-terminus kinase domain in the cytoplasm is regulated by phosphorylation.

The ER stress response includes suppression of translation (Ron 2002). The ER stress response also induces Bip and Chop transcription expression and *XBP1* mRNA splicing. All of these were shown to be increased by thapsigargin, a well-known ER stress stimulus. In contrast, usEPs did not induce these responses. Thus, the canonical ER stress response was not stimulated by usEPs. This suggests that under these conditions, usEPs does not cause protein unfolding in the ER. In further investigation of common gene expression induced by UV radiation and heat shock. both of these stress stimuli increase GADD45β RNA expression and decrease expression of *XIAP* and *BL2.* Heat shock increases expression of both *Hsp27* and *Hsp70.* Interestingly, usEP elicited none of these responses. Taken together, like unfolded protein response and response to UV radiation, usEPs inducestress responses as indicated by the PERK- and GNC-induced phosphorylation of eIF2ά and the decreased phosphorylation of 4E-BP. However, usEPs-induced stress is sensed and responded from characters that are distinct compared to heat, radiation, and the un-folded protein response. As the authors point out, there are additional studies to be carried out in these domains. Importantly, the authors indicate that usEP responses from 4E-BP1 suggests that mTORc1 could be a sensor of usEPs.

Morotomi-Yano et al. (2013) did not analyze stress levels that induced cell death, but threshold for stress-induced RCD is cell-type dependent. It has been shown that as the charging levels approach and exceed 1×10^{-2} Vs/cm RCD will occur by mechanisms that are cell type-dependent.

What aspects or components of usEPs induce cell death? usEP are considered to be non-thermal and this is the case when the pulse repetition rate is relatively slow. Most often units of usEP-induced cell death are discussed in J/cc despite the fact that these pulses are non-thermal. This is a carryover from considerations of conventional electroporation and radiation. Cellular responses to usEP are different than effects of conventional electroporation. In the presentations and discussions of usEP conditions, we use the charging unit of Vs/cm. This is based on the concept that usEP with short or fast rise times have intracellular effects (Schoenbach et al. 2001, 2004, 2009; Gowrishankar et al. 2006). These intracellular effects are believed to induce apoptosis and RCD mechanisms in a cell type and context-dependent manner. These intracellular effects contribute to immunogenic cell death (ICD) if they induce appropriate responses in dying cancer cells as will be discussed (Kepp et al. 2014).

9.4 Regulated Cell Death (RCD) Responses to usEPs

A simple electrical model for effects on biological cells predicted that electric fields with durations in the nanosecond range would interact with intracellular structures This was supported by the experimental finding that the 60 ns pulses with electric fields of 53 kV/cm modified intracellular granules of human eosinophils (Schoenbach et al. 2001). Shorter pulse duration (10 ns > 60 ns > 100 ns) had greater effects on intracellular Ca^{2+} release from the endoplasmic reticulum (Semenov et al. 2013).

For pulse rise times, shorter (15 ns) rise times had greater effects to decreasethe mitochondrial membrane potential and induce cell death than pulses with longer (150 ns) rise time (Beebe et al. 2012, 2013). These short duration, short (fast) rise time pulses that affect intracellular structures were predicted to induce apoptosis in Jurkat cells (Beebe et al. 2002, 2003; Gowrishankar et al. 2006) and regulated cell death when apoptosis is not the major mechanism of cell suicide in usEPs. These usEP with fast rise times are also suspected to induce immunogenic cell death (ICD) and induce host immune responses when some tumor types are treated with usEPs (Fig. 9.2).

When discussing cell death, it is key to remember that except for accidental cell death (ACD) by immediate injury, cell death is a process. Dying is a progression of events that generally begins with a stimulus that is stressful or toxic, proceeds through RCD-specific pathways, and ends in lifelessness.

Processes in apoptosis occur in three stages as initiation, effector, and execution phases. Induction is the rate-limiting step and depends on the stimulus and its intensity and the cellular context. Figure 9.3 shows this can be through the intrinsic pathway due to DNA damage, for example, or external factors such as UV light, growth factor withdrawal or other stresses leading to cytochrome c release, and caspase activation by apoptosome formation with APAF-1, cytochrome c and ATP. Apoptosis can also occur through the extrinsic pathway with engagement of death receptor activation by FAS, or TRAIL, for example, formation of the death induced signaling complex (DISC) and caspase activation. Each of these pathways uses different sets of induction proteins. The intrinsic pathway leads to cytochrome c release and initiator caspase-9 in the apoptosome and executioner caspase-3 activation while the extrinsic pathway activates initiator caspase 8, which can lead directly to executioner caspases or can

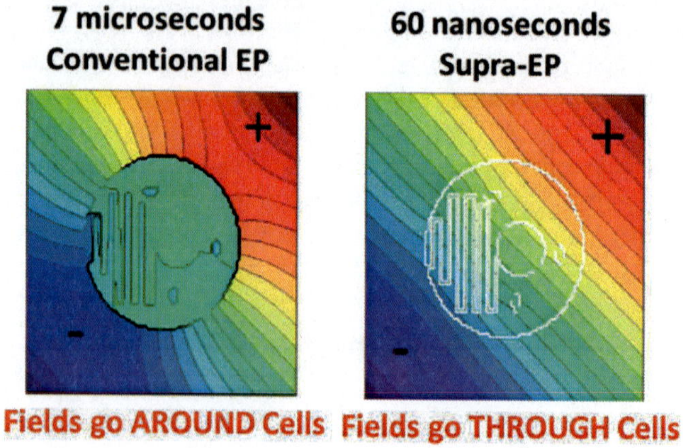

Fig. 9.2 Cellular responses to microsecond conventional electroporation versus nanosecond supra-electroporation. Slow pulses with 7 μs durations and rise times greater than 100 ns do not readily penetrate cells but go around them. In contrast, fast pulses with 60 ns durations and rise times less than 100 ns go through cells affecting intracellular organelles. Adapted from Gowrishankar et al. (2006)

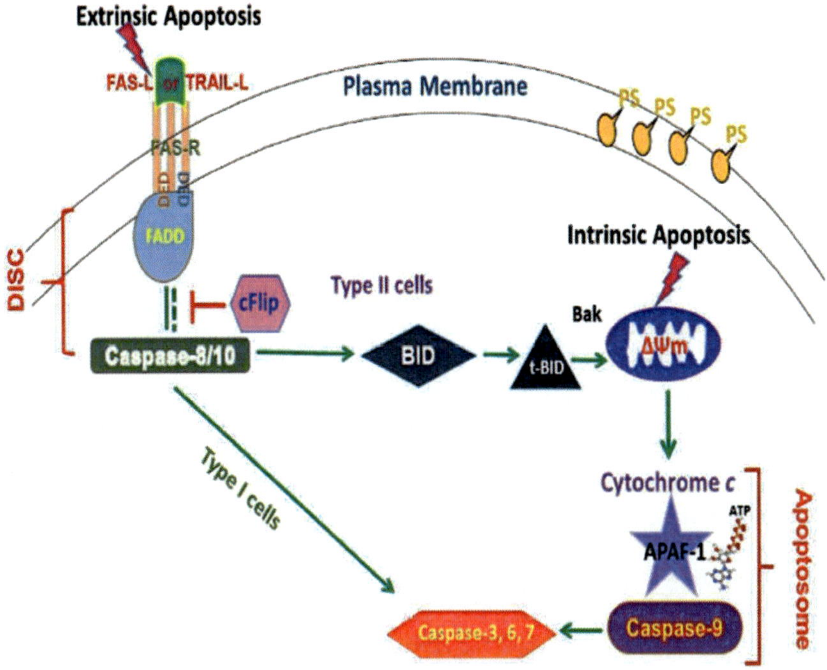

Fig. 9.3 Extrinsic and Intrinsic Apoptosis mechanisms. Extrinsic apoptosis occurs through death receptors such as the FAS receptor (Fas-R) or tumor necrosis factor (TNF-R) when the FAS or Trail ligand (FAS-L) or TNF-L binds to activate Caspase 8 at the death-induced signaling complex (DISC). In Type I cells caspase -8 directly activates caspase-3. In type II cells caspase 8 cleave BID, forming t-Bid, which leads to cytochrome c release and the formation of the apoptosome with the accumulation of APAF-1, ATP, and caspase-9, which is activated to activate caspase-3. Apoptotic events that occur intracellularly, such as acting on the mitochondria, occur through the intrinsic pathway with the release of cytochrome c, apoptosome formation, caspase 9 and caspase-3 activation. Modified from Beebe et al. (2013)

cleave BID and use the intrinsic pathway strategy with cytochrome c release and caspase activation with apoptosome formation. The activation of the caspases begins the effector phase. The steps from cytochrome c release to effector caspase activation are rapid. The final phase is the execution or degradation phase when caspase proteases start to disassemble the cell structure, and apoptotic morphology becomes apparent. There are feedback loops that continue the execution phase as more significant levels of cytochrome c are released, apoptosis inhibitors are inactivated, and caspases are exponentially activated as one caspase activates many more caspases. In vitro, under these conditions, cells will eventually lose membrane integrity. If the apoptosis process is observed at this late time, cells will have appeared to have died by ACD. In vitro, where phagocytosis does not occur, apoptosis may have been activated, but cells will eventually lose pasma membraneintegrity, appearing morphologically as necrosis if their molecular or morphologic histories were not investigated. This appearance of death after apoptosis is referred to as post-apoptotic necrosis. Thus, as indicated above, artificial in vitro environments can pose problems for the analysis of apoptosis and other RCD in general, because they exhibit overlapping features, and the final appearance is cells with ruptured membranes or necrosis. It is for this reason that morphologic characteristics of apoptosis in vitro can be misleading concerning RCD identification. However, in vivo, there is another set of phases to consider.

As shown in Fig. 9.3 for apoptotic cell death, this begins as extrinsic or intrinsic apoptosis by the formation of one of the caspase activation platforms or complexes. The extrinsic mechanism uses the DISC with FAD/TNF receptor, FADD/TRADD and initiator caspase-8 and the intrinsic mechanism uses the apoptosome with APAF-1, cytochrome c, and ATP to activate initiator caspase-9.

Extrinsic apoptosis begins at death receptors on the plasma membrane of cells, most often defined using the FAS and TRAIL receptors and their cognate ligands and formation of the DISC, a caspase activation platform activating initiator caspase-8 or caspase-10. Intrinsic pathway is initiated by intracellular events, such as stress or damage to organelles, but is also used in some cell types through the extrinsic pathway. As indicated above, this involves the mitochondrial release of cytochrome c, and formation of the other caspase activation platform with APAF-1 and ATP recruiting initiator caspase 9 and then caspase-3 activation.

Apoptosis activation at the DISC is assembled by *the homeotropic interaction* of domains on the proteins that form the complex. These proteins include an intracellular death domain (DD) of the FAS or TNF receptor that binds the DD of FAS-associated death domain or TNF receptor-associated death domain (FADD/TRAAD). At the other end of FADD/TRAAD are death effector domains (DED). They recruit and bind the DED of caspase-8. This forms the DISC, which activates caspase-8. C-FLICE inhibitory protein (c-FLIP)can also be recruited that prevents capsase-8/10 activation. The relative abundance of the DED-domain containing proteins appears to determine when caspases can be activated. For example, when inhibitory cFLIP is in low abundance, the probability that two caspase homodimers will combine is greater, and caspase-8/10 activation is more likely than when c-FLIP is in high abundance (Schleich et al. 2012).

What happens after the DISC is formed depends on the cell type. In some cells, called type I cells (thymocytes), caspase-8/10 activation leads directly to caspase-3 activation. In other cell types, called type II cells (hepatocytes), caspase-8 activation leads to cleavage of the protein BID forming a truncated protein called t-Bid that promotes cytochrome c release from the mitochondria, which activates the intrinsic apoptosis pathway. When there are sufficient quantities of caspase-8 and FADD for DISC formation, caspase-8 directly activates caspase 3. Alternatively, caspase-8 can cleave Bid and proceed through the intrinsic pathway. It has been suggested that the efficiency of caspase-8 cleavage of BID is the rate-limiting step in the involvement of death receptor TRAIL-induced apoptosis (Kantari and Walczak 2011).

Since the realization that cells died by regulated pathways was first defined as apoptosis, ten or twelve different RCD pathways have emerged, although they are interconnected. Tang et al. (2019) illustrated the chronology for RCD mechanisms from the discovery of apoptosis in 1972 up to 2018. An evolution of RCD mechanisms can also be discerned by reviewing the nomenclature committee's recommendations from 2009 (Kroemer et al. 2009) to 2018 (Galluzzi et al. 2018). Galluzzi and cell death recommendation of the nomenclature committee on cell death 2018 (Galluzzi et al. 2018) listed twelve different RCD mechanisms from morphological, biochemical, and functional assessments (Fig. 9.4).

Although we will not discuss all RCD pathways in detail, those pathways that are induced by usEPs will be discussed. Nevertheless, defining CD is a more complex exercise than originally conceived in the late twentieth century. However, to simplify classifications, it is expedient to define CD as "regulated" or "accidental". In the late 1990s, when CD responses to usEPs were first investigated, RCD was synonymous with apoptosis and accidental cell death (ACD) was equivalent to "necrosis". PCD and apoptosis were sometimes used interchangeably; however, PCD, as indicated in Fig. 9.1, is RCD that occurs as part of developmental programs or maintenance of

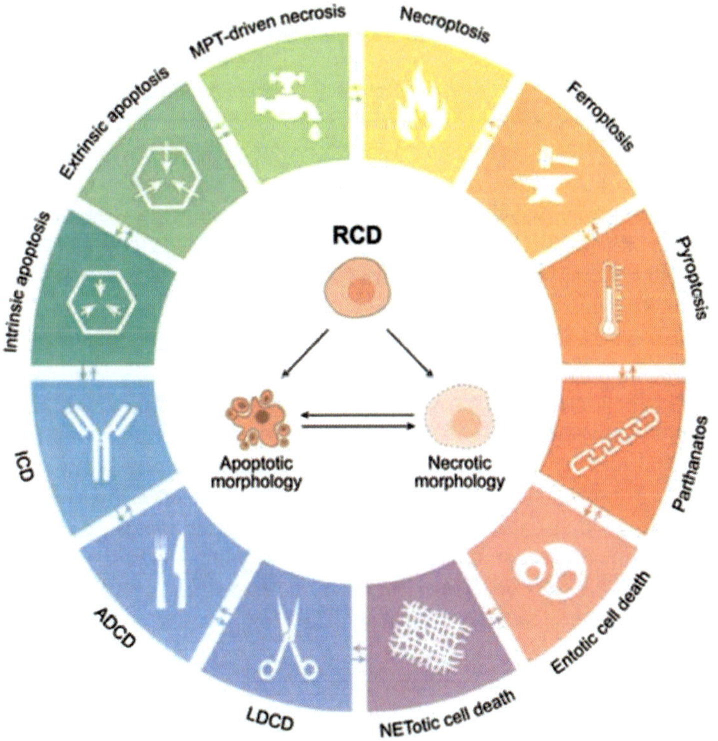

Fig. 9.4 Major cell death subroutines. Using molecularly oriented definitions of terms, RCD mechanisms include intrinsic apoptosis, extrinsic apoptosis, mitochondrial permeability transition (MPT)-driven necrosis, necroptosis, ferroptosis, pyroptosis, parthanatos, entotic cell death, NETotic cell death, lysosome-dependent cell death, autophagy-dependent cell death, immunogenic cell death, cellular senescence, and mitotic catastrophe (from Galluzzi et al. 2018)

normal homeostatic functions. Apoptosis is now known to be one of several RCD mechanisms (Fig. 9.4). Several non-apoptotic RCD mechanisms have been shown to be programmed as in "programmed necrosis" and distinct from primary necrosis or ACD. Thus, the term "necrosis" has historically referred to primary necrosis as non-regulated CD, which is now referred to as accidental cell death (ACD) with immediate loss of plasma membrane integrity. The presence of non-apoptotic regulated cell necrosis was generally missed because TUNEL staining was often considered to be a maker for apoptosis. However, TUNEL staining identifies apoptotic cells as well as non-apoptotic cells or "necrotic cells", some of which should have been considered regulated necrosis. TUNEL staining does not specifically identify apoptotic DNA, which is identified as DNA laddering on an SDS-PAGE gels. DNA laddering is a distinguishing characteristic feature of DNA degraded by caspase-activated DNAses (CAD), which is a specific event during apoptosis. CAD can only cleave DNA at regions that are available between or among histone nucleosomal cores that are 180–185 base pairs apart, giving a ladder appearance upon electrophoresis. TUNAL identifies random DNA fragments on gels. So, "regulated necrosis" pathways are other RCD programs as in non-apoptotic, caspase-independent CD. If CD is not immediate, it is highly likely to include some regulated aspects, either stress-related, pro- or anti-survival. While there are many RCD mechanisms, apoptosis will be the major one discussed in responses to usEPs in this chapter, primarily because it was the first RCD realized and is the considered the prototypical RCD mechanism (Napoletano et al. 2019). However, other RCD mechanisms are induced by usEP including necroptosis (Fig. 9.9), which includes programmed lytic CD by pore formation in the PM and parthanatos (Fig. 9.10), which is CD induced by PAR poly(ADP-ribose) formation on proteins related to excessive DNA damage.

Another RCD mechanism is characterized by caspase-1 activation and leads to programmed lytic CD. Although usEPs lead to increases in ROS (see the section on mitochondria), pyroptosis has not been shown to be induced by usEPs; but that may be because it has not been specifically targeted for identification. Like necroptosis, pyroptosis results in formation of protein-bearing plasma membrane pores and results in cell membrane lysis (Chen et al. 2016; Rathkey et al. 2018; Robinson et al. 2019). However, these RCD mechanisms are regulated pathways and differ from plasma membrane lysis due to accidental cell death. Pyroptosis is provoked by activation of inflammasome, which activates pro-inflammatory caspases (caspase-1/11) leading to the release of interleukin-1β [IL1β] and IL18 (Tang et al. 2019). These inflammatory signals are sensed by TLRs or Nod-like receptors (NLR) as pathogen-associated molecular patterns (PAMPs) or damage associated molecular patterns (DAMPs). They lead to upregulation of nuclear factor of κB (NF-κB) that triggers proinflammatory caspase-mediated pyroptosis (Broz and Dixit 2005). It is now appreciated that gasdermin D (GSDMD), a protein cleaved by pro-inflammatory caspases, is the fundamental component of pyroptosis. Pro-inflammatory caspases cleave GSDMD, which translocates to the cytosolic side of the plasma membrane, binds phospholipids and forms a pore that lyse the cell membrane. These processes are well documented (He et al. 2015; Kayagaki et al. 2015; Chen et al. 2016; Ding et al. 2016; Liu et al. 2016; Shi et al. 2017). Mitochondrial reactive oxygen species (mitoROS) appear to play an important role in IL-1β release and pyroptosis, mediated by both NLRP3 and GSDMD (Platnich et al. 2018). Caspase-1 exhibit redox-sensitive cysteine residues and can be modulated by ROS by reversible oxidation and glutathionylation (Meissner et al. 2008). Specific mechanistic understanding of the processes of in pyroptosis are developing (Robinson et al. 2019).

Cell death has also been associated with a process called autophagy. The term comes from self (auto) eating (phagy). It is generally an effort to preserve cell integrity under insults of stress as a resistance to starvation or exposure to toxic substances. It is also a means of housekeeping to eliminate defective cell structures of organelles, such as dysfunctional mitochondria, called mitophagy or quality control of defective proteins (Yonekawa and Thorburn 2013; Napoletano et al. 2019) Autophagy is a catabolic process that delivers dysfunctional cellular constituents to lysosomes for degradation and recycling. This is accomplished by forming a cellular compartment called that phagosome, a double membrane structure that fuses with lysosomes, which are acidic organelles containing nucleases, proteases, and lipases among other degrading proteins that hydrolyze and degrade the phagosome cargo. Autophagy connects with most, if not all, RCD mechanisms, especially apoptosis, which has been most studied. These are complex and complicated interactions. This is exemplified by autophagy roles in promoting cell death and protecting against cell death (Yonekawa and Thornburn 2013). While there have been many studies investigating these phenomena, there is still missing mechanistic understanding of these interactions. It has also been considered that there is an autophagic cell death process. This has been difficult to unravel because of the interactions with other RCD mechanisms and in many cases, the artificial methods of experimentation (Yonekawa and Thorburn 2013). Perhaps an easy way out of this discussion, although likely correct, is to consider that nature has conceived multiple ways to eliminate cells and perhaps has conserved autophagy through evolution as a means of survival in stressful times. In this scenario, autophagy attempts to survive, perhaps by interacting with apoptosis mechanisms, and upon failure depends apoptosis or other mechanisms for demise. What has been considered CD *by* autophagy is better defined as CD *with* autophagy.

Considered through an in vivo scope, there is the decision to induce the apoptosis program or another RCD mechanism, the actual suicide or disassembly of the cells, the engulfment of dying cells by phagocytes, and the final degradation of the apoptotic cell inside the phagocyte. So, in cases of PCD under developmental or homeostatic conditions, the cellular neighborhood never "sees" the dying cells lose their membrane integrity because the final stage of apoptosis in the PCD routine takes place inside a phagocyte; PCD never causes inflammation, which occurs when cells leak their content into their environment. However, in the case of usEP-induced apoptosis or other RCD mechanism induced in larger tumor masses that cannot be cleared by phagocytes, many cells will lose their integrity before they can be removed; it takes a week or more for a 4–6 mm tumor in mice to be cleared from the treatment site after elimination by usEPs.

9.5 Examples of usEP-Induced RCD

The earliest characterization of apoptosis in vivo was defined by morphological characteristics that differentiated cells were undergoing apoptosis from cells dying by primary necrosis or ACD. Apoptotic cells exhibited cytoplasmic shrinkage, chromatin condensation, nuclear fragmentation, PM blebbing, and formation of apoptotic bodies with intact membranes (Wyllie et al. 1972, 1973; Kerr 2002). However, because many different RCD mechanisms are known and have overlapping morphological characteristics, defining RCD based on morphological features is not recommended by the Nomenclature Committee on Cell Death (Galluzzi et al. 2015). There are several RCD routines that finish with ruptured cell membranes; however, these occur by regulated pathways with different upstream phenomena. Also, morphological definitions do not report or reveal the molecular events that lead to the final cell demise. It is these molecular details that are the most defining characteristics of RCD. It is most suitable to establish at least two different unambiguous molecular features for an appropriate definition of any RCD mechanism.

The early studies in the authors' laboratories indicated that usEPs killed biological cells, including cancer cells; however, they were not eliminated immediately. This delayed cell death suggested that the great majority of these cells did not die by what we now call ACD, although the common historical term for that condition was called necrosis or immediate loss of cell membrane integrity (ACD). The only other cell death process, known at the time was apoptosis, so identifying apoptosis markers was the focus of these early studies. Human Jurkat and HL-60 cells were commonly used cancer cells in the lab, so those were subjected to assays that would demonstrate the presence or absence of apoptosis markers such as active caspases, which were known to be the enzymes of apoptotic cell death (Porter and Janicke 1999; Budihardjo et al. 1999), and the release of cytochrome c from the mitochondria into the cytoplasm (Beebe et al. 2002, 2003).

To prove that a given response is specific to a given stimulus, it is crucial to demonstrate a stimulus-dose-dependent response. For usEP, the dose–response is generally considered as electric field- or pulse number-dependent effects. Although not evaluated in this way at the time of the studies, considered another way, the responses were considered as a charging effect in Vs/cm, which, as discussed earlier, is calculated as electric fields in volts/cm (E) times the pulse duration in seconds (τ) times the square root of the pulse number ($n^{1/2}$). This square root is based on the scaling law, where the square root of pulse number is in accord with the random walk theory (Schoenbach et al. 2009). So, for caspase activation, cytochrome c release, and annexin-V binding the responses were dependent of the pulse duration, which determined the charging effect of Vs/cm.

Caspase was assayed by catalytic assay (Beebe et al. 2002) that was established before active caspase kits were available (Parvathenani et al. 1998) and an indicator of active caspases z-VAD-fmk. Active caspase were shown to be present in an usEP-duration-dependent manner in the hours after treatment. So, for Jurkat cells treated with five pulses with durations of 10, 60, and 300 ns and electric fields adjusted to equalize energy density (~ 1.7 J/cc), (150, 60, and 26 kV/cm, respectively) it was found for that the 300 ns condition (1.7×10^{-2} Vs/cm) was greater than the 60 ns (8.1×10^{-3} Vs/cm) condition, which was greater than the 10 ns (3.4×10^{-3} Vs/cm) condition. These studies demonstrated that the usEP effects were due to charging effect rather than energy density effects. When cells were analyzed four hours after treatment, caspase-activity with the 10 ns condition 4 h after treatment was now similar to the 60 ns condition 2 h after treatment, indicating a time-dependent increases in usEP-induced apoptosis with the lower 10 ns condition. The 60 ns and 300 ns condition were saturated 2 h after treatment, so analyses at 4 h were not greater. One interpretation of these results is that usEP effects are dependent on the pulse duration, and this is correct; however, for the same energy density conditions the pulse duration determines the charging effects. Thus, these findings indicated that this effect of usEPs on apoptosis induction was due primarily to charging effects. The results also indicated that usEPs did not directly activate caspases but did this through effects that induced cytochrome c release from the mitochondria. (See section on usEP effects on mitochondria,cytochrome *c* release).

Cytochrome c release is another definitive marker of apoptosis, as illustrated in Fig. 9.3. Figure 9.5 shows the results from Jurkat cells treated with usEPs as the mitochondrial fraction, which includes cytochrome c oxidase as an integral mitochondrial protein is separated from the cytosol fraction before electrophoresis. The fractions show from left to right the mitochondrial pellet (MP) fraction containing both cytochrome c and cytochrome c oxidase followed by the cytosol fractions from the 0, 10, 60, and 300 ns treated cytosol fractions containing only cytochrome c. This indicates that the cytochrome c in the cytosol is not due to mitochondria contaminating the fraction, as noted in the absence of cytochrome c oxidase. The increase in cytochrome c is dependent on the pulse duration or the Vs/cm with the 60 ns and 300 ns fractions saturated with cytochrome c, like the caspase activity in the cell extracts discussed above. These results suggest that the mitochondria are one of the major targets for usEPs. This theory is addressed in the section on mitochondria.

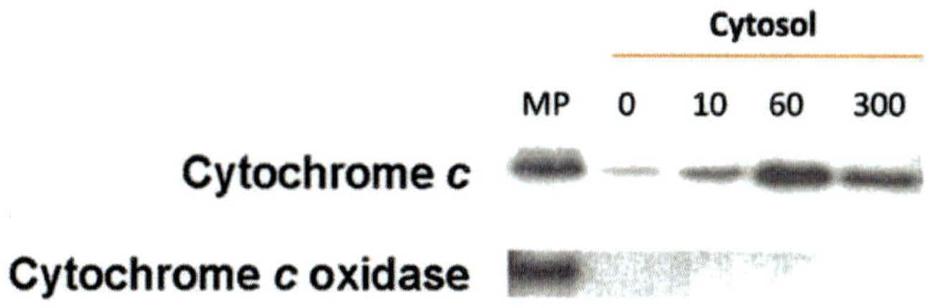

Fig. 9.5 usEPs induce cytochrome c release from Jurkat cells in a Vs/cm manner. Jurkat cells were treated with usEPs with 10 ns an energy density ~ 1.7 J/cc using three pulse with durations and electric fields of 10 ns 150 kV/cm, 60 ns 60 kV/cm, and 300 ns 26 kV/cm Two hours after treatment. The cytosol or mitochondrial fractions were prepared for SDS-PAGE by using standard procedures. Proteins (5–10 μg) were separated by 10% SDS-PAGE and transferred to PVDF membranes. Membranes were treated with a mouse monoclonal anti-cytochrome c antibody or a mouse monoclonal anti-bovine cytochrome c oxidase subunit IV, an integral mitochondria membrane-specific marker. Proteins were visualized by using enhanced chemiluminescence (from Beebe et al. 2003)

These cells also demonstrated usEPs increased Annexin-V binding on treated cell membranes, a known indication of apoptosis. Annexin-V binds to phosphatidylserine (PS) on the plasma membrane's external leaflet as it is externalized from the interior leaflet of the lipid bilayer. This effect was also dependent on the charging effect. It was later found that the usEP could "pull" PS from the inside to the outside leaflet of the plasma membrane as a physical means of PS externalization as opposed to an enzymatic mechanism (Vernier et al. 2006a, b). However, in these studies, PS externalization was attenuated by a caspase inhibitor, suggesting that at least some of the PS externalization was due to caspase activation. Nevertheless, it is now suggested that, to avoid this pulse-induced physical appearance of apoptosis-related PS externalization, this assay should be carried out at least 30 min after usEP treatment (Batista Napotnik et al. 2012). These studies also demonstrated that cytochrome *c* was released into the cytoplasm, indicating a likely intrinsic apoptosis mechanism. So, the cytoplasmic appearance of cytochrome c was also dependent on the charging effect on cells rather than the input of energy. This Vs/cm finding provided further support that the effects of usEPs are not due to thermal effects.

9.6 Use of Jurkat Cell Mutants to Demonstrate usEP-Induced Caspase-Dependent Apoptosis and Caspase-Independent RCD

The earlier studies with effects of usEP on Jurkat cell (Beebe 2003) were extending by analyzing Jurkat clones that were deficient in elements of the DISC, which is activated in the extrinsic apoptosis pathway and deficient in APAF-1, which is activated in the intrinsic apoptosis pathway (Ren et al. 2012). One clone used was also deficient in caspase-8, another deficiency in FADD, which are components of the DISC that activate caspase 8 in the extrinsic pathway. Figure 9.6 shows usEP effects on the clone that is deficient in APAF-1, which prevents the apoptosome formation and holoenzyme with cytochrome c and ATP to activate caspase-9 in the intrinsic apoptosis pathway (see Fig. 9.3).

Figure 9.6 shows viability electric field-dependent responses to the Jurkat wildtype clone that expressed APAF-1 as a critical component of the apoptosome (blue line) and the APAF-1 knock out clone without formation of an apoptosome (gold line). At lower electric fields there was a decrease in viability in the APAF- knock out clone compared to the wildtype clone. In the studies by Ren et al. (2012) and Beebe (2013), these lower electric fields also showed decreased capase-9 and -3 activity as expected because APAF-1 was not present to activate them. Thus, cell death was caspase-dependent, as indicated by the red line spanning 10 and about 40 kV/cm. As the electric fields were increased to 50 and 60 kV/cm, cell death (and caspase activation (Ren et al. 2012) did not rely on the presence or absence of the APAF-1 knock out clone -cell death was caspase-independent.

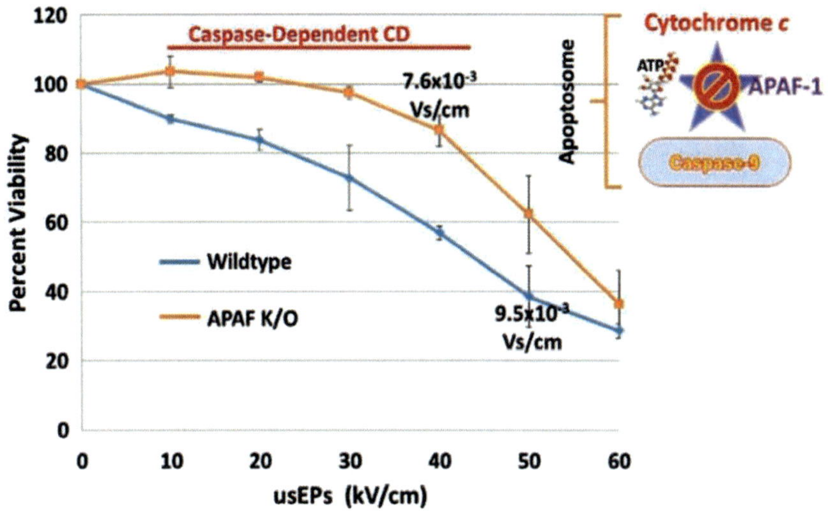

Fig. 9.6 usEPs induces caspase-dependent and -independent activation in Jurkat cells. – A wildtype Jurkat clone (light blue) and a Jurkat with APAF-1 knocked out clone (gold) were treated with ten 60 ns pulses with various electric field as indicated. Twenty-four hours post pulse cell viability was determine using the CellTiter-Glo luminescent assay (Promega), which determines ATP content. Volt-seconds/cm was determined as in the introduction to biological effect of usEPs (Schoenbach et al. 2009) The inset shows the apoptosome with cytochrome c, ATP, caspase-9, and APAF-knocked out (amended from Ren et al. 2012)

These studies shed new light on usEP-induced cell death in several ways. This study in Fig. 9.6 showed for the first time that usEPs could induce RCD caspase-dependent apoptosis and caspase-independent cell death depending on the electric field intensity. This is an important distinction because previously the presence of active caspases was only coincident with usEPs cell death; it could not be stated that cell death was dependent on active caspases until the absence of active caspases was shown attenuate cell death, which is shown here.

We analyzed Bid (22 kDa) and t-Bid (15 kDa) on quantitative western blots with the caspase-8- and FADD-deficient Jurkat clones. Surprisingly, t-Bid was present in these clones as well as the wildtype clone, meaning the usEPs induced Bid cleavage even when the upstream extrinsic apoptosis mechanisms were absent or present. Thus, Bid could be cleaved without formation of the DISC with FADD and activation of caspase-8, suggesting another mechanism induced by usEPs. Looking into this further and considering that usEPs induced increases in intracellular calcium (see Chap. 8 and Beebe et al. 2012, 2013), we considered the possibility that usEPs also activated Ca^{2+}-dependent proteases. We found that usEPs induced calpains, Ca^{2+}-activated proteases in addition to activating caspases. Thus, usEPs activate multiple RCD mechanisms that involve caspase, calpains and caspase-independent systems in Jurkat cells as well as in other cells.

Nevertheless, there was a threshold above which caspases were not needed to induce cell death. Furthermore, these studies demonstrated that usEP-induced apoptosis in Jurkat cells primarily through the intrinsic pathway with cytochrome c release because cell death was decreased when the intrinsic apoptosis activation platform APAF1 was absent. While this could occur through the extrinsic pathway using Bid/t-Bid to affect mitochondria but there were no effects on viability when caspase-8 or FADD were absent (Ren et al. 2012; Beebe 2013). These data provided further support that usEPs do not depend on the extrinsic apoptosis pathway to induce cell death. However, usEPs did activate another pathway involving Bid/t-Bid and Ca^{2+}-dependent and—independent calpains (Beebe et al. 2013), which may be part of a caspase-independent pathway.

Regarding caspase-dependence and independence mechanisms, more specifically, using ten pulses, 60 ns durations, and electric fields between 40 and 50 kV/cm as a dividing condition, caspase-dependent apoptosis is induced with $\leq 7.6 \times 10^{-3}$ Vs/cm and caspase-independent apoptosis is induced with $\geq 9.5 \times 10^{-3}$ Vs/cm. This study also shows that the IC50 for usEP in the wildtype is below 50 kV/cm or 9.5×10^{-3} Vs/cm and the IC50 for the APAF-1 knockout clone is greater than 9.5×10^{-3} Vs/cm. However, there is not a rigid delineation between the two mechanisms, and we can consider that through a middle part of the usEP "dose" that both mechanisms are operating. In any case, by invoking caspase-independent cell death, these finding provided evidence that usEPs can bypass defects in apoptosis such as an APAF-1 deficiency and still induce apoptosis. Consistent with this conclusion is the finding illustrated in the section on mitochondria (Chap. 8) that Jurkat clones over expressing Bcl-xl as a means to protect mitochondria had no effect on usEP-induced cell death compared to the wildtype clone (Fig. 9.13, mitochondria section). Analysis of the mitochondrial membrane potential ($\Delta\Psi$m) with the APAF-1 knock out clone in Fig. 9.7 provided additional evidence about caspase-dependent and-independent cell death.

The absence of the apoptosome and caspase activation does not affect the usEP-induced dissipation of the $\Delta\Psi$m. In Fig. 9.7, we used the same wildtype and APAF-1 knock-out clone that was used in Fig. 9.6. Unlike the wildtype clone, APAF-1 knock-out clone did not exhibit caspase-9 catalytic activity and therefore no caspase-3 catalytic activity (Ren et al. 2012). As indicated in Fig. 9.6, Jurkat clone exhibited usEP-induced caspase-dependent cell death at lower electric fields up to 40 kV/cm (shown in the solid blue line in Fig. 9.7), where cell death was absent or attenuated in the APAF-1 deficient clone (Fig. 9.6). Caspase-independent cell death occurred at higher electric fields greater than 40 kV/cm (dotted red line in Fig. 9.7), where the absence of APAF-1 did not matter and there were no differences in cell death between the two clones (Fig. 9.6; Ren et al. 2012; Beebe et al. 2013). While it appeared that usEP-induced RCD in Jurkat clones occurred through an intrinsic pathway with cytochrome *c* release and caspase activation, it was not clear if cytochrome *c* release and caspase activation occurred before or after the loss of $\Delta\Psi$m. Data in Fig. 9.7 shows that for cells exposed to 60 ns pulses at all electric fields, there were no differences between wildtype and APAF-1-deficient cells for usEP-induced dissipation of $\Delta\Psi$m. There was an electric field-dependent loss of $\Delta\Psi$m

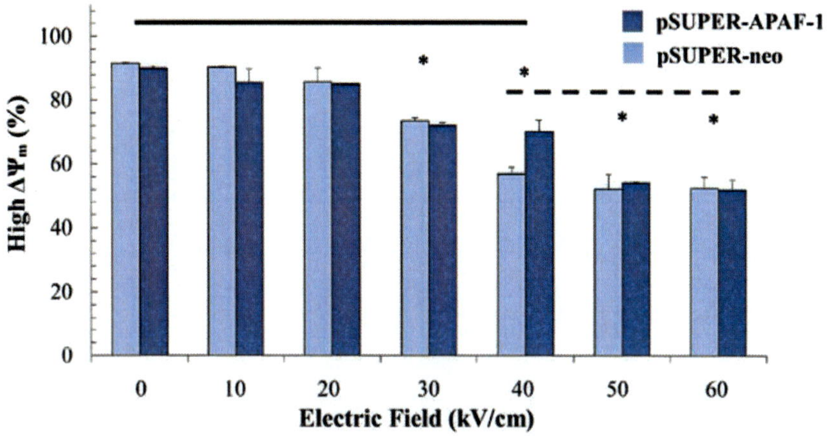

Fig. 9.7 Effects of usEPs on the ΔΨm of Jurkat that exhibit different RCD mechanisms—E6.1 Jurkat cells that exhibited APAF- 1 knocked out or were wildtype were loaded with TMRE to determine effects on the ΔΨm were treated with ten usEPs with durations of 60 ns and electric fields intensity of 50 kV/cm. Ten-15 min the ΔΨm, was determined by flow cytometry. The experiment represents three independent experiments (n = 3). + indicates statistically significant difference, $p < 0.05$ (Ruedlinger and Beebe, unpublished)

with as many as 50% of the cells with a low ΔΨm 10–20 min after pulsing. Loss of ΔΨm appeared to be independent of whether cells died by a caspase-dependent mechanism (below 40 kV/cm) or a caspase-independent mechanism (above 40–50 kV/cm). These data indicate that first, there is a loss of ΔΨm regardless of whether usEPs activate caspases or not. Second, the loss of ΔΨm likely occurs upstream of caspase activation, which suggests that usEP are inducing intrinsic cell death in these cells. Notice that usEPs at 30 kV/cm were just sufficiently high to cause a significant loss of ΔΨm and loss in cell viability (30%) in both clones (Fig. 9.6), yet the APAF-1 knock-out clone was still 100% viable 24 h later. Using these clones, caspase-dependent cell death reached 40% cell death in the wildtype and 30% cell death in the APAF-1 clone.

Figure 9.8 shows a study with usEP treatment of E4 squamous carcinoma cells demonstrated that RCD mechanisms were as complex or more complex than that observed in usEP treated Jurkat cells and apoptosis was more heterogeneous than a single pathway (Ren and Beebe 2011). By analyzing several cellular markers, electric field thresholds were identified for morphologic and molecular markers that typified cell responses to ten 300 ns usEPs. As electric fields were raised to 24 kV/cm (1.2×10^{-2}Vs/cm) and then above, there were increases in PI uptake, indicating plasma membrane permeabilization, and Annexin-V binding, indicating phosphatidylserine externalization. As the electric field was increased to 32 kV/cm (3.0×10^{-2} Vs/cm), there were increases in caspase activation and decreases in the mitochondrial membrane potential; however, cytochrome c release did not occur until after electric fields exceeded 50 kV/cm (4.7×10^{-2} Vs/cm). This hierarchy of cellular

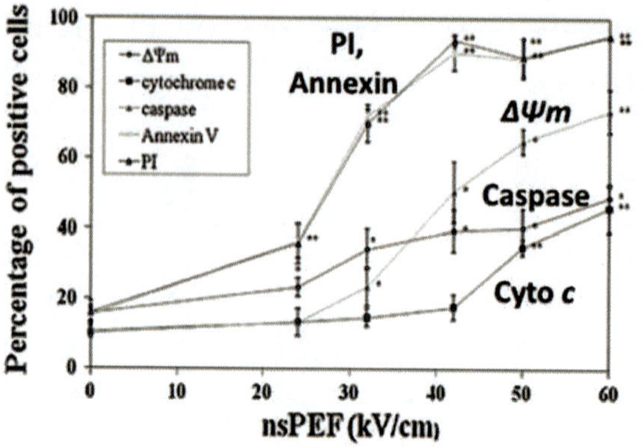

Fig. 9.8 Sensitivity thresholds for usEP effects on cell membranes, caspase activation and cytochrome c release. E4 cells were exposed to 10 pulses at 300 ns ranging from 0 to 60 kV/cm and percentage of cells that showed significant responses were analyzed by flow cytometry for decreases in the $\Delta\psi m$ with TMRE 15 min post pulse, active caspases with FITC-VAD-fmk 1 h post pulse, and cytochrome c using the InnoCyte (Method I) 1 h post pulse. FITC-Annexin-V and PI were added immediately after nsPEF exposure and incubated for 15 min before analysis by flow cytometry. After initial measurements, all responses were stable for a given electric field for at least 90 min. Values are presented as percent cells with positive responses. The graph represents the mean ± SE of three independent experiments. * $P < 0.05$ and ** $P < 0.01$ (from Ren and Beebe 2011)

changes occurred beginning first at the plasma membrane at lower electric fields and then with intracellular changes including cytochrome c release from mitochondria at the higher electric fields and charging effects in the usEP regimen. Also, there was caspase activation at electric fields that did not include cytochrome c release, suggesting that an extrinsic apoptosis pathway had been activated at these lower electric fields. Results from Jurkat cell apoptosis suggested intrinsic apoptosis because cytochrome c release occurred near the time that caspases were activated (Beebe et al. 2003). Extrinsic apoptosis pathways initially activated caspase-8, which could then activate caspase-3 directly and/or activate Bid, which would lead to cytochrome c release, which occurred late in the E4 cells. There was a time-dependent Bid cleavage that was partially dependent on calpain activation. Thus, in the midst of caspase activation, calpains-induced Bid cleavage to promote cytochrome c release that enhanced caspase-dependent apoptosis through the mitochondrial pathway. While this study with E4 cells investigated more usEP effector responses than the Jurkat cell study, it appeared that usEP-induced cell death could be more complex including both extrinsic pathways in E4 cells and intrinsic apoptosis suggested in the Jurkat study. It could also be considered that E4 cells were type I cells and Jurkat type II cells.

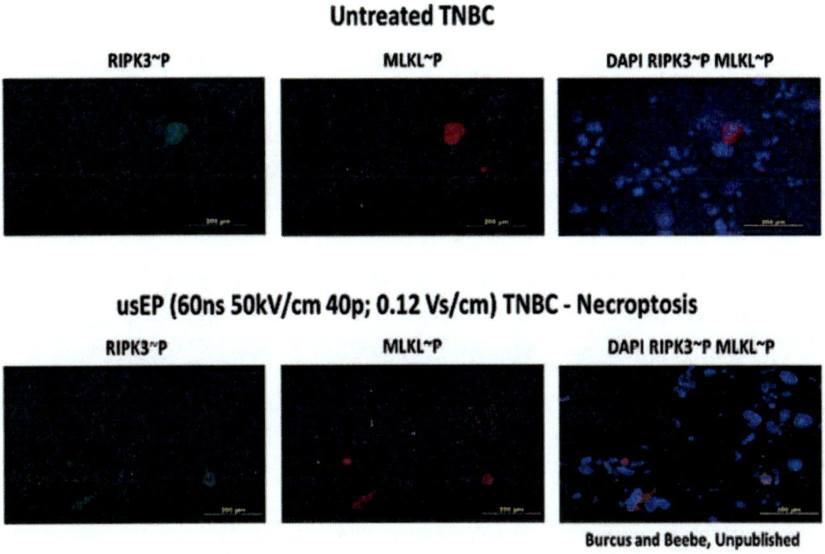

Fig. 9.9 Necroptosis- usEPs induce necroptosis in TNBC cells - HCC1937 TNBC cells were untreated (control, top panel) or treated with usEPs in cuvettes (0.1 mm) with forty pulses (1 Hz) with 60 ns durations and electric fields at 50 kV/cm (bottom panel). Cells were plated in chamber slides for 3 h, fixed and stained sequential with rabbit anti phospho-MLKL (red) and rabbit anti phospho-RIPK3 (green) and secondary goat anti-rabbit antibodies with the indicated colors. A typical 10× image is shown as represented in 3 independent experiments with 5–7 frames analyzed in each experiment. Significant increases occurred with usEPs ($p < 0.05$). Burcus and Beebe, unpublished

9.7 usEPs Induce RCD in a Cell Type-Dependent Manner

It was fortuitous that Jurkat, HL-60 and E4 cells were used for these initial studies since we were looking for apoptosis, because it is now known that usEPs do not induce apoptosis all cell types; usEP-induced apoptotic and specific RCD processes are cell type-dependent. This was clearly shown by comparing cell death characteristic between Jurkat cells and HeLa S3 cells (Morotomi-Yano et al. 2013). In these studies, the authors confirmed the presence of apoptosis in Jurkat cells using different assays as indicators then those used by Beebe et al. (2003), Ren et al. (2012). These authors used conditions for analyzing RCD mechanisms in the same range as those studies above (Beebe et al. 2002, 2003). They used durations of 80 ns, electric fields at 20 kV/cm with 20 pulses or 7.2×10^{-3} Vs/cm, with a roughly estimated usEP EC50 value of 5.1×10^{-3} Vs/cm, which closely matched Jurkat cell data above (6.0×10^{-3} Vs/cm) (Beebe et al. 2003; Ren et al. 2012). Using western blots with caspase-3-specific antibodies, they showed that usEPs induced caspase activation as indicated by the presence of smaller active caspase-3. They also showed the presence of DNA ladders that are typical of apoptosis when DNA is cleaved by caspase activated DNAases and that PARP1 was cleaved as a common apoptosis marker. In contrast,

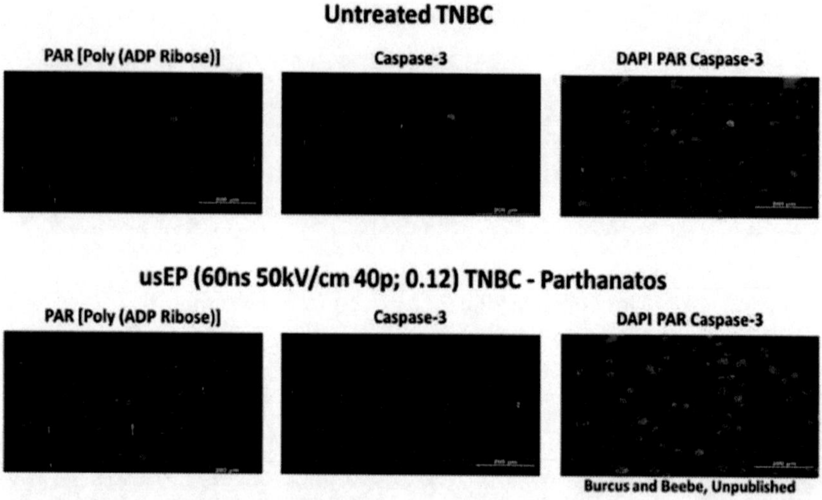

Fig. 9.10 Parthanatos- usEPs induce parthanatos in TNBC cells - HCC1937 TNBC cells were not pulsed (control, top panel) were treated in cuvettes (0.1 mm) with usEPs with forty pulses (1 Hz) with 60 ns durations and electric fields at 50 kV/cm (bottom panel). Cells were plated in chamber slides for 3 h, fixed and stained sequential with rabbit anti-[poly(ADP ribose)] (PAR) (red) and rabbit anti caspase-3 (green) and then counter stained with DAPI for localization of cell nuclei. A typical 10 × image is shown as represented in 3 independent experiments with 5–7 frames analyzed in each experiment. Significant increases occurred with usEPs for PAR, but not for caspase-3 ($p <$ 0.05). a distance of 200 μm is indicated. Burcus and Beebe, unpublished

HeLa S3 cells did not show either of these apoptosis markers, indicating that HeLa S3 cell death was not caused by activation of caspase proteases under their exposure conditions of 80 ns, 20 kV/cm and 40 shots or 1.0×10^{-2} Vs/cm. These cells were slightly more resistant to usEPs with roughly estimated EC50 of 9.5×10^{-3} Vs/cm. Interestingly, this condition induced caspase-independent apoptosis in Jurkat cells in the studies above (Ren et al. 2012). As a control, they demonstrated that HeLa S3 cells exposed to UV radiation (100 mJ/cm^2) could induce caspase activation and DNA laddering, proving that HeLa S3 cells could in fact express both of these apoptosis markers but not responses to usEPs.

In contrast to apoptosis induction, usEPs caused HeLa S3 cells to produce a significant increase in expression of PAR, which is due to the accumulation of protein modification by poly ADP-ribosylation (PAR). Although called necrosis by the authors, this RCD mechanism with increased levels of PAR is called parthanatos (David et al. 2009; Tang et al. 2019). This results from excessive activation of PARP-1, which is a chromatin-associated nuclear protein normally responsible for repairing single-stranded or double-strand DNA breaks. Recognizing DNA breaks, PARP1 uses ATP and nicotinamide adenine dinucleotide (NAD^+) to induce poly ADP-ribosylation and generating PAR polymers. RCD by parthanatos is caused by depleting NAD^+ and

ATP and by loss of the mitochondrial membrane potential ($\Delta\Psi$m). During apoptosis, PARP1 is inactivated by caspases since DNA repair is counter-intuitive to disassembling the genome. As shown in this study, PAR formation and parthanatos is independent of caspase activation and DNA laddering, both typical of apoptosis. Hyperactivated PARP1 binds to apoptosis inhibitory factor (AIF), which leads release from mitochondria and translocation to the nucleus. This study demonstrated that usEP-induced RCD is dependent on the cell type. While HeLa S3 cells clearly express apoptosis machinery, as shown by UV radiation, and share apoptosis operations that can lead to cell death, such as dissipation of $\Delta\Psi$m, this cell death is caspase-independent. Jurkat cells were also shown to undergo caspase-independent cell death at higher electric fields, that RCD mechanisms was not further investigated (Ren et al. 2012; Beebe et al. 2013).

usEP-induced RCD is dependent on Ca^{2+} and culture conditions

In a subsequent study Morotomi-Yano et al. (2014) continued these RCD studies in Jurkat and HeLa S3 cells using a wider range of usEP conditions in the presence and absence of the ubiquitous second messenger Ca^{2+}. Interestingly, they found that usEP-induced PAR formation was only present in the presence of Ca^{2+} and below 40 their 40-shot condition or about 1.0×10^{-2} Vs/cm. Above those conditions in the presence of absence of Ca^{2+} PAR was not expressed. In contrast, in the absence of Ca^{2+} but not in its presence at 60 shots (1.1×10^{-2} Vs/cm), active caspase-3 and PARP cleavage were observed. Thus, in these HeLa S3 cells, Par formation and parthanatos is a Ca^{2+}-dependent RCD at lower usEP conditions and apoptosis is present in the absence of Ca^{2+} at higher usEP conditions. When they analyzed apoptosis with and without Ca^{2+}, they found apoptosis was Ca^{2+}-independent; apoptosis was time-dependent in the presence and absence of Ca^{2+}. Finally, they showed that parthanatos was a usEP-induced RCD mechanism in HEK293 and K562 cells with up to 40–50 shots (1.1×10^{-2} Vs/cm) with no appearances of active caspase-3 or cleaved PARP, while apoptosis was the RCD mechanism in HL-60 cells with a threshold of about 30-shot or 8.7×10^{-3} Vs/cm, which confirmed the finding of Beebe et al. (2002). All of these finding are consistent with usEPs induced RCD in a Vs/cm-dependent manner and context-dependent.

In another valuable study of cell death mechanisms, usEP-induced cell death was analyzed human monocyte U937 cells, (Pakhomova et al. 2013). TheU937 cells were much more resistant to usEPs then Jurkat cells or HeLa S3 cells different from Jurkat. Rough estimations of the usEP EC for these cells are 7.3×10^{-2}, 9.5×10^{-3} and 5.1–6.0×10^{-3} Vs/cm, respectively; so Jurkat cell > 10 times more sensitive than U937 cells. This was later explained because usEPs induce FAS expression in Jurkat cells, which facilitated apoptosis, and U937 cells overexpress a cFLIP, a caspase action inhibitor that inhibited apoptosis (Estlack et al. 2014). Nevertheless, while the studies by Ren et al. (2012) demonstrated that RCD mechanisms depend on electric fields or Vs/cm, this study demonstrated that RCD also depended on the culture conditions, another example of context-dependent RCD. When U937 cells were treated with two different usEP conditions in cell culture media, there was very little caspase activity or cleaved PARP expressed. These cell swelled and formed blebs defined as necrotic-type blebs. In contrast when cells were treated with the same condition in the presence of sucrose to prevent swelling, both caspase activation and PARP cleavage were observed. Conditions included 300 ns, 7 kV/cm with 600 pulses or 5.1×10^{-2} Vs/cm and 60 ns, 30 kV/cm, with 50 pulses or 1.2×10^{-2} Vs/cm. As predicted by apoptosis condition induced by charging effects, 5.1×10^{-2} Vs/cm produced more caspase activation and PARP cleavage than 0.2×10^{-2} Vs/cm.

Pakhomova et al. (2014) extended their RCD studies by considering roles for Ca^{2+} in cell death process in U937 cells, CHO cells (Chinese hamster ovary), and BPAE cells (bovine pulmonary artery endothelial). As indicated above, when U937 cells were in media with no Ca^{2+} and sucrose, the sucrose appears to have prevented cell rupture by oncosis, yet the cells died later by apoptosis. When the media contain Ca^{2+}, there was a rapid (within 1.5 h) death of 90% of cells with a reduced level of caspase catalytic activation. Sucrose only partially protected the early oncosis response when Ca^{2+} was present. The authors concluded that either the pores were large enough for sucrose to enter the cell and thereby unable to prevent cell swelling or that Ca^{2+} caused an early RCD that was independent of an osmotic mechanism.

In order to evaluate the early permeability events altered by Ca^{2+}, the authors invoked an inventive approach used indium tin oxide (ITO)-coated slides, which provided an inert coating for cell adhesion and allowed the replacement of media on cells that were fragile after usEP treatment thereby avoiding centrifugation. The ITO coverslips exhibit high electrical conductance, which allows a uniform usEP exposure to the cells on an optically transparent slide. The usEP conditions on these slides is about 20-fold fewer pulses, or about 4.5 times lower Vs/cm (8.1×10^{-4} vs. 3.6×10^{-3}).

When Ca^{2+} was not present, the cells were spared the early cell death cell permeabilization, which the authors cell necrosis instead of oncosis, but the cells died later in the presence of caspase3/7 activation indicating apoptosis. However, when Ca^{2+} was present for the first 60–90 min after pulse treatment, there was an abrupt increase in propidium iodide uptake indicating membrane permeabilization in what the authors called necrotic cell death. This cell death was associated with growth of membrane blebs in a Ca^{2+}-independent and delayed osmotically-independent pore expansion.

Like many papers evaluating usEP-induced cell death, the absence of finding apoptosis as a RCD mechanism tends to summon a definition of necrotic cell death. This was the case in the study by Morotomi-Yano et al. (2013) while overlooking the presence of PAR, which indicates RCD by parthanatos, rather than a marker for necrosis. While the U937 cells did not die by a caspase- or PARP cleavage-mediated manner, these U937 cells did not lose membrane integrity until hours after usEP treatment. A delayed cell death is most often a RCD.

While the term "necrosis" does not tell us anything about how the cell dies, it is now known that these cells died by one or the other of two RCD mechanisms that result by forming plasma membrane pores from the inside of the cell by developing protein complexes that insert into the plasma membrane (Tang et al. 2019). The swelling cell death described in these usEP treated U937 cells is similar to ischemic cell death, which results by a failure of plasma membrane ion pumps to maintain an ionic equilibrium. This is defined as oncosis, which is derived from "ónkos", meaning swelling. Oncosis is also accompanied by increases in membrane permeability and blebbing, which as described by the authors, are blister-like, fluid-filled structures and empty of organelles (Majno and Joris 1995).

It is likely that the delayed increase in osmotic-independent permeability is due to one of the RCD mechanisms that generate the formation of pores in the plasma membrane by using intracellular protein to include increases in permeability as a means of regulated cell demise (Tang et al. 2019). These protein-containing pores are distinct from usEP-induced nanopores that were characterized with lipid channels properties and different from nanopores that can expand with higher usEP conditions that mimic pores produced by classical electroporation (Pakhomov et al. 2009; see section on usEP effects on plasma membranes). While osmotic increases in cell volume or oncosis and increases in cell permeability are typical of necrosis, when cell death is delayed, these RCD mechanisms have been called programmed or regulated necrosis.

Necroptosis is considered programmed necrotic RCD because it shows morphological characteristics like necrosis (Pasparakis and Vandenabeele 2015). Protein kinases RIPK1 and RIPK3 induce formation of what is called the necrosome, which is a signaling platform that recruits the pseudokinase MLKL. RIPK3 than phosphorylates MLKL causing conformational changes that stimulate binding to membrane phospholipids and formation of a MLKL oligomer that causes plasma membrane (Murphy et al. 2013; Dondelinger et al. 2014; Hildebrand et al. 2014). Some studies suggested that mitochondrial production of ROS (Vanden Berghe et al. 2010), activation of a mitochondrial serine/threonine protein phosphatase (PGAM5), and/or activation of the mitochondrial permeability transition pore (mPTP) may trigger necroptosis (Wang et al. 2012). While it is not clear how ROS leads to opening the mPTP, it has been suggested redox-sensing events upstream of RIPK1 and RIPK3 may be involved (Zhang et al. 2017). In addition, studies from the authors' laboratories indicate that usEPs can induce ROS, especially in the presence of Ca^{2+}. that may oxidize cyclophilin D, which opens the mPTP (see Fig. 9.9 in section on usEP effects on mitochondrial). As shown below, usEPs induce phosphorylation of RIPK3 and MLKL in triple negative breast cancer cells (TNB)C cells to induce necroptosis (Fig. 9.9).

usEPs induce necroptosis and parthanatos RCD programs in human triple negative breast cancer cells in vitro

Figure 9.9 shows that usEPs induce necroptosis in these human TNBC HCC-1937 cells. This is indicated by the co-localization of phosphorylated RIP3 (receptor-interacting serine/threonine-protein kinase 3) and phosphorylated MLKL (mixed Lineage Kinase Domain Like Pseudokinase). RIP3 is a member of the receptor interacting protein family of serine/threonine protein kinases and a component of the TNF complex. The interaction of two of these family members RIPK-1 and RIPK-3 for what is called the necrosome complex. This complex recruits MLKL, which is phosphorylated (Cho et al. 2009; He et al. 2009; Zhang et al. 2009) and ultimately oligomerizes and inserts into the plasma membrane with negatively charged phospholipids to form pores (Wang et al. 2014; Zhang et al. 2016; Robinson et al. 2019).

Figure 9.10 show that usEPs induce parthanatos in human TNBC cells HCC1937 TNBC cells as indicated by the significant increase in staining for PAR. However, like mouse 4T1-luc and rat N1-S1 cells, these TNBC cells do not respond to usEPs with caspase activation. Interestingly, Figs. 9.9 and 9.10 shows that usEPs induces necroptosis and parthanatos in HCC-1937 TNBC cells.

Given that these RCD mechanisms are caused by specific events, it is possible to suggest some likelihood of how usEPs induced these RCD mechanisms in these TNBC cells. Parthanatos is a poly [ADP-ribose] polymerase 1 (PARP1)-dependent, caspase-independent RCD mechanism; usEP-cells did not express active casapse3/7. PARP-1 is a protein associated with chromatin and is involved in repairing single- and double-stranded breaks in DNA. Upon recognizing DNA damage, activated PARP1, which exhibits (ADP-ribosyl)-transferases activity, transfers ADP-ribose residues as highly negatively charged polymers from NAD^+ onto proteins, mainly histones at the sites of DNA damage and also PARP1 itself. These ribosylated or PAR proteins function as scaffolds for DNA repair enzymes and are recognized as a parthanatos marker (Wei and Yu 2016; Pascal 2018; Alemasova and Lavrik 2019). The mechanism of parthanatos is binding of hyperactivated PARP1 to apoptosis-inducing factor (AIF) and its release from the mitochondria and translocation to the nucleus where it results in fragmenting DNA into large (20–50 kb) fragments (Andrabi et al. 2008; Wang et al. 2011). The major cause of DNA damage induce through parthanatos is oxidative stress-induced. Key features of parthanatos are common in many usEP-treated cell types including caspase independence, decrease in mitochondrial membrane potential ($\Delta\Psi m$), ROS production, Ca^{2+} dependence, $\Delta\Psi m$, independence of the cytoprotective effects of BCL2 proteins (Robinson et al. 2019) The role of ROS, Ca^{2+} and BCL2 in usEP-induced effects are discussed in the section of mitochondria and DNA damage is discussed in the section on the nucleus. usEPs were also shown to induce parthanatos in HEK293 and K562 cells (Morotomi-Yano et al. 2014) without caspase activation.

So, from these two series of experiments with human reagents for apoptosis, parthanatos and necroptosis, it becomes clear that the same cell types can die by more than one mechanism, and using human HCC1937 TNBC cells, caspase is not an active RCD mechanism, but the latter two are active RCD mechanisms. We could ask the question, is it possible for more than one cell death mechanism to be functioning in the same cell type; this is not an unreasonable possibility given that cells have not been specifically equipped to respond to usEPs like they are for other dangerous stimuli like heat, cold and radiation. However, looking closely at these two experiments and considering some morphological characteristics of these to RCD programs (Tang et al. 2019), it appears likely that these two cell death programs are not occurring in the same cells. Parthanatos does not occur with cell swelling. Figure 9.10 shows that parthanatos positive cells do not appear to be swollen. In contrast, cells that die by necroptosis do swell and it appears that cells that exhibit co-localization of RIPK3 and MLKL appear swollen. However, morphology was not a focus of this study. Since the electric fields are relatively homogeneous within parallel plate cuvettes, differences in induction of RCD is not likely due to responses to different electric fields. Nevertheless, under limiting conditions, cells in the S-phase exhibited greater membrane integrity and maintained cytoskeletal structure more than cells that were not synchronized to S-phase (Hall et al. 2007). Although these studies are done with low passage numbers, it is likely there are clonal variations in cell cultures that could account for different responses to usEPs or other cell death inducing stimuli. Still, Jurkat cell exhibited caspase-dependent and caspase-independent mechanisms that likely crossed over at some electric fields, and both Jurkat and E4 cells exhibited activation of caspases and calpains. So, it is possible that there are instances of cross-overs between or among RCD mechanisms.

Necroptosis can be induced by a number of stimuli including FAS ligand (Holler et al. 2000) and TLRs-3 and -4 (He et al. 2011). TLRs function to connect the innate and adaptive immune systems by recruiting mechanism that activate transcription factors including NF-κB and interferon (IFN)-regulatory factors (IRFs) that lead to innate immune responses (Kawasaki and Kawai 2014).

Figure 9.11 shows TLRs are in a family of pattern recognition receptors (PRRs) that recognize danger-associated molecular patterns or DAMPs, which are molecules released from dead and dying cells. Pathogen-associated microbial patterns or

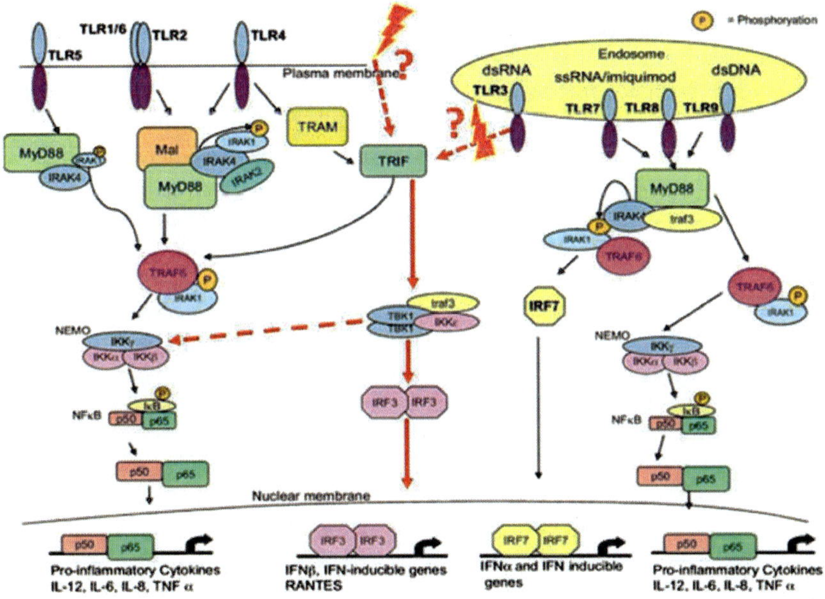

Fig. 9.11 TLR signaling pathways and downstream effector molecules. TLRs are shown with agonist, downstream mediators, and functional response elements (amended with possible NPS From O'Neill et al. 2009)

PAMPs are expressed by microbes. Antigen presenting cells (APCs) like dendritic cells (DCs) express these PRR that lead to the activation of adaptive immunity. TLR and other PRR initiate inflammatory responses as part of host defense systems. Other PRRs include nucleotide-binding oligomerization domain (Nod)-, leucine-rich repeat–containing receptors (NLRs), C-type lectin receptors (CLRs), RIG-I-like receptors (RLRs), and AIM-2 like receptors. (Wu and Chen 2014; Iwasaki and Medzhitov 2015). Another interesting system is the cGMP-cAMP synthase (cGAS)-STING mechanism, which is a DNA sensor that activates the innate immune system by producing the second messenger cGAMP, which activates the endoplasmic receptor (ER) adaptor protein STING (Burdette and Vance 2013; Li and Chen 2018).

Possible role for Toll-like receptors (TLR) in usEP-induced RCD

In other studies, DCs, usEPs and LPS appear to activate different pathways in mouse monocyte-derived dendritic cells (moDCs). Response to usEPs and LPS were differentiated in moDCs because LPS increased and usEPs decreased expression of activation markerCD86; LPS increased and usEPs had no effect on expression of activation marker CD40; usEPs increased and LPS had no effect on activation marker CD68 expression. Knowing LPS uses TLR-4, usEPs most likely use, at least in part, the TLR-3.

Figure 9.11 shows at least 11 isotypes of TLRs. Increasing evidence indicates that administration of TLR agonists enhances humoral and cellular immunity. TLR are expressed on the plasma membrane like TLR-1, -2, -4, -5, and -6. In contrast, TLR-3, -7, -8, and -9 are expressed on endosomes and/or endoplasmic reticulum so their agonists must be internalized, primarily by antigen presenting cells (APCs). While the Figure is complex, importantly notice that TLR-4 and TLR-3 signal through TRAF3-TBK1-IRF3, which is unique among TLR signaling; TLR9 signals though MyD88-TRAK1/4-IRF7. Both pathways activate IFN inducible genes and NFκB mechanisms.

Taking data from Figs. 9.9 and 9.10 with TNBC cells and studies with moDCs presented above, it is not unreasonable to suggest that usEPs could cause release of mitochondrial DNA, which would be senses by the cell as a mimic of the presence a of microbial PAMPs. Although it is academic, it could also be sensed as a DAMP since it is possible the cell would respond to a misplaced pieces of DNA in the cytoplasm as a danger signal. Either pathogen or danger signal, mitochondrial DNA would be sensed by an endosomal/cytoplasmic receptor such as TLR-3 as opposed to an extracellular TLR-4, which senses pathogen patterns such as LPS before they enter the cell. In fact, the data suggests an intrinsic paradigm for cells responding to intracellular molecular patterns as opposed to extracellular molecular patterns.

In the examples of the TLR cell PRR responses to DAMPs and PAMPs, the innate immune system is highlighted as the first responders to invasion by pathogens or mutiny by cancer. The innate immune cells not only identify danger signals and pathogen molecular patterns, but they also alert the adaptive immune system for assistance by secreting cytokines that recruit and alert adaptive immune cells. Innate cells called antigen presenting cells (APCs), which include dendritic cells (DCs), also phagocytize pathogens or cancer antigens, digest them, and present them to T-cells as possible antigens on major histocompatibility complex (MHC) I and MHC II molecules. In this way T-cells develop and express specific, high affinity T-cell receptors that very specifically recognize those antigens presented to them by APCs and use them to identify and eliminate cells that express those antigens. But rather than go into exactly how these T-cell receptors develop and how adaptive immune cells function to eliminate cancer, the discussion will focus on the recent findings that usEPs can actually activate innate and adaptive immune cells in the host to help eliminate cancer. Interestingly, that immune cells can act against cancer cells is only a relatively recent insight.

In a series of studies in the authors' lab, we asked whether usEPs could induce activation of TLRs on cancer cells or immune cells. Such a cellular response would suggest that usEPs could be sensed by cells like they respond to bacteria, viruses or to other danger signals perceived identified by their PRRs. A likely possibility would be activation of TLRs if usEPs caused mitochondrial DNA to be released into the cytoplasm. As discussed in the section on mitochondria, the mitochondria are of bacterial ancestral origin, so it is possible that appearance of mitochondrial DNA in the cytoplasm, which could be sensed as foreign under normal conditions, could appear as a mimic of bacterial DNA, and activate TLR. If this were the case, a response to usEPs through TLRs could induce necroptosis (Fig. 9.12).

usEPs activate a unique TLR-3-like pathway. Administration of TLR agonists enhances humoral and cellular immunity (He et al. 2011; Kawasaki and Kawai, 2014). TLRs signaling pathways are possible sites of usEPs on the plasma membrane or at intracellular sites on endosomes. Although some TLRs including TLR-4 are on the plasma membrane, TLR-3, -7,- 8, -9 are on endosomes and/or endoplasmic reticulum. usEPs can affect intracellular structures. Based on scheme in Fig. 9.11, usEPs appears to act through a TLR3-like pathway, using TRAF3-TBK1-IRF3, (tumor necrosis factor receptor-associated factor 3- TNF-alpha/NF-kappa B (TANK) binding kinase 1 - interferon regulatory factor 3),which is unique among TLR signaling, but not likely through a TLR-4 pathway. usEPs (60 ns 50 kV/cm, 50 pulses, 0.02 Vs/cm) induce transient IRF3 phosphorylation (at 30 min, but not at 120 min) (panels C), but has no effect on stimulator of interferon genes (STING) phosphorylation (panels A). This suggests usEPs activate TLR pathways through a unique TLR signaling mechanism that is *MyD88-independent*, TRAF3-TBK1-IRF3 pathway. The central signaling node in this pathway is the TANK (TRAF-associated NFκB kinase), which is a non-canonical IKK kinase that phosphorylates the nuclear factor kB (NFκB) known for its role in innate immune antiviral response. However, TBK1 also regulates anti-tumor immunity, cell proliferation, apoptosis and autophagy (Helgason et al. 2013). usEPs signal through this kinase to phosphorylate and activate IRF3, which mediates downstream signaling to gene transcription mechanisms that set IFN-inducible genes in motion through the JAK-STAT pathway.

In summary, RCD mechanisms are numerous and complex. Many of their morphologic characteristics overlap, making morphology a risky method for specific identification of RCD mechanisms. Two RCD mechanisms end with plasma membrane pores formed from protein complexes inside cells. These can give appearance of necrotic

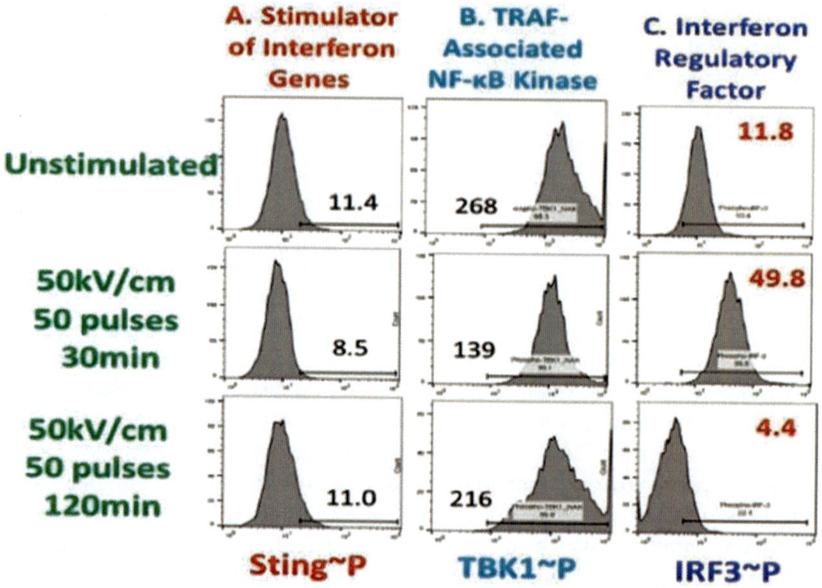

Fig. 9.12 usEPs activates a TLR3-like pathway in human breast cancer cells. Human triple negative breast cancer cells (HCC1937) were treated with usEPs as indicated or unstimulated, labeled 30 min and 120 min after treatment with fluorescent antibodies for phosphorylated proteins including Sting, TBK1 and IRF3 and then analyzed by flow cytometry. The data represents a typical experiment. (Lassiter and Beebe, unpublished)

cells with lysed plasma membranes, but differ from ACD, which is immediate but with the same end morphologies. usEPs have been shown to induce several RCD mechanisms including apoptosis, necroptosis and parthanatos. However, studies have not been idesigned to use markers for all RCD mechanisms. It seems unlikely that usEPs induce ACD, except in a small population of cells that are exposed to extremely high electric field conditions. Generally, when cells in vitro are exposed to greater than 0.01 Vs/cm, some RCD mechanisms in a cell type specific manner are highly likely. It is hypothesized that RCD mechanisms are caused by charging events induced by electric fields that affect both the plasma membrane and intracellular structures and functions.

References

Alemasova EE, Lavrik OI (2008) Poly(ADP-ribosyl)ation by PARP1: reaction mechanism and regulatory proteins. Nucleic Acids Res 47(8):3811–3827

Andrabi SA, Dawson TM, Dawson VL (2008) Mitochondrial and nuclear cross talk in cell death: parthanatos. Ann NY Acad Sci 1147:233–241

Batista Napotnik T, Wu YH, Gundersen MA, Miklavčič D, Vernier PT (2012) Nanosecond electric pulses cause mitochondrial membrane permeabilization in Jurkat cells. Bioelectromagnetics 33(3):257–264

Beebe SJ, Fox PM, Rec LH, Buescher ES, Somers K, Schoenbach KH (2002) Nanosecond pulsed electric field (nsPEF) effects on cells and tissues: apoptosis induction and tumor growth inhibition. IEEE Trans Plasma Sci 30:286–292

Beebe SJ, Fox PM, Rec LJ, Willis EL, Schoenbach KH (2003) Nanosecond, high-intensity pulsed electric fields induce apoptosis in human cells. FASEB J 17:1493–2145

Beebe SJ, Chen YJ, Sain NM, Schoenbach KH, Xiao S (2012) Transient features in nanosecond pulsed electric fields differentially modulate mitochondria and viability.PLoS One 7(12):e51349

Beebe SJ, Sain NM, Ren W (2013) Induction of cell death mechanisms and apoptosis by nanosecond pulsed electric fields (nsPEFs). Cells 2:136–162

Broz P, Dixit VM (2005) Inflammasomes: mechanism of assembly, regulation and signalling. Nat. Rev. Immunol. 2016; 16, 407–420

Budihardjo I, Oliver H, Lutter M, Luo X, Wang X (1999) Biochemical pathways of caspase activation during apoptosis. Annu Rev Cell Dev Biol 15:269–290

Burdette DL, Vance RE (2013) STING and the innate immune response to nucleic acids in the cytosol. Nat Immunol 14:19–26

Chen X et al (2016) Pyroptosis is driven by non-selective gasdermin-D pore and its morphology is different from MLKL channel-mediated necroptosis. Cell Res 26:1007–1020

Cho YS et al (2009) Phosphorylation-driven assembly of the RIP1-RIP3 complex regulates programmed necrosis and virus-induced inflammation. Cell 137:1112–1123

Conradt B, Wu YC, Xue D (2016) Programmed cell death during caenorhabditis elegans development. Genetics 203(4):1533–1562

D'Arcy MS (2019) Cell death: a review of the major forms of apoptosis, necrosis and autophagy. Review Cell Biol Int 43(6):582–592. https://doi.org/10.1002/cbin.11137

David KK, Andrabi SA, Dawson TM, Dawson VL (2009) Parthanatos, a messenger of death. Front Biosci (landmark Ed) 14:1116–1128

He WT, Wan H, Hu L, Chen P, Wang X, Huang Z, Yang ZH, Zhong CQ, Han J (2015) Gasdermin D is an executor of pyroptosis and required for interleukin-1β secretion. Cell Res 25(12):1285–1298

Ding J, Wang K, Liu W, She Y, Sun Q, Shi J, Sun H, Wang DC, Shao F (2016) Pore-forming activity and structural autoinhibition of the gasdermin family. Nature 535(7610):111–116

Dondelinger Y, Declercq W, Montessuit S et al (2014) MLKL compromises plasma membrane integrity by binding to phosphatidylinositol phosphates. Cell Rep 7(4):971–981

Estlack LE, Roth CC, Thompson GL 3rd, Lambert WA 3rd, Ibey BL (2014) Nanosecond pulsed electric fields modulate the expression of Fas/CD95 death receptor pathway regulators in U937 and Jurkat Cells. Apoptosis 19:1755–1768

Fulda S, Gorman AM, Hori O, Samali A (2010) Cellular stress responses: cell survival and cell death. Int J Cell Biol 2010:214074

Galluzzi L, Bravo-San Pedro JM, Vitale I, Aaronson SA, Abrams JM, Adam D, Alnemri ES, Altucci L, Andrews D, Annicchiarico-Petruzzelli M, Baehrecke EH, Bazan NG, Bertrand MJ, Bianchi K, Blagosklonny MV, Blomgren K, Borner C, Bredesen DE, Brenner C, Campanella M, Candi E, Cecconi F, Chan FK, Chandel NS, Cheng EH, Chipuk JE, Cidlowski JA, Ciechanover A, Dawson TM, Dawson VL, De Laurenzi V, De Maria R, Debatin KM, Di Daniele N, Dixit VM, Dynlacht BD, El-Deiry WS, Fimia GM, Flavell RA, Fulda S, Garrido C, Gougeon ML, Green DR, Gronemeyer H, Hajnoczky G, Hardwick JM, Hengartner MO, Ichijo H, Joseph B, Jost PJ, Kaufmann T, Kepp O, Klionsky DJ, Knight RA, Kumar S, Lemasters JJ, Levine B, Linkermann A, Lipton SA, Lockshin RA, López-Otín C, Lugli E, Madeo F, Malorni W, Marine JC, Martin SJ, Martinou JC, Medema JP, Meier P, Melino S, Mizushima N, Moll U, Muñoz-Pinedo C, Nuñez G, Oberst A, Panaretakis T, Penninger JM, Peter ME, Piacentini M, Pinton P, Prehn JH, Puthalakath H, Rabinovich GA, Ravichandran KS, Rizzuto R, Rodrigues CM, Rubinsztein DC, Rudel T, Shi Y, Simon HU, Stockwell BR, Szabadkai G, Tait SW, Tang HL, Tavernarakis N, Tsujimoto Y, Vanden Berghe T, Vandenabeele P, Villunger A, Wagner EF, Walczak H, White E, Wood WG,

Yuan J, Zakeri Z, Zhivotovsky B, Melino G, Kroemer G (2015) Essential versus accessory aspects of cell death: recommendations of the NCCD 2015. Cell Death Differ 22(1):58–73

Galluzzi L, Vitale I, Aaronson SA, Abrams JM, Adam D, Agostinis P, Alnemri ES, Altucci L, Amelio I, Andrews DW et al (2018) Molecular mechanisms of cell death: recommendations of the Nomenclature Committee on Cell Death

Gowrishankar TR, Esser AT, Vasilkoski Z, Smith KC, Weaver JC (2006) Microdosimetry for conventional and supra-electroporation in cells with organelles. Biochem Biophys Res Commun 341(4):1266–1276

Green DR (2005) Apoptotic pathways: ten minutes to dead. Cell 121(5):671–674

Hall EH, Schoenbach KH, Beebe SJ (2007) Nanosecond pulsed electric fields have differential effects on cells in the S-phase. DNA Cell Biol 26(3):160–171

He S, Wang L, Miao L, Wang T, Du F, Zhao L, Wang X (2009) Receptor interacting protein kinase-3 determines cellular necrotic response to TNF-alpha. Cell 137(6):1100–1111

He S, Liang Y, Shao F, Wang X (2011) Toll-like receptors activate programmed necrosis in macrophages through a receptor-interacting kinase-3-mediated pathway. Proc Natl Acad Sci USA 108:20054–20059

Helgason E, Phung QT, Dueber EC (2013) Recent insights into the complexity of Tank-binding kinase 1 signaling networks: the emerging role of cellular localization in the activation and substrate specificity of TBK1. FEBS Lett 587(8):1230–1237

Hildebrand JM, Tanzer MC, Lucet IS et al (2014) Activation of the pseudokinase MLKL unleashes the four-helix bundle domain to induce membrane localization and necroptotic cell death. Proc Natl Acad Sci USA 111(42):15072–15077. https://doi.org/10.1073/pnas.1408987111

Holler N, Zaru R, Micheau O, Thome M, Attinger A, Valitutti S, Bodmer JL, Schneider P, Seed B, Tschopp J (2000) Fas triggers an alternative, caspase-8-independent cell death pathway using the kinase RIP as effector molecule. Nat Immunol 1(6):489–495

Iwasaki A, Medzhitov R (2015) Control of adaptive immunity by the innate immune system. Nat Immunol 16(4):343–353

Kawasaki T, Kawai T (2014) Toll-like receptor signaling pathways. Front Immunol 5:461

Kayagaki N, Stowe IB, Lee BL, O'Rourke K, Anderson K, Warming S, Cuellar T, Haley B, Roose-Girma M, Phung QT, Liu PS, Lill JR, Li H, Wu J, Kummerfeld S, Zhang J, Lee WP, Snipas SJ, Salvesen GS, Morris LX, Fitzgerald L, Zhang Y, Bertram EM, Goodnow CC, Dixit VM (2015) Caspase-11 cleaves gasdermin D for non-canonical inflammasome signalling. Nature 526(7575):666–671

Kepp O, Senovilla L, Vitale I, Vacchelli E, Adjemian S, Agostinis P, Apetoh L, Aranda F, Barnaba V, Bloy N et al (2014) Consensus guidelines for the detection of immunogenic cell death. Oncoimmunology 3:e955691

Kerr JF (1971) Shrinkage necrosis: a distinct mode of cellular death. J Pathol. 105(1):13–20

Kerr JFR (2002) History of the events leading to the formulation of the apoptosis concept. Toxicology

Kerr JF, Searle J (1972) The digestion of cellular fragments within phagolysosomes in carcinoma cells. J Pathol 108(1):55–58

Kerr JF, Wyllie AH, Currie AR (1972) Apoptosis: a basic biological phenomenon with wide-ranging implications in tissue kinetics. Br J Cancer 26(4):239–257

Kantari and Walczak (2011) Caspase-8 and bid: caught in the act between death receptors and mitochondria. Biochim Biophys Acta 1813(4):558–563

Kroemer G, Galluzzi L, Vandenabeele P, Abrams J, Alnemri ES, Baehrecke EH, Blagosklonny MV, El-Deiry WS, Golstein P, Green DR, Hengartner M, Knight RA, Kumar S, Lipton SA, Malorni W, Nuñez G, Peter ME, Tschopp J, Yuan J, Piacentini M, Zhivotovsky B, Melino G (2009) Nomenclature Committee on Cell Death 2009. Classification of cell death: recommendations of the Nomenclature Committee on Cell Death 2009. Cell Death Differ 16(1):3–11

Kumar V, Abbas AK, Aster JC, Fausto N (2010) Robbins and Cotran pathologic basis of disease, 8th ed. Saunders/Elsevier, Philadelphia, PA, pp. 12–41. ISBN 978-1416031215

Li T, Chen LJ (2018) The cGAS–cGAMP–STING pathway connects DNA damage to inflammation, senescence, and cancer. J Exp Med 215(5):1287–1299

Linkermann A, Skouta R, Himmerkus N, Mulay SR, Dewitz C, De Zen F, Prokai A, Zuchtriegel G, Krombach F, Welz PS, Weinlich R, Vanden Berghe T, Vandenabeele P, Pasparakis M, Bleich M, Weinberg JM, Reichel CA, Bräsen JH, Kunzendorf U, Anders HJ, Stockwell BR, Green DR, Krautwald S (2014) Synchronized renal tubular cell death involves ferroptosis. Proc Natl Acad Sci USA 111(47):16836–16841

Liu X, Zhang Z, Ruan J, Pan Y, Magupalli VG, Wu H, Lieberman J (2016) inflammasome-activated gasdermin D causes pyroptosis by forming membrane pores. Nature 535(7610):153–158

Maier P, Hartmann L, Wenz F, Herskind C (2016) Cellular pathways in response to ionizing radiation and their targetability for tumor radiosensitization. Int J Mol Sci 17(1):102

Majno G, Joris I (1995) Apoptosis, oncosis, and necrosis. An overview of cell death. Am J Pathol 146(1):3–15

Meissner F, Molawi K, Zychlinsky A (2008) Superoxide dismutase 1 regulates caspase-1 and endotoxic shock. Nat Immunol 9(8):866–872

Morotomi-Yano K, Oyadomari S, Akiyama H, Yano K (2012) Nanosecond pulsed electric fields act as a novel cellular stress that induces translational suppression accompanied by eIF2α phosphorylation and 4E-BP1 dephosphorylation. Exp Cell Res 318(14):1733–1744

Morotomi-Yano K, Akiyama H, Yano K (2013) Nanosecond pulsed electric fields induce poly(ADP-ribose) formation and non-apoptotic cell death in HeLa S3 cells.Biochem Biophys Res Commun 438(3):557–562

Morotomi-Yano K, Akiyama H, Yano K (2014) Different involvement of extracellular calcium in two modes of cell death induced by nanosecond pulsed electric fields. Arch Biochem Biophys 555–556:47–54

Murphy JM, Czabotar PE, Hildebrand JM et al (2013) The pseudokinase MLKL mediates necroptosis via a molecular switch mechanism. Immunity 39(3):443–453. https://doi.org/10.1016/j.immuni.2013.06.018

Napoletano F, Baron O, Vandenabeele P, Mollereau B, Fanto M (2019) Intersections between Regulated Cell Death and Autophagy. Trends Cell Biol 29(4):323–338

Neutelings T, Lambert CA, Nusgens BV, Colige AC (2013) Effects of mild cold shock (25 °C) followed by warming up at 37 °C on the cellular stress response. PLoS ONE 8(7):e69687

O'Neill LA, Bryant CE, Doyle SL (2009) Therapeutic targeting of Toll-like receptors for infectious and inflammatory diseases and cancer. Pharmacol Rev 61(2):177–197

Pakhomov AG, Bowman AM, Ibey BL, Andre FM, Pakhomova ON, Schoenbach KH (2009) Lipid nanopores can form a stable, ion channel-like conduction pathway in cell membrane. Biochem Biophys Res Commun 385:181–186

Pakhomova ON, Gregory BW, Semenov I, Pakhomov AG (2013) Two modes of cell death caused by exposure to nanosecond pulsed electric field. PLoS One 8(7):e70278

Pakhomova ON, Gregory B, Semenov I, Pakhomov AG (2014) Calcium-mediated pore expansion and cell death following nanoelectroporation. Biochim Biophys Acta 1838(10):2547–2554

Parvathenani LK, Buescher ES, Chacon-Cruz E, Beebe SJ (1998) Type I cAMP-dependent protein kinase delays apoptosis in human neutrophils at a site upstream of caspase-3.J Biol Chem 273(12):6736–6743

Pascal JM (2018) The comings and goings of PARP-1 in response to DNA Damage—review. DNA Repair (amst) 71:177–182

Pasparakis M, Vandenabeele P (2015) Necroptosis and its role in inflammation. Nature 517:311–320

Porter AG, Jänicke RU (1999) Emerging roles of caspase-3 in apoptosis. Cell Death Differ 6(2):99–104

Platnich JM, Chung H, Lau A, Sandall CF, Bondzi-Simpson A, Chen HM, Komada T, Trotman-Grant AC, Brandelli JR, Chun J, Beck PL, Philpott DJ, Girardin SE, Ho M, Johnson RP, MacDonald JA, Armstrong GD, Muruve DA (2018) Shiga Toxin/lipopolysaccharide activates caspase-4 and gasdermin D to trigger mitochondrial reactive oxygen species upstream of the NLRP3 inflammasome. Cell Rep 25(6):1525-1536.e7

Rathkey JK, Zhao J, Liu Z, Chen Y, Yang J, Kondolf HC, Benson BL, Chirieleison SM, Huang AY, Dubyak GR, Xiao TS, Li X, Abbott DW (2018) Chemical disruption of the pyroptotic

pore-forming protein gasdermin D inhibits inflammatory cell death and sepsis. Sci Immunol 3(26):eaat2738
Ren W, Beebe SJ (2011) An apoptosis targeted stimulus with nanosecond pulsed electric fields (nsPEFs) in E4 squamous cell carcinoma. Apoptosis 16(4):382–393
Ren W, Sain NM, Beebe SJ (2012) Nanosecond pulsed electric fields (nsPEFs) activate intrinsic caspase-dependent and caspase-independent cell death in Jurkat cells. Biochem Biophys Res Commun 421:808–812
Robinson N, Ganesan R, Hegedűs C, Kovács K, Kufer TA, Virág L (2019) Programmed necrotic cell death of macrophages: focus on pyroptosis, necroptosis, and parthanatos. Redox Biol. 26:101239
Ron D (2002) Translational control in the endoplasmic reticulum stress response. J Clin Invest 110(10):1383–1388
Schleich K, Warnken U, Fricker N, Oztürk S, Richter P, Kammerer K, Schnölzer M, Krammer PH, Lavrik IN (2012) Stoichiometry of the CD95 death-inducing signaling complex: experimental and modeling evidence for a death effector domain chain model. Mol Cell 47(2):306–319
Schoenbach KH, Beebe SJ, Buescher ES (2001) Intracellular effect of ultrashort electrical pulses. Bioelectromagnetics 22:440–448
Schoenbach KH, Joshi RP, Kolb JF, Chen N, Stacey M, Blackmore PF, Buescher ES, Beebe SJ (2004) Ultrashort electrical pulses open a new gateway into biological cells. Proc of the IEEE 92:1122–1137
Schoenbach KH, Joshi RP, Beebe SJ, Baum CE (2009) A scaling law for membrane permeabilization with nanopulses. IEEE Trans Dielectrics Electrical Insul 16:1224–1235
Semenov I, Xiao S, Pakhomova ON, Pakhomov AG (2013) Recruitment of the intracellular Ca^{2+} by ultrashort electric stimuli: the impact of pulse duration. Cell Calcium 54(3):145–150
Shi J, Gao W, Shao F (2017) Pyroptosis: gasdermin-mediated programmed necrotic cell death. Trends Biochem Sci. 42(4):245–254
Stockwell BR, Friedmann Angeli JP, Bayir H, Bush AI, Conrad M, Dixon SJ, Fulda S, Gascón S, Hatzios SK, Kagan VE, Noel K, Jiang X, Linkermann A, Murphy ME, Overholtzer M, Oyagi A, Pagnussat GC, Park J, Ran Q, Rosenfeld CS, Salnikow K, Tang D, Torti FM, Torti SV, Toyokuni S, Woerpel KA, Zhang DD (2017) Ferroptosis: a regulated cell death nexus linking metabolism, redox biology, and disease. Cell 171(2):273–285
Tang D, Kang R, Berghe TV, Vandenabeele P, Kroemer G (2019) The molecular machinery of regulated cell death. Cell Res. 29(5):347–364
Vanden Berghe T, Vanlangenakker N, Parthoens E, Deckers W, Devos M, Festjens N, Guerin CJ, Brunk UT, Declercq W, Vandenabeele P (2010) Necroptosis, necrosis and secondary necrosis converge on similar cellular disintegration features. Cell Death Differ 17(6):922–930
Vernier PT, Sun Y, Gundersen MA (2006a) Nanoelectropulse-driven membrane perturbation and small molecule permeabilization. BMC Cell Biol 7:37
Vernier PT, Ziegler MJ, Sun Y, Gundersen MA, Tieleman DP (2006b) Nanopore-facilitated, voltage-driven phosphatidylserine translocation in lipid bilayers–in cells and in silico. Phys Biol. 3:233–247
Wang Y, Kim NS, Haince JS, Kang HC, David KK, Andrabi SA, Poirier GG, Dawson VL, Dawson TM (2011) Poly(ADP-ribose) (PAR) binding to apoptosis-inducing factor is critical for PAR polymerase-1–dependent cell death (parthanatos). Sci Signal 4(167):ra20
Wang H, Sun L, Su L, Rizo J, Liu L, Wang LF, Wang FS, Wang X (2014) Mixed lineage kinase domain-like protein MLKL causes necrotic membrane disruption upon phosphorylation by RIP3. Mol Cell 54(1):133–146
Wang Z, Jiang H, Chen S, Du F, Wang X (2012) The mitochondrial phosphatase PGAM5 functions at the convergence point of multiple necrotic death pathways. Cell 148(1–2):228–243
Wei H, Yu X (2016) Functions of PARylation in DNA damage repair pathways. Genomics Proteomics Bioinform 14(3):131–139
Wyllie AH, Kerr JF, Currie AR (1972) Cellular events in the adrenal cortex following ACTH deprivation. J Pathol 106(1):Pix

Wyllie AH, Kerr JFR, Currie AR (1980) Cell death: the significance of apoptosis. Int Rev Cytol 68:251–306

Wu J, Chen ZJ (2014) Innate immune sensing and signaling of cytosolic nucleic acids. Annu Rev Immunol 32:461–488

Zhang DW et al (2009) RIP3, an energy metabolism regulator that switches TNF-induced cell death from apoptosis to necrosis. Science 325:332–336

Zhang J, Yang Y, He W, Sun L (2016) Necrosome core machinery: MLKL. Cell Mol Life Sci 73(11–12):2153–2163

Zhang Y, Su SS, Zhao S, et al (2017) RIP1 autophosphorylation is promoted by mitochondrial ROS and is essential for RIP3 recruitment into necrosome. Nat Commun 8:14329. https://doi.org/10.1038/ncomms14329

Chapter 10
Model Rate Equation Evaluation of an Extrinsic Apoptotic Pathway

Ravi Joshi

Abstract Apoptosis is one of the most complex signaling pathways, and can be triggered by a number of factors, radiation, chemotherapeutic drugs, electric field application, etc. The apoptotic pathways can generally be divided into signaling via the death receptors (extrinsic pathway) or the mitochondria effects (intrinsic pathway), and each pathway implies caspases as effector molecules. The role and extent of electric field modifications to the apoptotic pathways remains unclear. Here in this chapter, we attempt to qualitatively probe some of the issues pertaining to apoptotic cell death based on simple model simulations. The objectives of the present discussions are: (i) to determine if a pulse-number threshold might suitably apply to cell killing by nano-second, high-intensity pulses, and (ii) to assess whether the intrinsic or the extrinsic pathway is more dominant following electric pulsing.

10.1 Introduction

Despite the potential applications of electrical pulsing to biomedical engineering, the biophysical details and mechanisms of electrically triggering cellular apoptotic pathways are not well understood. From a high-level, systems standpoint, cell death can be viewed as being caused by the inactivation of one (or several) critical enzymatic sub-systems or process pathways. Such inactivation should be regarded as a stochastic event given the variability in growth-cycles of exposed cells, their age, individual orientations and locations relative to the pulsing electrodes, differences in sizes and shapes, and other heterogeneities (e.g., concentrations and molecular states) that can affect various reaction rate constants. Assuming that the inactivation times are random variables, and that the cellular system consists of several components of which the most severe defect leads to biological failure; then the Weibull distribution becomes an appropriate descriptor for cell survivability (Martz and Waller 1982; Smith 1991; van Boekel 2002). Survivability based on such a Weibull distribution yields a threshold followed by either a concave, convex or an exponential fall-off based on the characteristic Weibull parameters. However, such analytical descriptions ignore the underlying physical details.

S. J. Beebe et al., *Ultrashort Electric Pulse Effects in Biology and Medicine*, Series in BioEngineering, https://doi.org/10.1007/978-981-10-5113-5_10

From a biophysical standpoint, two pathways to cellular apoptosis are recognized. An "extrinsic pathway" which is instigated at the cellular plasma membrane through death receptors; and the "intrinsic pathway," which responds to intracellular cues through mitochondrial-initiated processes. The former pathway can be triggered by extracellular death signals linked with CD95 molecules and tumor necrosis factor (TNF)-related apoptosis-inducing ligands. Apoptosis (which is important in maintaining homeostatic balance (Krammer 2000; Horvitz 1999; Jacobson et al. 1997) can be activated by deprivation of survival signals, genetic or toxicological damage (Brune 2000), *or through external electrical pulsing*. A common observation in response to these stimuli is the activation of caspases, a group of cysteine proteases that serve as the main initiators and effectors of apoptosis. For example, death receptors, such as CD95, enable the binding or clustering of molecules to form the death inducing signaling complex (Green 2000; Peter and Krammer 2003). Trimerization of death receptors upon cross-linking recruits procaspase-8. Subsequent proteolytic activation through several cleavage steps (based on proximity interactions) leads to caspase-8 (C8) activation. A series of biochemical reactions are then set into motion leading to apoptosis.

In some cells, activation of caspase-8 in large quantities leads to the activation of other caspases, including the executioner caspase-3 that ultimately leads to apoptosis. In other cells, the amount of caspase-8 generated at the DISC could be small, and the activation cascade would not propagate directly, but instead could be amplified via the mitochondria. This process is initiated by the cleavage of Bid by caspase-8, followed by the translocation of the truncated Bid (tBid) to the mitochondria, which induces the release of proapoptotic molecules such as cyt c and Smac/DIABLO to the cytoplasm. Cytochrome c release closely depends on the opening of mitochondrial permeability transition pores (MPTPs).

Thus, the mitochondria-dependent activation (intrinsic pathway) involves cytochrome c release from mitochondria induced by stress, irradiation, or inflammation (Budihardjo et al. 1999; Li et al. 1997). In the more recent electrical pulsing context (Schoenbach et al. 2008), it is postulated that the externally applied electric field modulates the transmembrane mitochondrial potential. This can directly open the mitochondrial permeability transition pore (Marzo et al. 1998), leading to cytochrome *c* release, with subsequent downstream apoptotic events. There is also a linkage between the electric field-initiated events at the plasma membrane and the mitochondrial processes. For example, caspase-8 has been shown to cleave the Bcl-2 family member Bid, causing release of cytochrome c. A mathematical model of this cascade leading to the activation of caspase-3 and the eventual DNA fragmentation and cell death was recently discussed (Bagci et al. 2006).

The role and extent of electric field modifications to the apoptotic pathways is unclear. Hard experimental data is lacking, in part due to the complexity of the biophysical mechanisms and the underlying competition between various processes. However, possible consequences of the applied field from a biophysical standpoint can be conjectured. This should help drive and focus experimental studies on specific hypothesis. Consequences of the applied voltage include the stimulation of CD95 and recruitment of procaspase-8 through enhanced dipole-interactions. Another possibility is that membrane poration by the external field leads to molecular diffusion through the pore bottleneck as observed with phosphatidylserine externalization (Hu et al. 2005). Such molecular transport towards local pores (in keeping with the fluid mosaic model (Singer and Nicolson 1972), should increase the proximity between molecules and enhance their interactions.

Quantitative modeling of the overall electric field-induced cellular apoptosis is an extremely difficult task. Details of the various biomechanisms, their temporal dynamics and mutual interactions are not known, and many of the parameters and rate constants have not been measured. Given these difficulties, a less ambitious and fairly simplified systems-level approach is taken here. The current modeling objectives are limited to the following; (a) ascertain whether a pulse-number threshold emerges for cell killing by the nanosecond, high-intensity pulses. Such a result would lend a qualitative biophysical explanation of the experimental observations. (b) Gauge the relative importance of the extrinsic mechanism relative to the intrinsic, mitochondrial pathway. This become germane, given that the ultrashort (nanosecond) pulsing might be expected to have a stronger effect on intracellular organelles (Schoenbach et al. 2007). (c) Lend qualitative support to the hypothesis that cell apoptosis is due to discrete events initiated at specific sites. By extrapolation, use of multi-prong electric pulsing systems would then be more effective in cell killing (or in triggering cellular electro-biochemistry) due to the variable field orientations and the possibility to affect multiple targets.

Time-dependent kinetics of the caspases and the various molecular species within the apoptotic pathway, can be simulated using a rate-equation model (Bagci et al. 2006); in part because of the availability of published rate parameters. This model contains a system of 31 coupled rate equations for the temporal evolution of the various concentrations. The initial equilibrium concentrations were determined by running the time-dependent rate-equation simulation until steady state. This model includes the growth of Bid due to Caspase 8 activation, followed by a C3-Bid positive feedback process, wherein increases in [C3] lead to enhanced cleavage of Bid, which in turn feed into enhancing the [C3] concentrations through release of cytochrome c and caspase-9 activation.

The role of the externally applied electric field from a modeling standpoint, is to trigger C8 release from specific sites within the plasma membrane. On the assumption that C8 release occurs at discrete sites on the plasma membrane, one can expect different cells to exhibit varying degrees of C8 activation depending on their orientation with respect to the external field. Our notion of discrete sites and specific death-signaling domains ties in with emerging evidence that lipid microenvironments on the cell surface, known as lipid rafts, may be critically involved in the ultimate cell fate (Simons and Toomre 2000; Holthius and Levine 2005). Death-inducing signaling receptor (DISC) molecules are known to be located at membrane rafts and act as the linchpins from which apoptosis signals are launched (Mollinedo and Gajate 2006). From a modeling standpoint, assignment of random caspase-8 release concentrations to each cell through a stochastic, Monte Carlo type implementation, can take this discrete nonuniformity into account. Hence, for our simulations (Song et al. 2010), the caspase-8 concentration released in each cell was assigned a value "*r* [Cmax]," with "*r*" being a random number and [Cmax] a preset concentration upperbound. Clearly, a large caspase-8 release can be expected to lead to irreversible increases in caspase-3, and predictions of cell death. In keeping with previous values (Bagci et al. 2006), initial concentrations of all compounds in our simulation were taken to be zero, except for procaspase-3, procaspase-9, Apaf-1, Bid, Bax, Bcl-2, inhibitor of apoptosis (IAP), and cytochrome *c*_mito, which were all set to 4.0 nM.

The activation and release of C8 by the voltage pulse takes time since a sequence of processes such as the formation of the death-inducing signaling complex (Kischkel et al. 1995), clustering followed by cleavage of C8 etc. are involved. Consequently, the physics-based model needs to account for a finite time delay in the availability of caspase-8 following the external pulsing. Furthermore, in response to an electric field impulse, the concentration of C8 must gradually rise as a function of time and eventually saturate. This aspect, not treated by the Bagci model, was included based on a CD95 activation approach (Bentele et al. 2004). As a result, these time-dependent C8 concentrations, following an excitation electrical pulse, were used as input to the rate-equation model for caspases.

10.2 Single Cell Simulation Results

Calculation results are presented for a single cell based on the above approach. Figure 10.1a, b show caspase-3 concentrations as a function of time starting from two different assumed levels of C8. At the lower 0.01 nM caspase-8 concentration, a decay in C3 over time is predicted; while increased values of C3 with time are seen in Fig. 10.1b for the higher 0.1 nM caspase-8. These results are indicative of a bi-stability in the apoptotic process and the requirement of a critical C8 level for cell death induction. For the higher 0.1 nM caspase-8 value, appreciable C3 activation is predicted to occur roughly in the 1–2 h time frame. This time scale turns out to be approximately in keeping with experimental data that gauges caspase activity and gathers intracellular data. The simulations suggest that if more C8 were to be activated (e.g., either due to a higher voltage pulse or through multiple pulsing), then the probable outcome for cell death in the overall population would be enhanced.

To gauge the response of C8 over time to multiple pulsing and evaluate possible cumulative effects, simulations were next carried out with low, but repetitive, 0.4 pM caspase-8 injections. Simulation results for the time-dependent [C3] concentration are shown in Fig. 10.2 in response to 1 and 10 pulses (at 1 Hz), respectively. The injections were taken to begin after 10,000 s (~2.7 h) from the initial time to allow the system to reach an initial steady state. The 10-pulse case points towards unstable cell behavior as reflected through the [C3] increase, despite the low caspase-8 injection levels.

The above calculations were based on instantaneous caspase-8 release following a voltage pulsing event for a cell. However, as already mentioned, the caspase-8 concentration should be taken to rise gradually in a time-dependent fashion. Using

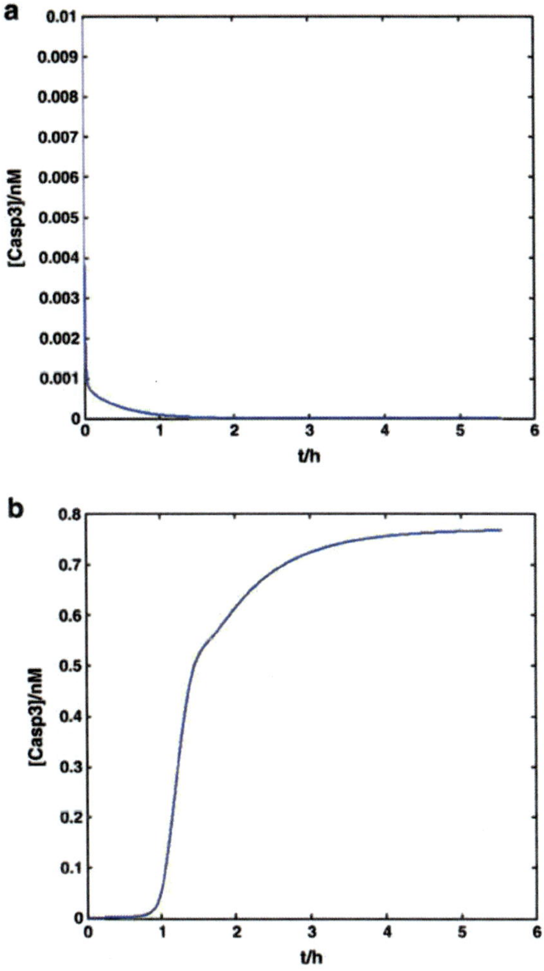

Fig. 10.1 Single cell simulation results for the caspase-3 concentration ([Casp3]/nM) versus time in hours (t/h) for two different starting caspase-8 values. **a** Caspase-8 value of 0.01 nM, and **b** initial caspase-8 value of 0.1 nM (After Song et al. 2010)

the Bentele model (Bentele et al. 2004), the normalized concentrations of caspase-8 and the blocked DISC states obtained through numerical simulations are shown in Fig. 10.3. As expected, both increase over time. However, a saturating trend is obtained for the concentration [C8]. This is due to a regulatory mechanism associated with the role of c-FLIP, which efficiently blocks caspase-8 activation at the DISC. The characteristic time delay seen in Fig. 10.3 is roughly 5 min. This is quite small compared to much longer period spanning hours over which C3 activation and eventual cell death occurs. This thus confirms that to a good approximation, biochemical triggering by an electric pulse can be taken to be a very rapid process.

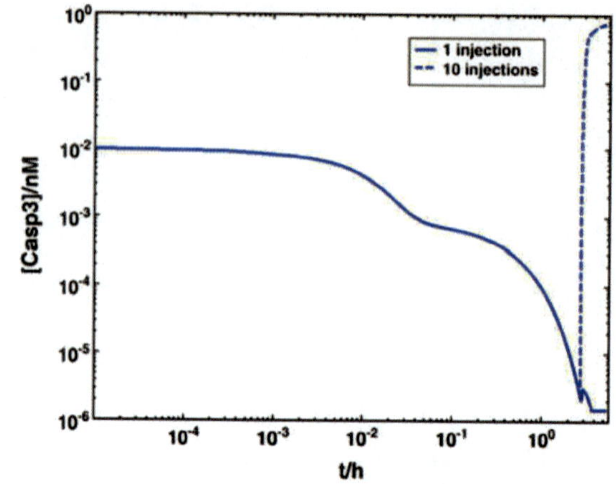

Fig. 10.2 Predicted logarithm of concentration C3 ([Casp3]/nM) versus time in hours (t/h) following 1- and 10-injections at 1 Hz of 0.4 pM caspase-8 in a single cell. The injections were taken to begin after 10,000 s (After Song et al. 2010)

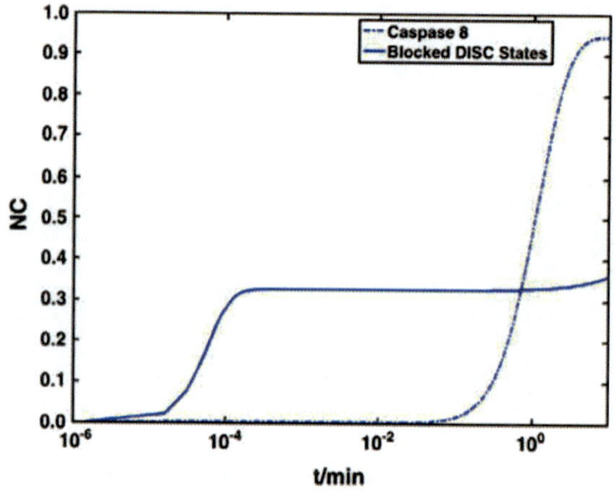

Fig. 10.3 Single cell simulation results for the temporal evolution of the normalized concentration (NC [nM/nM]) for caspase-8 and blocked DISC states versus time in minutes (t/min). The normalization is with respect to the maximum value of 0.2 nM for caspase-8 (After Song et al. 2010)

10.3 Predictions of Ensemble Population Dynamics

Having discussed some aspects of single cell response, ensemble simulations were carried out next to include the statistical variability inherent in actual experiments. Thousand cells were used, and the C8 injection in each cell was randomly varied

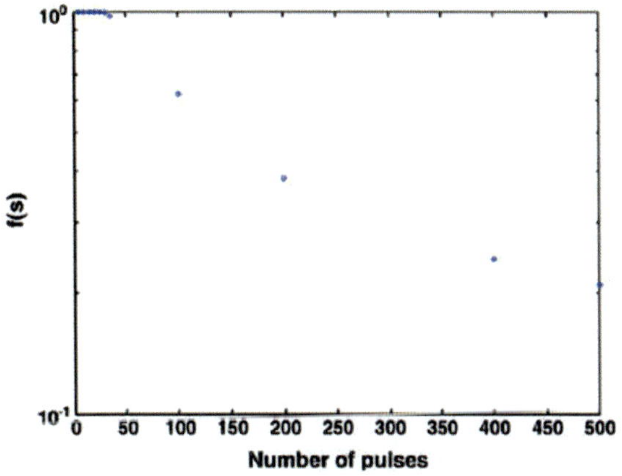

Fig. 10.4 Predicted logarithm of cell-survival ratio (Log f(s)) versus pulse number for 1000 cells. Survival ratio $f(s) = N(s)/N_{tot}$ with N(s) the number of surviving cells and N_{tot} the total cell number (After Song et al. 2010)

between zero and a maximum value (C_{max}) by assigning random numbers. Predictions of cell survival versus pulse number for low level C8 injection (with $[C_{max}] =$ 0.04 pM) are shown in Fig. 10.4. The injections, corresponding to the pulse numbers, were set at a frequency of 1 Hz, in keeping with the experimental data (Pakhomov et al. 2004). Figure 10.4 clearly shows a threshold effect, and virtually all cells are predicted to survive until about 50 pulses are incident on the cells. Beyond the threshold, a near-exponential fall-off in survival is predicted, which again is in keeping with the nanosecond electric pulsing experiments. However, at the very large pulse numbers, a shift towards a somewhat saturating trend can be seen in Fig. 10.4. Physically, this outcome arises from our choice of a fixed set of random numbers for each C8 injection at the 1 Hz frequency. In other words, those cells assigned a relatively high random number, were always associated with higher C8 activation, leading to their more likely (and faster) outcomes of cell death. Those with low random numbers continued to survive. This constant random number assignment to the cells is a likely scenario under in vivo conditions using electrical pulsing that is fixed in location and orientation.

In any case, results of Fig. 10.4 are roughly in keeping with the experimental cell-survival data. In particular, a pulse-number threshold is predicted followed by a near-exponential fall-off. The analysis is more insightful and based on biophysics instead of the statistical treatments that invoke survival distributions. Also, the intrinsic pathway is shown to be much weaker as compared to the extrinsic mechanism for electric pulse induced cell apoptosis. In addition, delays of about an hour are predicted for detectable molecular concentration increases following electrical pulsing. Furthermore, since cell killing depends on the amounts of caspase-8 activation and cytochrome c released, differing survival thresholds are to be expected between different cell types. For instance, our previous experimental reports on electric pulsing of cells showed a relative lack of PS externalization in B16 cells, whereas Jurkat cells have shown strong PS effects (Schoenbach et al. 2008). This underscores inherent differences in cell parameters and their responses to external stimuli.

A feature of this rate-equation modeling discussed in this chapter is the predicted delay of about an hour in the time taken for molecular concentration increases following electrical pulsing. This delayed biochemical process in response is similar to a recent report on cellular dye update upon electric pulsing (Kennedy et al. 2008). The experiments by Kennedy et al. identified two distinctive electroporative uptake signatures: an initial almost negligible dye absorption immediately after the pulse exposure, followed by a high-level, accelerating uptake at much longer times. This trend is generally also in line with earlier reports (for example, Neumann et al. 2008) indicating multi-stage evolution following electrical pulsing.

A number of aspects were ignored in our simple treatment. For example, it was assumed in the simulations that a chosen amount of cytochrome c release would result from each electrical pulsing. This assumes that the cytochrome c outflow is roughly similar to field-induced intracellular calcium release (Joshi et al. 2007), and that its concentration within the mitochondria is large enough to discount depletion effects. Similarly, it was assumed that in the multiple-pulsing scenario, the continued presence of strong external fields would not alter the rates and that the reaction kinetics would continue to follow the parameters of the Bagci model. The role of reactive oxygen species (ROS) generation by the electric pulsing has also been ignored in the present treatment. Reports in the literature (Wartenberg et al. 2008) suggest that application of electric fields could elevate ROS. This, in turn, would alter apoptotic destruction through the ROS-mediated pathways involving oxidative stress (Kannan and Jain 2000; Zamzami et al. 1997). For example, ROS is known to induce dissociation of cytochrome c from cardiolipin on the inner mitochondrial membrane, and facilitate its release via mitochondrial permeability transition-dependent mechanisms. Last but not least, our numerical implementation of the simple multiple-pulsing scenario assumed a fixed amount of caspase-8 activation repeatedly in response to the external pulsing. However, quite conceivably, there could be a depletion effect and the CD95 initiated process might not produce as much C8 after multiple pulses. However, the activation could physically arise from different discrete sites within the plasma membrane. A multitude of such sites would alleviate the depletion issue and provide an electrically-induced effect after each electrical pulse. This hypothesis of multi-targets and sites would be in keeping with the observed enhancements in cellular apoptotic killing and gene delivery through multi-electrode systems (Jaroszeskia et al. 1999). Such multi-electrode systems would provide sufficiently high electric fields at different locations and orientations within cells for greater bio-effects (Heller et al. 2010).

References

Bagci EZ, Vodovotz Y, Billiar TR, Ermentrout GB, Bahar I (2006) Bistability in apoptosis: roles of Bax, Bcl-2, and mitochondrial permeability transition pores. Biophys J 90:1546–1559

Bentele M, Lavrik I, Ulrich M, Stofier S, Heennann DW, Kalthoff H, Krammer PH, Eils R (2004) Mathematical modeling reveals threshold mechanism in CD95- induced apoptosis. J Cell Biol 166:839–851

Brune B (2000) Nitric oxide and apoptosis in mesangial cells. Kidney Int 61:786–789

Budihardjo I, Oliver H, Lutter M, Luo X, Wang X (1999) Biochemical pathways of caspase activation during apoptosis. Annu Rev Cell Dev Biol 15:269–290

Green DR (2000) Apoptotic pathways: paper wraps stone blunts scissors. Cell 102:1–4

Heller R, Cruz Y, Heller LC, Gilbert RA, Jaroszeski MJ (2010) Electrically mediated delivery of plasmid DNA to the skin, using a multielectrode array. Hum Gene Ther 21:357–361

Holthius JCM, Levine TP (2005) Lipid traffic: floppy drives and a super-highway. Nat Rev 6:209–220

Horvitz HR (1999) Genetic control of programmed cell death in the nematode Caenorhabditis elegans. Cancer Res 59:1701–1706

Hu Q, Joshi RP, Schoenbach KH (2005) Simulations of nanopore formation and phosphatidylserine externalization in lipid membranes subjected to a high intensity, ultrashort electric pulse. Phys Rev E 72:031902/1–10

Jacobson MD, Weil M, Raff MC (1997) Programmed cell death in animal development. Cell 88:347–354

Jaroszeski MJ, Gilbert R, Nicolau C, Heller R (1999) In vivo gene delivery by electroporation. Adv Drug Deliv Rev 35:131–137

Joshi RP, Nguyen A, Sridhara V, Hu Q, Nuccitelli R, Schoenbach KH (2007) Simulations of intracellular calcium release dynamics in response to a high-intensity, ultra-short electric pulse. Phys Rev E 75:041920/1–10

Kannan K, Jain SK (2000) Oxidative stress and apoptosis. Pathophysiology 7:153–163

Kennedy SM, Ji Z, Hedstrom JC, Booske JH, Hagness SC (2008) Quantification of electroporative uptake kinetics and electric field heterogeneity effects in cells. Biophys J 94:5018–5027

Kischkel FC, Hellbardt S, Behrmann I, Germer M, Pawlita M, Krammer PH, Peter ME (1995) Cytotoxicity-dependent APO-1(Fas/CD95)-associated proteins form a death-inducing signalling complex (DISC) with the receptor. EMBO J 14:5579–5588

Krammer PH (2000) CD95's deadly mission in the immune system. Nature 407:789–795

Li P, Nijhawan D, Budihardjo I, Srinivasula SM, Ahmad M, Alnemri ES, Wang X (1997) Cytochrome c and dATP-dependent formation of Apaf-1/caspase-9 complex initiates an apoptotic protease cascade. Cell 91:479–489

Martz HF, Waller RA (1982) Bayesian reliability analysis. Wiley Interscience, New York

Marzo I, Brenner C, Zamzami N, Jürgensmeier JM, Susin SA, Vieira HLA, Prévost MC, Xie Z, Matsuyama S, Reed JC, Kroemer G (1998) Bax and adenine nucleotide translocator cooperate in the mitochondrial control of apoptosis. Science 281:2027–2031

Mollinedo F, Gajate C (2006) Fas/CD95 death receptor and lipid rafts: new targets for apoptosis-directed cancer therapy. Drug Resist Updat 9:51–73

Neumann E, Toensing K, Kakorin S, Budde P, Frey J (2008) Mechanismof electroporative dye uptake by mouse B cells. Biophys J 74:98–108

Pakhomov AG, Phinney A, Ashmore J, Walker K III, Kolb JF, Kono S, Schoenbach KH, Murphy MR (2004) Characterization of the cytotoxic effect of high-intensity, 10-ns duration electrical pulses. IEEE Trans Plasma Sci 32:1579–1586

Peter ME, Krammer PH (2003) The CD95 death-inducing signaling complex and beyond. Cell Death Differ 10:26–35

Schoenbach KH, Hargrave B, Joshi RP, Kolb JF, Nuccitelli R, Osgood C, Pakhomov A, Stacey M, Swanson RJ, White J, Xiao S, Zhang J, Beebe SJ, Blackmore PF, Buescher ES (2007) Bioelectric effects of intense nanosecond pulses. IEEE Trans Dielectr Electr Insul 14:1088–1107

Schoenbach KH, Xiao S, Joshi RP, Camp JT, Heeren T, Kolb JF, Beebe SJ (2008) The effect of intense subnanosecond electrical pulses on biological cells. IEEE Trans Plasma Sci 36:414–422
Simons K, Toomre D (2000) Lipid rafts and signal transduction. Nat Rev 1:31–39
Singer SJ, Nicolson GL (1972) The fluid mosaic model of the structure of cell membranes. Science 175:720–731
Smith RL (1991) Weibull regression models for reliability data. Reliab Eng Syst Saf 34:55–77
Song J, Joshi RP, Beebe SJ (2010) Cellular apoptosis by nanosecond, high-intensity electric pulses: model evaluation of the pulsing threshold and extrinsic pathway. Bioelectrochemistry 79:179–186
van Boekel MAJS (2002) On the use of the Weibull model to describe thermal inactivation of microbial vegetative cells. Int J Food Microbiol 74:139–159
Wartenberg M, Wirtz N, Grob A, Niedermeier W, Hescheler J, Peters SC, Sauer H (2008) Direct current electrical fields induce apoptosis in oral mucosa cancer cells by NADPH oxidase-derived reactive oxygen species. Bioelectromagnetic 29:47–54
Zamzami N, Hirsch T, Dallaporta B, Petit PX, Kroemer G (1997) Mitochondrial implication in accidental and programmed cell death: apoptosis and necrosis. J Bioenerg Biomembr 29:185–193

Chapter 11
Thermal Effects in Bioelectrics

Karl H. Schoenbach

Abstract Electroporation is considered a nonthermal process, and when used in medical therapies great care is taken to ensure that temperature effects due to Joule heating are avoided by limiting the average electrical power of the electroporation pulse train. However, moderate temperature increases, far below those which are used in hyperthermia therapies, have shown to increase safety and efficacy of electroporation-based therapies. This chapter provides an introduction into the heating mechanisms by applied electrical pulses, Joule heating and heating due to dielectric relaxion, followed by a short introduction of external heat sources, with emphasis on infrared light sources. The second part describes basic thermal effects on cells, particularly on cell membranes. This is followed by an overview of recent in vitro and in vivo studies on the effects of thermal assistance in electrotherapies with emphasis on its application in cancer treatments. Moderate heating of tumors in combination with electrical pulse treatments was found, for a wide range of pulse durations, to cause a significantly higher rate of complete tumor regression compared to the use of pulsed electric fields alone.

11.1 Introduction

Nanosecond pulsed electric field effects are considered nonthermal effects, that means that the only parameters that determine the outcome of bioelectric studies and applications are the electrical parameters such as pulse amplitude, pulse duration, pulse shape, pulse number and repetition rate. However, considering that cells and tissues are defined by a finite resistivity and permittivity, the interaction of electric fields with these targets always leads to heating, be it resistive heating (Joule heating), or heating caused by dielectric relaxation. In most of the studies, scientists have made every effort to limit the increase in temperature particularly for the case that multiple pulses need to be applied to obtain a desired effect. This can be done by reducing the repetition rate, or increasing the time between pulses, respectively. The thermal losses due to heat conduction and convection between subsequent pulses allow the average temperature to be kept at the desired low level.

S. J. Beebe et al., *Ultrashort Electric Pulse Effects in Biology and Medicine*,
Series in BioEngineering, https://doi.org/10.1007/978-981-10-5113-5_11

However, a moderate, controllable increase in temperature might actually help to provide a better outcome of a desired process or treatment. We will show in the following that a moderate temperature increase, either through Joule heating or dielectric relaxation, or by using external heating sources, can be used to support the nonthermal effects of pulsed electric fields, e.g. by reducing the need for large electric fields or large number of pulses in medical treatments. This will be documented by presenting experimental results for bioelectric applications with electrical pulses with durations ranging from milliseconds to hundreds of picoseconds.

Before describing such experiments, we will briefly introduce the main mechanisms for heat generation by electrical pulses in cell suspensions and tissues: Joule heating, and the localized heating of cell structures, particularly cell membranes, by means of dielectric relaxation. Also, a short introduction of the use of external heating sources, microwaves and infrared radiation will be given. This will be followed by a discussion of the temperature effects on single cells, particularly on their membranes, and on tissues. In the last section, we will demonstrate the advantages of controlled, moderate heating for medical applications.

11.2 Heating Mechanisms in Bioelectrics

11.2.1 Joule Heating

Joule heating, or resistive heating, respectively, is based on the transfer of momentum from ions, which are accelerated in the electric field, to molecules, charged or neutral, through elastic collisions. This causes a random motion of particles, known as thermal motion, which is superimposed to the directed motion of the ions along the field lines, the so-called drift motion. It is generally assumed that almost all the acquired energy of the ions from the electric field is converted into thermal energy. In this case, the increase in temperature, ΔT, caused by a square pulse with an electric field amplitude, E, and a duration τ, is:

$$\Delta T = \sigma E^2 \tau / \rho c \tag{11.1}$$

where ρ is the density (mass/volume), c is the specific heat capacity, and σ is the electrical conductivity (the inverse of resistivity). Assuming that the density of a cell suspension in a cuvette is approximately that of water ($\rho = 1$ g/cm^3) and the specific heat capacity, c, is 1 cal/g K (1 cal = 4.2 J), we can estimate the temperature increase in the medium in a cuvette when an electrical pulse is applied. With an electrical conductivity of the cell suspension of $\sigma = 10$ mS/cm, typical for many of the commercially available solutions, in a 2 mm cuvette, a square pulse of 6 kV and 100 ns duration causes an increase in temperature by 0.22 K.

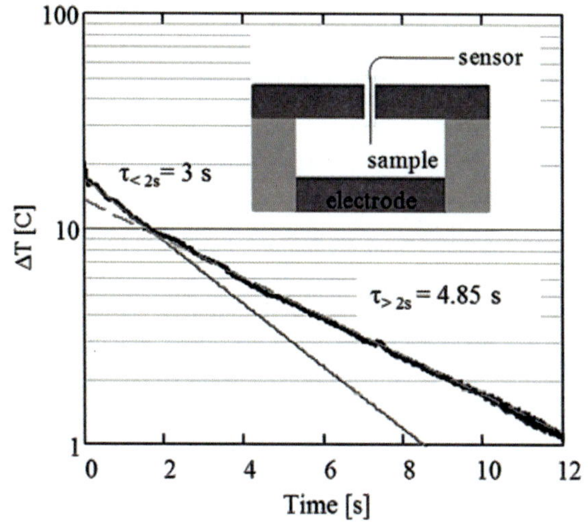

Fig. 11.1 Measured temperature decay time constant (bold) of the medium in an exposure chamber. The temperature by the time the probe was placed in the exposure chamber was 40 °C, and it decayed to 23 °C. The initial thermal decay time constant (<2 s) was measured to be 3 s, while it was increased to 4.85 s at times beyond 2 s (Camp et al. 2012)

The temperature decreases after the pulse and reaches its base level after a sufficient time. The recovery time is determined by the heat loss processes, mainly through thermal conduction. Figure 11.1 shows the temporal development of the temperature in an exposure chamber after quickly heating the medium in the center between the electrodes to a temperature of 40 °C (Camp et al. 2012). It consists of two parts, an initial fast temperature drop with a 1/e time constant of 3 s, followed by a somewhat slower decay with a time constant of 4.85 s. The initial decay is likely determined by thermal conduction through the planar electrodes, the following, slower decay likely by convection.

Most bioelectric studies as well as the therapeutic applications require multiple pulses. This causes an accumulative effect with respect to temperature. In the experiment (Fig. 11.1), the temperature after each pulse at 1 Hz is reduced by only 30%. The temperature increase caused by N consecutive pulses would therefore be the product of the temperature increase caused by a single pulse, T_0, and $N \times 0.7$. Using the data in the previous example ($E = 30$ kV, $\tau = 100$ ns, $\sigma = 10$ mS/cm) and assuming $N = 100$, the temperature increase after pulsing would be about 15 °C, a substantial increase in temperature.

It needs to be noted that the thermal time constants are determined by the geometry of the exposure chamber (cuvette) used in the experiment, as well as the thermal properties of the suspension. In other configurations and with other suspensions, the time constants are certainly different, and so will be the thermal effect caused by trains of pulses. Temperature measurements in tissues have shown that the temperature increase cannot be neglected, but it seems less severe than that in cuvettes. Only

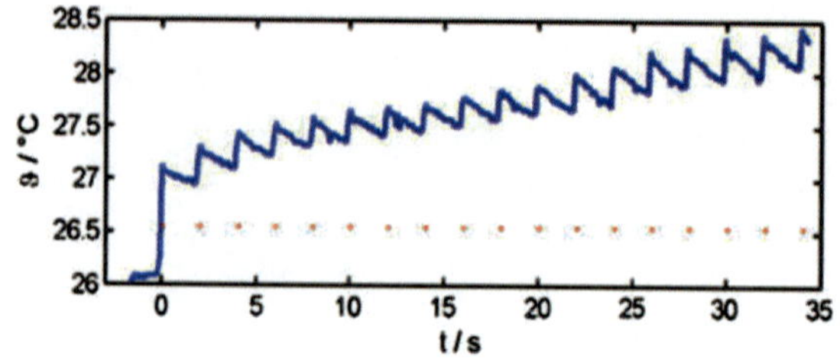

Fig. 11.2 Temperature increase inside the tumor due to 2.2 kV/300 ns pulses with 2 s interpulse spacing. The maximum temperature was reached after about 40 pulses. At the end of the experiment after 100 pulses, 29.4 °C was reached. The dots depict the time of pulse application. The higher temperature increase for the first pulse was attributed to the drop in voltage with subsequent pulses because of the voltage divider between the pulse-forming network and the skin fold (Pliquett and Nuccitelli 2014)

minor temperature increases were observed using a micro-thermocouple (Pliquett and Nuccitelli 2014) (Fig. 11.2). The conclusion, based on these results, as expressed by the authors, was "Since the temperature rise in B-16 mouse melanoma tumors due to equally spaced ($\Delta t = 2$ s) 300 ns-pulses with $E = 40$ kV/cm usually does not exceed $\Delta T = 3$ K at all parts of the skin fold between the electrodes, a hyperthermic effect on the tissue can be excluded."

Whereas the consideration of temperature increase through Joule heating is valid for pulses in the millisecond and microsecond range and even for nanosecond pulses with durations large compared to 10 ns, the temperature increase by using single–digit, nanosecond pulses and subnanosecond pulses requires to consider a second type of heating: heating though dielectric relaxation.

11.2.2 Heating Due to Dielectric Relaxation

The heating process is based on the effect of rapidly changing electric fields on permanent dipoles in suspensions or tissues. At low frequencies the dipoles follow easily the electric field and almost all of them is polarized, exhibiting a large capacitance. When the frequency increases beyond a certain value, the dipoles cannot follow the electric field entirely and the capacitance decreases, so they display a short-range, oscillatory motion. At the same time dielectric losses in the medium increase because the frequent reorientation of the dipoles causes collisions with neighboring ones, which generates a large amount of thermal energy (concept of microwave heating).

In order to utilize this process, the Fourier spectrum of the pulsed electric fields needs to extend into the frequency range of the dispersion of cell components. Effects of the dielectric relaxation of the membrane become important in the range of tens to hundreds of MHz (Kloesgen et al. 1996) while the dielectric relaxations of water molecules and dissolved ions, which are the main constituents of cytoplasm and cell exterior, begin in the GHz range. The lower frequency range for membrane

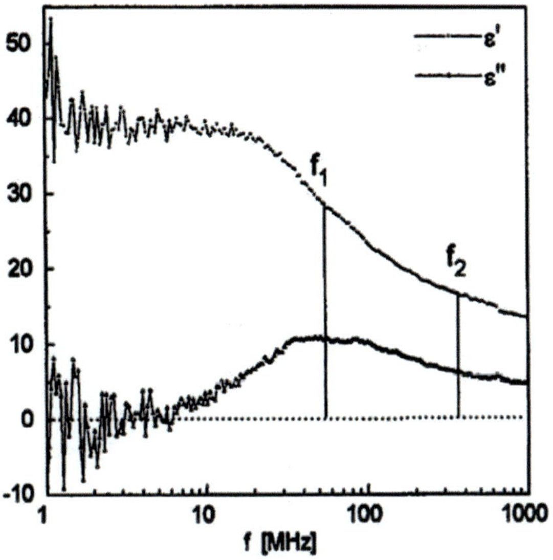

Fig. 11.3 The real (ε′) and the imaginary part (ε″ = σ/2πf) of the dielectric relaxation of a DMPC stack as a function of the applied rf frequency 1–1000 MHz at T = 308.5 ± 0.1 K. The dispersion frequency f_1 can be attributed to the diffuse thermal rotational motion of the phosphatidylcholine headgroups whereas the dispersion contribution that is characterized by the relaxation frequency f_2 corresponds to the bound water (Kloesgen et al. 1996)

effects is attributed to the diffuse thermal rotational motion of the headgroups of membrane lipids. Figure 11.3 shows the results of dielectric spectroscopy studies on the phospholipid membrane systems of dimyristoylphosphatidylcholine (DMCP) (Kloesgen et al. 1996).

Taking the membrane relaxation into consideration and including the effect of saline dipoles, Kotnik and Miklavcic (2000) evaluated the effect of dielectric relaxation on the power dissipation in cells. It was shown that the increases in the power dissipation in the medium (P_e) and the cytoplasm (P_i) increase dramatically in the GHz range, but the effect on dipoles in the membrane (P_m) occurs, as expected, already in the MHz range.

The data for τ_{em1}, τ_{em2}, and τ_{re} are listed in Table 11.1. Note that ν_{rm1} and ν_{re} correspond to f_1 and f_2 in Fig. 11.3.

In order to obtain approximate values for pulse durations where the effect of dielectric relaxation needs to be considered, the Fourier spectrum of rectangular pulses with different pulse duration is superimposed to the power spectrum. For simplicity square wave pulses are considered only. The spectral envelope of a square wave pulse with duration τ_p is constant up to a corner frequency of $\omega_c = 2/\tau_p$ and then decreases by 1/ω (Schoenbach 2010). In Fig. 11.5, the Fourier spectra of rectangular pulses of 10 μs, 100 ns and 1 ns were superimposed to the spectral power distributions in Fig. 11.4. It can be seen that pulses in the single-digit nanoseconds

Table 11.1 Parameters of dielectric relaxation of cytoplasm, membrane, and extracellular medium (Kotnik and Miklavcic 2000)

Parameter	Symbol	Value	Reference
Dielectric relaxation of cytoplasm and extracellular medium			
First relaxation time	τ_{re}	6.2×10^{-12} s	Büchner et al. (1999)[a]
First relaxation step	$\Delta\varepsilon_e$	5.9×10^{-10} As/Vm	Büchner et al. (1999)[a]
Dielectric relaxation of membrane			
First relaxation time	τ_{rm1}	3.0×10^{-9} s	Klösgen et al. (1996)
First relaxation step	$\Delta\varepsilon_{m1}$	2.3×10^{-11} As/Vm	From Klösgen et al. (1996)[b]
Second relaxation time	τ_{rm2}	4.6×10^{-10} s	Klösgen et al. (1996)
Second relaxation step	$\Delta\varepsilon_{m2}$	7.4×10^{-12} As/Vm	From Klösgen et al. (1996)[b]

[a]Permittivity of 0.154 M NaCl at 35 °C
[b]Scaled by $\varepsilon_m/\varepsilon = 0.125$, where $\varepsilon_m = 4.4 \times 10^{-11}$ As/Vm is the static permittivity of lipid bilayer (see Table), and $\varepsilon = 3.5 \times 10^{-10}$ As/Vm is the static permittivity of the multilamellar sample used by Klösgen et al. (1996)

begin to affect the power dissipation in cell membranes and therefore lead to local heating of membranes (Fig. 11.5).

Based on these modeling results, it can be concluded that shorter pulses, in the picosecond range, likely have major effects on the membrane temperature. Longer pulses, on the other hand, are mostly causing heating of the medium through Joule heating, as discussed in the previous section.

Croce et al. (2010), and Salimi et al. (2013) have studied the effects of dielectric relaxation caused by short electrical pulses on the membrane temperature more quantitatively. The model by Croce et al. (2010), which considered heat diffusion, allowed

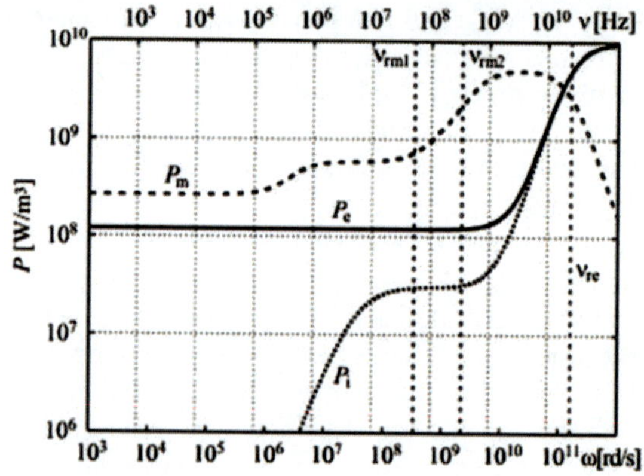

Fig. 11.4 Power deposition into the membrane (P_m), the cytoplasm (P_i), and the medium (P_m) dependent on frequency for an electric field of ~ 150 V/cm. The three dotted verticals correspond to the relaxation frequencies; from left to right: $\nu_{rm1} = 1/\pi\tau_{rm1}$, $\nu_{rm2} = 1/\pi\tau_{rm2}$ and $\nu_{re} = 1/\pi\tau_{re}$ (Kotnik and Miklavcic 2000)

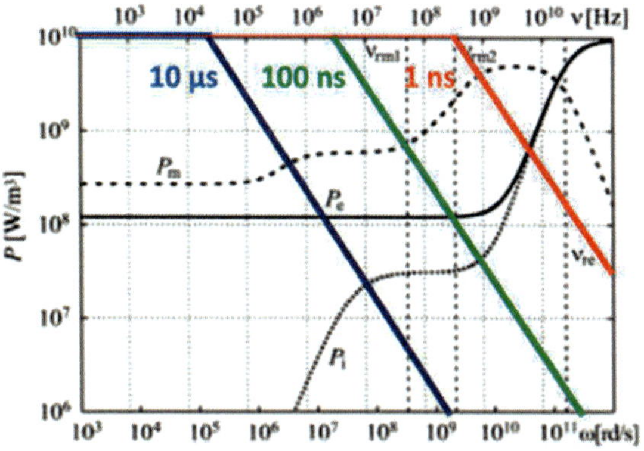

Fig. 11.5 Spectral distribution of three square-wave pulses of 10 μs, 100 ns and 1 ns duration superimposed to the power spectrum as shown in Fig. 11.4

the computation of the temporal development of the average membrane temperature for ultrashort pulses. Figure 11.6 shows the results for a 1 ns pulse compared to that of a 10 ns pulse. Clearly, the reduction of the pulse duration into the single digit range

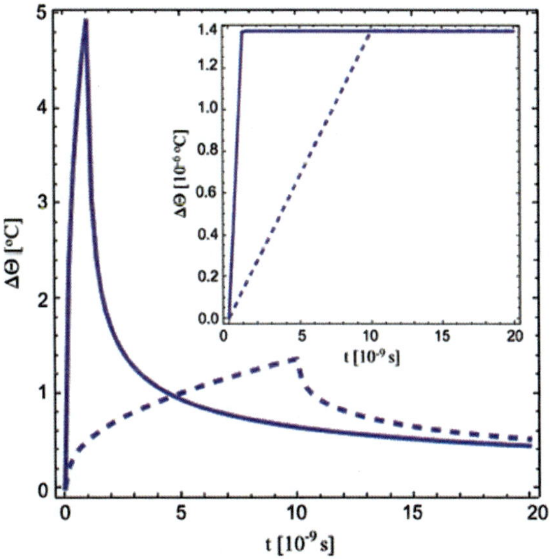

Fig. 11.6 Increase of (average) membrane temperature versus time, for an incident (rectangular) pulse with a specific absorption dose of 1 J/kg, applied at t = 0. (Full line) Pulse width = 1 ns. (Dashed line) Pulse width = 10 ns. The corresponding temperature increase in the cytoplasm is shown in the inset (Croce et al. 2010). A SAR of 1 J/kg corresponds to about 10 kV/cm for a 1 ns pulse

has major impact on the membrane temperature. This is consistent with the conclusion obtained from considering the effect of dielectric relaxation in the frequency domain (Fig. 11.5).

Effects of the membrane dielectric relaxation on electroporation caused by nanosecond pulses have been studied by Salimi et al. (2013). The main result of their theoretical study using a finite element model was that the process of electroporation occurs faster when membrane relaxation is considered, and that the required electric field is significantly reduced—an important consequence of moderate heating in bioelectric studies and applications.

11.2.3 External Heating

Whereas the previous sections described heating by pulsed electric fields only, there is certainly the possibility to control the temperature of a suspension or tissue externally. There is a multitude of options, ranging from heating by thermal conduction via a heater in physical contact with the biological target, to radio waves, microwaves, millimeter waves, and optical heating. Most of the in vitro thermal studies in bioelectrics rely on contact thermal sources, such as heaters or chillers. In vivo studies and applications require generally external heating through radiation sources. We will in the following focus on infrared sources, such as IR lasers and IR LEDs. Such optical sources allow the use of optical fibers, and consequently will not distort the applied electric fields in the target.

An example for such a device, integrated in a pulse delivery system, is shown in Fig. 11.7. Through an optical fiber, a semiconductor IR laser was used to irradiate the area between two or more electrodes. Since the emission from the fiber end is determined by the fiber aperture, the irradiated area can be varied by varying the distance of the fiber end. The closer the fiber end is placed to the irradiated surface,

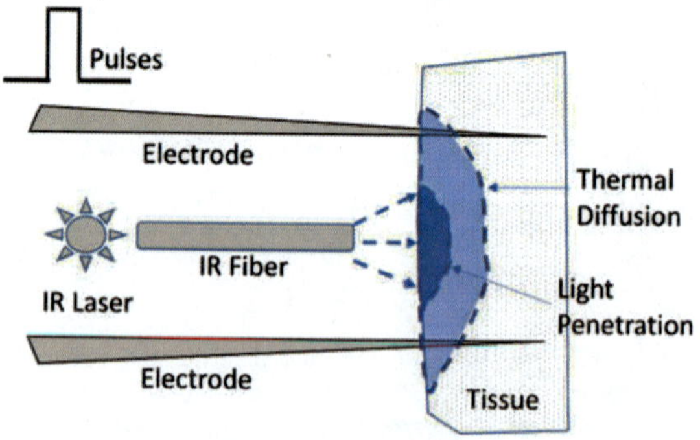

Fig. 11.7 Schematic of an optical irradiation system centrally located in an electroporation device with four needle electrodes. Modifying the distance between target and fiber allows changing the spot size of the infrared radiation on the tissue surface

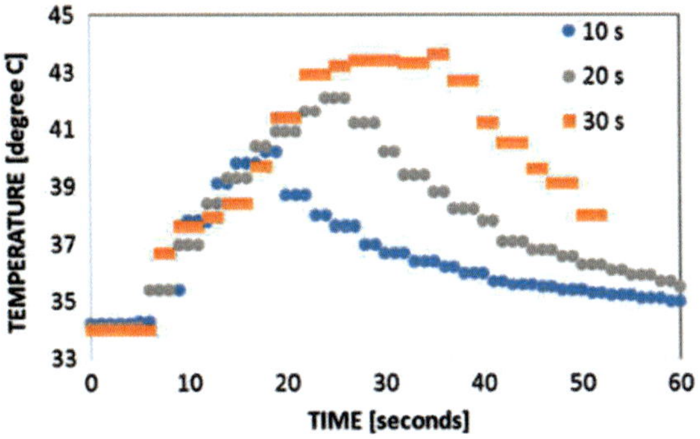

Fig. 11.8 At the optical power level of 8 W, a peak temperature of 42 °C was reached when the laser was on for about 20 s. The rise of the temperature was for this power level measured as 0.6 °C/s

the smaller is the irradiated area. The advantage of this system is the possibility to assist electroporation in positions where the electric field alone will not suffice. In the system shown in Fig. 11.7, the optical fiber is centrally located between the electrodes, wherein the electric field has a minimum for the case of needle electrodes and therefore thermal assistance is most helpful in compensating for the reduced bioelectric effects at this position.

With such a system the target temperature can be controlled before and even during a treatment. Temperature measurements on the skin of a guinea pig used in animal studies irradiated by an 8 W IR laser at a wavelength of about 1 μm showed that the temperature at the target can be varied on a time scale of seconds (Fig. 11.8), reducing the probability for a hyperthermia effects because of the reduced thermal dose at times of minutes or even seconds.

It needs to be mentioned that for electrotherapy assisted by optical irradiation, optical fibers with diameters of less than one millimeter can easily be included in catheters and can be used to irradiate e.g. tumors in deeper lying tissue.

11.3 Thermal Effects on Cells

What are the thermal effects on cells and tissue? Most of the responses focus on hyperthermia, which is a procedure of raising the temperature of tissue to such levels that cell death occurs. It is used as a sole or an adjunctive procedure in cancer treatment. The thermal dose required for cell death combines time of exposure with a constant temperature (Fig. 11.9). The longer the time of exposure the less temperature increase is required for cell death. In extreme cases, using a laser to heat cells for 300 μs with a temperature increase of up to 160 K did not cause cell death (Simanovskii et al. 2005).

A quantitative method to obtain information on the thermal dose required to induce cell death is based on the solution of the Arrhenius integral:

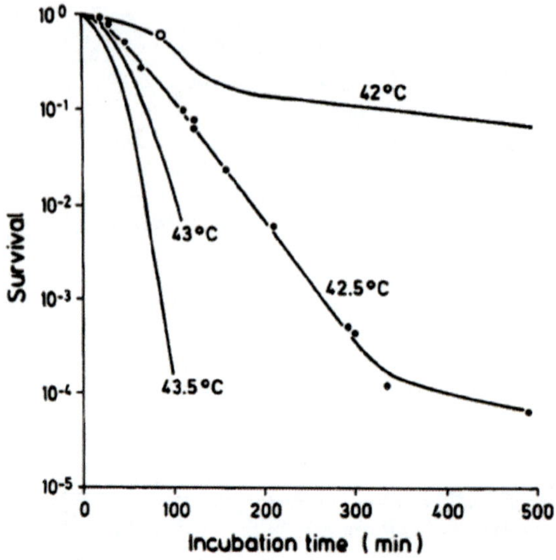

Fig. 11.9 Cell survival versus incubation time (Hildebrandt et al. 2002) with temperature as parameter

$$\Omega = \int A \exp\{-E_a/RT(t)\}dt \tag{11.2}$$

where Ω describes the proportion of cells killed, collapsed blood vessels or protein molecules coagulated; R is the universal gas constant, E_a is the activation energy, and A is a frequency factor. Activation energy and frequency factors are listed in the Table 11.2 (Sherar et al. 2000).

Using the data for cell death as listed in the table to integrate the temperature, T, over time, t, provides information on the percentage of dead cells when a temperature pulse is applied over a given time (Eq. 11.2).

The temperature distribution in tissue can be related to the amplitude and duration of electrical pulses by using Pennes' Bioheat equation:

$$\rho C \frac{\partial T}{\partial t} + \nabla \cdot (-k \nabla T) = \rho_b C_b \omega_b (T_b - T) + \sigma(T) E2 \tag{11.3}$$

Table 11.2 Activation energies and frequency factors for thermal damage in tissue (Sherar et al. 2000)

Process	Activation energy, E_a (J mol^{-1})	Frequency factor, A(s^{-1})
Cell death	5.064×10^5	2.984×10^{80}
Microvascular blood flow stasis	6.67×10^5	1.98×10^{106}
Protein coagulation	2.577×10^5	7.39×10^{37}

where ρ the tissue density, c the specific heat capacity, and σ the conductivity (dependent on temperature). The term $\rho_b C_b \omega_b (T_b - T)$ describes the losses due to blood perfusion. The temperature changes with time in a specific tissue volume (first term) is due to the in- or out-flow of thermal energy (second term), the loss of heat due to blood perfusion (third term), and due to Joule heating (fourth term). An example for the temperature distribution between two needle electrodes, calculated by means of the Pennes' Bioheat equation is shown in Fig. 11.10 (bottom) (Garcia et al. 2011).

Cell death through hyperthermia is by no means the only effect caused by increased temperature. Quoting a passage from a paper by Wust et al. (2002): "Various targets in the cell affected by rises in temperature have been found, such as membranes, the cytoskeleton, synthesis of macromolecules, and DNA repair. The expression of several genes can be upregulated or downregulated by heat, for example, the family of heat-shock proteins (HSP). Several other temperature-dependent interactions have been found, regulating molecular functions such as apoptosis, the cell cycle, and DNA repair." We will focus in the following on thermal effects on cell membranes, since these cell components play a major role in bioelectric research and medical applications based on nsPEF.

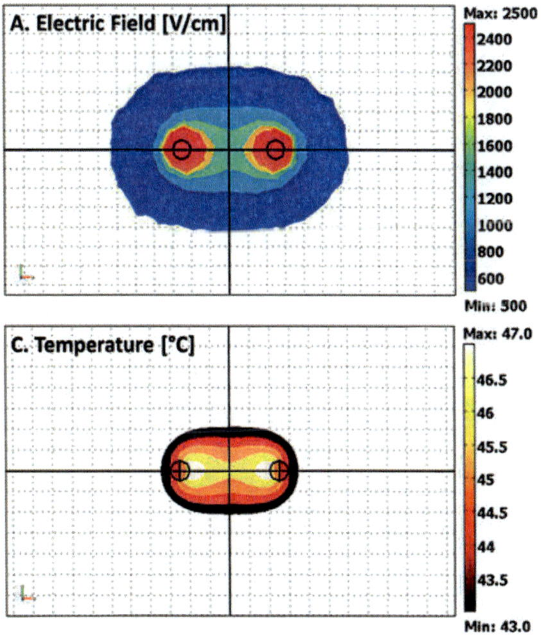

Fig. 11.10 top: Electric field distribution in a two-needle electrode pulse delivery system. The voltage applied to the needles with a gap of 0.5 cm was 1 kV. bottom: Temperature distribution after 80, 50 μs pulses at 1 Hz. The grid resolution in the distributions is 1.0 mm (Garcia et al. 2011)

11.3.1 Thermal Effects on Cell Membranes

Membranes undergo changes in their structure when heated. Simple phospholipid bilayers below a transition temperature, T_M, are in a gel phase. The membrane melts above the transition temperature and transits into a liquid disordered phase with increasing fluidity (Marrink et al. 2005). This is schematically shown in Fig. 11.11.

The transition temperature from gel-like to fluid-like is dependent on the composition of the bilayer (Table 11.3). Lipid bilayers do not need to be composed of a single type of lipid. Most membranes are a complex of mixture of different lipid molecules. Depending on the position in the cell membrane, different transition temperatures can be expected in individual cells.

Atomic force microscopy was used to study the bilayer changes with temperature in a mica-supported bilayer. An example, in this case, a DPPC (dipalmitoylphosphatidylcholine) bilayer, with increasing temperature is shown in Fig. 11.12

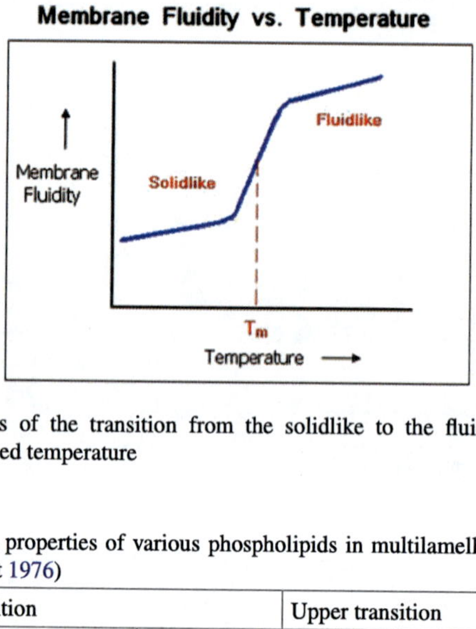

Fig. 11.11 Schematics of the transition from the solidlike to the fluidlike structure of cell membranes with incresed temperature

Table 11.3 Transition properties of various phospholipids in multilamellar aqueous suspension (Mabrey and Sturtevant 1976)

Lipid	Lower transition			Upper transition		
	T_{m1} (°C)	H1 (kcal mol^{-1})	Cooperative unit (molecules)[a]	T_{m2}(°C)	H2 (kcal mol^{-1})	Cooperative unit (molecules)[a]
DLPC	–	–	–	−1.8	1.7_0	980
DMPC	14.2	1.0_0	280	23.9	5.4_4	330
DPPC	35.3	1.8_3	290	41.4	8.7_4	260
DSPC	51.5	1.8_5	160	54.9	10.6_2	130
DMPE	–	–	–	49.5	5.8_0	140

[a]The size of the cooperative unit is particularly affected by impurities and other unrecognizable influence. The values given are maximum observed values

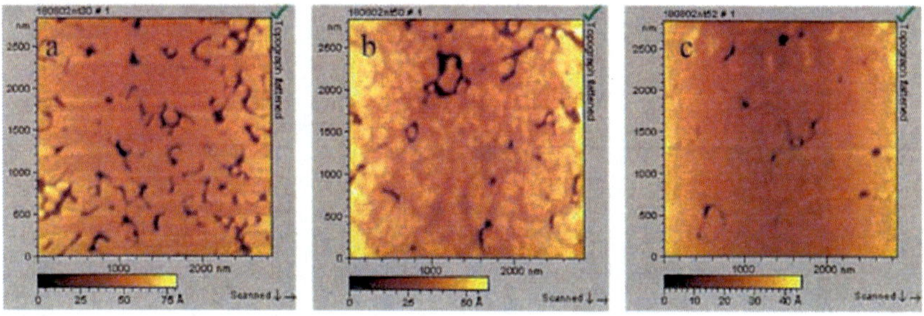

Fig. 11.12 AMF topography mages showing the phase transition in DPPC bilayer upon heating. **a** Heating DPPC bilayer to room temperature, 22 °C, **b** heating to 50 °C, **c** heating to 52 °C (Leonenko et al. 2004)

(Leonenko et al. 2004). When the temperature is increased from 22 °C (Fig. 11.12a) to 50 °C (Fig. 11.12b) and then to 52 °C (Fig. 11.12 c) the bilayer transits from a gel phase into a fluid phase. The transition occurs at 42–52 °C, leading to the formation of lower (thinner) domains. The coexistence of two domains, with thicknesses of 4.2 ± 0.4 nm and 3.3 ± 0.3 nm is visible in Fig. 11.12b. At 52 °C the lower domain becomes dominant. This process was found to be reversible when the temperature was reduced from the highest values to room temperature (22 °C).

Rather than studying the effect of increasing temperature on the fluidity of cell membranes, the effect of chilling cells (cell lines V-79 and B16F-1) down to 4 °C was recorded using electron paramagnetic resonance. With lower temperature the cell membranes became less fluid, with higher order parameters in the three types of domains, and higher proportion of domains with highest order parameter (Kandušer et al. 2008).

An important observation in studies with variable temperature is the thinning of the bilayer, when heated beyond a phase transition. In experiments where the temperature was increased from room temperature to 50 °C, the bilayer thickness was found to be considerably reduced (Leonenko et al. 2004). The measured thickness and the estimated values from theoretical fit show a reduction in thickness from 5.5 to 3.6 nm for fully hydrated DPPC bilayer (Table 11.4) and 4.5–3.3 nm for a fully hydrated DOPC bilayer, a change by more than 50%.

Time Domain Dielectric Spectroscopy (TDDS) measurements of the electrical parameters of cells, in this case, Hepa 1–6 cells, showed a considerable increase in conductivity (Table 11.5) of the cell membrane (Camp et al. 2012). The conductivity increased by almost a factor of four when the temperature was increased from 25 to 47 °C. The relative permittivity also increased from about 14–18. The drastic change in conductivity could in part be explained by reduction in membrane thickness, a parameter which was not included in the evaluation of the TDS data.

It is reasonable to assume that such changes in membrane structure, particularly the domain structure and thickness, will have effects on the electroporation or nanoporation, respectively. This is a correct assumption as will be shown in the following. Just considering the reduction in thickness, it is reasonable to assume that

Table 11.4 Comparison between the length L, measured by force analysis, the indention corresponding to the force of 1 nN, and the height given by AFM for a fully hydrated DPPC bilayer (Leonenko et al. 2004)

DPPC	Thickness of bilayer, estimated from theoreticalfit L, in nm	Indentation for 1 nN, in nm	Thickness of bilayer, measured by AFM cross section, in nm	Thickness for bilayer, literature data, in nm
$T = 65$ °C	2.5	4	2.5	–
$T = 50$ °C	3.3	4	3.6	3.6
$T = 36$ °C	3.6	5	4.4	–
$T = 22$ °C	4.5	5.8	5.5	4.7

Table 11.5 Conductivity of cytoplasm, medium and plasma membranes of Hepa 1–6 cells at 25 °C, 37 °C, 42 °C, and 47 °C (Camp et al. 2012)

Conductivity						
Temperature (°C)	Membrane (10^{-5} S/m)		Cytoplasm (10^{-1} S/m)		Medium (S/m)	
	Avg	Std	Avg	Std	Avg	Std
25	2.44	0.28	3.79	0.08	1.37	0.09
37	4.05	0.13	4.53	0.07	1.74	0.17
42	5.13	0.24	4.73	0.27	1.88	0.13
47	9.21	0.21	5.1	0.23	2.03	0.16

increasing the temperature will cause increased electric fields in the membrane, and consequently lower the voltage of pulses which is required to achieve electroporation. Reducing the temperature, on the other hand, will reduce the probability that electroporation occurs (Kandušer et al. 2008).

11.4 How Heating Affects Electroporation

11.4.1 Modeling Results

That the probability for electroporation increases with temperature is evident in the Molecular Dynamics (MD) simulations of a lipid layer with constant electric field strength of 0.5 V/nm at 25 °C and 47 °C. The high background electric field is typically used in MD simulations to probe poration (Song et al. 2011). It serves as an accelerated test of the pore formation process, since low electric fields would take inordinately long simulation time. The simulation results in Fig. 11.13 are in agreement with the experimental trends in that for an electric field of 0.5 V/nm no pore formation occurs at 25 °C. However, at 47 °C, there is a clear indication of pore formation at 0.75 ns. The MD results, by showing snapshots of membrane

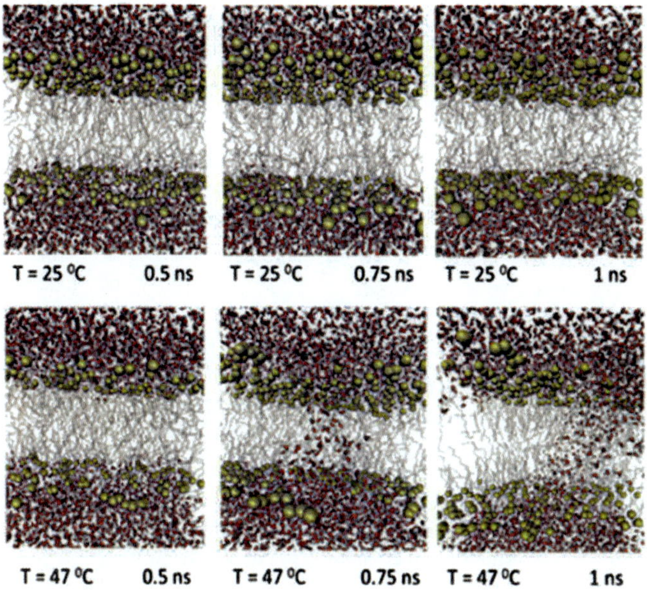

Fig. 11.13 Molecular dynamic simulations of a lipid bilayer exposed to a constant electric field strength of 0.5 V/nm for 1 ns at 25 °C and 47 °C (Camp et al. 2012)

poration at two temperatures, point to the possibility that relatively modest variations in temperature can result in different degrees of electrically stimulated bio-effects. The results underscore the role of temperature in facilitating and accelerating the poration process. A similar trend toward a synergism involving both thermal and electrically driven biophenomena was reported (Song et al. 2011).

11.4.2 Experimental Results—Joule Heating

Most of the experiments performed to explore temperature effects in bioelectric studies have relied on varying the temperature by means of external heating. Studies where the Joule heating caused by the applied electrical pulses was utilized are rather scarce. The reason is that researchers working on electroporation generally try to avoid heating in order to make sure that the observed effects are only due to electrical fields.

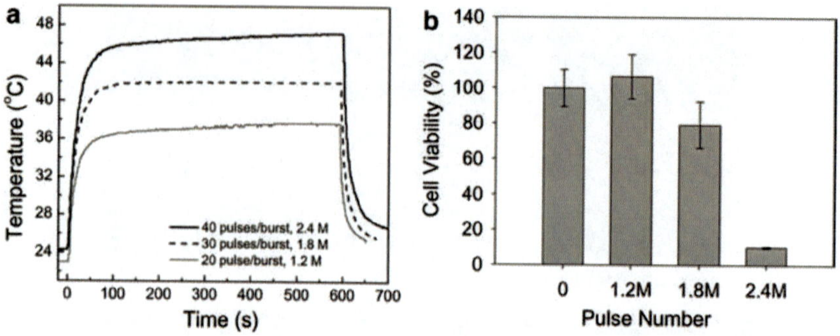

Fig. 11.14 **a** Left: Temporal development of the temperature in a suspension with increasing number of pulses. The pulse duration was 200 ps, and the pulse amplitude was 20 kV/cm. **b** Right: the viability, measured by the WST assay 18 h after exposure, at different pulse numbers, corresponding to different suspension temperatures as shown in Fig. a

One experiment where Joule heating was used as a means to increase the temperature in a cell suspension into the 40 °C range has been performed in the subnanosecond range (Xiao et al. 2011). With 200 ps long pulses, operated in a burst mode, at an amplitude of 20 kV/cm and a repetition rate of 10 kHz, it required 2.4×10^6 pulses, corresponding to a time of 10 min, to increase the temperature from room temperature to about 46 °C (Fig. 11.14a). With a pulse amplitude of just 20 kV/cm the pulses were unlikely to generate much of a drop in viability based on electrical effects alone. However, with increasing temperature a decrease in viability was observed already when the temperature reached values of about 42 °C. When the temperature reached levels above 45 °C, the cell viability dropped to 10% (Fig. 11.14b).

11.4.3 Experimental Results—External Heating or Cooling Based on Thermal Conduction

An experiment where the thermal effect on cell permeabilization was studied by cooling the suspension, rather than heating confirmed the hypothesis that the temperature induced changes in the cell membrane affect the degree of membrane permeabilization. In this experiment, cooling was obtained by placing the cuvette with the suspension on ice, and the temperature was reduced from 37 to 4 °C. As a result, the cell membrane permeability decreased by more than a factor of two as shown in Fig. 11.15 (Kandušer et al. 2008).

Moderate changes in temperature affect the viability of cells when exposed to electrical pulses but may also be used to improve the efficiency of gene electrotransfer. Studies on this topic using the human Keratinocyte cell line, HaCaT, were performed by placing a cuvette containing the suspension into an oil bath to achieve median temperatures of 40 °C, 43 °C, and 45 °C (Donate et al. 2015). In order to determine

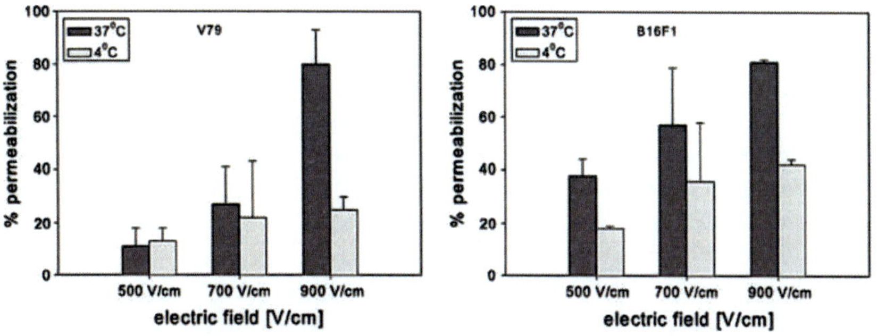

Fig. 11.15 Cell membrane permeabilization in cell line V79 and B16 F1 at 4 °C and 37 °C, measured by bleomycin uptake. A train of 8 pulses with a pulse duration of 100 μs at a repetition frequency of 1 Hz was applied (Kandušer et al. 2008)

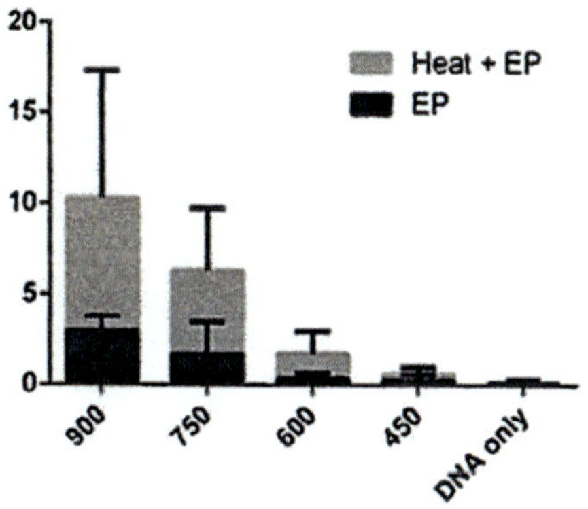

Fig. 11.16 Increase in GFP expressing events of the heat pretreatment over electroporation (Donate et al. 2015)

the efficiency of gene delivery, the Green Fluorescent Protein (GFP) expression was determined by means of flow cytometry. Increasing the temperature to 43 °C (Heat + EP) caused a strong increase in GFP expressing events (Fig. 11.16). The study showed that moderate increases in temperature allow to maximize gene expression or, when focusing on increased viability, the voltage can be reduced by 30%.

11.4.4 Experimental Results—External Heating Including Radiative Sources

The cell temperature in the two experimental studies (Kandušer et al., 2008; Donate et al. 2015) as listed in Sect. 11.4.3 was controlled by thermal conduction from a static cooling or heating source. More recently there are reports on several attempts using microwave or optical heating, to explore the effect of temperature on gene therapy and cell permeabilization. They address thermally assisted electroporation using millisecond pulses as well as nanosecond and subnanosecond pulses.

Studies on gene electrotransfer performed by heating the suspension (Donate et al. 2015) through heat conduction were expanded in in vivo experiments with guinea pigs (Donate et al. 2016) using an IR laser. In these experiments the laser was used to illuminate and heat the skin of the guinea pigs. Such a system, as shown in Fig. 11.7, allowed to illuminate the skin at any time before, during, or after the electroporation. The study demonstrated that increased temperature during electrotransfer maintained or even increased expression with a reduction in the applied electroporation voltage. The temperature in these studies were set at 43 °C for 10 s allowing adequate time for pulsing at a range of voltages (50–100 V) for luciferase experiments and for GFP experiments.

Whereas the thermally assisted gene electrotransfer studies were performed with pulses in the millisecond range, in more recent studies the bioelectric effects of microsecond pulses with thermal assistance was explored (Edelblute et al. 2017). However, instead on gene therapy, the focus was on irreversible electroporation (IRE), a nonthermal tumor ablation technology. The studies showed that that moderate heating significantly increased the efficiency of this tumor treatment—in this case for pancreatic carcinoma (Edelblute et al. 2017). Using either a heat block (in vitro studies), or a pulse delivery system (in vivo studies) where the temperature was controlled with a laser optic fiber placed at the center of a four-needle electrode (Fig. 11.7), the tumor ablation zone could be considerably enlarged by heating the target to 43 °C (Fig. 11.17). Comparing the results of in vivo IRE treatments without heating and that with heating the tissue to 43 °C (Moderate Heating IRE: MHIRE) showed that moderate heating significantly extended median survival and achieved a high rate of complete tumor regression. Nearly 60% of tumor bearing mice treated with MHIRE were tumor free at the completion of the study, whereas tumor regression was not observed in the IRE groups (Fig. 11.18).

The observed increase in efficiency of the tumor ablation was attributed to the measured increased conductivity of the tumor tissue when heated. This is an effect which was also observed for cells in a medium as shown in Table 11.5 (Camp et al. 2012). According to Ohm's law, at constant electric field and pulse duration, the reduced resistance of the tumor tissues allows the electrical energy deposition in the tumor tissue to increase, and correspondingly to cause a larger ablation.

Such ablation studies were extended into the nanosecond pulse regime (Edelblute et al. 2018). In in vitro studies it was shown that moderate heating at 43 °C for 2 min significantly enhanced nanosecond pulse-induced cell death of KLN205 murine squamous carcinoma cells as evidenced by propidium iodide uptake. Furthermore, the ablation zone in a 3-dimensional cell-culture pulsed at 43 °C was 3 times larger than in cells exposed to nanosecond pulse stimulation at room temperature.

Similar to the previous in vivo studies (Donate et al. 2016; Edelblute et al. 2017), pulse stimulation electrodes with an integrated optical fiber, as shown in Fig. 11.7, was used to explore the effect of thermally assisted tumor ablation, but this time with nanosecond pulses. The thermally assisted nanopulse treatment of murine ectopic

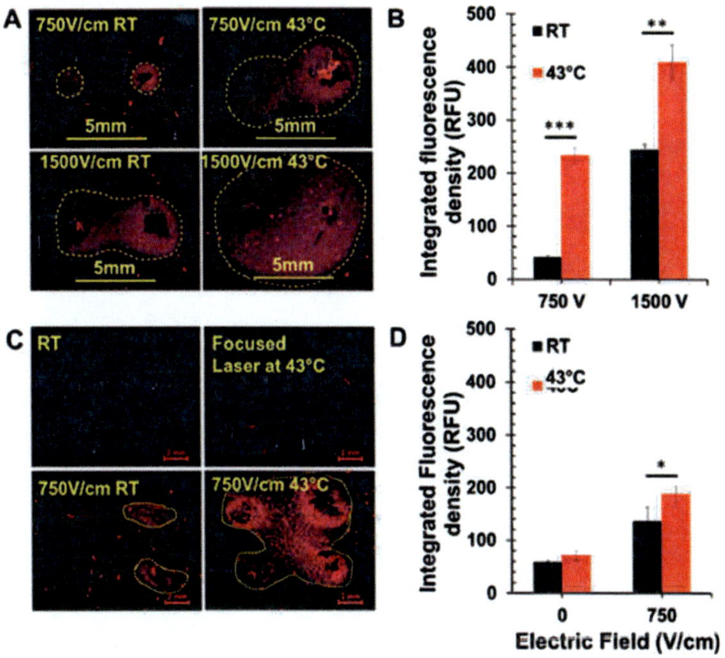

Fig. 11.17 Enlargement of ex vivo tumor ablation zone by IRE with moderate heating. A 3D agarose gel Pan02 tumor model was treated by IRE with a two-needle (**a**, **b**) or a four-needle (**c**, **d**) electrode. Area with red color was zone of dead cells indicated by Propidium Iodide (PI) staining. IRE parameters: pulse duration 100 μs, frequency 1 Hz, pulse number 80 and applied electric fields 750 V/cm or 1500 V/cm. RT: room temperature; 43 °C: samples preheated at 43 °C by heat block (**a**, **b**) or laser (**c**, **d**). 5 mm or 2 mm: Scale bar. Integrated fluorescence density, as a measure of cell death, was calculated with ImageJ software (Edelblute et al. 2017)

squamous cell carcinoma in mice extended significantly overall survival, delayed tumor growth, and achieved a high rate of complete tumor regression. With 600, 100 ns pulses at 5 Hz, at an electric field of 9.8 kV/cm, the median overall survival was 22 days without moderate heating but reached over 63 days for treatments with moderate heating. Nearly 69% (11 of 16) of tumor-bearing mice were tumor free at the completion of the study, whereas complete tumor regression was not observed in the control groups and the groups treated with pulses at 9.8 kV/cm but without moderate heating.

A more detailed experimental analysis of the effect of IR laser irradiation on tissue in a nanosecond treatment system as shown schematically in Fig. 11.7 was done on pig skin because of its similarities to human skin. The measurements of the optical transmission through the pig skin showed a strong increase in transmission with time (Fig. 11.19). This is an important observation, since the change in the transmission through the skin affects the temperature distribution in the subcutaneous tissue which is the main target of electrical pulses.

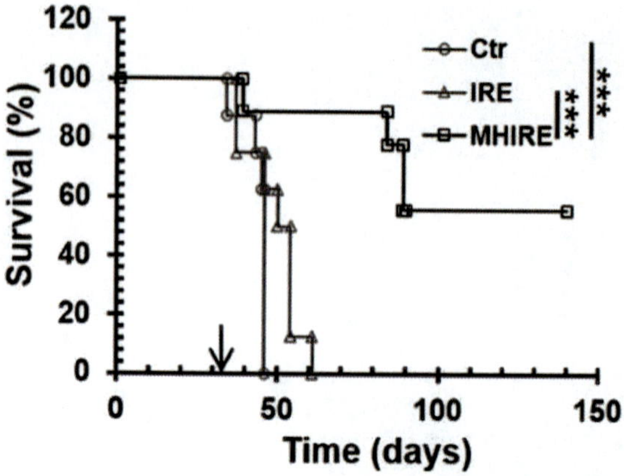

Fig. 11.18 Kaplan-Meier survival curves of mice treated with IRE or MHIRE. Pan02 pancreatic tumors with the size of 8–10 mm were treated with IRE or MHIRE at day 31 indicated by arrow. IRE parameters: pulse duration 100 μs, frequency 1 Hz, pulse number 90 and applied electric fields 2000–2500 V/cm. Ctr: no treatment (n = 8); IRE: treated with IRE (n = 9) (Edelblute et al. 2017)

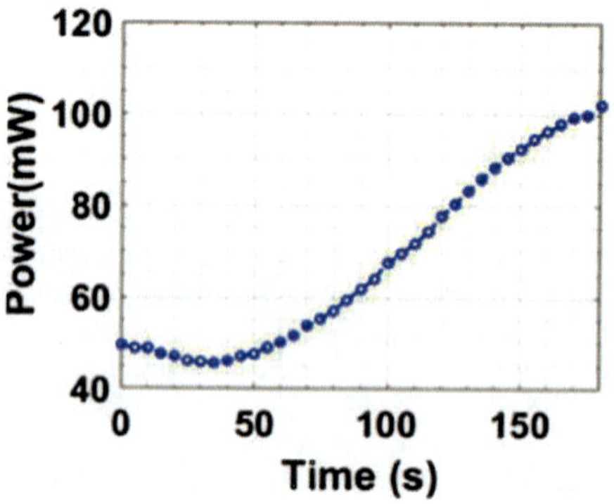

Fig. 11.19 Temporal dependence of the increase in the transmission of laser light at 980 nm at constant power of 2.3 W through a 3.9 mm thick pig skin tissue. It shows that the transmission doubles for a radiation time of 3 min (Hornef et al. 2020)

A treatment system, as shown in Fig. 11.20, which takes these changes into account was used for in vivo studies of the effects on lung squamous cell (KLN205) tumors in mice. The electrical pulse parameters were 100 ns, applied 600 times at a frequency of 5 Hz. The pulse electric field amplitude in the tumor was 14 kV/cm and 22 kV/cm. Treatments with pulses of 14 kV/cm amplitude without thermal assistance were below the threshold of tumor ablation. Also, IR laser irradiation which caused an increase in tumor temperate to a constant level of 43 °C did not show an effect on tumor ablation. However, nanosecond pulses with moderate heating at temperatures below the pain level was found to have a significant synergistic effect, resulting in overall tumor regression up to 50% (Fig. 11.21) similar to results obtained

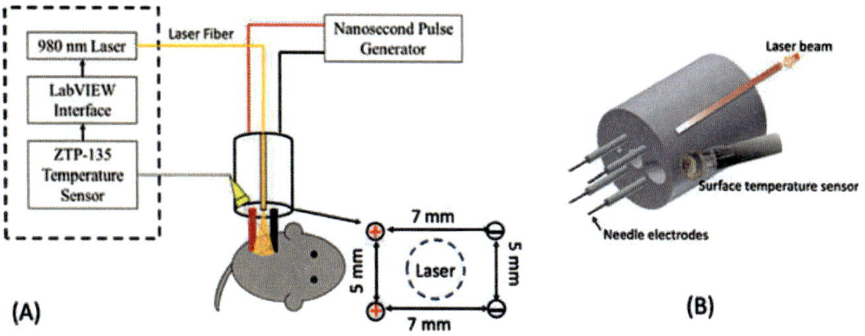

Fig. 11.20 **a** Treatment system for the in vivo study. The temperature reading on the surface of the skin by means of a thermopile, as shown in figure b, allows to adjust the laser power by means of the LabVIEW interface. **b** View of the needle electrode array with integrated laser fiber optic and surface temperature sensor (Hornef et al. 2020)

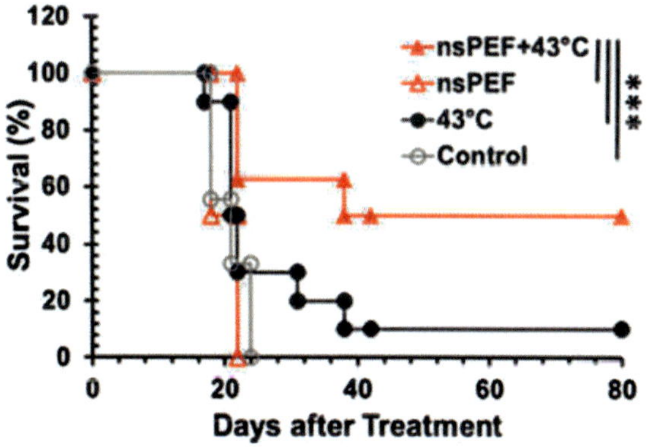

Fig. 11.21 Kaplan Survival Curves of in vivo treatment. Mice were treated on day 0 with 14 kV/cm nanosecond pulses alone, with laser heating to 43 °C for one minute alone, combined nanosecond treatment and laser heating or neither one (control) (Hornef et al. 2020)

with microsecond pulses on pancreatic carcinoma (Edelblute et al. 2017). Increasing the electric field at the same temperature, 43 °C, to 22 kV/cm, caused increased tumor shrinking, as shown in Table 11.6, but did not show a significant survival improvement.

Thermally assisted pulsed studies not just covered the microsecond and nanosecond pulse regime, but also extended into the subnanosecond pulse regime (Camp et al. 2012). Microwave heating with the suspension placed in a cuvette as shown in Fig. 11.22 provided a clear indication of the effect of increased temperature on electropermeabilization. A suspension with HEPA 1–6 cells was exposed to 2000 pulses at an electric field of 84 kV/cm at different suspension temperatures. When the temperature was increased beyond the physiological temperature to 42 °C and

Table 11.6 Comparison of treatment effectiveness by nanosecond pulses of 14 kV/cm and 22 kV/cm, respectively, with and without moderate heating

nsPEF Treatment Condition:	Tumor Size at Day	Statistics	Complete	Statistics
(100 ns, 5 Hz, 600 pulses)	11 (Mean ± SE)	(T-test)	Regression	(LogRank)
14 kV/cm	195.1 ± 7.1 (mm^3)	p = 0.137	0% (0/8)	$p < 0.001$
14 kV/cm with MH	129. ± 12.6 (mm^3)	p = 0.137	50% (4/8)	$p < 0.001$
22 kV/cm	121.2 ± 10.0 (mm^3)	p = 0.056	62.5% (5/8)	$p > 0.05$
22 kV/cm with MH	47.9 ± 7.4 (mm^3)	p = 0.056	75% (6/8)	$p > 0.05$

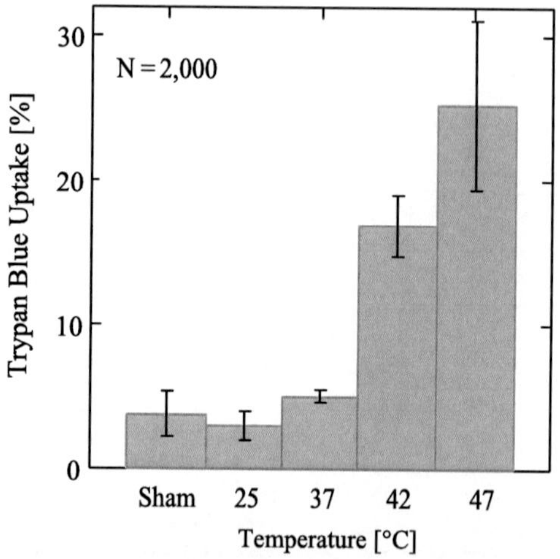

Fig. 11.22 Trypan blue uptakes in 4 h after exposure to 2000, 200 ps pulses at different temperatures. The measured electric field was kept constant at 84 kV/cm at room temperature. The data were based on three to five experiments each (n = 3–5) (Camp et al. 2012)

47 °C, Trypan Blue uptake, an indicator for cell permeabilization, increased by a factor of 3 at 42 °C and a factor of 5 at 47 °C (Fig. 11.22).

In a separate study where the temperature was controlled and no pulses were applied, keeping the cell temperature at a constant level of 42 °C did not cause an increase in permeabilization over the time of observation (100 min) (Fig. 11.23). Even at 47 °C, it took more than 10 min to thermally induce cell death, which was longer than the time it took to obtain the results shown in Fig. 11.22.

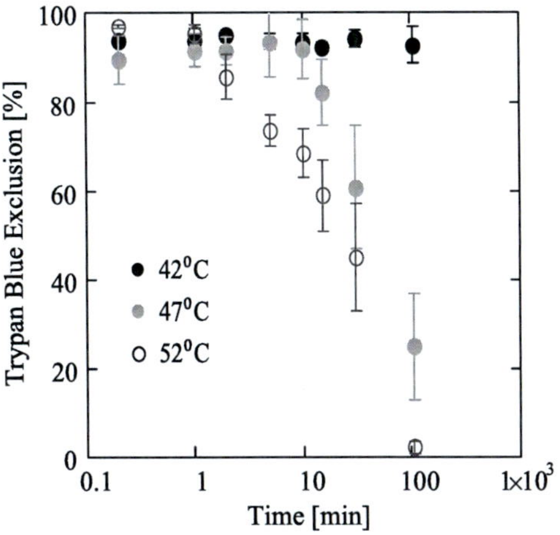

Fig. 11.23 Trypan blue exclusion versus exposure time of cells placed in a water bath of 42 °C, 47 °C, and 52 °C (Camp et al. 2012)

11.5 Summary

Short time heating of cells and/or tissue to moderate temperatures above the physiological temperature, but significantly below the thermal dose threshold for hyperthermia treatments, allows us to reduce the energy required for electroporation. This has been demonstrated with studies performed over a wide range of electrical pulse parameters, from the millisecond to the subnanosecond range, and has been shown to be a synergistic effect. Using thermal assistance in electrotherapies which rely on electroporation therefore allows us to scale down on pulse amplitudes, pulse durations, number of pulses, and consequently to reduce the stress on patients.

Whereas the outcome of studies on thermally assisted electroeffects does not depend on the method of heating, heating by nonionizing radiation allows a fast rise in temperature and consequently minimizes the time for which the tissue is exposed to the elevated temperature. Therefore, the probability for hyperthermia is decreased even further. Although noncoherent radiation in the millimeter or microwave range can be used to obtain fast rising temperatures, infrared radiation sources such as light emitting diodes (LEDs) and even lasers have the advantage that the light can be channeled through thin, dielectric optical fibers, which do not interfere with the electrical circuit.

Infrared radiation also allows the accurate local deposition of heat to support bioelectric effects in areas with lower electric field strength, which exists in the center area between the needles—see e.g., Fig. 11.7. While lasers and LEDs can be used for irradiation of skin, as described in the millisecond and microsecond studies, the fibers which guide optical energy to the target also can be integrated in catheters.

Minimally invasive procedures with catheter electrodes and laparoscopic procedures could be a next step. Local heating of electrically treated tumors would then be only a small addition to the electrical procedure performed at the treated tissue, but with likely benefits as shown in this study.

References

Buechner R, Hefter GT, May PM (1999) Dielectric relaxation of aqueous NaCl solutions. J Phys Chem A 103:1±9

Camp JT, Jing Y, Zhuang J, Kolb J, Beebe SJ, Song J, Joshi RP, Xiao S, Schoenbach KH (2012) Cell death induced by subnanosecond pulsed electric fields at elevated temperatures. IEEE Trans Plasma Sci 40:2334–2347

Croce RP, De Vita A, Pierro V, Pinto IM (2010) A thermal model for pulsed EM field exposure effects in cells at nonthermal levels. IEEE Trans Plasma Sci 38:149–155

Donate A, Burcus N, Schoenbach K, Heller R (2015) Application of increased temperature from an exogenous source to enhance gene electrotransfer. Bioelectrochemistry 103:120–123

Donate A, Bulysheva A, Edelblute C, Jung D, Malik MA, Guo S, Burcus N, Schoenbach KH, Heller R (2016) Thermal assisted *in vivo* gene electrotransfer. Current Gene Therapy 18:83–89

Edelblute CM, Hornef J, Burcus NI, Norman T, Beebe SJ, Schoenbach K, Heller R, Jiang C, Guo S (2017) Controllable moderate heating enhances the therapeutic efficacy of irreversible electroporation for pancreatic cancer. Sci Rep 7, Article number: 11767

Edelblute CM, Guo S, Hornef J, Yang E, Jiang C, Schoenbach K, Heller R (2018) Moderate heat application enhances the efficacy of nanosecond pulse stimulation for the treatment of squamous cell carcinoma. Technol Cancer Res Treat 17:1533033818802305

Garcia PA, Rossmeisl JH Jr, Neil RE II, Ellis TL, Davalos RV (2011) A parametric study deliniating irreversible electroporation from thermal damage based on a minimally invasive intracranial procedure. BioMed Eng OnLine 10:34

Hildebrandt B, Wust P, Ahlers O, Dieing A, Sreenivasa G, Kerner T, Felix R, Riess H (2002) The cellular and molecular basis of hyperthermia. Crit Rev Oncol/hematol 43:33–56

Hornef J, Edelblute CM, Schoenbach KH, Heller R, Guo S, Jiang C (2020) Thermal analysis of infrared irradiation-assisted nanosecond pulsed tumor ablation, nature research. Sci Rep 10:5122

Kandušer M, Šentjurc M, Miklavčič D (2008) The temperature effect during pulse application on cell membrane fluidity and permeabilization. Bioelectrochemistry 74:52–57

Kloesgen B, Reichle C, Kohlsmann S, Kramer KD (1996) Dielectric spectroscopy as a sensor of membrane headgroup mobility and hydration. Biophys J 71:3251–3260

Kotnik T, Miklavcic D (2000) Theoretical evaluation of the distributed power dissipation in biological cells exposed to electric fields. Bioelectromagnetics 21(5):385–394

Leonenko ZV, Finot E, Ma H, Dahms TES, Cramb DT (2004) Investigation of temperature-induced phase transitions in DOPC and DPPC phospholipid bilayers using temperature-controlled scanning force microscopy. Biophys J 86:3783–3793

Mabrey S, Sturtevant JM (1976) Investigation of phase transitions of lipids and lipid mixtures by sensitivity differential scanning calorimetry. Proc Natl Acad Sci U S A 73:3862–3866

Marrink SJ, Risselada J, Mark AE (2005) Simulation of gel phase formation and melting in lipid bilayers using a coarse grained model. Chem Phys Lipid 135:223–244

Pliquett U, Nuccitelli R (2014) Measurement and simulation of Joule heating during treatment of B-16 melanoma tumors in mice with nanosecond pulsed electric fields. Bioelectrochemistry 100:62–68

Salimi E, Thomson DJ, Bridges GE (2013) Membrane dielectric dispersion in nanosecond pulsed electroporaration of biological cells. IEEE Trans Diel Electr Insul 20:1256–1265

Schoenbach KH (2010) Bioelectric effects of intense nanosecond pulses. In: Pakhomov AG, Miklavcic D, Markov MS (eds) Advanced electroporation techniques in biology and medicine. CRC Press, Taylor and Francis Group, Bota Racon, FL, pp 19–49

Sherar MD, Moriarty JA, Kolios MC, Chen JC, Peters RD, Ang LC, Hinks RS, Henkelman RM, Bronskill MJ, Kucharcyk W (2000) Comparison of thermal damage calculated using magneticresonance thermometry, with magnetic resonance imagingpost-treatment and histology, after interstitial microwavethermal therapy of rabbit brain. Phys Med Biol 45:3563–3576

Simanovskii D, Sarkar M, Irani A, O'Connell-Rodwell C, Contag C, Schwettman A, Palanker D (2005) Cellular tolerance to pulsed heating. In: SPIE Proceedings, Laser-Tissue Interactions XVI, vol 5695, BIOS 2005

Song J, Joshi RP, Schoenbach KH (2011) Synergistic effects of localtemperature enhancements on cellular responses in the context of high-intensity, ultrashort electric pulses. Med Biol Eng Comput 49:713–718

Wust P, Hildebrandt B, Sreenivasa G, Rau B, Gellermann J, Riess H, Felix R, Sclag PM (2002) Hyperthermia in combined treatment of cancer. Lancet Oncol 3(8):487–497

Xiao S, Guo S, Vasyl N, Heller R, Schoenbach KH (2011) Subnanosecond electrical pulses cause membrane permeabilization and cell death. IEEE Trans Biomed Eng 58:1239–1245

Chapter 12
Synergy Between Electric Pulse and Thermal Effects

Ravi Joshi

Abstract Generally, in studies of the bioelectric effects of nanosecond pulses, thermal effects are not considered. While this is certainly true for the delivery of single or a small number of pulses, developments in medical treatment based on tissue ablation have been based pulse trains applied at a high repetition rate. In fact, temperature increases may trigger thermally activated bioeffects. This suggests technological opportunities for electro-manipulation that takes advantage of synergisms between thermal and electrically driven processes. Even with modest temperature changes, large thermal gradients could be established, which in itself can lead to additive electric field creation. The focus in this chapter is on thermal aspects and the possible synergies with electrical stimulation for bio-effects.

12.1 Introduction

Most of the literature on the application of high-intensity ultrashort cellular pulses (e.g., Joshi et al. 2005, Song et al. 2011; Pierro et al. 2014; Croce et al. 2010; Camp et al. 2012), has ignored possible thermal effects because of the small (nanosecond) duration of the applied voltage pulses. One potential effect of local temperature enhancements is the increased fluidity of fatty acid tails within the phospholipid bilayer which would facilitate membrane poration (Song et al. 2011). Here we argue that issues associated with local heating should not be entirely dismissed. The importance of locally driven phenomena (such as sectional heating) cannot be overstated, even if globally averaged changes over the entire cell may remain negligible. As an example, rise of temperature induces localized calcium release (termed blips from single channels or puffs from a small collection of channels) can lead to global cellular signaling through a regenerative feedback mechanism (Swillens et al. 1999). More importantly, even if the temperature changes remained small and well controlled (to avoid non-local collateral damage), it is entirely conceivable that *substantial local temperature gradients could be created.*

This issue of differential power dissipation in biological cells, which leads to temperature gradients, was first treated by the Slovenian group (Kotnik and Miklavčič 2000). It was shown that power dissipation within the membrane, becomes much

S. J. Beebe et al., *Ultrashort Electric Pulse Effects in Biology and Medicine*, Series in BioEngineering, https://doi.org/10.1007/978-981-10-5113-5_12

more pronounced in the MHz and lower GHz regions. In theory, a temperature differential across the outer cell membrane could be maintained by having an electromagnetic pulse-train of relatively short duration of ~ 5 ns (Croce et al. 2010). They demonstrated that temperature differentials across the outer cell membrane in the 5–10 K range were possible. For example, it was shown based on an equivalent lumped-circuit continuum analysis, that ultrashort high-intensity pulses could produce fast localized heating at the cell membrane, with the cytoplasm temperature being essentially unaffected. Physically, this consequence arises from large differences between the conductivity of cell membranes and the cytosol. Peaks of membrane temperature between 1 °C and 5 °C for single 10 ns and 1 ns pulses, respectively, were predicted (Croce et al. 2010). For a typical 5 nm cell membrane, temperature gradients ranging from 0.2×10^9 to 10^9 K/m were reported (Croce et al. 2010). Such possibilities of setting up thermal gradients also arise in the context of pulsed laser irradiation of cells (Garner et al. 2016), and radio-frequency or alternating current electric excitations (Garner et al. 2013). Other specific ways to create temperature gradients include the use of nanoparticles in the vicinity of tumor cells, and subjecting the system to electromagnetic radiation (Chen and Zhang 2006; Richardson et al. 2006), with thermal gradients as large as 10^8 K/m being reported a few years ago (Govorov et al. 2006). With laser excitation, temperature gradients of the order of 10^6 K m^{-1} can routinely be obtained in experiments (Jiang et al. 2009).

The thermoelectric effect is a potentially relevant phenomenon associated with thermal gradients, and involves the development of an electric field. Molecular dynamic (MD) simulations in water have shown that a thermal gradient of 10^{10} K/m can result in the creation of an internal electric field of about 10^8 V/m (Bresme et al. 2008). It was demonstrated that a thermal gradient polarizes water in the direction of the gradient, leading to a non-negligible electrostatic field whose origin lies in the water reorientation under nonequilibrium conditions. More specifically, the hydrogen atoms (or more generally the smaller of the two species in a polar molecule) begins pointing preferentially towards the cold region. Based on simple continuum mechanics, the electrostatic field (E) and the driving temperature gradient (dT/dr) are related as: $E \sim (1 - 1/\varepsilon_r)\,(dT/dr)/T$, where ε_r is the dielectric constant. Hence, the electric field induced by the temperature gradient should be larger for high dielectric constant materials such as water. It is therefore very plausible that the creation of such fields driven by temperature gradients could enhance electroporation in biological cells. This aspect, however, has not been probed to the best of our knowledge. A more subtle point concerns the lifetime of the temperature gradient, and this can be understood from the following argument. When a thermal gradient is applied, the homogeneous system evolves to reduce the entropy production, which is made possible by creating an internal polarization. The heat transfer, which is dominated by intermolecular interactions and molecular polarization, is then reduced (Muscatello et al. 2011). Thus, the polarized system as a whole becomes a less effective heat conductor, implying that any local temperature increase and spatial gradient created would likely be sustained for a longer duration.

In this chapter, we focus on both the thermal heating in the presence of an external electric field to gauge synergistic effects, and also discuss thermal gradients as drivers for electroporative enhancements, even though the actual temperature values might not have changed appreciably from their equilibrium levels. Thus, aspects such as phase changes or protein denaturization are ignored assuming that the absolute temperature changes due to such ultrashort pulses are minimal, around 1–5 °C at most (Pakhomov et al. 2003). Such a study is particularly relevant to high-intensity, nanosecond pulses which are inherently nonthermal in nature, but could establish large thermal gradients.

12.2 Simulation Results with Heating and Electric Pulsing

Analysis of the cellular response to electrical pulses requires an evaluation of the time- and spatial-dependent voltage and current distributions within the cell. This voltage is the source for pore opening and other cellular bio-responses. A possible method to modeling is to represent the electrical characteristics of the cells by a distributed equivalent circuit. Dynamic changes in cell parameters need to be included for self-consistency, since application of the transmembrane voltage can lead to membrane conductance increases over time associated with electroporation.

Details of the numerical distributed circuit approach for computing the spatio-temporal voltages and current flows in cells have already been as discussed in detail earlier in another chapter. Essentially, the entire cell is broken up into discrete segments of resistor–capacitor (RC) circuit elements. The computational region here was taken to be a sphere that included the cell cytoplasm, its plasma membrane, and the surrounding cell-suspension medium. Application of the Kirchhoff current law at each discretized node, then yielded a set of N coupled linear equations in the N node voltages.

Local temperature increases due to dissipation of the electrical energy would be governed by the following thermal diffusion equation: $\delta T/\delta t = (k_T \nabla\ T + E.J)/(\rho_M c_M)$, where T is the temperature, k_T the thermal conductivity of the bio-segment, E and J denote the local electric field and current density, respectively, through the membrane. Also, ρ_M is the density, and c_M the membrane specific heat. Over time scales much shorter than those of heat diffusion across the cell (e.g., ~ 200 ns), membrane temperatures will essentially evolve adiabatically. The early times are an important temporal regime since they correspond to intervals during or immediately following the application of electrical pulses. Events triggered by the external fields are being initiated during this time, and so the presence of locally elevated temperatures would work to influence the overall outcomes and kinetic rates, thereby providing synergy to the field-driven processes.

The distributed circuit model was applied to a spherical cell of radius 10 μm consisting of cytoplasm ($\varepsilon = 60\ e_0$, $\sigma = 0.13$ S/m), outer medium ($\varepsilon = 80\ e_0$, $\sigma = 0.6$ S/m), and a 5 nm thick plasma membrane ($\varepsilon = 8\ e_0$, $\sigma = 5.3 \times 10^{-6}$ S/m). The

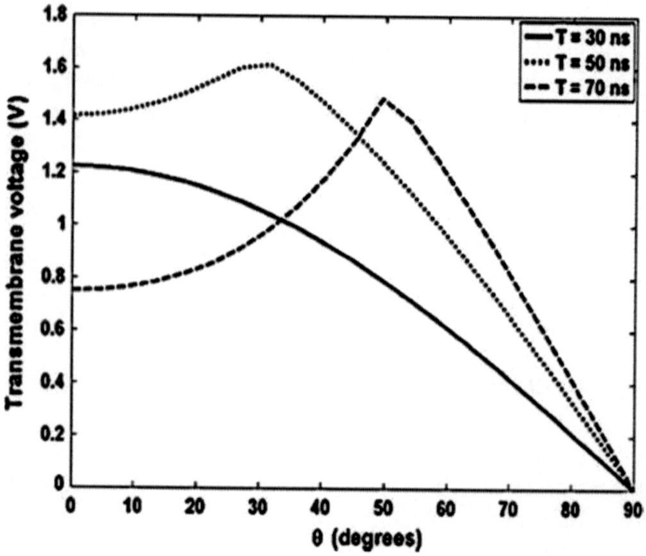

Fig. 12.1 Simulation results for the transmembrane potential versus angular position due to a trapezoidal pulse having a 6 ns rise- and fall-times, and a 60 ns ON time. Values at the three different time instants are shown (after Song et al. 2011)

transmembrane potential as a function of angular location at different time instants is shown in Fig. 12.1.

This result of Fig. 12.1 was in response to a rectangular 60 kV/cm pulse with a 6 ns rise time, a 60 ns ON time and a 6 ns fall time. At the earliest 30 ns snapshot, the potential across the membrane at the poles (i.e., $\theta = 0°$) is predicted to be above 1.2 V. However, as poration at the polar caps initiates, the localized conductivity increases. This leads to a slight lowering of the transmembrane voltage at the polar region as seen in the 50 ns curve. Finally, the 70 ns curve reveals an overall expansion of the porated region around the polar cap.

Changes in local conductivity due to membrane poration were computed, and the values shown in Fig. 12.2. Large increases by orders of magnitude are predicted over time starting at the $\theta = 0°$ location, with the highest conductivity of about 0.007 S/m occurring at the "poles" of the spherical cell. Based on the time-dependent conductivity and transmembrane voltage, the power dissipation density could then be computed.

These power density values are shown in Fig. 12.3 as a function of time for five different angular locations. The initial power dissipation is negligible since the membrane acts as a virtual open circuit with the displacement current being the dominant mechanism. At later times (~ 40 ns and beyond), the conduction current develops as pores begin to form and increase both in density and radial size. Due to the nonlinear increases in conductance and pore densities over time, the power density attains a peak, and then begins to reduce. The decrease can be associated

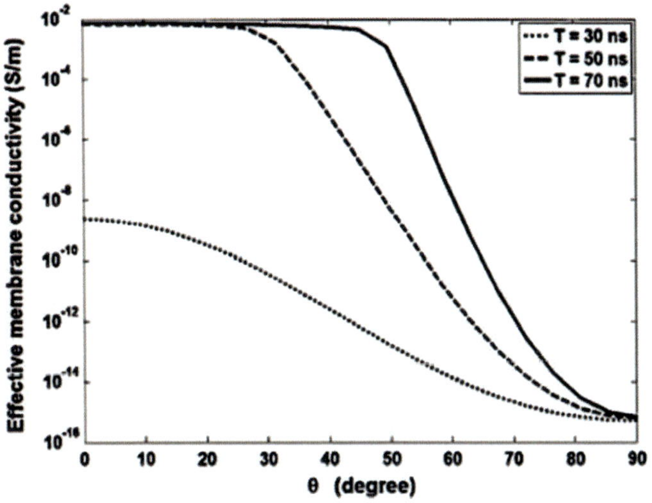

Fig. 12.2 Calculated membrane conductivity versus angular position due to the electrical pulse of Fig. 12.1. Plots at the three different time instants of 30, 50, and 70 ns are shown (after Song et al. 2011)

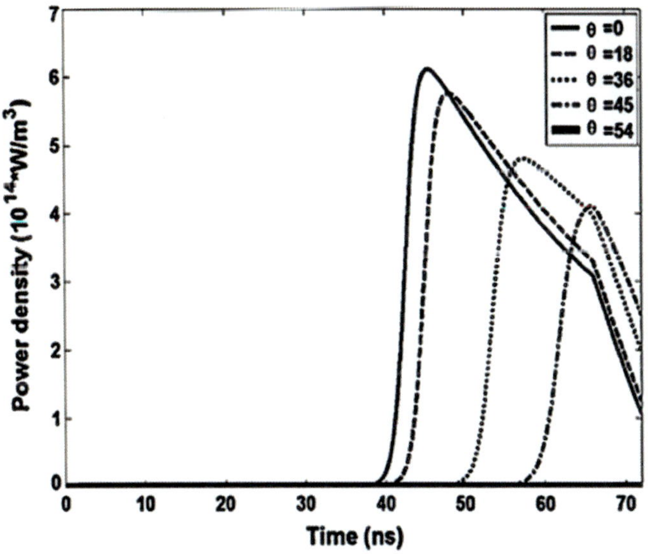

Fig. 12.3 Calculated dissipated power density at the membrane as a function of time. Five different angular locations are shown (after Song et al. 2011)

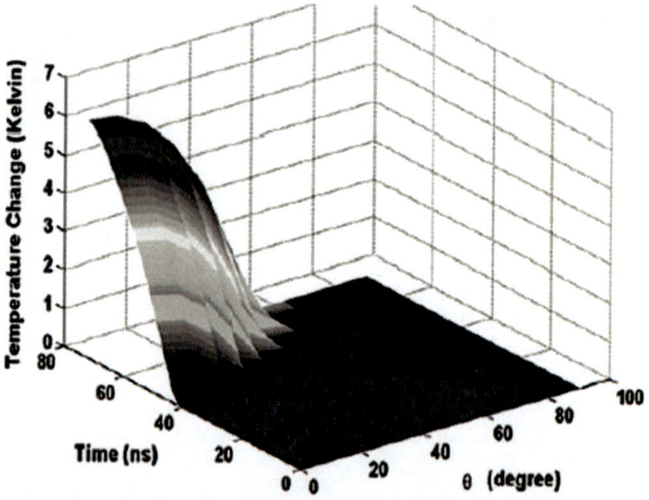

Fig. 12.4 Results for the membrane temperature change as a function of time at different angular locations (after Song et al. 2011)

with a collapse of the local transmembrane potential as the pores effectively produce an electrical "short".

Based on the power density, changes in temperature as a function of time and angular location were obtained next. Values for the density ρ_M and membrane specific heat c_M were chosen to be 10^3 kg/m^3 and 2×10^3 J/kg/K, respectively, in keeping with reports (Croce et al. 2010). Figure 12.4 shows the temperature profile. Interestingly, the largest temperature increase, which occurs at the poles (i.e., $\theta = 0°$), is predicted to have a value of about 6 K. However, a rapid fall-off is seen in Fig. 12.4 away from the poles, with near negligible temperature changes predicted for much of the cell. So, while a near-thermalized situation seems to prevail in an averaged sense, localized heating could be a factor at specific sites. For example, local cellular temperature increases at the membrane could work to energetically enhance procaspase-8 cleavage, and hence promote cellular apoptosis.

As a simple step towards showing that the localized hot spots produced by exposure of cells to ultrashort pulses could then work to accelerate bio-physical processes, we use poration as a demonstrative example. Poration is a well-known cellular effect initiated at the plasma membrane (Neumann et al 1989; Weaver and Chizmadzhev 1996). Heat-induced events at membranes include unfolding of spectrin, redistribution of proteins, and denaturation (Davio and Low 1982; Ivanov 1999). Though these latter processes are relatively slow, application of repetitive pulses could be used as a technique to bring about such temperature-driven changes as well.

Many of the bioprocesses are difficult to model and quantify due to lack of data and parameter values. However, membrane poration can be simulated with a high degree of accuracy based on the MD technique. Hence here, the possible role of localized temperature elevations on electroporation was probed as a simple, yet pertinent, example. Results of MD simulations are shown in Fig. 12.5a, b for a patch of DPPC membrane. The primary objective was to see if temperature increases would foster electro-poration. MD simulations were carried out with a 0.5 V/nm background electric field—one at a temperature of 313 K and the other at 316 K. The trends were consistent, and as seen in the 1 ns snapshot of Fig. 12.5a (which represents a typical simulation outcome), no pore is predicted to form for the 313 K case. However, the simulation at 316 K shows pore formation at the same time instant. Water nanowires connecting the extra- and intra-cellular regions through the membrane are also apparent. The result underscores the role of temperature in facilitating and accelerating the poration process. These results, by showing snapshots of membrane poration at two specific temperatures, point to the possibility that relatively modest variations in temperature generated by exposing cells to electrical pulses, can result in different degrees of bio-effects.

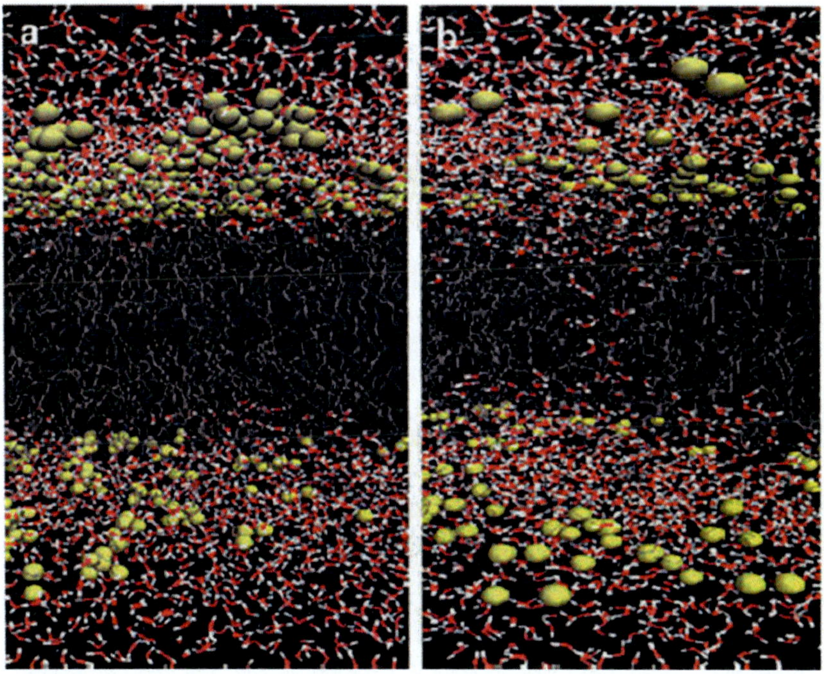

Fig. 12.5 Molecular Dynamics snapshots of the electroporation process in a dipalmitoyl-phosphatidyl-choline (DPPC) membrane under the E field of 0.5 V/nm. Snapshots at 1 ns for two different temperatures of: a 313 K, and b at a temperature of 316 K. A nano-wire forms at the higher temperature (after Song et al. 2011)

Data pertaining to temperature effects (non-electrical in origin) on the mitochondrial membrane are already available (Nijhuis et al. 2006). Experiments on HL60 cells seem to indicate a thermal induction of apoptosis launched by mitochondrial membrane processes involving the Bcl-2 family (Narita et al. 1998). Hence, it is plausible that electrical pulse-induced localized mitochondrial membrane heating could add to several biological outcomes, and so further studies seem warranted. In this regard, experiments on liver cancer cells (Camp et al. 2012) showed enhanced rate of trypan blue uptake (which is indicative of cell death) upon exposure to subnanosecond high electric field pulses, when the temperature was raised above 37 °C. The exposure of Hepa 1–6 cells to 2000 pulses of 200 picosecond duration and electric field amplitudes exceeding 80 kV/cm induced cell death in almost 30% of the cells when the temperature was increased to 47 °C for the time of the pulsing. On the other hand, for temperatures at 37 °C and below, the same exposure to pulsed electric fields did not show any measurable effect. Even for the maximum elevated temperature of 47 °C, thermal effects were not found to cause fatalities for the time of exposure, which was, for 2000 pulses at a repetition rate of 7–9 pulses per second, on the order of 5 min. The use of elevated temperature together with the effect of multiple pulses, was the likely cause for the observed high death rate of the Hepa cells. This hypothesis is in keeping with the results of molecular dynamics simulations presented earlier in this section.

An even more recent report involves the irradiation assisted nanosecond tumor ablation of squamous cancers (Hornef et al. 2020). This recent study showed that nanosecond pulsed electric field cancer therapy may be improved further with the assistance of moderate heating of the target. Heating of the tumor cells was achieved in an integrated manner through a feedback-looped heating system that used a 980-nm fiber optic laser. Synergistic effects between the nanosecond pulsed electric field and the moderately elevated temperature at the target were observed, resulting in enhanced overall survival tumor regression up to 50% in the treatment of lung squamous cell cancer in mice.

12.3 Role of Thermal Gradients

The dynamics of pore formation and water entry into a lipid bilayer membrane with thermal gradients, in response to an externally applied static electric field, was studied (Song et al 2017). Simulations were based on the molecular-dynamics (MD) scheme using the GROMACS package (Berendsen et al. 1995) with a 2 fs time step. Different temperatures were assigned to various regions to simulate the appropriate temperature gradients. The lipid membrane was taken to comprise of dipalmitoyl-phosphatidyl-choline (DPPC) molecules. A constant electric field was used for each simulation, while the top and bottom surfaces of the membrane were assigned two separate fixed values. The water-membrane system contained 6323 water molecules and 128 DPPC lipid molecules in a 6.9 nm × 7.4 nm × 7.0 nm simulation box. This

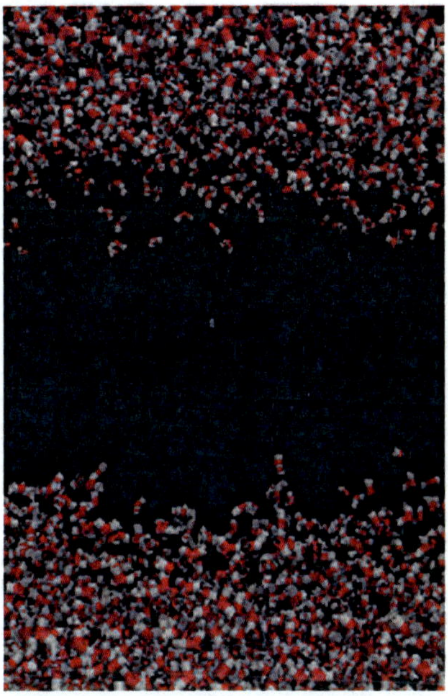

Fig. 12.6 A 7 ns snapshot of DPPC membrane section subject to a constant 0.4 V/nm external electric field. A uniform temperature of 295 K was assumed (after Song et al. 2017)

membrane system was charge-neutral and represents a homogeneous section of a simple lipidic membrane system.

A 7 ns snapshot for a DPPC membrane set at a uniform temperature of 295 K is shown in Fig. 12.6. The central portion represents the membrane, while the water molecules are seen at the top and bottom on the *trans* and *cis* faces. This fixed temperature case serves as a benchmark test case. A high electric field of 0.4 V/nm was applied for these simulations. This high field provides an accelerated test of the bio-physical process of pore formation, since low electric fields would otherwise take inordinately long simulation times. As seen in the 7 ns snapshot of Fig. 12.6 (which represents a typical outcome from a total of six simulation runs starting with different initial velocities), no pore is predicted to form in the membrane. A very slight protrusion at the bottom of the membrane near the middle represents a small and insignificant excursion of the water molecules, but with no beginnings of real pore formation.

To check whether any differences in the outcome may arise due to a temperature gradient, the MD simulations were performed once again on the same geometry, but with the top and bottom membrane surfaces kept at temperatures of and 300 K and 295 K, respectively. A snapshot from the MD simulations at 7 ns showing the molecular structure of the DPPC and water system is given in Fig. 12.7. A nanopore is clearly seen and a water wire is seen to connect the water reservoirs at the top and bottom. The entry of water molecules is slightly stronger at the top, which was the

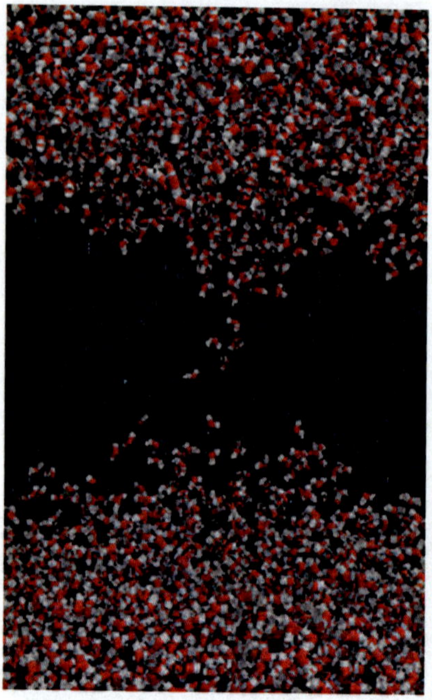

Fig. 12.7 A 7 ns snapshot of DPPC membrane section and the adjoining water molecules subject to a constant 0.4 V/nm external electric field. The membrane temperatures at the top and bottom were maintained at 300 K and 295 K, respectively (after Song et al. 2017)

positive (anode) side and suggests a polarity dependence. This aspect was previously discussed by our group (Hu et al. 2013). It is associated with the molecular stacking of the water network fashioned by the strong external electric field. Basically, in the case of freely rotating molecules, there is competition between energy reduction and the loss of orientational entropy upon alignment which is described by the well-known Langevin equation. For liquid water, orientations of water molecules are also subject to angle restrictions associated with hydrogen bonding and a strong tendency to minimize the loss of hydrogen bonds at the interface (Du et al. 1993; Luzar et al. 1985). The hydrogen bonding between water molecules favors near-parallel dipole orientations relative to the membrane surface upon electric field application. Hence, in order to optimize hydrogen bonding, angular distributions of water molecules relative to the pore walls would be biased against orientations in which the hydrogen atoms point toward the circular walls. This then means that the water would likely enter from the outside at the anode end and move more easily from the inner membrane at the opposite cathode side. The prediction of a faster observable poration from the anode side has indeed been reported experimentally (Vernier et al. 2004).

The difference in pore formation with temperature is again underscored by the results presented in Fig. 12.8a. These MD simulations were carried out at a slightly higher field of 0.45 V/nm and the two snapshots shown were obtained at the longer

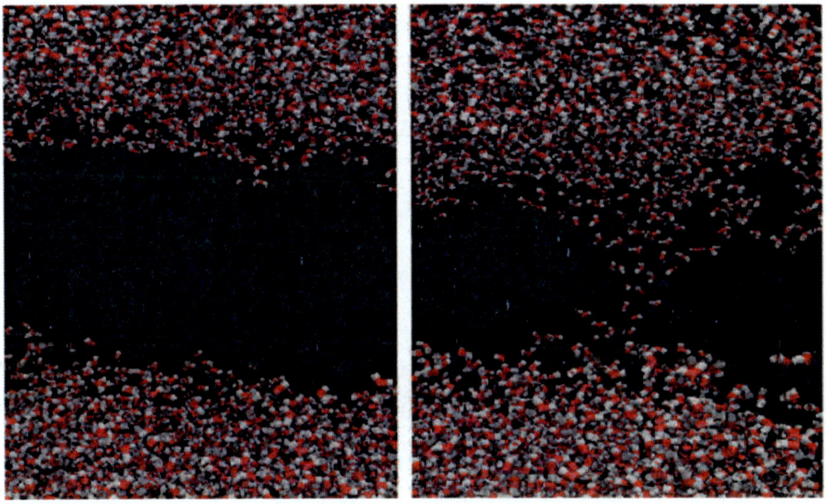

Fig. 12.8 A 10 ns snapshot of a DPPC membrane and the adjoining water molecules at a constant 0.45 V/nm external electric field. **a** Constant temperature of 295 K, and **b** different membrane temperatures at the top and bottom of 300 K and 295 K, respectively (after Song et al. 2017)

10 ns instant. Results shown in Fig. 12.8a were obtained at a constant temperature of 295 K; while for the case shown in Fig. 12.8b, the membrane temperatures at the top and bottom were chosen to be at 300 K and 295 K, respectively. Despite the longer simulation time of 10 ns (as compared to 7 ns for Fig. 12.6) and the higher electric field, no pore formation is evident in Fig. 12.8a. However, in Fig. 12.8b, a nanopore is predicted. The initiation is once again from the upper anode end. Given the 5 nm membrane thickness, this result predicts that the 2×10^9 K/m temperature gradient would synergize electroporation.

The entire mechanism of electric field creation due to internal thermal gradients in water perhaps requires a more detailed explanation and analysis, beyond just numerical MD simulations. This process is best understood in terms of system energy minimization. Assuming near local equilibrium, the solvation energy "G_{sol}" (i.e., the single particle free enthalpy) of a single sphere of diameter "d" is given by: $G_{sol} = -k_B T s d^2$, where k_B is the Boltzmann constant, T the absolute temperature, and "s" a positive constant (Eastman 1926, 1928). Incidentally, the solvation energy G_{sol} is related to the Soret coefficient S_T as: $S_T = [1/(k_B T)]\ \delta G_{sol}/\delta T$, which then leads to: $S_T \sim -s\, d^2$. This Soret effect (also referred to as thermo-migration, or the Ludwig-Soret effect), describes the mass flow induced by a temperature gradient in a fluid (Wiegand 2004), and leads to the following equilibrium equation: $\nabla c + cST\, \nabla T = 0$, with "$c$" being the species concentration.

For the polar water molecules, one can crudely represent this as a dumbbell system having a dipole with two hard spheres of diameters "d_1" and "d_2". Considering such water molecules in a system having a spatial temperature gradient, the overall solvation energy for the molecular system would then be given as: $G_{sol} = -k_B s[T_1 {d_1}^2$

$+ T_2 d_2^2]$, where $T_{1,2}$ are the local temperatures at the two sphere 1,2. Physically, the length-scale of temperature variation is much larger than the length of the molecular dumbbell. Mathematically, these temperatures can be expressed in terms of the mean temperature at the midpoint as: $T_{1,2} = T \pm (L/2)\{(T_1 - T_2)/L\} \equiv T \pm \Delta$, where $L \sim d_1 + d_2$ is the length of the dumbbell, and $T \sim (T_1 + T_2)/2$ is the average temperature in the vicinity of the water molecule. Denoting d_1 to be the diameter of the larger oxygen molecule, and the ratio of the diameters $d_1/d_2 = R$ (i.e., $R > 1$), one gets: $G_{sol} = - k_{B}s\ d_2^2\ [T\ (R^2 + 1) + \Delta\ (R^2 - 1)]$. Clearly for minimizing the energy G_{sol}, one requires $\Delta > 0$. Hence, the temperature T_1 of the molecular system on the side containing the larger oxygen atom would be larger. More generally the smaller of the two species system in the polar molecule would begin pointing preferentially towards the colder region, in the presence of temperature gradients. It may also be pointed out that the effect would be stronger for polar molecules having a larger dipole separation distance L.

In addition to the separation L, the temperature can also be expected to play a role. More generally, the dipoles could be oriented at an angle θ with respect to the applied field direction. In this case then, the average term $\langle\cos(\theta)\rangle = \int_0^{\pi} \{\cos(\theta)\ \sin(\theta)\ \exp[-G_{sol}/(k_B T)]\ d\theta\}/\{\int_0^{\pi} \sin(\theta) \quad \exp[-G_{sol}/(k_B T)]d\theta\}$ would have to be considered. In such a case, lower temperatures would provide a greater likelihood of orientation with the electric field. Also, at higher electric field values, the dipoles could be better aligned and even form a possible lattice structure to enhance the Soret effect. Furthermore, if the outer membrane of a biological cell were subject to electric field pulsing (with or without any additional optical or RF excitation), the outer side would be the hottest. For the cell region facing the anode (e.g., the anodic polar cap), the electronegative oxygen atom of the polar water system would consequently be near the outer membrane, while the hydrogen would be a bit further away. This local polarized field would enhance the total field near the anode pole, and drive cellular electroporation more strongly. This is thus the essence of the additive synergy between a thermal gradient and an applied electric field.

The thermally-induced polarization and development of an associated local electric field can therefore, be seen to be a secondary and synergistic effect, while the primary driver for electroporation is the applied voltage. In fact, it is the electric fields in this case that would cause cellular heating, which would then lead to a differential thermal gradient. The thermal gradient would aid and assist in the electroporation process. In principle, poration at lower electric fields could occur, provided a significantly higher temperature difference is present to drive the process and achieve a comparable result. However practically, there would be two issues with this route: (a) Elevating and then maintaining high temperature gradients would be difficult, and (b) phase transitions could easily occur at the higher temperatures or thermally-activated detrimental cellular effects might be initiated. As a simple example though, simulations at a lower 0.35 V/nm field were carried out with a 15 K thermal gradient, with the outer membrane kept at 310 K. A small "water-wire", can be seen forming in Fig. 12.9 which is a 6.5 ns snapshot. Two different aspects are made obvious from

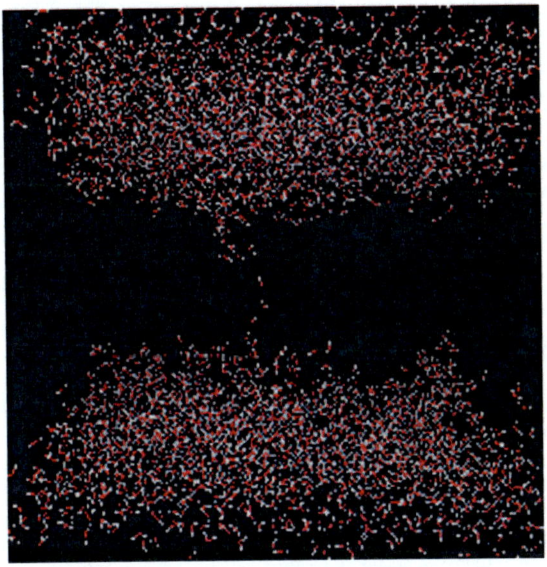

Fig. 12.9 A 6.5 ns snapshot of a DPPC membrane and the adjoining water molecules at a constant 0.35 V/nm external electric field. Different membrane temperatures at the top and bottom of 310 K and 295 K, respectively, were used (after Song et al. 2017)

this result. First, it is possible to electroporate at lower electric fields if higher temperature gradients are used. However, differential values of only up to 10 K have been reported (Pierro et al. 2014), and so maintaining high temperature differentials across membranes would not be easy or practical. Second, the phase transition temperature of DPPC is about 314 K (Leonenko et al. 2004), and so this sets another limit on the differential temperature. Though other much lower electric fields (for example, 0.25 V/nm) were tried for the MD simulations with at most a 313 K temperature for the outer membrane surface (i.e., a 18 K thermal differential), no nanopore formation was observed despite running the simulations up to 15 ns.

In any event, the development of an electrical field due to temperature gradients opens up possibilities for triggering various bio-physical processes for a variety of applications. Here, the role of temperature gradients in facilitating electroporation has been probed. However, other forms of excitation (such as the use of infrared lasers) could also be used to create thermal gradients based on selective and spatially-dependent absorption.

References

Berendsen JC, van der Spoel D, van Drunen R (1995) GROMACS - A Message-passing parallel molecular-dynamics implementation. Comput Phys Commun 95:43–56

Bresme F, Lervik A, Bedeaux D, Kjelstrup S (2008) Water polarization under thermal gradients. Phys Rev Lett 101: 020602/1–4.

Camp JT, Jing Y, Zhuang J, Beebe SJ, Song J, Joshi RP, Schoenbach KH (2012) Cell death induced by subnanosecond pulsed electric fields at elevated temperatures. IEEE Trans Plasma Sci 40:2334–2347

Chen W, Zhang J (2006) Using nanoparticles to enable simultaneous radiation and photodynamic therapies for cancer treatment. J Nanosci Nanotechnol 6:1159–1166

Croce RP, De Vita A, Pierro V, Pinto IM (2010) A thermal model for pulsed EM field exposure effects in cells at nonthermal levels. IEEE Trans Plasma Sci 38:149–155

Davio SR, Low PS (1982) Characterization of the calorimetric C-transition of the human erythrocyte membrane. Biochemistry 21:3575–3582

Du Q, Superfine R, Freysz E, Shen YR (1993) Vibrational spectroscopy of water at the vapor/water interface. Phys Rev Lett 70:2313–2316

Eastman E (1926) Thermodynamics of non-isothermal systems. J Am Chem Soc 48:1482–1493

Eastman E (1928) Theory of the Soret effect. J Am Chem Soc 50:283–291

Garner AL, Deminsky M, Neculaes VB, Chashihin V, Knizhnik A, Potapkin B (2013) Cell membrane thermal gradients induced by electromagnetic fields. J Appl Phys 133:214701/1-11

Garner AL, Neculaes VB, Deminsky M, Dylov DV, Joo C, Loghin ER, Yazdanfar S, Conway KR (2016) Plasma membrane temperature gradients and multiple cell permeabilization induced by low peak power density femtosecond lasers. Biochem Biophys Rep 5:168–174

Govorov AO, Zhang W, Skeini T, Richardson H, Lee J, Kotov NA (2006) Gold nanoparticle ensembles as heaters and actuators: melting and collective plasmon resonances. Nanoscale Res Lett 1:84–90

Hornef J, Edelblute CM, Schoenbach KH, Heller R, Guo S, Jiang C (2020) Thermal analysis of infrared irradiation-assisted nanosecond-pulsed tumor ablation. Sci Rep 10:5122

Hu Q, Zhang Z, Qiu H, Kong M, Joshi RP (2013) Physics of nanoporation and water entry driven by a high-intensity, ultrashort electrical pulse in the presence of cellular hydrophobic interactions. Phys Rev E 87:032704/1-9

Ivanov IT (1999) Investigation of surface and shape changes accompanying the membrane alteration responsible for the heat-induced lysis of human erythrocytes. Colloids Surf B 13:311–323

Jiang HR, Wada H, Yoshinaga N, Sano M (2009) Manipulation of colloids by a nonequilibrium depletion force in a temperature gradient. Phys Rev Lett 102: 208301/1-5

Joshi RP, Hu Q, Schoenbach KH, Beebe SJ (2005) Simulations of transient membrane behavior in cells subjected to a high-intensity ultrashort electric pulse. Phys Rev E 71:031914/1-9

Kotnik T, Miklavčič D (2000) Theoretical evaluation of the distributed power dissipation in biological cells exposed to electric fields. Bioelectromagnetics 21:385–394

Leonenko ZV, Finot E, Ma H, Dahms TES, Cramb DT (2004) Investigation of temperature-induced phase transitions in DOPC and DPPC phospholipid bilayers using temperature-controlled scanning force microscopy. Biophys J 86:3783–3793

Luzar A, Svetina S, Zeks B (1985) Consideration of the spontaneous polarization of water at the solid/liquid interface. J Chem Phys 82:5146–5154

Muscatello J, Romer J, Sala J, Bresme F (2011) Water under temperature gradients: polarization effects and microscopic mechanisms of heat transfer. Phys Chem Chem Phys 13:19970–19978

Narita M, Shimizu S, Ito T, Chittenden T, Lutz RJ, Matsuda H, Tsujimoto Y (1998) Bax interacts with the permeability transition pore to induce permeability transition and cytochrome c release in isolated mitochondria. Proc Natl Acad Sci USA 95:14681–14686

Neumann E, Sowers AE, Jordan CA (1989) Electroporation and electrofusion in cell biology. Plenum Press, New York

Nijhuis EHA, Poot AA, Feijen J, Vermes I (2006) Induction of apoptosis by heat and c-radiation in a human lymphoid cell line; role of mitochondrial changes and caspase activation. Int J Hyperth 22:687–698

Pakhomov AG, Doyle J, Stuck BE, Murphy MR (2003) Effects of high power microwave pulses on synaptic transmission and long term potentiation in hippocampus. Bioelectromagnetics 24:174–181

Pierro V, De Vita A, Croce RP, Pinto IM (2014) Membrane heating in living tissues exposed to nonthermal pulsed EM fields. IEEE Trans Plasma Sci 42:2236–2244

Richardson HH, Hickmaan ZN, Govorov AO, Thomas AC, Zhang W, Kordesch M (2006) Thermooptical properties of gold nanoparticles embedded in ice: characterization of heat generation and melting. Nano Lett 6:783–788

Song J, Garner AL, Joshi RP (2017) Effects of thermal gradients created by electromagnetic fields on cell membrane electroporation probed by molecular dynamics simulations. Phys Rev Appl 7:024003

Song J, Joshi RP, Schoenbach KH (2011) Synergistic effects of local temperature enhancements on cellular responses in the context of high-intensity, ultrashort electric pulses. Med Biol Eng Comput 49:713–718

Swillens S, Dupont G, Combettes L, Champeil P (1999) From calcium blips to calcium puffs: theoretical analysis of the requirements for interchannel communication. Proc Natl Acad Sci 96:13750–13755

Vernier PT, Sun Y, Marcu L, Craft CM, Gundersen MA (2004) Nanoelectropulse-induced phosphatidylserine translocation. Biophys J 86:4040–4048

Weaver JC, Chizmadzhev Y (1996) Theory of electroporation: a review. Bioelectrochem Bioenerg 41:135–160

Wiegand S (2004) Thermal diffusion in liquid mixtures and polymer solutions. J Phys Cond Matt 16:R357–R379

Chapter 13
Synergy Between Electric Pulsing and Shock Waves for Cell Poration

Ravi Joshi

Abstract The application of pulsed electric fields and discharges can lead to the generation of shockwaves in aqueous media. The interaction of shockwaves with biological cells and membranes is a relative recent area of research, with applications ranging from water purification and desalination, bacterial decontamination, and extracorporeal shock wave lithotripsy. Here, the possible role and impact of shockwaves on membrane pore creation is discussed. Model results based on Molecular Dynamic simulations are presented. The results also touch upon synergistic benefits of a scenario that combines the application of shockwave with an electric field.

13.1 Introduction

In addition to electrical stimulation, the interaction of shock waves with biological cells has been a topic of active research (Choubey et al. 2011; Santo and Berkowitz 2014; Steinhauser and Schmidt 2014). Shockwaves move faster than the speed of sound, and can carry high energy in an abrupt fashion. Uses of shock waves include extracorporeal shock wave lithotripsy (ESWL) to treat kidney stones below 20 mm in size, or destruction of cancer cells (Anpilov et al. 2004). In particular, the collapse of nanobubbles generates water 'nanojets' at very high speeds, which can penetrate membranes to cause poration within picosecond time scales. Controlled shock waves thus have the potential for medical use, since they increase the permeability of cell membranes (Espinosa et al. 2013) through pressure-induced poration. In principle, this then allows the entry of various macromolecules, drugs, and genetic material into the cell (Ahmed et al. 2015). The effect of shock waves can be enhanced by engineering the collapse of bubbles near the membrane vicinity (De Cock et al. 2015; Kobayashi et al. 2011) leading to permanent cell poration (Choubey et al. 2011; Santo and Berkowitz 2014; De Cock et al. 2015; Lentacker et al. 2014). It was also demonstrated, based on their all-atom molecular dynamics (MD) simulations, that a shock wave can induce nanobubble collapse to form a strong nanojet (Choubey et al. 2011).

S. J. Beebe et al., *Ultrashort Electric Pulse Effects in Biology and Medicine*,
Series in BioEngineering, https://doi.org/10.1007/978-981-10-5113-5_13

It would seem entirely possible that the combined application of shock waves and external electric fields could have a synergistic effect. The dual input could enhance the poration speed, or lead to a larger nanopore size, or help reduce time delays and/or the electric field magnitudes necessary for poration. Yet such synergistic effects have not been analyzed to date. Quantitative predictions in this regard would serve as a very useful and practical instrument towards optimizing poration techniques for drug and molecular delivery. Another important point is that electric pulses tend to porate cells at both poles facing the electrodes (Schoenbach et al. 2001). Furthermore, electric field magnitudes tend to fall off rapidly from the location of the electric stimuli. Hence, generating useful bio-effects in cells and tissues that are further removed from electrode contact sites might not be as effective if one solely relied on electric stimulation. Nanobubble facilitated sonoporation (Abdalkader et al. 2017), on the other hand, can be quite site specific and an efficient noninvasive alternative for cellular drug and gene delivery. Furthermore, shockwaves (sonoporation) could be used as a 'pretreatment' of cell membranes in the electroporation process. The advantages of such a dual strategy would be to either: (i) help reduce the voltage (and electric field) requirements on the external power circuitry, with the accompanying benefit of reduced collateral damage, or (ii) result in bigger pores at the same external field for faster delivery of larger molecules. This dual approach also opens up the parameter space and enables more directed focus on specific sites, or cells that are harder to porate. As an example of the latter, U937 cells were found to be more resilient to electric pulsing than Jurkats (Ibey et al. 2010), and so a dual strategy might prove more useful in this context. This technique could even be extended to tissues or larger areas such as a tumor mass. Here, we discuss results of numerical evaluations for the combined effect of both an external field and an incident shock wave of cellular poration based on molecular dynamics (MD) simulations.

13.2 Molecular Dynamics Simulation Results

Simulations results based on MD technique showing the effect of nanobubble generated shockwaves incident on cell membranes are discussed in this context. Snapshots of the Dipalmitoylphosphatidylcholine (DPPC) membrane lipid bilayer along with the surrounding water molecules obtained from the MD simulations (Hu et al. 2019, 2018) are shown in Fig. 13.1. The pressure wave was initiated by assigning velocities of -4 km s^{-1} to the topmost two layers of water molecules. The negative sign simply designates the downward direction. Figure 13.1a shows the t = 0 ns snapshot which is the initial structural configuration. The nanobubble was set at the top-center location with a diameter taken to be 4 nm. The choice of a small-radius nanobubble was based on the following known qualitative features (Alheshibri et al. 2016): (a) The small nanobubbles are very stable and can exist in aqueous media for long periods. Hence, it would be more likely to such smaller entities would exist in situations involving sonoporation. (b) The surface area of a given total volume of bubbles is inversely proportional to their radius. (c) The buoyancy of larger bubbles causes them

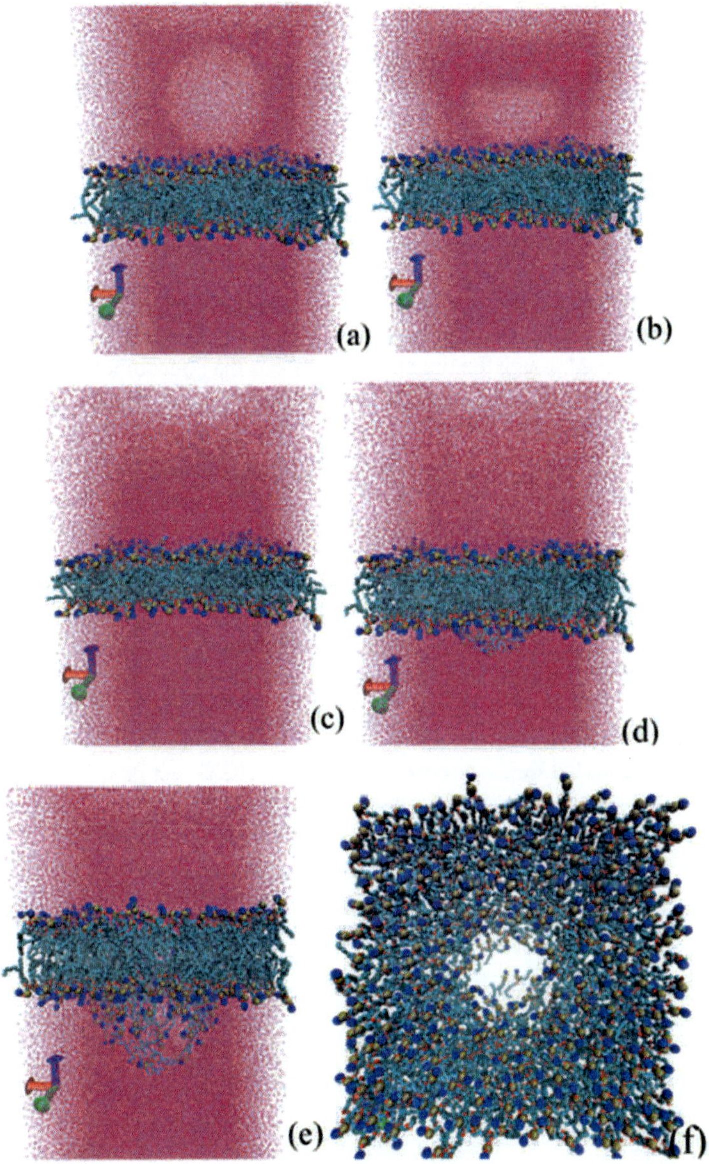

Fig. 13.1 Snapshots of the lipid bilayer with 512 DPPC molecules and 172,302 SPC water molecules obtained from simulations. The diameter of the nanobubble at the top-center was set at 4 nm. **a** The initial t = 0 ps snapshot, **b** result at t = 1.1 ps with the bubble starting to collapse, **c** configuration at t = 3.6 ps showing disappearance of the bubble, **d** snapshot at t = 10.6 ps showing the nanojet pushing lipids away from the membrane as a pore starts to form, **e** pore formation at membrane at t = 23.4 ps, and **f** the top view of the pore at the 23.4 ps instant (after Hu et al. 2018)

to rise to surfaces making it difficult to place them near, but separated from, cells and tissues. (d) Smaller bubbles have higher pressure and will function more effectively as point sources allowing them to deliver stronger pressure intensities to cells and small segments of tissues. In any event, as shown in our results of Fig. 13.1a at about 1.1 ps, the downward moving water molecules incident on the bubble are predicted to cause its collapse. This scenario is shown in Fig. 13.1b. The t = 3.6 ps snapshot of Fig. 13.1c reveals the disappearance of the bubble. The impact of the nanojet on the lipid membrane structure, however, is relatively minimal. At a much later time of 10.6 ps, the nanojet can be seen pushing the lipids away from the membrane as a pore starts to form. The lipid movement though is not very discernable. Figure 13.1e shows pore formation in the membrane patch at t = 23.4 ps. The nanojet can be seen to have passed through the lipid bilayer, and has the effect of forcing molecular ejections from the bottom layer. Some of the DPPC molecules are seen to be loosely connected and dangling from the bottom part of the membrane. The top view shows the pore more clearly at the 23.4 ps instant in Fig. 13.1f. These results thus highlight the role of nanojets on pore formation in biological membranes, as already reported in the literature, and validate our numerical implementation.

Though a nanojet can form pores, these would likely tend to shrink and eventually close. This follows from the energetics at the atomic level (Bennett et al. 2014; Tarek 2005), or more simply from considerations of the continuum elastic response (Helfrich 1973). Details concerning such dynamic pore evolution after nanojet termination were obtained from MD simulations. The result showing a 4 ns snapshot after nanojet termination is given in Fig. 13.2a. The pore cross-section is somewhat elliptical with a major axis of about 5.3 nm and the minor axis dimension of 2.1 nm as indicated by the arrows on the figure. The aspect ratio is thus roughly 2.52. The side view at the 4 ns instant (Fig. 13.2b) shows a lipid section detached from the membrane. This detachment was the result of the impingement of the water molecules from the nanojet. At a later t = 8 ns time, the pore size is seen to shrink. The dimensions were reduced to 3.4 nm by 2.3 nm as shown by the arrows in Fig. 13.2c. Furthermore, the lipid segment is predicted to move back towards the cell membrane. This is easily seen from the corresponding side view at t = 8 ns in Fig. 13.2d. The intermolecular interactions and the hydrophobic nature of the lipid molecules help drive this outcome.

Application of an electric field can serve to not only keep the pores open, but also produce controlled expansion. This aspect of the pore dynamics is presented through the results of Fig. 13.3. Both a shockwave and an external 0.2 V nm^{-1} electric field were applied as the external driving inputs. The shockwave was launched initially, and as with the previous case, the pressure wave was initiated by assigning velocities of – 4 km s^{-1} to water molecules at the topmost layers. The electric field was then turned on after passage of the pressure disturbance, taken to be at the t = 0 instant. The simulation result for the membrane along with a pore is shown in Fig. 13.3a at t = 4 ns. The pore is predicted to have dimensions of about 3.5 nm by 3.9 nm as indicated by the arrows in Fig. 13.3a. With the external field still applied, the nanopore is predicted to expand and grow to a size of 4.5 nm by 5.1 nm in Fig. 13.3b by the 8 ns time instant. Thus, an electric field driven pore expansion is clearly seen.

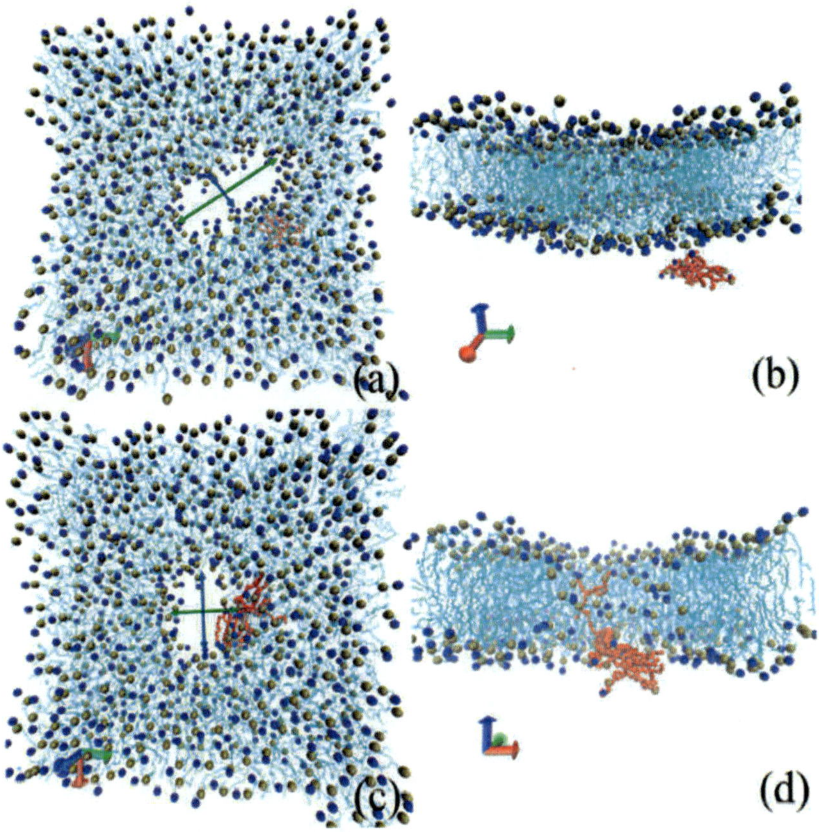

Fig. 13.2 MD simulations results. **a** After 4 ns from nanojet termination, the pore has dimensions of 5.3 nm by 2.1 nm. **b** The side view at the 4 ns instant showing a lipid piece (in red) detached from the membrane. **c** At t = 8 ns, the pore size is seen to shrink to 3.4 nm (green arrows) by 2.3 nm (blue arrows). **d** Finally, the corresponding side view at t = 8 ns with the lipid segment seen to move back to the cell membrane (after Hu et al. 2018)

In this context, it may also be mentioned that a fairly high electric field was applied for these simulations. This high field provides an accelerated test of the biophysical process of pore formation, since low electric fields would otherwise take inordinately long simulation times. Molecular Dynamic approaches typically always use such (or even much higher) field values (Tarek 2005; Tieleman et al. 2003).

The role of the pressure wave, and in particular, the possibility of a limiting threshold behavior on the velocity of the incident pressure wave was probed next. The shock wave was initiated by applying a smaller velocity of -3 km s^{-1}. At this lower velocity, no distinct pore formation was predicted, as seen from the snapshot of Fig. 13.4a. However, the central portion of the membrane can be seen to be heavily deformed, and presents a low lipid density region. Upon subsequent application

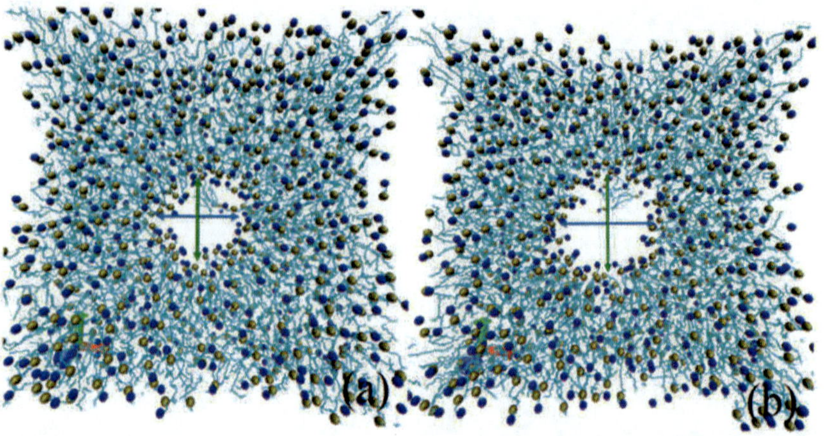

Fig. 13.3 Results with a shockwave and an external 0.2 V nm^{-1} electric field applied after the shockwave. **a** At the t = 4 ns instant, the pore has a dimension of about 3.5 nm by 3.9 nm; **b** after 8 ns, the pore is seen to expand to 4.5 nm by 5.1 nm clearly due to electric field application (after Hu et al. 2018)

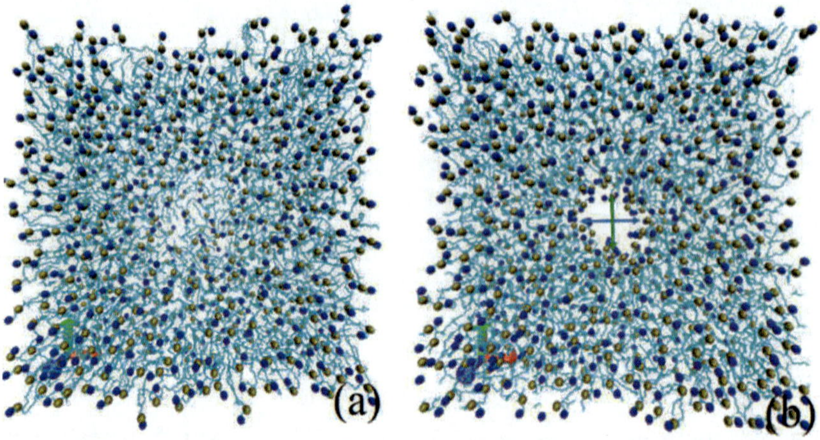

Fig. 13.4 Simulation with both a shockwave having lower velocity and a subsequent 0.2 V nm^{-1} field. **a** A t = 48 ps snapshot following a shock wave with a starting velocity of − 3 km s^{-1}. No pore is predicted, but the central portion of the membrane under the nano-bubble is heavily deformed. **b** A t = 4 ns snapshot after a 0.2 V nm^{-1} electric field is turned on following the shock wave. A pore with 2.4 nm by 3.0 nm is formed (after Hu et al. 2018)

of a 0.2 V nm − 1 electric field, poration was predicted as shown in Fig. 13.4b. The t = 4 ns snapshot (with the zero time taken from the start of the electrical stimulation) shows a pore with dimensions of roughly 2.4 nm by 3.0 nm. This results thus underscores the role of acoustic pressure in 'softening' sites on the membrane for subsequent electroporation. It may be added, that membrane poration of a DPPC bilayer within 4 ns by a 0.2 V nm^{-1} external field is unusually short, and most reports have predicted either much higher fields or longer application times as being necessary for pore formation (Joshi and Schoenbach 2010; Hu et al. 2013; Tarek 2005). Hence, the present result underscores the point that the shockwave helps pore formation at a lower threshold.

Molecular Dynamics (MD) simulations of poration with multiple nano-bubbles were also carried out, and the results discussed next. Since nanobubbles might be pre-existing or can be created by external means, one would practically expect a distribution of these entities, rather than the occurrence of an isolated bubble. One can envision three separate scenarios: (a) separated and non-overlapping bubbles, (b) bubbles with partial overlap above the membrane region, and (c) serial bubbles located in tandem above the membrane. Figure 13.5 shows MD results obtained starting with two 3 nm bubbles side to side (i.e., in a parallel configuration) at the same height from the membrane patch. There was no overlap between the two bubbles.

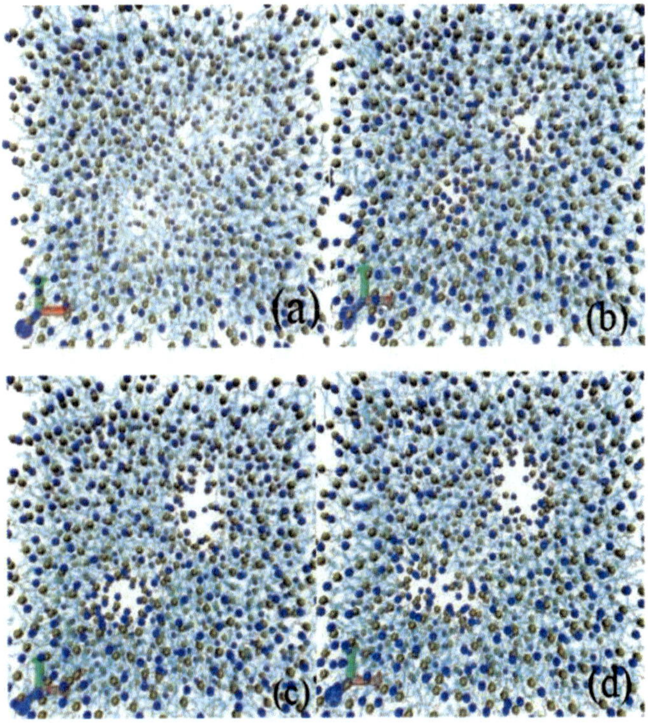

Fig. 13.5 Snapshots obtained from simulation with both a shockwave with the − 4 km/s initial velocity for two separated bubbles and a subsequent 0.2 V/nm field from the start of the electrical pulse. **a** A t = 0 ps snapshot. **b** A t = 1 ns snapshot showing the beginnings of two nanopores. **c** Results at the 2 ns instant with more prominent pores. **d** The state at t = 4 ns, with the two pores seen to be dissimilar in shape and size (after Hu et al. 2019)

The initial velocity of the water molecules was taken to be − 4 km/s (a representative value associated with the shockwave), and an electric field of 0.2 V/nm was applied *after passage of the shockwave.* The snapshots of the lipid membrane patch at time instants of 0 ns, 1 ns, 2 ns, and 4 ns from the start of the electric pulsing, are shown in Fig. 13.5a–d, respectively. At 1 ns, the beginnings of two nanopores can be seen in Fig. 13.5b. The result at 2 ns reveals the pores becoming more prominent and better defined. The state at t = 4 ns shown in Fig. 13.5d clearly reveal two distinct of dissimilar shape and size. The dimensions of the left pore in Fig. 13.5d were 2.8 nm × 1.6 nm, while that on the right had dimensions of 4.7 nm × 2.8 nm. It is thus apparent that within a relatively short time of 4 ns, two nanopores spaced quite close to each other can be formed. This scenario of multiple nanopores close to each other, is in keeping with the notion of ″supraporation″. It is, therefore, expected that the use of multiple bubbles will likely enhance the pore density and hence, lead to stronger material transport into cells. Furthermore, it may be noted that almost all other MD simulations of electroporation in the literature have only yielded a single nanopore in a membrane patch! The difference here in Fig. 13.5 was the presence of two separated bubbles which led to two nanojets incident on the membrane, and represents a new result. Hence, this combined strategy clearly seems to offer the high pore-density advantage.

References

Abdalkader R, Kawakami S, Unga J, Higuchi Y, Suzuki R, Maruyama K, Yamashita F, Hashida M (2017) The development of mechanically formed stable nanobubbles intended for sonoporation-mediated gene transfection. Drug Deliv 24:320–327

Ahmed SE, Martins AM, Husseini GA (2015) The use of ultrasound to release chemotherapeutic drugs from micelles and liposomes. J. Drug Target 23:16–42

Alheshibri M, Qian J, Jehannin M, Craig VSJ (2016) A history of nanobubbles. Langmuir 32:11086–11100

Anpilov AM, Barkhudarov EM, Christofi N, Kop'ev VA, Kossyi IA, Taktakishvili MI, Zadiraka YV (2004) The effectiveness of a multi-spark electric discharge system in the destruction of microorganisms in domestic and industrial wastewaters. J. Water Health 2:267–277

Bennett WFD, Sapay N, Tieleman DP (2014) Atomistic simulations of pore formation and closure in lipid bilayers. Biophys J 106:210–219

Choubey A, Vedadi M, Nomura K, Kalia RK, Nakano A, Vashishta P (2011) Poration of lipid bilayers by shock-induced nanobubble collapse. Appl Phys Lett 98:023701/1–4

De Cock I, Zagato E, Braeckmans K, Luan Y, de Jong N, De Smedt SC, Lentacker (2015) Ultrasound and microbubble mediated drug delivery: acoustic pressure as determinant for uptake via membrane pores or endocytosis. J Control Release 197:20–28

Espinosa S, Asproulis N, Drikakis D (2013) Chemotherapy efficiency increase via shock wave interaction with biological membranes: a molecular dynamics study. Microfluid Nanofluid 16:613–622

Helfrich W (1973) Elastic properties of lipid bilayers: theory and possible experiments. Z Naturforsch C 28:693–703

Hu Q, Hossain S, Joshi RP (2018) Analysis of a dual shock-wave and ultrashort electric pulsing strategy for electro-manipulation of membrane nanopores. J Phys D 51:285403

Hu Q, Zhang L, Joshi RP (2019) Numerical evaluations of membrane poration by shockwave induced multiple nanobubble collapse in presence of electric fields for transport through cells. AIP Adv 9:045006

Hu Q, Zhang Z, Qiu H, Kong M, Joshi RP (2013) Physics of nanoporation and water entry driven by a high-intensity, ultrashort electrical pulse in the presence of cellular hydrophobic interactions. Phys. Rev. E 87:032704/1-9

Ibey BL, Pakhomov AG, Gregory BW, Khorokhorina VA, Roth CC, Rassokhin MA, Bernhard JA, Wilmink GW, Pakhomova ON (2010) Selective cytotoxicity of intense nanosecond-duration electric pulses in mammalian cells. Biochim Biophys Acta 1800:1210–1219

Joshi RP, Schoenbach KH (2010) Bioelectric effects of intense, ultrashort electric pulses. Crit Rev Bio-Med Eng 38:255–304

Kobayashi K, Kodama T, Takahira H (2011) Shock wave-bubble interaction near soft and rigid boundaries during lithotripsy: numerical analysis by the improved ghost fluid method. Phys Med Biol 56:6421–6440

Lentacker I, De Cock I, Deckers R, De Smedt SC, Moonen CTW (2014) Understanding ultrasound induced sonoporation: definitions and underlying mechanisms. Adv Drug Deliv Rev 72:49–64

Santo KP, Berkowitz ML (2014) Shock wave interaction with a phospholipid membrane: coarse-grained computer simulations. J Chem Phys 140:054906/1-7

Schoenbach KH, Beebe SJ, Buescher ES (2001) Intracellular effect of ultrashort electrical pulses. Bioelectromagnetics 22:440–448

Steinhauser MO, Schmidt M (2014) Destruction of cancer cells by laser-induced shock waves: recent developments in experimental treatments and multiscale computer simulations. Soft Matter 10:4778–4788

Tarek M (2005) Membrane electroporation: a molecular dynamics simulation. Biophys J 88:4045–4053

Tieleman DP, Leontiadou H, Mark AE, Marrink SJ (2003) Simulation of pore formation in lipid bilayers by mechanical stress and electric fields. J Am Chem Soc 125:6382–6383

Chapter 14
Probing Potential for Cellular Stimulation by Time-Varying Magnetic Fields

Ravi Joshi

Abstract Membrane pore creation by applying voltage pulses has been used for various applications including gene electrotransfer, electrochemotherapy, and tissue ablation. Electric pulse stimulation, however, requires insertion of electrodes into tissue, which may not always be conducive or convenient. On the other hand, time-varying magnetic fields can also induce time-dependent electric fields based on Lenz's law and Maxwell's equations, and thus be an alternative to creating trans-membrane voltages across membranes. Pulsed magnetic fields would be a contact-less, noninvasive technique allowing clinicians to affect any target within the body. This chapter discusses the concept of magnetically induced voltages, presents model results, with further elaboration of electromagnetic bio-stimulation.

14.1 Introduction

Exposure of biological cells and tissues to magnetic fields has not been studied as much as to electric fields. The first successful stimulation of nerves using magnetic fields, was in 1982 (Polson et al. 1982). Since then, pulsed magnetic fields are used in transcranial magnetic stimulation (TMS). The latter has been effective in the treatment of depression (Epstein and Davey 2002), seizures (Anninos et al. 1999) and Parkinson's disease (Cotelli et al. 2008). However, progress in experimental techniques has resulted in the burgeoning development of new approaches to target and observe the effects of magnetic fields at the intracellular and molecular levels (Cho et al. 2012; Qin et al. 2016; Dobson 2008).

When exposed to a time-varying magnetic field, an excitable tissue can be stimulated by an electric current, which induces a potential across cell membranes. Such induced voltages can lead to physiological changes, such as opening of the mitochondrial transition pore (Weaver 2003), calcium release from internal stores (Scarlett et al. 2009), induction of mitochondria-dependent apoptosis (Beebe et al. 2003), changes to ATP synthesis (Feng et al. 2008), and metabolism (Valdez et al. 2006).

Consequently, analysis of magnetic field effects is of significance in understanding bio-functional changes, potential treatment possibilities for neurological diseases, and to ascertain the dose dependencies and potential threshold limits. Also, while

S. J. Beebe et al., *Ultrashort Electric Pulse Effects in Biology and Medicine*, Series in BioEngineering, https://doi.org/10.1007/978-981-10-5113-5_14

medical applications based on electric fields require direct application via electrodes inserted into tissue, pulsed magnetic fields would allow treatment without invasive electrodes. *This advantage could lead to an expansion of bioelectric treatments by allowing clinicians to affect any target within the body.* Since time-varying currents can induce magnetic fields and give rise to AC electric fields based on Lenz's law and Maxwell's equations, it becomes germane to probe magnetic stimulation in the context of cell function manipulation.

14.2 Simple Analysis

An analysis into the time-dependent development of electric fields at cell membranes due to an externally applied magnetic field is presented based on numerical simulations. The parameter space for the magnetic stimulation for bringing about membrane poration are discussed. Our results focus on the time-dependent magnetic vector potential, and the resulting transmembrane potential (TMP). This thus constitutes a feasibility study into the possibility of generating suitable TMPs for cellular poration via AC currents for a contactless approach.

Figure 14.1 shows the basic geometry used in this modeling effort. A ring of radius "*a*" is shown to carry an AC current *I(t)*, with the co-ordinate system (*x,y,z*) having its origin at the center of the ring. This current carrying ring in Fig. 14.1 is displaced by distances of c_y and c_z (along the y- and z-axes) from the center of a spherical cell. For concreteness, the induced electric field (E_{ind}) and transmembrane potential (TMP) at this chosen cell are assumed to be of focal interest to the present calculations. Calculations of these quantities here involved the following sequence. (i) First, the magnetic vector potential produced by the AC current is

Fig. 14.1 Schematic of a current carrying coil above biological cells

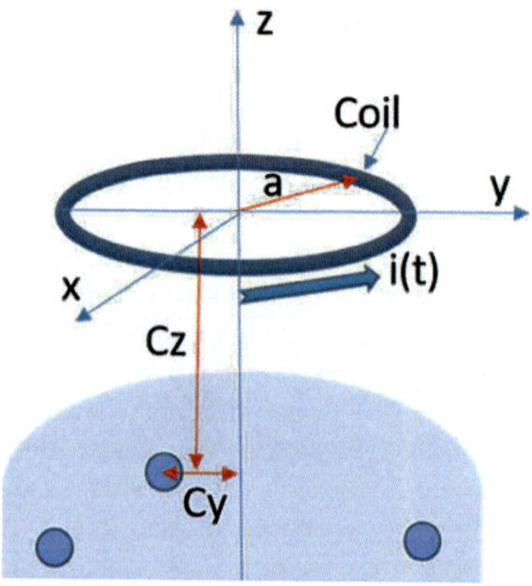

first calculated. The expressions are well-known, especially in the reference frame centered at the current carrying ring. In spherical coordinates, $A_r = A_\theta = 0$, while the third component takes the form (Simpson et al. 2001):

$$A_\varphi = \frac{\mu_0 NI(t)a}{4\pi}\int_0^{2\pi}\frac{\cos(\varphi)d\varphi}{\sqrt{r^2 + a^2 - 2aR\sin(\theta)\cos(\varphi)}}, \tag{14.1}$$

where N represents the number of turns of the circular coil and R the radial distance between a field point and a differential element on the ring. Equation (14.1) can alternaticely be cast in terms of the complete elliptic integrals of the first and second kind [K(m) and E(m), respectively] as:

$$A_\varphi = \frac{\mu_0 NI(t)a}{\pi\sqrt{a^2 + x^2 + y^2 + z^2 + 2a\sqrt{x^2 + y^2}}}\frac{(2 - m)K(m) - 2E(m)}{m}, \tag{14.2}$$

where $m = \frac{4a\sqrt{x^2+y^2}}{a^2+x^2+y^2+z^2+2a\sqrt{x^2+y^2}}$. (ii) Next, expressions for the magnetic vector potential are obtained in a reference frame at the center of the spherical cell. This is done by first converting $\boldsymbol{A}$ to a Cartesian frame centered at the ring, and then, a translation to the cell center. Thus, the components of $\boldsymbol{A}$ in the Cartesian frame become:

$$A_x = -\sin(\Phi)A_\Phi = -\left[y/\left(x^2 + y^2\right)^{1/2}\right]A_\Phi, \tag{14.3a}$$

$$A_y = \cos(\Phi)A_\Phi = \left[x/\left(x^2 + y^2\right)^{1/2}\right]A_\Phi, \tag{14.3b}$$

$$\text{and, } A_z = 0. \tag{14.3c}$$

Upon translation to a new rectangular system centered at the cell, the components become:

$$A_{x'} = -\left[\left(y' - c_y\right)/\left(x'^2 + \left(y' - c_y\right)^2\right)^{1/2}\right]A_\Phi, \tag{14.4a}$$

$$A_{y'} = \left[x'/\left(x'^2 + \left(y' - c_y\right)^2\right)^{1/2}\right]A_\Phi, \tag{14.4b}$$

$$\text{and, } A_{z'} = 0, \tag{14.4c}$$

Finally, converting the co-ordinates back to the spherical system centered at the cell, one obtains the following expressions for the magnetic vector potential:

$$A_{r'} = A_{x'}\sin(\theta')\cos(\Phi') + A_{r'}\sin(\theta')\sin(\Phi'), \tag{14.5a}$$

$$A_{\theta'} = A_{x'}\cos(\theta')\cos(\Phi') + A_{y'}\cos(\theta')\sin(\Phi'), \tag{14.5b}$$

$$\text{and, } A_{\Phi'} = -A_{x'}\sin(\Phi') + A_{y'}\cos(\Phi'). \tag{14.5c}$$

The radial component of the total electric field induced (**E**) is then given by: $E_r = -j\omega A_r{}' - \text{gradient}(V)$. The scalar potential *V* appears due to charge accumulation that appears from the application of a time-varying magnetic field (Stratton 1941). Expressions for the scalar potential are well known, and for a spherical geometry, can be expressed as:

$$V(r,\theta,\phi) = D_0 r^2\sin(\theta)\cos(\phi), \text{ for } R_+ < r < \infty \tag{14.6a}$$

$$V(r,\theta,\phi) = \left(C_1 + D_1/r^2\right)\sin(\theta)\cos(\phi), \text{ for } R_- < r < R_+ \tag{14.6b}$$

$$\text{and,} V(r,\theta,\phi) = C_2\sin(\theta)\cos(\phi), \text{ for } 0 < r < R. \tag{14.6c}$$

where D_0, D_1, C_1, and C_2 are constants that can be determined from the continuity in potential and current density, while R_- and R_+ are the inner and outer radii of the cell membrane. The condition on potentials at the two interfaces ($r = R_-$ and $r = R_+$) yields:

$$D_0/R_+^2 = C_1 R_+^2 + D_1/R_+^2, \tag{14.7a}$$

$$\text{and } C_2 R_- = C_1 R_- + D_1/R_-^2. \tag{14.7b}$$

Similarly, continuity between the normal component of the current density across either side of the two interfaces yields:

$$S_0\left[-j\omega B_0 C/2 + 2D_0/R_+^3\right] = S_1\left[-j\omega B_0 C/2 + 2D_1/R_+^3 - C_1\right], \tag{14.8a}$$

$$S_1\left[-j\omega B_0 C/2 + 2D_1/R_-^3 - C_1\right] = S_2\left[-j\omega B_0 C/2 - C_2\right], \tag{14.8b}$$

where $S_0 = \sigma_0 + j\omega\varepsilon_0$, $S_1 = \sigma_1 + j\omega\varepsilon_1$ and $S_2 = \sigma_2 + j\omega\varepsilon_2$. Hence, based Eqs. (14.7) and (14.8), the scalar potential in all the regions (i.e., intracellular, membrane and extracellular) can be uniquely determined. Using the magnetic vector potential from Eq. (14.5), the entire induced electric field at the membrane is then known.

14.3 Some Simulation Results

The input excitation current waveform that was assumed to feed a bundle of 25 turns of coil of radius “a” is shown in Fig. 14.2. It consisted of a narrow trapezoidal pulse having a 50 ns ON time, and 10 ns rise- and fall-times.

Some simulation results that have been reported (Hu et al. 2010) are given and discussed next. Details of the various parameters used for the modeling are provided in Table 14.1. Figures 14.3 and 14.4 show the results for the magnetic vector potential A_r for different values of the parameters c_y and c_z. The value of A_r is seen to exhibit a minima when the separation parameter (c_y) equaled the coil radius (a), as seen in Fig. 14.3. Furthermore, the magnetic vector potential A_r is predicted to reduce as shown in Fig. 14.4, as the distance of the coil center from the cell increased.

Ultimately, the success of applying a magnetic field (or the driven source current in the coils) depends on the ability to porate cell membranes. The threshold for such poration is known to occur at a transmembrane potential (TMP) of about 1 V. Hence, it is germane to quantify the TMP predictions for such magnetic stimulation. Results for the TMP as a function of angle across the surface (φ) and time is shown in Fig. 14.5a at the 105 ns instant. Due to the oscillating nature of the applied waveform, both positive and negative TMP values are predicted, Furthermore, the peak magnitudes can be seen to exceed the 1 V threshold, implying there is potential for successful poration by such magnetic stimulation. The voltage profile at a later time of 165 ns is shown in Fig. 14.5b. The peak values, as expected, are predicted to be at 90 degree.

The development of the TMP can perhaps be better understood by looking at the equipotential profiles. Such a result is shown in Fig. 14.6 at four different times. These were at 105 ns (when the input pulse reaches the half way of the rising edge), then at the 130 ns instant (during the input pulse on time), a snapshot after 165 ns (when the input pulse reached the half way of the rising edge), and an equipotential surface at 180 ns (10 ns after the input pulse was zero). The induced voltage depends

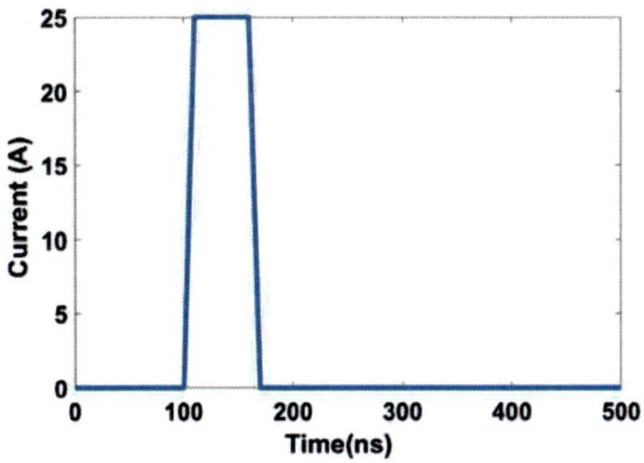

Fig. 14.2 The current waveform taken to excite the 25-turn coils of radius “a” to create the time-dependent magnetic field

Table 14.1 List of parameters used in the simulations

Parameter	Value
Radius of coil a	0.15 m
Number of turns in the coil	25
Cell radius ("a")	10 microns
Membrane thickness	5 nm
Coil axis-cell distance c_y	0.05 m
Coil center to the cell plane c_z	0.08 m
Membrane permittivity	7 ε_0
Suspension permittivity	80 ε_0
Cytoplasm permittivity	80 ε_0
Membrane conductivity	10^{-7} S/m
Suspension conductivity	0.2 S/m
Cytoplasm conductivity	0.02 S/m
Pulse ON time	50 ns
Pulse rise time	10 ns
Pulse fall time	10 ns
Time step	1 ps
Simulation time	500 ns
Pulse peak value	500 A
Sampling frequency	1 THz

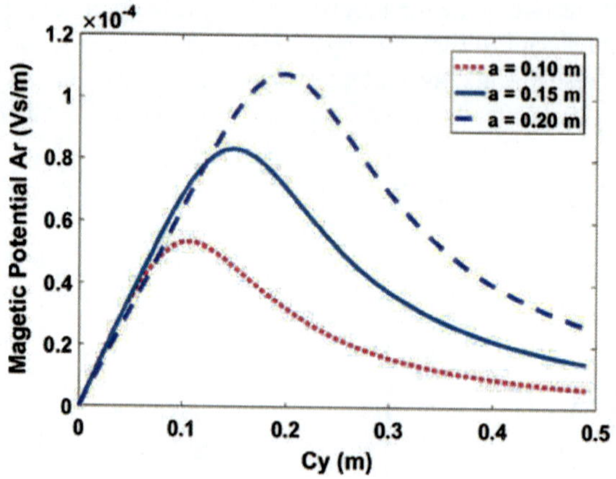

Fig. 14.3 Magnetic vector potential variation with separation between the centers of the cell and coil for different coil radii

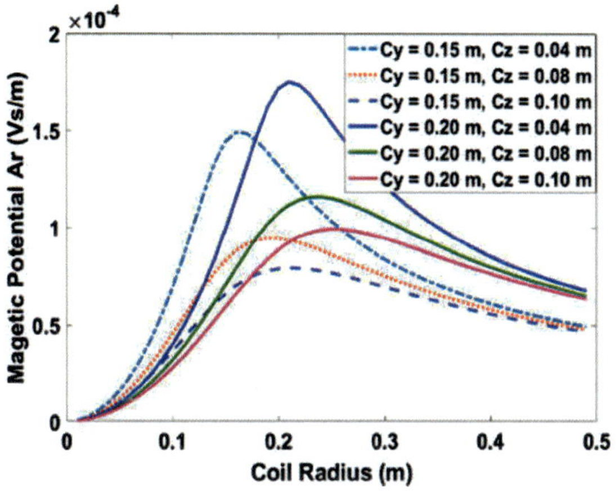

Fig. 14.4 Magnetic vector potential variation with coil radius for different c_y and c_z distances

on rate of change of flux reversal (i.e., the flip in the flux direction), and hence is guided by the rising- or falling-edge of the current pulse. This aspect is emphasized through the sketch of Fig. 14.7.

14.4 Brief Discussion Relating to Electromagnetic Bio-stimulation

Current pulses which can produce time-varying magnetic fields, have been shown to be capable of inducing electric potential at cell membranes, for in vitro stimulation. The transmembrane potential generation was based on Maxwell's principle of voltage induction from time-varying magnetic fields. The advantage of such stimulation is that it presents the possibility of contactless operation by relying on "action at a distance". The TMP development in the present situation is akin to AC-like electric stimulation based on the H-FIRE concept demonstrated by the Davalos group (Arena et al. 2011). The strongest response is predicted when lateral distance between cell and coil center would equal the coil radius. Furthermore, symmetry points and angular dependencies different from electric stimulation have been shown to exist. Thus, this could be a complementary technique to electric stimulation. Also, in the present modality, the TMP peaks depend on the gradient in A_r, and which would be at locations different from the gradients in the electric potential. The next step would be to couple the induced TMP to membrane poration as a function of time. This could be analyzed based on Smoluchowski-based approaches (Joshi and Schoenbach 2010; Neu and Krassowska 1999; Kotnik et al. 2012).

An aspect that has not been studied here, but could be important in the future, is the possibility of using time-dependent magnetic fields over longer time frames to

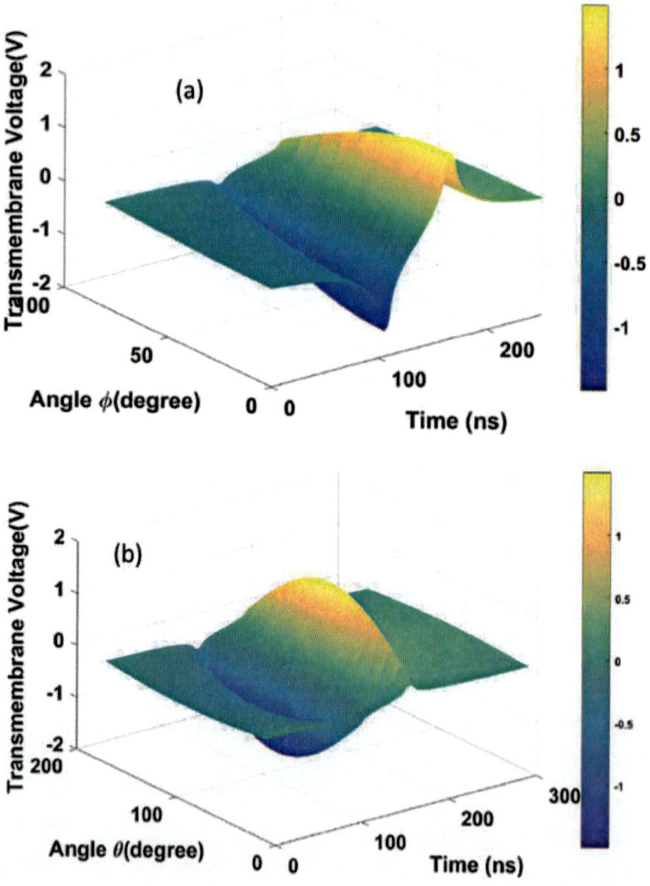

Fig. 14.5 TMP variation with angular displacement and time for as spherical cell at different times. **a** A 105 ns snapshot, and **b** a result at 165 ns

induce localized heating. Magnetic nanoparticles (MNPs) would enhance local fields and concentrate the energy absorption in desired areas. Besides, such MNPs would be able to permeate a tumor mass because of their small size. Hence ideally, and combined use of magnetic stimulation, with MNPs that can attach to tumor specific binding agents could facilitate localized hyperthermia-based killing of unhealthy cells with minimal collateral damage. The role of temperature on electroporation has already been discussed in another chapter, and related simulation results shown. A qualitative picture of the above idea is shown in Fig. 14.8.

The use of magnetic nanoparticles (MNPs) would raise the localized permeability leading to stronger absorption of incident electromagnetic radiation and elevate the magnetic field intensities. The latter, at least for time-varying excitation, would then lead to stronger electric fields and create larger transmembrane potentials (TMPs).

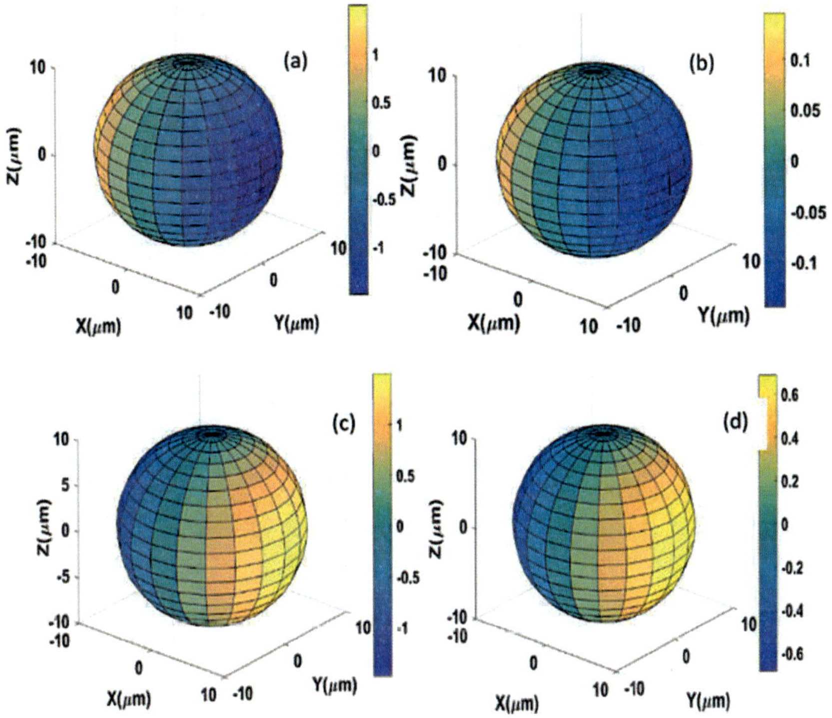

Fig. 14.6 TMP across the cell surface at four different time instants of: **a** 105 ns, **b** 130 ns, **c** 165 ns, and **d** 180 ns

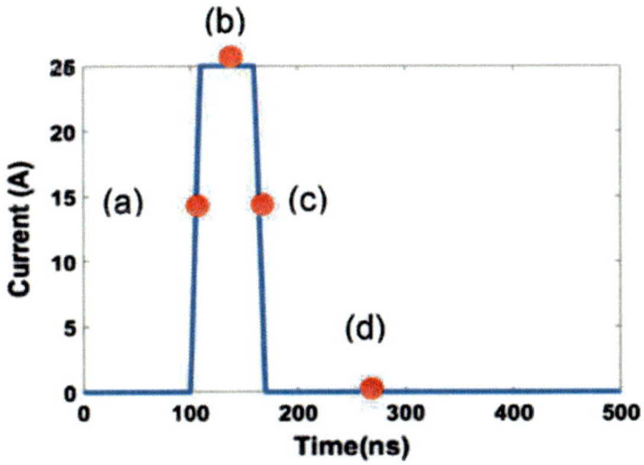

Fig. 14.7 The profile for the time dependent input current exciting the coil

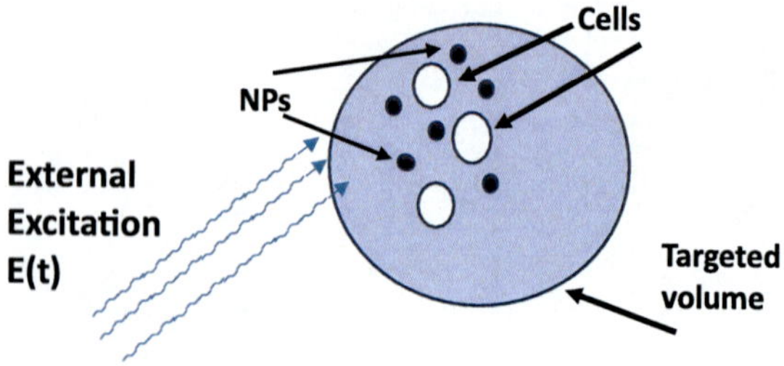

Fig. 14.8 A schematic of possible localized nanoparticle (NP) heating through the selective absorption of incident electromagnetic radiation. The elevated temperatures at sites in close proximity to cells could lead to enhanced electroporation and even cell death or damage

Thus, a nanoparticle-based treatment of cells by external electromagnetic radiation would entail the injection of NPs to load them up and alter the material properties in an environment local to the cells. The effective permeability in a mixture of MNPs and extra-cellular fluid near a cell would be modulated to yield an effective value (μ_{eff}) as:

$$\mu_{eff} = \mu_s \left(\frac{\mu_p + 2\mu_s + 2f(\mu_p - \mu_s)}{\mu_p + 2\mu_s - f(\mu_p - \mu_s)} \right). \tag{14.9}$$

The heating around cells due to such MNPs is shown in Fig. 14.8 and the schematic below. As is well known, temperature enhancements at cell membranes can alter biological function, increase reaction rates, and has even been shown to lead to enhance electroporation (EP). Results of enhancements in the TMP from calculations are shown in Fig. 14.9.

The model previously discussed in Eq. (14.5), along with the following relation:

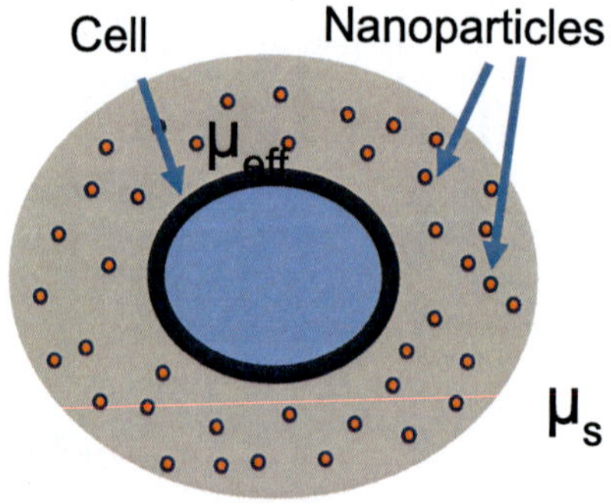

Fig. 14.9 A qualitative sketch of a localized NP distribution near a cell to enhance the effective permeability, and hence increase heat-related bio-effects

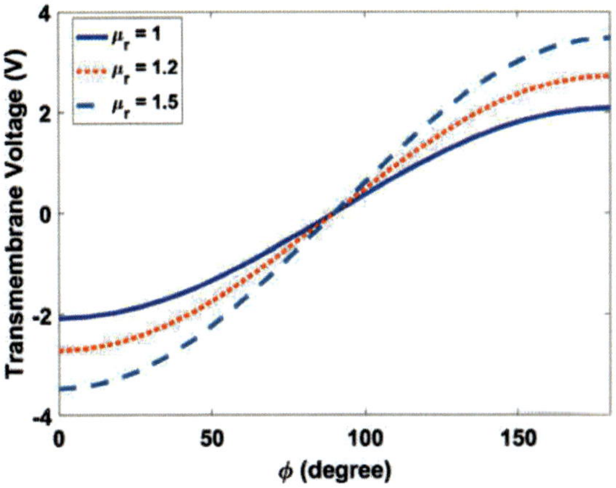

Fig. 14.10 Dependence of the transmembrane potential (TMP) on the local permeability

$$\nabla \cdot \mathrm{A} + \frac{1}{\mathrm{c}^2}\frac{\partial \varphi}{\partial \mathrm{t}} = 0, \tag{14.10}$$

yielded the dependence of the TMP on the effective permeability as shown in Fig. 14.10.

Usually the problem in this area of localized temperature enhancements is that the heat capacity of the NPs is not very large. Hence, despite preferential energy absorption at the NP sites, the temperature increases may remain modest. However, a relatively recent report (Albarqi et al. 2019) that used Cobalt- and manganese-doped, iron oxide nanoparticles (CoMn-IONP) encapsulated in biocompatible PEG–PCL (poly(ethylene glycol)-*b*-poly(ε-caprolactone))-based nanocarriers, seem to indicate large increases in heat capacity. As a result, the concentration of Fe needed from the NPs went down from about 1700 mg/Kg to roughly 6 mg/Kg. This reduction promises to favorably impact localized heating, while reducing the dangers of injecting foreign material into the bio-system.

References

Ahmed M, Brace CL, Lee FT Jr, Goldberg SN (2001) Principles of and advances in percutaneous ablation. Radiology 258:351–369

Albarqi HA, Wong LH, Schumann C, Sabei FY, Korzun T, Li X, Hansen MN, Dhagat P, Moses AS, Taratula O (2019) Biocompatible nanoclusters with high heating efficiency for systemically delivered magnetic hyperthermia. ACS Nano 13:6383–6395

Anninos PA, Tsagas N, Jacobson JI, Kotini A (1999) The biological effects of magnetic stimulation in epileptic patients. Panminerva Med 41:207–215

Arena CB, Sano MB, Rossmeisl Jr. JH, Caldwell JL, Garcia PA, Rylander MN, Davalos RV (2011) High-frequency irreversible electroporation (H-FIRE) for non-thermal ablation without muscle contraction. BioMed Eng OnLine 10:102/1-20

Beal MF (1998) Mitochondrial dysfunction in neurodegenerative diseases. Biochim Biophys Acta 1366:211–223

Beebe SJ, Fox PM, Willis EL, Schoenbach KH (2003) Nanosecond, high intensity pulsed electric fields induce apoptosis in human cells. FASEB J 17:1493–1495

Berg J, Tymoczko J, Stryer L (2002) Biochemistry. Freeman& Company, New York

Cho MH, Lee EJ, Son M, Lee JH, Yoo D, Kim JW, Park SW, Shin JS, Cheon J (2012) A magnetic switch for the control of cell death signaling in in vitro and in vivo systems. Nat Mater 11:1038–1043

Cotelli M, Manenti R, Cappa SF, Zanetti O, Miniussi C (2008) Transcranial magnetic stimulation improves naming in Alzheimer disease patients at different stages of cognitive decline. Eur J Neurol 15:1286–1292

Dobson J (2008) Remote control of cellular behaviour with magnetic nanoparticles. Nat Nanotechnol 3:139–143

Epstein CM, Davey K (2002) Iron-core coils for transcranial magnetic stimulation. J Clin Neurophysiol 19:376–381

Feng HL, Yan L, Cui Y (2008) Effects of repetitive transcranial magnetic stimulation on adenosine triphosphate content and microtubule associated protein-2 expression after cerebral ischemia-reperfusion injury in rat brain. Chin Med J (engl) 121:1307–1312

Hu Q, Joshi RP, Miklavcic D (2020) Calculations of cell transmembrane voltage induced by time varying magnetic fields. IEEE Trans Plasma Sci 48:1088–1095

Joshi RP, Schoenbach KH (2010) Bioelectric effects of intense, ultrashort electric pulses. Crit Rev Bio-Med Eng 38:255–304

Kotnik T, Kramar P, Pucihar G, Miklavcic D, Tarek M (2012) Cell membrane electroporation- Part 1: the phenomenon. IEEE Electr Insul Mag 28:14–23

McBride HM, Neuspiel M, Wasiak S (2006) Mitochondria: more than just a powerhouse. Curr Biol 16:R551-560

Neu JC, Krassowska W (1999) Asymptotic model of electroporation. Phys Rev E 59:3471–3482

Polson MJ, Barker AT, Freeston IL (1982) Stimulation of nerve trunks with time-varying magnetic fields. Med Biol Eng Comput 20:243–244

Qin S, Yin H, Yang C, Dou Y, Liu Z, Zhang P, Yu H, Huang Y, Feng J, Hao J, Deng L, Yan X, Dong X, Zhao Z, Jiang T, Wang HW, Luo SJ, Xie C (2016) A magnetic protein biocompass. Nat Mater 15:217–226

Scarlett SS, White JA, Blackmore PF, Schoenbach KH, Kolb JF (2009) Regulation of intracellular calcium concentration by nanosecond pulsed electric fields. Biochim Biophys Acta 1788:1168–1175

Schapira AH, Gu M, Taanman JW, Tabrizi SJ, Seaton T, Cleeter M, Cooper JM (1998) Mitochondria in the etiology and pathogenesis of Parkinson's disease. Ann Neurol 44:S89-98

Simpson J, Lane J, Immer C, Youngquis R (2001) Simple analytic expressions for the magnetic field of a circular current loop. NASA/TM-2013-217919

Stratton JA (1941) Electromagnetic theory. McGraw-Hill, New York

Valdez LB, Zaobornyj T, Boveris A (2006) Mitochondrial metabolic states and membrane potential modulate mtNOS activity. Biochim Biophys Acta 1757:166–172

Weaver J (2003) Electroporation of biological membranes from multicellular to nanoscales. IEEE Trans Dielectr Electr Insul 10:754–776

Chapter 15
Pulsed Power Generators

Shu Xiao

Abstract This chapter presents an overview of commonly used pulsed generators in bioelectrics. The concept of pulsed power is discussed in Sect. 15.1. One of the key components of ultrashort pulse generator is the switch that is capable of closing in time of nanoseconds or subnanoseconds. A variety of switches, including gas spark gap switches and MOSFETs, can be used (Sect. 15.2). Besides the switches, the circuits that are often used in producing ultrashort pulses are discussed in Sect. 15.3. Finally, when ultra-high voltage pulses are needed, an air-core transformer or a Marx generator can be used, which are discussed in Sect. 15.4.

15.1 Introduction

Electrical pulses causing diverse biological effects are produced by pulsed power generators, which have discrete components that are capable of handling high voltages and high output currents. Although generators have different operating principles, they generally are configured in a structure that includes a charger, an energy storage, a switch, and a load (Fig. 15.1). A pulse generator works in the way of "slow charging and fast discharging". At the beginning, the charger pumps either a DC current or a pulsed one to the energy storage, which could be a capacitor, an inductor, or the combination of both. Upon the completion of charging, turning on the switch releases the stored energy to the load. The time for discharging could be three orders of magnitude shorter than charging. For example, charging a capacitor typically could take 1 ms, but discharging it could just be as short as 1 μs. As such, a gain of current or voltage is achieved, so the pulse's instantaneous power is amplified compared to the average power in the charging process, albeit the total energy remains approximately the same.

In bioelectrics applications, the load to absorb the pulse energy is a volume of cell medium or a bulk of tissue. As the load is resistive in nature, the type of energy storage determines the pulse waveform. If the energy storage is a capacitor, the pulse amplitude rises almost instantaneously and decays exponentially with a duration governed by a time constant, RC (R: the load resistance, C: the capacitance of the energy storage). If the energy storage is an inductor, a similar waveform is generated,

S. J. Beebe et al., *Ultrashort Electric Pulse Effects in Biology and Medicine*, Series in BioEngineering, https://doi.org/10.1007/978-981-10-5113-5_15

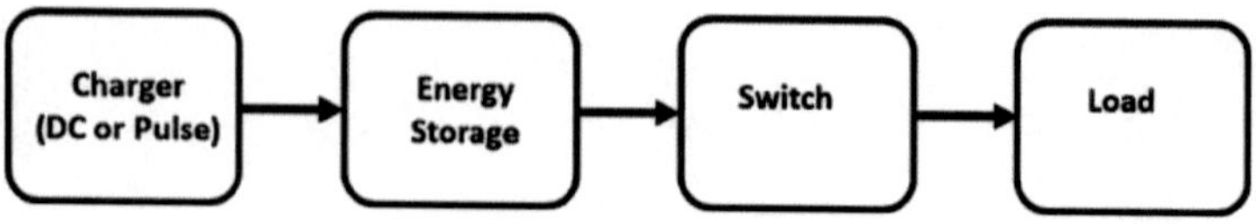

Fig. 15.1 The block diagram of a pulsed power system. The charger can be a DC voltage source or a pulsed source. The energy storage can be a capacitor, an inductor, or the combination of both. A switch allows the stored energy to transfer to a load, which, in bioelectrics applications, can be cell medium or a bulk of tissue

and the amplitude decreases with a time constant, L/R (L: the inductance). When the energy storage is the combination of capacitor(s) and inductor(s), an oscillatory waveform or a square waveform can be produced depending on the circuit parameters. No matter what type the energy storage is, a turn-on switch plays a critical role, which can be a vacuum switch, a gas switch, or a solid-state switch. If the pulse width is to be adjusted, a fully controlled solid-state switch capable of turning on within a given time interval should be used, although the current level may be limited. In applications where the output current is critically high, gas switches are preferred and, in this case, the pulse width is generally fixed.

15.2 High Voltage Switches

In generating high voltage pulses, the switch performance is central to the pulse rise and fall time, pulse duration, and repetition rate. In the nanosecond range, spark gap switches and MOSFETs are the most frequently used switches. Spark gap switches are fast in switching speed, inexpensive, and easy-to-build, although they have shortcomings such as turn-on jitter, voltage error, and short lifetime due to the deterioration of electrode condition. For bioelectrics applications, though the problem of switching error exists in a single pulsed treatment, but the pulse frequency and the pulse number are generally low, so the switch performance remains relatively steady within each treatment. Therefore, the benefits of spark gap switches outweigh the shortcomings, making them the top choice of high voltage switches. Since a spark gap is only capable of turning on, it is mostly incorporated into a transmission line (or a pulse forming network) that generates pulses with a fixed pulse duration. On the other hand, MOSFETs can be programed to turn on and off by a low voltage gate signal, offering flexibility of varying the pulse duration. The drawback is they have either a low voltage or a low current constraint. A high-power application therefore requires connecting MOSFETs in series to increase the total switching voltage or in parallel to increase the total current. But stacking MOSFETs increases the current path and therefore the inductance, making the circuit slower. When it comes to the choice of a switch for generating nanosecond pulses, it is preferable to use spark gap switches for short nanosecond pulses (< 100 ns), whereas MOSFET switches are for long nanosecond pulses (> 100 ns) with the benefit of pulse width adjustment.

15.2.1 Gas Spark Gap Switches

Spark gaps are normally filled with atmospheric-pressure air or dense gas. They are typically constructed in the sphere-sphere, cylinder-cylinder, sphere-plane, or cylinder-ring configurations, which result in different degrees of electric field homogeneity and switching performances. The switching process is well understood and normally starts from free electrons caused by cathode emission or external source. These free electrons, accelerated by the electric field in the electrode gap, acquire sufficient energy and cause collisional ionization of gas molecules or atoms. New electrons and ions are created. The electron multiplication is referred to as the Townsend avalanche process (or the Townsend first process). The number of electrons increases exponentially ($n = n_0e^{\alpha d}$, $e^{\alpha d} \gg 1$) over a distance d. The Townsend's first coefficient, α, is to measure the number of electrons produced by an electron per unit length in the direction of electric field. Besides collisional ionization, photoionization and (or) other ionization mechanisms, such as that through metastable atoms, also contribute to the overall ionization process. In accompany with ionizations, there are deionization processes such as attachment or recombination that reduce the number of electrons.

If the electric field lasts sufficiently long, positive ions can reach the cathode and cause more electrons to emit to the electrode gap due to ion bombardment, a process known as the Townsend secondary process. The secondary electron emission marks the onset of breakdown of the gap. As such, the breakdown is associated with a formative time lag, which involves the time for electron multiplication by collisions and electron emission at the cathode by positive ion impact. The breakdown is a statistical process which is sensitive to the electrode material, electrode condition, and gas purity.

The formative time lag and jitter can be greatly reduced by applying voltages much greater than the static breakdown voltage, which allows for a higher degree of ionization in the gap. In this case, the Townsend first process transitions to the streamer mode, in which intense ionization and excitation of gas molecules occur in front of the avalanche head. For which, αd reaches ~ 18–20 (Raizer 1997). While the Townsend processes are mostly invisible, streamers are brightly luminous with photoionization and photoemission involved. Secondary luminous structures can be seen stemming from the primary streamer structure. Streamer breakdown can be so fast that no motion of positive ions is involved. As a result, the formative time lag is determined by the electron transit time across the gap, instead of the positive ion transit time.

The breakdown voltage of a uniform gap depends on the gas species, and the product of pressure and gap distance (*pd*). The dependence is shown in the Paschen's Law. In high power gas spark gaps, three commonly-used gases are air, N_2 and Sulfur Hexafluoride (SF6). Among these, SF6 has the highest breakdown strength owing to its strong ability to attach free electrons to form negative ions. These ions are much less mobile and therefore effectively drive the electrons out of the avalanche process. The Paschen's curves for air, N_2, and sulfur hexafluoride (SF_6) are shown in

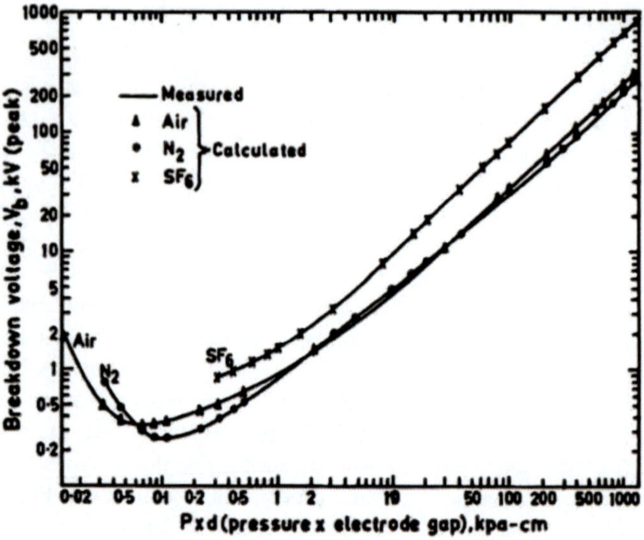

Fig. 15.2 The breakdown curves (Paschen's curves) for air, N_2, and SF_6 (Husain and Nema 1982)

Fig. 15.2 (Husain and Nema 1982). For a 1 cm gap, the breakdown voltage of air is ~ 30 kV, which is approximately the same as that of N_2. But for SF6, the breakdown voltage almost becomes tripled. The Paschen's curve can be explained based on the Townsend mechanism (Kuffel 2000). But at very high *pd* and low *pd*, the breakdown voltage deviates from the Paschen's curve. The breakdown mechanism shifts from the Townsend mechanism to the streamer mechanism for high *pd* and to the vacuum breakdown mechanism for low *pd*.

Once a conduction channel forms either through the Townsend mechanism or the streamer mechanism, more current can flow, forming a self sustaining conductive channel. This process is accompanied by temperature increase and thermal ionization. The ionization further increases the number of electrons, resulting in more Joule heating of the (gas) channel, which in turn increases the pressure and the radius of the channel. As soon as the voltage across the electrodes collapses entirely, a spark is formed, and the gap switch closes entirely. The expanded channel acts like a piston on the remaining gas and produces a shock in the gas (Braginskii, 1958).

The pulses generated by closing a spark gap switch in a pulsed power generator are largely determined by the spark gap performance. The characteristics of a gas switch can be understood from the equivalent circuit as shown in the Fig. 15.3, where $L_{channel}$, C_{sw} and $R_{channel}$ are the inductance, capacitance, and dynamic resistance of the switch channel.

Many investigators conducted research on the resistance of the spark gap and proposed different models (Engel et al. 1989; Martin et al. 1993). Due to the complex nature of the spark gap physics, there has not been a consensus model. It was pointed out though that the conductivity of the channel σ in the nanosecond time scale is not

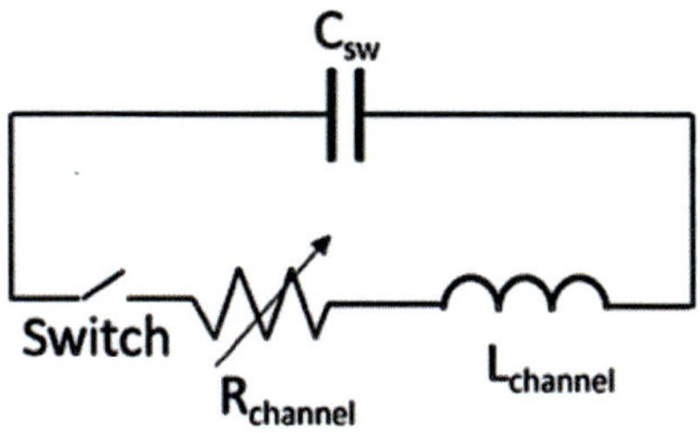

Fig. 15.3 Equivalent Circuit of a gas spark gap switch

well modeled by a constant electrical conductivity (Hussey et al. 1999). The variation of conductivity was also observed in a spark gap used to generate bipolar nanosecond pulses (Xiao et al. 2018). But for a given instant, if the channel conductivity σ is known, the resistance of the spark can be estimated from a cylindrical conductive column:

$$R(t) = \frac{d}{\sigma \pi a(t)^2}, \tag{15.1}$$

where d is the length of the spark channel and $a(t)$ is the radius. For a small gap (d = 2 mm) with a hold-off voltage of several kV, using typical values a = 0.1 mm and σ=2 × 10^4S/m (Martin et al. 1993), R is calculated as ~ 3.2 Ω. This suggests that the switch resistance is not an issue for high impedance pulse generators. However, for a low impedance generator (a few Ω), the switch resistance is not small and the voltage loss across the switch is no longer negligible.

The inductance associated with a spark gap was investigated in Persephonis et al. (1992). Their major conclusions are: (1) any increase or decrease of the inductance associated with the switch corresponds to shrinking or thickening of the spark channel; (2) the inductance increases proportionately with gap distance at a constant pressure; and (3) the increase of pressure leads to an increase in inductance at a fixed electrode distance.

The switch inductance of a spark gap was obtained with a good approximation by treating the conduction channel as a coaxial transmission line of inner radius a(t) and outer radius a_0 (Levinson et al., 1979):

$$L(t) = 2 \times 10^{-7} d \ln\left[\frac{a_0}{a(t)}\right] \tag{15.2}$$

For practical purpose, the inductance per unit length of a spark gap can be treated approximately as (Lehr et al. 1998):

$$L_s = 10^{-6}(H) \tag{15.3}$$

Equation 15.3 suggests that a 5 mm-gap approximately carries an inductance of 5 nH. This poses a 100 ps delay for a 50 Ω load (L/R time). For a subcentimeter gap, it is therefore reasonable to rule out the role of switch inductance as the stray inductance of the wiring between the load and the generator can easily exceed tens of nH.

The capacitance of a spark gap switch depends on the inter-electrode distance and the electrode surface area. The gap capacitance is the main source of switch current during the channel formation and provides the energy for the onset of breakdown, resulting in a fast closing of the switch (Frost et al. 1993). In conjunction with the fast risetime, the capacitance is responsible for the prepulse before a main pulse observed in the load.

A spark gap switch breaks down randomly on the electrode surface and the delay time of a spark gap switch is statistical, so a spark gap switch naturally introduces voltage variation. In order for a switch to close on command at relatively stable voltages, several methods can reduce the jitter: (1) applying high voltage charging pulses with reduced pulse duration and letting the gap break down at over-voltages; (2) using trigger methods by employing a third electrode such as in a trigatron or a mid-plane spark gap. A third electrode that generates seed electrons helps to reduce voltage scattering (Frey et al. 2009); and (3) illuminating the electrode surface or electrode gap with a high-power laser.

The repetition rate of a spark gap is in the range of 1 Hz to several kHz. It depends on the gas recovery, i.e., how fast it takes to cool down the spark channel after a switching event and so the gas density and temperature return to their ambient levels. Gases, such as hydrogen, that have high thermal diffusivity tend to have a fast recovery time. A hydrogen spark gap was shown to exhibit 100-μs recovery time (Moran and Hardesty 1991), which suggests a potential repetition rate of 10 kHz.

15.2.2 Liquid Switches

Liquid switches, normally water switches or oil switches, were used in ultrafast, very high voltage, and low impedance pulse generators (Martin 1992; Woodworth et al. 2005; Heffernan et al. 2005). In the final stage of a pulse generator, they can be used as a peaking switch to sharpen the risetime of a slow pulse from the pulse forming line. Working in the self-breaking mode, they were used to generate pulses with nanosecond or even subnanosecond risetimes (Huiskamp et al. 2015).

The dielectric strength of a liquid depends on the electrode condition, electrode polarity, liquid purity, and most importantly, the duration of the applied pulse, which can be described in an empirical scaling relationship. For example, the breakdown strength of water, E (in MV cm^{-1}), between a spherical and a plane electrode, where the spherical electrode is the anode, is described by Schoenbach et al. (2008a)

$$E = \alpha 0.23 A^{-0.058} t^{-1/3} \tag{15.4}$$

where t is the effective stress time in μs, A is the electrode area within 90% of the peak electric field in cm^2, and α is a constant determined by the electrode geometry.

Shorter stress time or the pulse time increases the water breakdown strength. For a 10 μs, the breakdown strength of water is 500 kV/cm, but increases to 3.5 MV/cm for a 10 ns pulse.

15.2.3 *MOSFET Switches*

A MOSFET (metal-oxide-semiconductor field-effect transistor) consists of either two P-type regions that are separated by an N-type region or two N-type regions separated by a P-type region. Two of the contacts of the MOSFET, the drain and the source, are attached to each of the two P-type (or N-type) regions. A third contact, the gate, is attached to the intervening N-type (or P-type) region but separated from it by an oxide layer (Fig. 15.4a). The voltage applied to the gate modulates the conductivity between the source and the drain by the field effect. The planar structure is applicable in the low voltage, MOS digital circuits. A vertical architecture is better suited to the high voltage and high-power switching. One example, an n-type MOSFET, is shown in Fig. 15.4b. Between the source and the drain, several layers with varying doping densities are configured through the drain-source gap. In the switch-off state when no trigger voltage is applied to the gate, the junction between the p region and the n-Epi (n-type epitaxial) layer is reverse biased and the voltage drop occurs mainly across this junction. Since the epitaxial layer growth technique is capable of atomic layer deposition, the n-Epi layer thickness can be controlled to produce a high hold-off voltage and yet a small on-state resistance. When the switch is turned on with a positive voltage applying fto the gate, electrons near the source in the n^+ region are attracted to the gate, resulting in an electron channel formed between the n^+ region and the p region. The number density of electrons increases and becomes greater than that of holes, thus population inversion in the p region occurs. The electrons diffuse to the n-Epi layer region and divert from the horizontal direction to the vertical

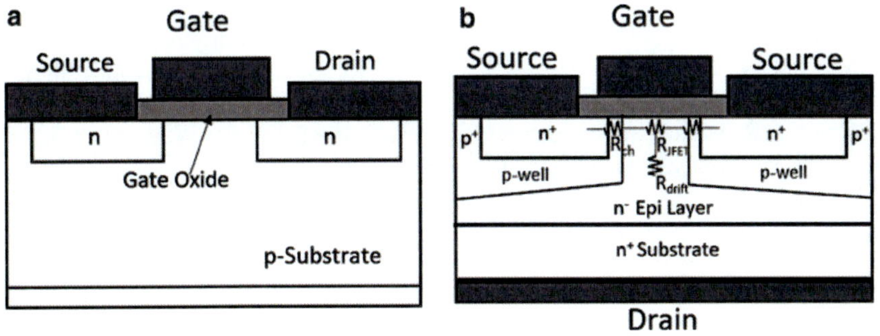

Fig. 15.4 A MOSFET switch. **a** A planar MOSFET. **b** A vertical power MOSFET

direction before flow to the drain. The current conduction between the source and the drain is then controlled by the electrons in the p region, similar to that in a Junction Field Effect Transistor (JFET).

Power MOSFETs have several parasitic capacitances, which are produced by adjacent locations of differently doped semiconductor materials. They include: the capacitance formed between the gate and the drain, C_{gd}, the capacitance formed between the gate and the source, C_{gs}, and the capacitance between the drain and the source, C_{ds}. For the circuit designer's reference, these parameters are typically found in the datasheet as C_{iss}, C_{oss}, and C_{rss} (Batarseh, 2011). For power MOSFETs, $C_{iss} \cong C_{gs}$ and $C_{rss} \cong C_{gd}$. In the power switching, C_{gs} needs to be charged to a threshold voltage for a switch to turn on. Therefore, its value affects the speed of the switching and a large current injection is warranted. For a nanosecond turn-on time, if a C_{gs} (1800 pF) needs to be charged to 10 V within 1 ns, the current needs to be 18 A. When it comes to turn-off, C_{gs} must be discharged before the MOSFET stops completely, and this process normally causes a turn-off delay. On the other hand, C_{gd} causes a feedback from the output to the input in the switching process, which makes the gate circuit susceptible to high dV/dt stress. Both C_{gs} and C_{gd} vary nonlinearly with the drain voltage and differ from their static values. The third capacitance, C_{ds}, is relatively small compared to the other two and does not affect the switch turn-on and turn-off characteristics as much.

MOSFETs are unipolar device with either electrons or holes as the majority charge carriers. The resistive set of parasitic components cause several voltage drops in the current flow (Godignon et al. 2017): the voltage drop across the Ohmic contact, R_c, the inversion layer channel, R_{CH}, the JFET resistance under the gate area, R_{JFET}, the drift resistance in the epitaxial layer, R_{drift}, and the substrate resistance, R_{sub}. The voltage across the drain and the source is $V_{ds} = I_{ds} * R_{ds}$, where $R_{ds} = R_c + R_{CH} + R_{JFET} + R_{drift} + R_{sub}$. The major resistive component however is the epitaxial layer resistance R_{EPI} ($R_{JFET} + R_{drift}$), which becomes much higher than the rest as the drain-source voltage increases (Barkhordarian 1997). Reducing the turn- on resistance R_{ds}, essentially R_{EPI}, allows for reduction of the switching loss. This has become one of the major thrusts in the development of high power MOSFETs. Wide Band Gap Materials (WBGM), such as SiC, GaN, or diamonds, can be made very thin to reduce R_{EPI} and meanwhile the WBGM MOSFETs can hold much higher voltages than can conventional Si MOSFETs.

The parasitic inductances are caused by the MOSFET leads. They are small and in the range of several nH. Coupled with external circuit inductances, they may produce a fast ringing in the gate-source voltage and the drain-source current when the switch turns on and off.

15.2.4 Switches for Subnanosecond Pulse Generators

Subnanosecond electric pulses are generated by sharpening nanosecond or longer pulses that are typically implemented by a peaking circuit in a step-up transformer, a Marx generator, or a PFL. This approach in general involves ultrafast sharpening switches, which can be solid state switches including Step Recovery Diodes (SRD), avalanche transistors, fast ionization dynistors (FID), or high-pressure gas/liquid switches. Solid state switches usually have low hold-off voltages (< 1 kV), because they rely on narrow PN junctions. In order to utilize these switches for high voltage generation, they are stacked in series. Gas or liquid spark gap switches on the other hand are much more suited to generating high voltage pulses and a single gap can hold tens to hundreds of kilovolts.

Step Recovery Diode

A step recovery diode (SRD) is also called fast recovery diode or snap-off diode (Ivanov et al. 2016; Ng 2009). An SRD has a special doping profile so that the injected carriers are close to the vicinity of the PN junction, which results in a rapid turn-off under reverse bias. Under the forward quiescent bias, the minority carriers are stored in the PN depletion layer. Under the reverse bias, the minority carriers are fully drained, causing the reverse current to interrupt rapidly and form a fast transient (Brown and Martin 1987). The SRD works when it is switched from the forward bias to the reverse bias.

Avalanche Transistors

An avalanche transistor (either NPN or PNP) creates an avalanche process in the base–collector depletion region, which is reverse biased (Miller and Ebers 1955; Rein and Zahn 1975). At the avalanche breakdown voltage, the region has a high field which facilitates the impact ionization. When electrons injected from the emitter drift towards the collector, they collide with the material lattice and form new electrons and holes. With a mechanism analogous to the avalanche process in a gaseous discharge, this impact ionization process produces ultrafast multiplication of charge carriers. The transistor is quickly turned on and therefore it can be used to generate subnanosecond pulses (Rein and Zahn 1975) or nanosecond pulses (Krishnaswamy et al., 2007).

Fast Ionization Dynistors (FID)

The operating principle of Fast Ionization Dynistors is the fast propagation of ionization wave inside multiple semiconductor layers, such as an $n^+p^inp^+$ structure (Fig. 15.5) (Grekhov et al. 2005). At the resting condition, the switch is biased with a static voltage, which allows the formation of a depletion region in the reverse biased, p^in region. The electric field is increased but is kept lower than the static breakdown field. When the switch is applied with a fast pulse that has a higher voltage than the static breakdown voltage, ionization occurs in the n region due to the electric field set up by the displacement current. Holes start traveling towards the super-high field region in the p^in junction in the form of an ionization wave, which increases the conductivity of the n region. Meanwhile, the n^+p^i and p^+n junctions are ionized due to the injection of electrons and holes into p^i and n-base layer. With plentiful charge carriers throughout the switch, the dynistor is turned on. The typical switching time is 0.2 ns with a delay of ~ 2 ns. The FID switches are stacked for the high-voltage

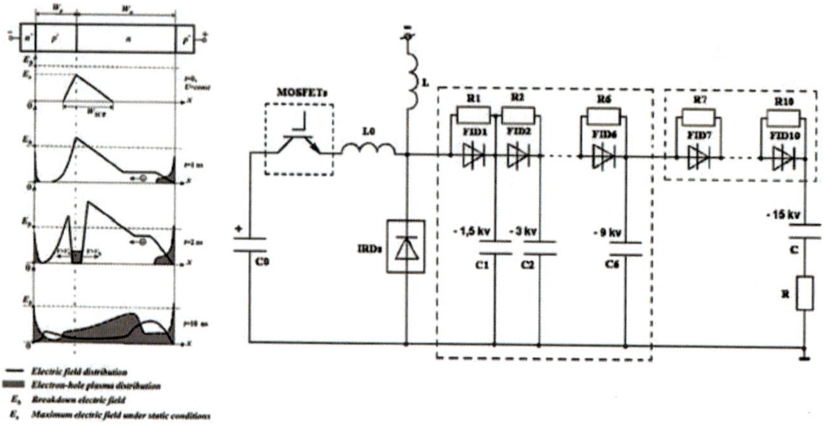

Fig. 15.5 A field ionization dynistor and a stacking circuit (Grekhov et al. 2005)

operation (Fig. 15.5). An opening switch stack, Inverse Recovery Diodes (IRDs), creates the fast trigger voltage (2 ns) needed to cause the first FID (FID1) to turn on due to overvoltage. The sequential discharging of capacitors precharged with higher voltages turn on all the FIDs eventually and so a subnanosecond pulse is formed and delivered to the load resistor R.

Gas and Liquid Switches

A sharpening (or peaking) switch can close within a fraction of 1 ns when it is in the overvolted condition, under which the applied electric field is much higher than the static breakdown field. Several MV/cm is typical. In a high-pressure gas sharpening switch, the applied pulse is of the order of the formative time for electron avalanching (Pincosy 1992). The electric field is so high that the electrodes have a large current emission. Diffuse discharge caused by the copious source of initiating electrons leads to very fast breakdown times. The stray inductance of the switches needs to be minimized, as such millimeter gaps are often used (Lehr et al. 1998). Because the switch gap is so short and the capacitance effect becomes apparent, a pre-pulse is usually observed due to the increased displacement current.

In liquid sharpening switches, oil is preferred as the switching medium. One benefit is that it has a dielectric constant (2.25) close to polypropylene (2.3), which is a commonly used material in the pulsed power systems. The continuity of dielectric constant can avoid the interfacial field enhancement and decrease the probability of unwanted dielectric breakdown. The other benefit is that the high dielectric strength of oil allows for high voltage switching with a short gap. The breakdown of the switch is associated with streamer formation and propagation, a mechanism generally accepted in the liquid breakdown. Electronic processes occur inside low-density, gas bubble-like streamers (Kolb et al. 2008) and lead to the formation of a spark channel.

In an oil switch used for a high voltage impulse radiator (Baum et al. 2004), the oil gap has a distance of 0.76 mm which allows a very high voltage (800 kV) to sharpen to a risetime of 130 ps. Like water switch, oil switch for repetitive operation also needs flow to remove the debris in the electrode gap.

15.3 Basic Pulsed Power Circuits

15.3.1 Combination of a Capacitor and an Inductor

An RLC circuit with a switch is one of the fundamental circuits in pulsed power systems (Fig. 15.6a). This circuit usually forms the final stage of a pulse generator. The prior stage can be a Marx generator or a Tesla transformer working as a high voltage pulse charger. The circuit parameters (R, L, and C) determine the ultimate output waveform. Assuming the capacitor is fully charged by t = 0 and the switch closes at that time, the voltage equation becomes

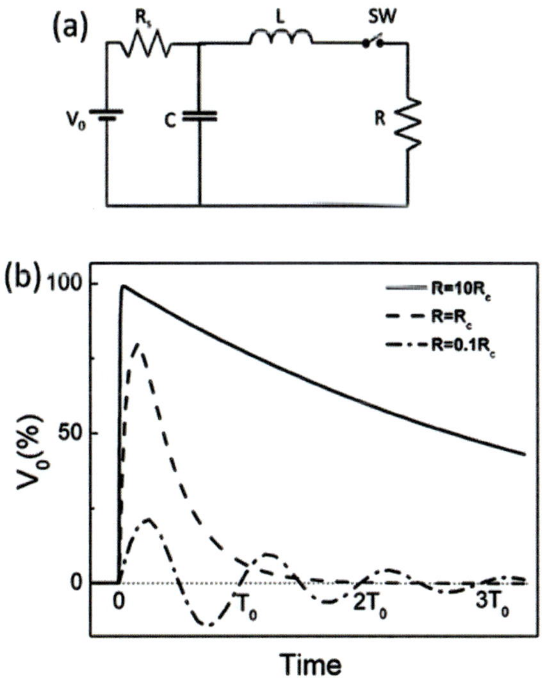

Fig. 15.6 **a** An RLC Circuit that is charged by a constant voltage (V_0). **b** The voltage across the load resistor has a waveform depending on the circuit parameters. Varying R and keeping Z_c (=2 $\sqrt{L/C}$) unchanged results in different waveforms

$$L\frac{di(t)}{dt} + Ri(t) + \frac{1}{C}\int_{0}^{t} i(\tau)d\tau - V_0 = 0 \tag{15.5}$$

Using the Laplace transform, this equation can be solved for the voltage and current in the load resistor. The solution is defined by the initial condition of the circuit. There is no current flow in the inductor and resistor before the switch closes. In addition, immediately after the switch is closed, the impedance of the inductor is higher than that of the resistor. Therefore,

$$L\left(\frac{di(t)}{dt}\right)\Big|_{t=0^+} = V_0 \tag{15.6}$$

where $t = 0^+$ refers to the infinitesimally short time after the switch is closed. The load voltage v_R is a combination of an oscillatory term due to the presence of L and C and an attenuated term due to the dissipation in R James Nilsson (2018). Depending on the speed of the dissipation with respect to the period of the oscillation, v_R is categorized into three forms: underdamped ($(R/2L)^2 < 1/LC$, Eq. 15.7), critically damped ($(R/2L)^2 = 1/LC$, Eq. 15.8), and overdamped forms ($(R/2L)^2 > 1/LC$, Eq. 15.9) (Fig. 15.6b):

$$v_R(t) = \frac{V_0 R}{L\omega_d} e^{-\alpha t}\sin(\omega_d t) \tag{15.7}$$

$$v_R(t) = \frac{V_0}{L} t e^{-\alpha t} \tag{15.8}$$

$$v_R(t) = \frac{V_0 R}{L\omega_d} e^{-\alpha t}\sinh(\omega_d t) \tag{15.9}$$

where $\alpha = \frac{R}{2L}$, $\omega_0 = \frac{1}{\sqrt{LC}}$ and $\omega_d = |\omega_0^2 - \alpha^2|^{1/2}$.

Using a fully controlled switch, a square waveform can be generated and in this case the critically damped condition is preferred. The RC time constant is chosen to be longer than the pulse width across the load. The pulse width is thus controlled by the switch turn-on time. One example is a MOSFET switch used in an RC pulse generator.

15.3.2 *Pulse Forming Line and Pulse Forming Network*

A transmission line, or pulse forming line (PFL), connected in Fig. 15.7a is another common type of energy storage in a pulsed power system. It is a cost-effective way to generate a square pulse. In a charged PFL, two waves of opposite directions coexist:

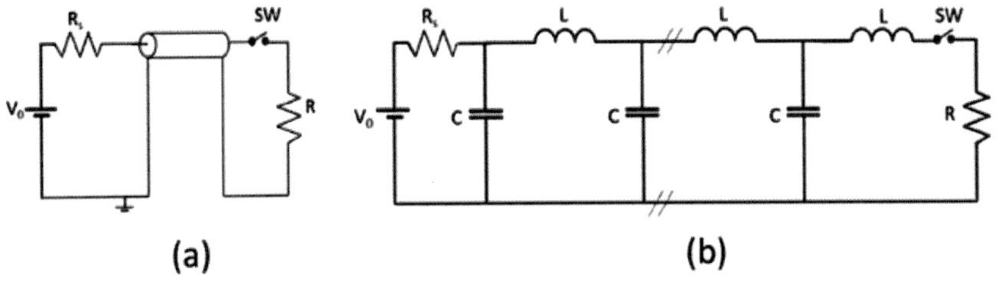

Fig. 15.7 A PFL is charged by a DC source and discharges to a load after closing the switch SW. **a** The line is a coaxial cable. **b** The PFL consists of individual inductors and capacitors

$$V = V_{01}e^{-jkz} + V_{02}e^{+jkz} \tag{15.10}$$

$$I = I_{01}e^{-jkz} + I_{02}e^{+jkz} \tag{15.11}$$

where $k = \omega\sqrt{LC}$ is the wave propagation constant; e^{-jkz} represents the phase of the positive wave and e^{+jkz} the negative wave; Here it is assumed that the line is loss-free, so k is a real number; L is the inductance per unit length of the PFL and C is the capacitance per unit length. For a coaxial cable that has a dielectric with a dielectric constant, $\varepsilon_r\varepsilon_0$, an inner conductor radius *a* and an inner radius of the outer conductor b, $L = \mu_0 \ln(b/a)/2\pi$, $C = 2\pi\varepsilon_r\varepsilon_0/\ln(b/a)$. When the PFL is fully charged to the same voltage as the charger, V_0, there is no net current flow in the line. As the charger can be a DC power supply, which means $k \to 0$, so the voltage relationship holds: $V_{01} = V_{02} = V_0/2$, which makes the total voltage V between the inner conductor and outer conductor V_0. The total current I is therefore zero, meaning $I_{01} = -I_{02}$.

Now if the switch in the line is closed, the positive wave with the amplitude of $V_0/2$ travels to the load. A pulse is seen and the energy is absorbed by R without reflection, assuming that R matches the PFL impedance:

$$R = Z_0 = \sqrt{L/C} \tag{15.12}$$

In the meantime, the negative wave travels away from the load. It is reflected entirely at the charging resistor R_s and becomes directed toward the load with its polarity remaining the same. Until the reflected wave returns to the load, it finishes a round trip (2 *l*). The time it takes thus determines the actual pulse width:

$$T_p = 2l/c = 2l\sqrt{\varepsilon_r\varepsilon_0\mu_0} \tag{15.13}$$

where *l* is the line length and *c* is the wave travelling speed. For a typical coaxial cable, the dielectric material is polyethylene (PE), which has $\varepsilon_r = 2.25$. Thus, a 1 m cable creates a T_p of 10 ns.

When the load resistance is not equal to the cable impedance, a reflection occurs for the wave arriving at the load, which results in an extension of the pulse duration. The reflection coefficient is:

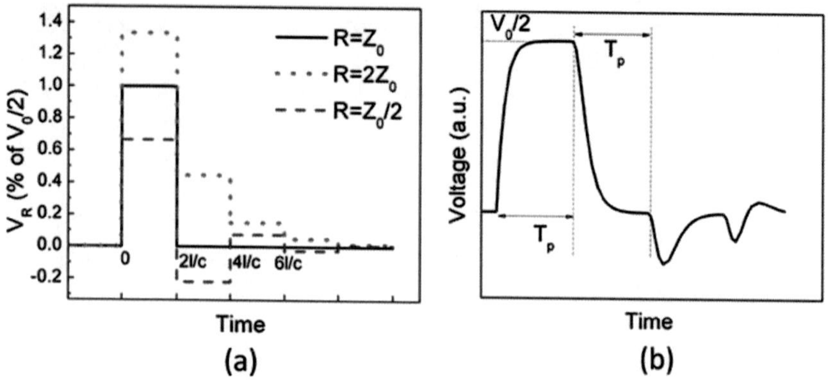

Fig. 15.8 Waveforms produced by a PFL pulser. **a** The load mismatching to the PFL results in different waveforms. **b** An inductance added to the load causes slow rise and fall time of the pulse as well as tails

$$\Gamma = (R - Z_0)/(R + Z_0) \tag{15.14}$$

The transmission coefficient is:

$$T = 2R/(R + Z_0) \tag{15.15}$$

For $R > Z_0$, the waveform consists of steps of pulses decreasing in amplitude. The first amplitude of the pulse is greater than $V_0/2$. For $R < Z_0$, the waveform is bipolar and alternates in every T_p. The first amplitude is lower than $V_0/2$ and the waveform dies down after several T_p (Fig. 15.8a).

In practice, the pulses generated from a PFL are not perfect square. They appear trapezoidal with finite risetime and fall time as well as some tails (Fig. 15.8b). This pulse imperfection is due to the stray inductance and resistance associated with the switching process as well as the inductance of the connection wires among the PFL, the switch, and the load. These inductances are high impedance to the travelling wave and cause it to partially reflect back. Therefore, they limit the rise of the load current. From the energy perspective, the inductances need to be charged in the risetime. After the inductances are fully charged, they are in principle shorted and do not exhibit any impedance. So the impedance matching condition of the PFL and load is met and the load absorbs the incident wave from the PFL. This occurs when the pulses appear flat. At the end of the pulses, the inductances release the energy in the fall time. After another round trip time (T_p), tails caused by the inductance reflection ensue.

Pulse generation using a PFL is convenient if the cable length and its size are not an issue. When a compact system is needed, a pulse forming network (PFN) consisting of individual inductors (L) and capacitors (C) is a better option (Fig. 15.7b). This is usually the case for producing microsecond or longer pulses. The inductance and capacitance are connected periodically, exactly the same way as in the equivalent circuit of a transmission line. The pulse width of a PFN is

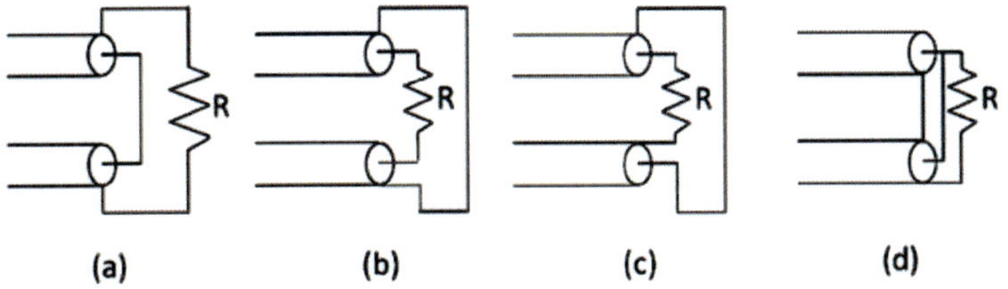

Fig. 15.9 Two transmission lines are connected in series (**a–c**) and in parallel (**d**)

$$T_p = 2n\sqrt{LC} \tag{15.16}$$

where n is the number of L or C. For a matched load, $R = \sqrt{L/C}$, the output voltage is half the charging voltage.

In the PFL (or PFN) generators, it is beneficial to have several lines connected in parallel or in series to form a single PFL in order to match different load conditions. For example, a cuvette loaded with a high-conductivity medium has a low resistance, approximately near 10 Ω. Therefore, five, 50 Ω lines in parallel can be used as a single PFL. If the load consists of a pair of electrodes submerged in a cell medium, it may exhibit a higher impedance, say 200 Ω. Accordingly, four, 50 Ω lines in series can be used as the PFL. Although one can still use a single 50 Ω line as the PFL to drive the 200 Ω load, for the purpose of impedance matching, an extra matching resistor is needed to lower the load resistance. Comparatively speaking, a high impedance PFL (200 Ω) produces faster risetime and better pulse shape than a low impedance PFL (50 Ω), if both PFLs involve the same switch and stray inductances. For a high impedance PFL, one can use a low power but high speed switch, for example, a MOSFET, DE275-102N06A (IXYS, CA), which was originally designed for RF power generation but instead used for generating 10 ns pulses (Frey et al. 2006).

The serial connections of two transmission lines can be made by connecting the inner conductors, the outer conductors, or connecting the outer and the inner conductors (Fig. 15.9a–c). The load resistance is twice the characteristic impedance of the transmission line. For the parallel connection, the outer conductors are connected, and the inner conductors are connected respectively (Fig. 15.9d). The load resistance is half the characteristic impedance of the transmission line.

15.3.3 Blumlein Line Generator

The Blumlein line generator evolves from a PFL. It overcomes the disadvantage of a PFL that only generates half of the charging voltage ($V_0/2$) and enables the full charging voltage across the load (V_0). The load is placed in the breakout section at the middle of a transmission line. Meanwhile, the inner conductor and insulating dielectric are kept intact. The load can be either grounded or floating (Fig. 15.10a, b) depending on whether the switch is grounded. It is important to distinguish these

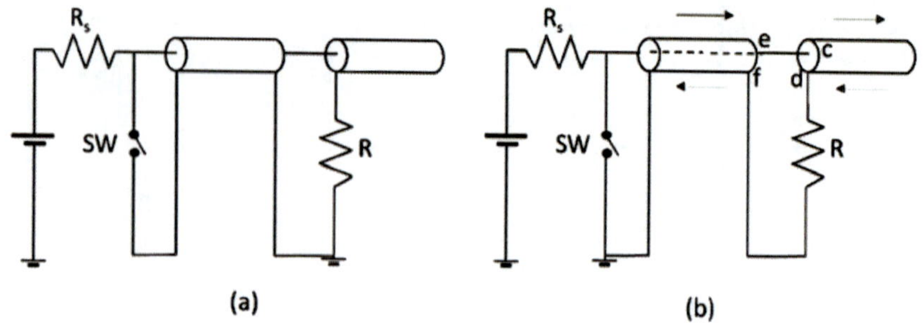

Fig. 15.10 A Blumlein line generator. **a** The load resistor is grounded; **b** the load resistor is floating

two scenarios, as one requires the switch to be grounded and the other does not. A floating switch is triggered from a driver circuit that is isolated from the ground by a transformer or an optoisolator. To measure the voltage drop across a floating load, two probes should be placed at both sides of the load. Each voltage is measured with respect to the actual ground. The absolute voltage drop is obtained by subtracting the voltages, which can be done in a digital oscilloscope.

When a Blumlein line is fully charged and the switch is open, no net current flows although two waves with half of the charging voltage coexist, as described by Eqs. (15.10) and (15.11). At the left end where the switch is located (Fig. 15.10), the reflection coefficient is one, the left-directed wave experiences a 100% reflection and then becomes the right-directed wave with the same amplitude. Likewise, the right-directed wave becomes the left-directed wave at the right end where the line is open, which also causes a 100% reflection. Because of the superposition of these waves, everywhere in the cable has a voltage of V_0. At the load, the reflections on both sides (cd and ef planes, Fig. 15.10b) are not unity. For the convenience of analysis, let's assume R is twice the characteristic impedance ($R = 2Z_0$). On the right side (plane cd), the waves in the cable still consist of two waves in both directions. The left-directed wave originates from the right end's reflection with the voltage of $V_0/2$. The right wave includes two parts: one being the transmitted wave from the resistor's left side (ef plane) and the other being the reflection of the left wave in cd plane:

$$V_0/2 * 1/2 + V_0/2 * 3/2 * 1/3 = V_0/2 \tag{15.17}$$

Note that the reflection coefficient in cd plane is 1/2 and the transmission coefficient in ef plane is 3/2. The voltage transmitted from the ef plane has a 2/3 drop across the resistor and the rest (1/3) goes to the transmission line. In total, $V_{cd} = V_0$.

At the left side of the resistor (ef plane), the left wave in total is still $V_0/2$, which consists of a reflected wave ($V_0/2*1/2$) and a transmitted wave from the right wave ($V_0/2$ *3/2*1/3). The right wave has a voltage of $V_0/2$. So the total voltage $V_{ef} = V_0$. As a result, the net voltage drop across the load, V_{cd}-V_{ef}, is zero.

When the switch is closed, the voltage balance across the load is disrupted. As the left end is shorted, which results in a right-directed wave ($-V_0/2$) in the left cable

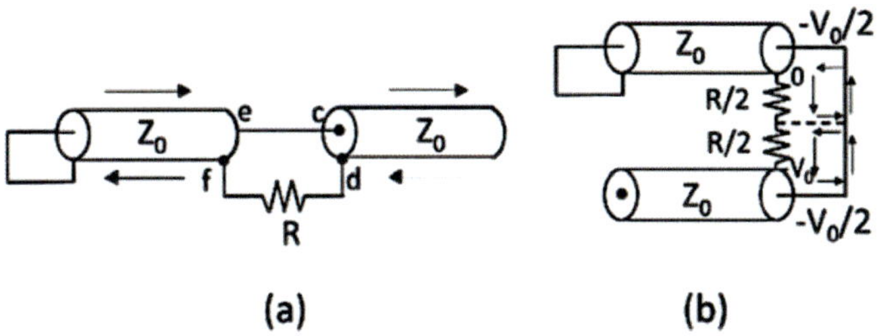

Fig. 15.11 A Blumlein line with a shorted side after closing the switch. **a** the load is a resistor; **b** the load splits into two matched resistors ($R/2 = Z_0$)

(Fig. 15.11a). When it arrives at the ef plane, the reflected wave is $-V_0/2*1/2$, so the total $V_{ef} = -V_0/2 * 3/2$. Meanwhile, the transmitted voltage from the ef plane include

$$V_{cf} = -V_0/2 * 3/2 \tag{15.18}$$

$$V_{cd} = -V_0/2 * 3/2 * 1/3 = -V_0/2 * 1/2 \tag{15.19}$$

So the voltage drop across the load caused by the left wave ($-V_0/2$) is

$$V_{df} = -V_0/2 * 3/2 * 2/3 = -V_0/2 \tag{15.20}$$

Note that V_{cf}, V_{cd}, and V_{df} satisfy Kirchhoff's law.

Simultaneously, the left-wave ($+V_0/2$) in the right cable encounters an impedance discontinuity in the cd plane. The following voltage relations hold:

$$V_{cd} = V_0/2 + V_0/2 * 1/2 \tag{15.21}$$

$$V_{ef} = V_0/2 * 3/2 * 1/3 \tag{15.22}$$

According to Kirchhoff's law, the voltage across the load resistor caused by the left wave ($+V_0/2$) is

$$V_{fd} = V_0/2 * 3/2 * 2/3 \tag{15.23}$$

Note that $V_{df} = -V_{fd}$. Including both the drops (Eqs. 15.20 and 15.23) caused by the left and right waves, the total voltage drop V_{df} across the load becomes $-V_0$. If point f is grounded, a negative pulse that has an amplitude of $-V_0$ is measured at point d. Although the resistance seems to create a discontinuity for both left and right cables, the reflected wave in each cable is canceled by the transmitted wave from the other cable because of the opposite polarity.

The pulse duration of a Blumlein line generator is given by the total length of the Blumline line:

$$T_p = 2l/c \tag{15.24}$$

This relationship can be understood from the concept of one Blumline line being equal to two PFLs connected in series. The matched condition requires that the load resistance is twice the characteristic impedance. The load resistance ($R = 2Z_0$) can be split into half and each provides a matched resistance for the PFLs (Fig. 15.11b), so there is no reflection for the wave that travels to the load. A fictitious line connection can be inserted between the midpoints of the load resistor and the inner conductor line. Since the current in each PFL is opposite and cancels each other, the fictitious line is physically not needed. The pulse across the load resistor disappears until the wave completes the round trip in each PFL, which takes as long as T_p.

A Blumlein line generator is typically constructed with coaxial cables or lumped capacitors and inductors. In the case where the generator size is critical, a polar liquid, such as water, can be used in a Blumlein line. Because water has a large dielectric constant (80) and a high breakdown strength, it is good for compact energy storage. A water-based PFL is commonly used in low impedance, high power, and high energy pulse generation. Water is also used as a switching medium together with the water PFL, which allows for a continuous impedance profile from the PFL to the load. A high voltage (tens of kV), low impedance (10 Ω), water Blumlein line generator is shown in Fig. 15.12 (Sun et al. 2007). It's compact size and tight integration of energy storage and switch allowed for a fast pulse risetime (a few ns). However, for such generator, there are some restrictions: (1) a pulse charging is needed in order for a high voltage output; (2) to expedite the switch recovery for high repetition rate switching, a water flow is needed and adds to the complexity of the generator; and (3) for generating high energy pulses, the electrode erosion occurs rapidly. All these shortcomings limit the use of this generator to a laboratory setting.

15.3.4 Pulse Generators with MOSFET Switches

A. *Unipolar Pulse Generation*

Using a MOSFET switch, nanosecond pulses can be generated in a resistor–capacitor circuit (Fig. 15.13a, b). Each switch is triggered by an isolation transformer, which is driven by one or more low voltage MOSFET drivers. As a voltage spike could occur across the gate and source, a Zener diode is placed at the transformer output to clamp the voltage. The voltage spike has several causes. The first is coupling between the drain and gate through the capacitance C_{gd}, and C_{gs} (Fig. 15.13c). As such, the gate voltage spike is determined by the capacitive divider ratio (C_{gd}/C_{gs}) and dV/dt, as well as the gate drive impedance. Note that a wide-band-gap (WBG) device, which has a low internal Miller capacitance, can inherently reduce capacitive

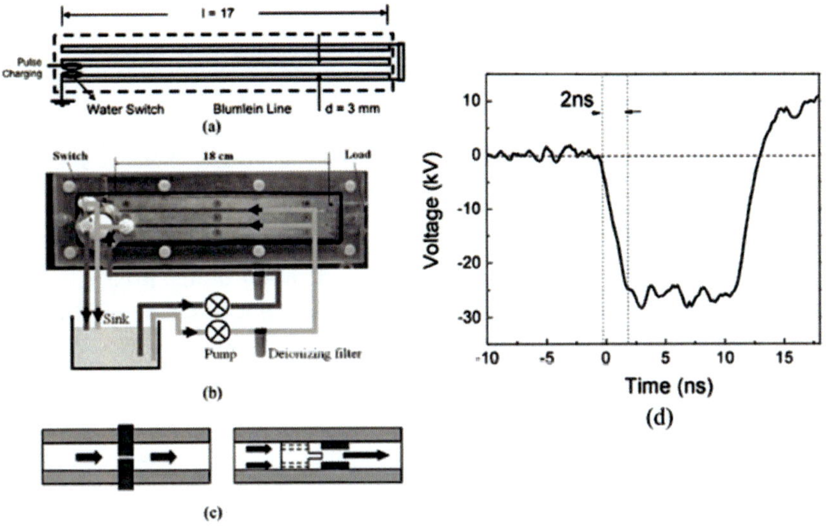

Fig. 15.12 A 10 ns, 10 Ω Blumlein generator with water as energy storage and switching medium. **a** Cross-section; **b** photograph of the generator with a flow system; **c** Cross-section of electrodes; Left: pin-pin electrodes with transverse flow. Right: coaxial electrodes with axial flow; **d** output waveform (Sun et al. 2007)

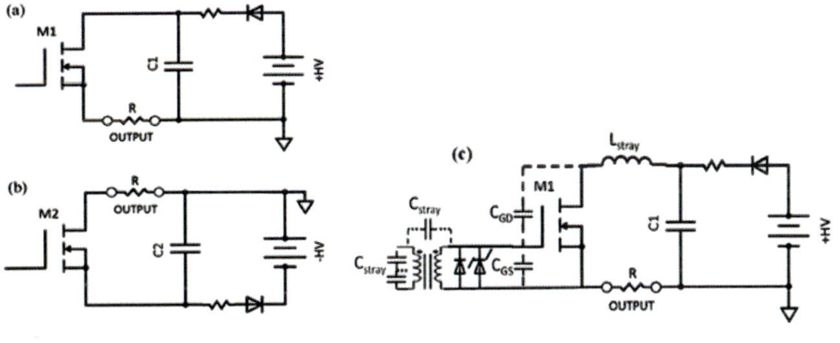

Fig. 15.13 A N-type MOSFET switch is used to generate **a** positive pulses and **b** negative pulses. **c** The gate is triggered by an isolating transformer (Ryan et al. 2018)

coupling. Still, one must minimize parasitic gate-drain board capacitance, which can be done by carefully avoiding parallel footprints. The second cause is uncoupled stray inductance, which is introduced by extra wiring in the trigger loop. A tight loop, which can be implemented on a well-laid PCB, is critical to minimizing these inductances. The third likely cause of a voltage spike is stray capacitance in the transformer. This capacitance lies in inter-winding and can be minimized by using just a few turns of winding. Considering the factors capable of producing exceeded voltage between the gate and source, emplacing a Zener diode can prevent breakdown in the gate-source region. Additionally, one can place another general-purpose diode in parallel to suppress the negative voltage swing during switch turn-off. Note that a large capacitance circuit beneficially reduces such voltage but may reduce speed as a tradeoff.

The load, in principle, is connected to a voltage source (the pre-charged capacitor), so the pulse duration is determined by the switched-on time of the period. In the case of a mid-range load impedance, the load voltage will drop, mainly due to the short RC time constant governed by the load resistance (R) and the capacitance (C). As a MOSFET's turn-on resistance is usually a fraction of 1 Ω, the load resistance dictates the total time constant. In the case of a high load impedance (~ 1 kΩ), the pulse duration may not be indicated by the driver's turn-on time, but instead by the appearance of a tail. This is because the MOSFET switch must deplete its charges of both C_{gs} and C_{gd} before returning to the "off" state. The duration of the tail derives from the extended time constant and is associated with a slow C_{gd} discharging time.

In cases where a load has a very low resistance (< 10 Ω), such as with blood samples, the turn-on resistance can matter and create a substantial voltage drop across the switch, possibly resulting in a significantly lower load voltage than the charging voltage. Multiple switches can, however, be paralleled to lower turn-on resistance. In such a case, the inductance (L) that exists due to the circuit loop wiring must ideally be minimized. The inductance not only affects pulse rise time, but also creates ringing in the leading edge (and falling edge) of the waveform as the circuit operates in the under-damped mode. Figure 15.14a shows a pulse with a duration of 80 ns (full width at half-maximum). The corresponding pulse spectrum extending to 10 MHz is shown in Fig. 15.14b. Although the charging voltage was 1 kV, the actual voltage was measured as 900 V and the rise time approximately 30 ns (10–90%).

B. *Biphasic Pulse Generation*

The switch-capacitor structure can be made into modules. Two such modules (Fig. 15.15a) were configured to produce biphasic pulses. One module has a positive

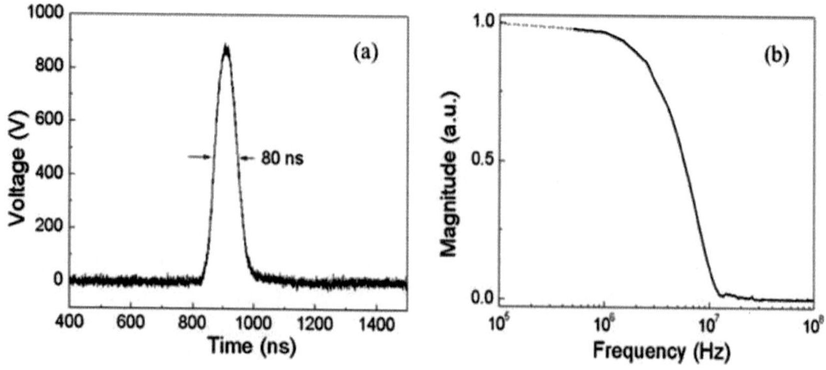

Fig. 15.14 The shortest pulses generated from a single phase 1 kV pulse generator, shown in both the time (**a**) and frequency (**b**) domains

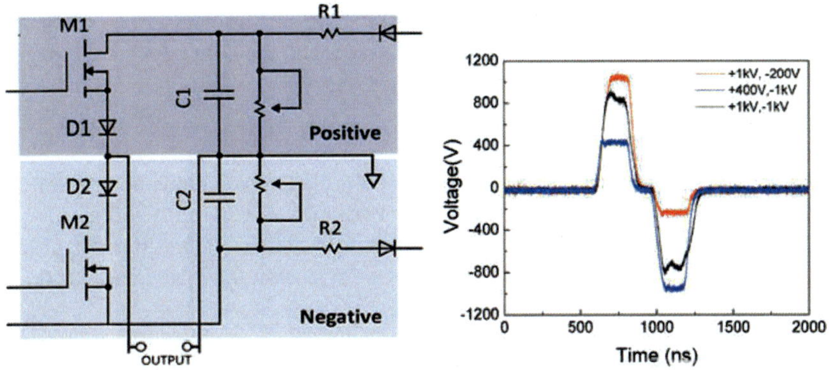

Fig. 15.15 Combining the positive and negative modules creates a biphasic pulse generator. Biphasic pulses were produced by combining two fundamental modules to feed the same load. The figure shows three waveforms for charging voltages: + 1 kV, − 200 V (red), ± 1 kV (black), and + 400 V, − 1 kV (blue). The pulse duration for the two modules was set equally at 200 ns

charging voltage and the other has a negative one. While the switches and capacitors are identical, the switch connection in the negative module is drain-to-ground (and source-to-negative) charging, which is opposite to that in the positive module. The two modules drive the same load resistor. The diode D_1 (D_2) is used to block the reverse current arising from the co-phase outputs in the multiphase configuration. Because of the limitation of the switch's maximum hold-off voltage, the output pulse magnitude is < 1 kV. The pulse width can be adjusted as short as 80 ns and up to any length above that, limited by the RC time constant. The advantage of this pulse generator is that the circuit is very simple, but highly robust. Because each module is independently triggered and charged, almost any set of biphasic pulses, symmetric or asymmetric, can be created. The voltage and pulse duration can be differentially adjusted. One can also trigger just one module so that either a positive or negative pulse is delivered. A delay of any length can be inserted in any interval. This configuration can serve most applications that require low voltage, nanosecond pulses.

When the charging voltages for C_1 and C_2 were set to ± 1 kV, the actual voltage produced across the load resistor (100 Ω) was approximately ± 800 V; the output voltage did not reach the full value of the charging voltage (Fig. 15.15b). When one of the charging voltages was lowered, however (for example, to 400 V), the other output reached its full value (− 1 kV). The positive output could also reach 1 kV if the negative charging voltage were lowered to a voltage less than − 800 V (for example, − 200 V). This is due to the coupling of the positive module and the negative module through the switches' parasitic capacitances, as discussed in Xiao et al. (2018). When setting the minimum interval to trigger the biphasic pulses, a delay time generally should be given to avoid a short circuit. However, the circuit still functions if the minimum delay is zero. This is because the switches must go through the rise and fall phase with some transient resistance, preventing them from being completely on simultaneously. As a result, the actual high voltage biphasic pulses may not have the precise pulse durations as the trigger pulses. Rather, they can be off by 30 ns, as determined by the overlap time of the switches' turn-on and turn-off phases.

C. *Multiphasic Pulse Generation*

The modularized design allows a pulse generator to be flexible in generating multiple pulses that have alternating polarities. For example, as the schematic shows in Fig. 15.16, four modules consisting of two sets each of identical positive and negative modules can generate multiphasic pulses to drive the same load. Diode D_1 (D_3)

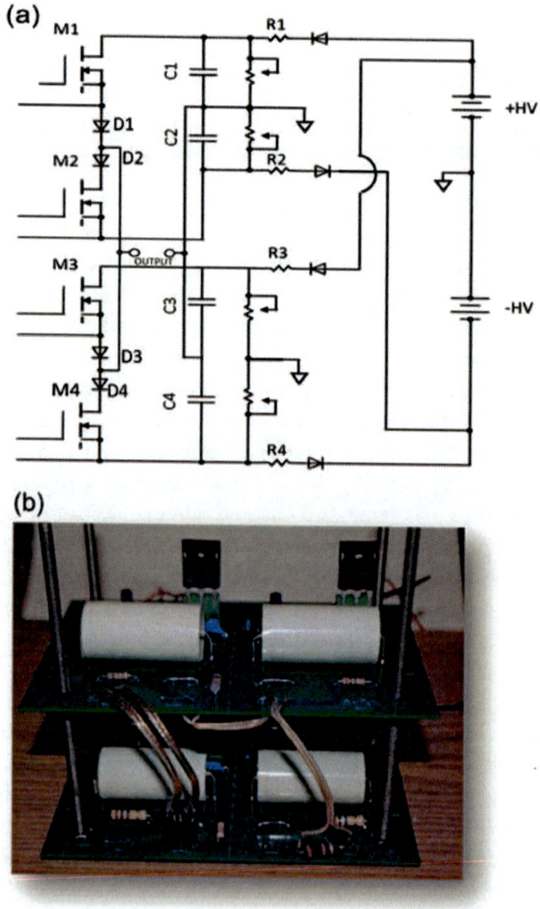

Fig. 15.16 Multiphasic pulses can be produced by combining **a** four fundamental modules, M1-M4, to supply the same load. Each module can be charged with different voltages and is responsible for producing one phase of the pulse. **b** Photograph of the custom-built modules based on the circuit in Fig. 15.5a, which generate a multiphasic wave with two positive phases and two negative phases (Ryan et al. 2018)

blocks the current from C_3 (C_1) and the diode D_2 (D_4) blocks the current from C_4 (C_2). This avoids coupling among the co-phase modules. This circuit could be a powerful way to produce very high repetition rate RF pulses. Three multiphasic pulses with single-phase pulse widths of 80 ns, 600 ns and 1000 ns from this configuration are shown in Fig. 15.17. Their pulse spectra are shown in Fig. 15.17d. When the phase duration changes, the center frequency shifts. This generator also can vary the pulse duration from phase to phase, creating asymmetric, bipolar pulses. If using four separate chargers, it has the ability to vary the pulse amplitude from phase to phase.

D. *Stacking MOSFETs for High Voltage Switching*

A recent advance in wide-bandgap devices has extended solid state switches to the high current and high voltage region. Commercially available MOSFET switches can handle current up to 300 A or voltage up to 10 kV, but there is still no single switch that has both a high current and high voltage rating. To extend either the voltage or current capability, either a parallel or a serial operation is needed. There are at least two types of stacking configurations distinguished from their triggering circuits. One is to equip each switch with a separate triggering circuit through a transformer or an optoisolator. The other circuit is to trigger the switches sequentially, starting from the switch that is source-grounded.

D.1 A High Voltage, Nanosecond Pulse Generator with Separate Triggers

When MOSFETs are connected in series for switching high voltages, in parallel with the MOSFET's drain and source an RC snubber circuit is needed. This circuit mainly protects the switch from overvoltage in the transient phase (either turning on or off). The RC snubber circuit also helps to protect the switch in case of misfiring or when the switch is not fully turned on. Also in parallel with the switches are stacking high voltage Zener diodes. They are used to create a crowbar protection in the case of misfiring or any overvoltage across the switch. The rule of thumb in choosing the resistance and capacitance is that the RC time constant should be on the same order of the switch turn-on time:

$$R_s C_s \cong t_{on}. \tag{15.25}$$

In a design example (Ryan et al. 2018), the switch turn-on time, t_{on}, was found to be ~ 30 ns, so the snubber resistance R_s was chosen to be 150 Ω and the capacitance, C_s, to be 200 pF.

In a biphasic pulse generator that consists of two generators for two opposite pulses, as Fig. 15.17 shows, five switches were stacked in each of the positive phase and negative phase. That, in principle, allowed pulses of ± 5 kV to be generated. The corresponding circuit (Fig. 15.18a) indicates that each switch was controlled by a driver (TC4422, Microchip Technology), which was powered from an isolated DC power module. In total, ten such DC modules were used for both phases. To trigger the drivers, optoisolators were used to create differential pulses to turn on the drivers. These optoisolators were also powered by the same isolated DC power modules.

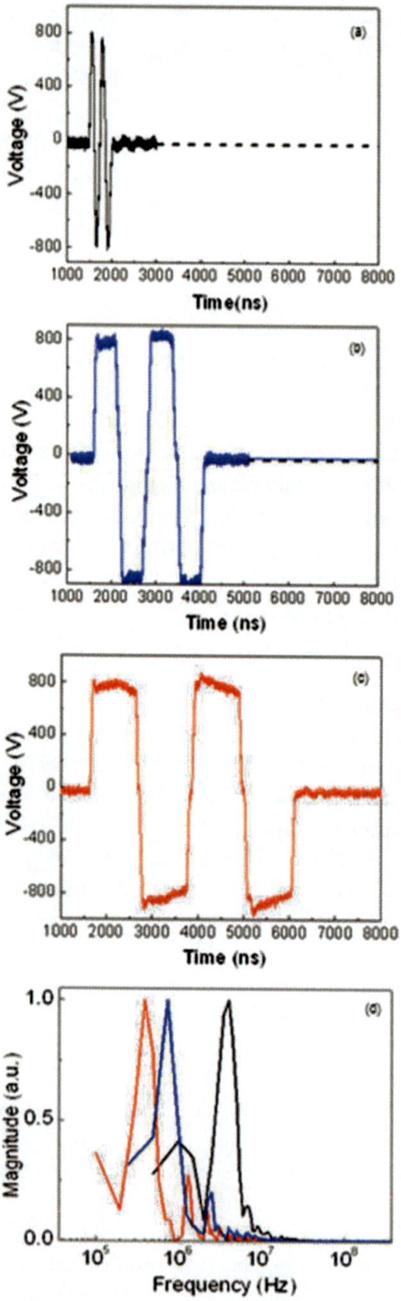

Fig. 15.17 Multiphasic pulses consisting of four phases of alternating polarities and their spectra. The phase widths are **a** 80 ns, **b** 500 ns and **c** 1000 ns. **d** The center frequency shifts as the phase width changes. The transient time from the positive to negative phase is 78 ns for (a), and 150 ns for (b) and (c)

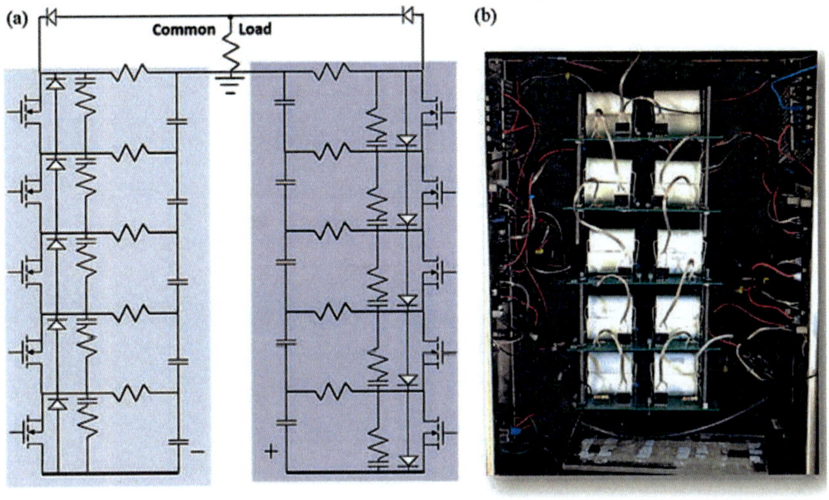

Fig. 15.18 A high voltage biphasic pulse generator with independently-controlled outputs. **a** circuit diagram negative and positive phase elements are highlighted in light grey and dark grey boxes, respectively. Each phase can be charged up to 5 kV, determined by **b** the five stacking switches

By comparison, a typical uniphasic pulse with a 5 kV peak voltage is shown in Fig. 15.19a for a load resistance of 100 Ω. The pulse duration can be varied as narrow as 200 ns. For the biphasic pulses (Fig. 15.19b), the output pulse has a maximum positive and negative peak of 4.5 kV corresponding to a 5 kV charging voltage. A 10% of voltage loss was measured, which can be attributed to the mutual coupling across stacks (Xiao et al. 2018).

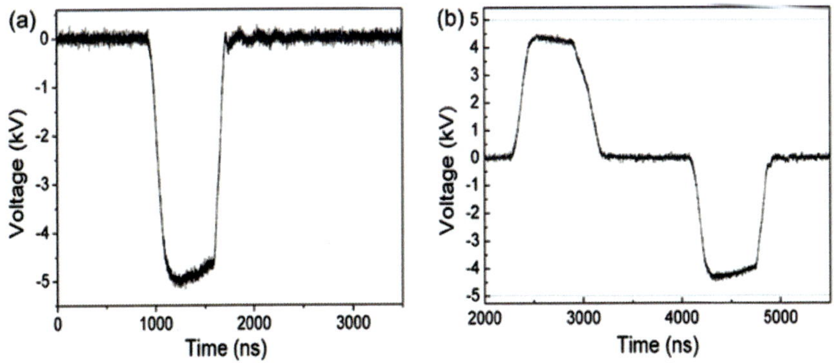

Fig. 15.19 An output voltage for a 5 kV, 600 ns **a** uniphasic pulse and **b** biphasic pulse. The charging voltages were ± 5 kV. A voltage loss (10%) was observed for both voltages

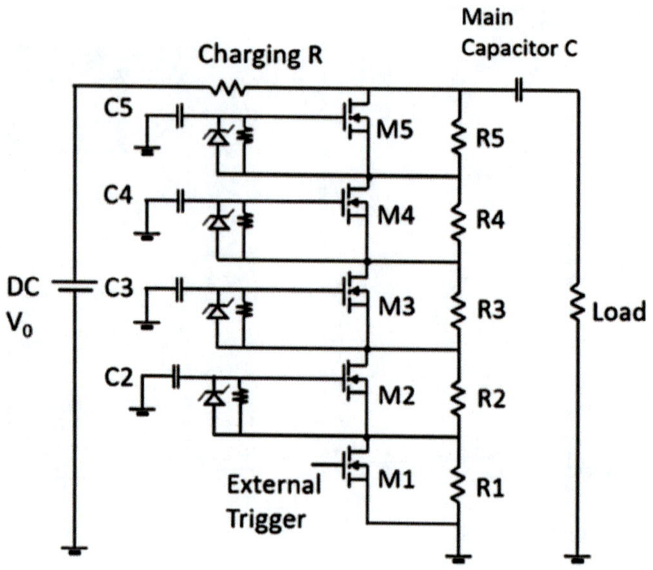

Fig. 15.20 A 5-Stage MOSFET circuit

D.2 A High Voltage, Nanosecond Pulse Generator with One Trigger

A 5-stage MOSFET stacking circuit with only one trigger is shown in Fig. 15.20. The MOSFET M1 is source-grounded and driven externally by a MOSFET driver. The gates of the remaining MOSFETs (M2 through M5) are connected to the trigger capacitors (C2–C5). Zener diodes are placed across the gates and sources of all MOSFETs except M1 to clamp the trigger voltage. The states of these switches depend on the charging and discharging of these capacitors. When a DC power supply is connected to the circuit, the trigger capacitors are charged through the biasing resistors (R1-R5) and Zener diodes.

The switches are triggered as soon as the capacitors begin to discharge. When M1 is turned on, the voltage drop across M1 is zero. The capacitor, C2, which has previously been charged, now finds a path to discharge to M2's gate to source capacitance C_{gs2}. The voltage across the gate to source of M2 is given by Baker and Johnson (1992), Baker and Johnson (1993)

$$\Delta V_{gs2} = \Delta V_{d1} \cdot \frac{C2}{C2 + C'_{gs2}} \tag{15.26}$$

where C'_{gs2} is the effective gate to source capacitance of M2, given as $C'_{gs2} \cong C_{gs2} + A_{V2}C_{gd2}$. The gain A_{V2} is determined by the ratio of the change in drain-source voltage to the gate-source voltage change of a MOSFET. The increase of ΔV_{gs2} results in M2 being turned ON and hence the voltage across M2 now drops to zero. The capacitor C3 likewise discharges to C_{gs3} through M2 and M1. The sequential discharges of the remaining capacitors lead to the closing of the entire switch stack. Note that the capacitances for later stages are not needed as large as C2 because they are charged to higher voltages.

Ideally, all five switches should be turned on instantaneously to prevent overvoltage from occurring across some of the switches. However, the switches' finite turn-on time causes a delay from M1 to M5, which could put the top switches at the risk of overvoltage when lower switches were already turned on. Having a snubber circuit for each switch can potentially avoid that problem. In the turn-off phase, it is possible that the bottom switch is turned off sooner than the rest switches, which can cause the bottom switch to be overvolted if the main capacitor retains a large percentage of its charges. Hence, a safe strategy to protect the bottom switch is to discharge the main capacitor entirely to the load before the bottom switch (M1) is set to open. In this case, all five switches can recover much earlier than the main capacitor is recharged to the full voltage, V_0. This strategy means that this stacking circuit can be used as a closing switch, typically found in a PFL or PFN generator.

15.4 High Voltage Boosters

15.4.1 *Transformer*

A transformer is usually used as a charging circuit for a PFL equipped with a self-closing switch (Katsuki et al. 2001; Lim et al. 2011; Reed 1988; Rohwein 1979). High voltage pulses are then generated upon the switching firing. A typical transformer consists of two windings wound around a transformer core, which is usually made of a magnetic material that has a high permeability and helps to confine and guide the magnetic flux. In high voltage applications, an air core is used to avoid the core saturation. As the secondary winding is electrically isolated from the primary winding, its voltage can be either positive or negative simply by switching the ground connection. A differential voltage can be obtained if the output is not grounded but instead kept floating.

In an ideal transformer with a perfect flux coupling, the leakage flux set up from the primary winding is negligible compared to the common flux that links the primary winding and the secondary winding. The voltage transformation ratio (voltage gain) is directly related to the number of turn ratio. A practical transformer, however, has an imperfect coupling due to the flux leakage manifested in the leakage inductance (Fig. 15.21). The transformer also has a winding resistive loss and core losses including eddy current loss and hysteresis loss. For high frequency operation (or the pulse mode), the electric coupling between the primary and the secondary windings, reflected in the capacitance C_{ps}, lowers the voltage gain.

Neglecting the coupling capacitance and the resistive loss of the transformer, considering the coupling coefficient M, the voltage across the load capacitor C_2 is described as

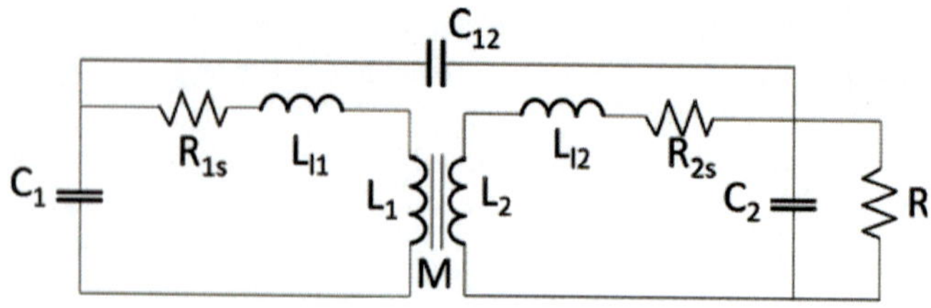

Fig. 15.21 The equivalent circuit of a practical transformer

$$V_2 = \frac{Vk}{C_2\sqrt{L_1L_2}} \frac{\cos(\omega_b t) - \cos(\omega_a t)}{\omega_a^2 - \omega_b^2} \tag{15.27}$$

where $\omega_a^2, \omega_b^2 = \frac{1}{2(1-k^2)}\left(\frac{1}{L_1C_1} + \frac{1}{L_2C_2}\right) \pm \frac{1}{2(1-k^2)}\sqrt{\left(\frac{1}{L_1C_1} + \frac{1}{L_2C_2}\right)^2 - \frac{4(1-k^2)}{L_1C_1L_2C_2}}$ and $k = \frac{M}{\sqrt{L_1L_2}}$.

The waveform is determined by the difference between oscillations in the primary and the secondary circuit (Smith 2002). The optimal coupling coefficient, k, is determined by $k_n = \frac{2n-1}{2n^2-2n+1}$ where n is a positive integer representing the sequential voltage excursion. Because the first voltage swing (n = 1, k = 1) will not reach the maximum in the air core transformer due to poor coupling, the maximum voltage is seen on the second voltage excursion (n = 2, k = 0.6). Therefore, k needs to be tuned by controlling the magnetic flux between the primary and secondary, which is strongly influenced by the distance between the coils.

A more practical method can be used to the design the desirable waveform, which is done by focusing on the times of the first and second voltage peaks and the maximum theoretical gain. The location of the first and second voltage peak are:

$$\tau_1 = \frac{\pi\omega}{2} \text{ and } \tau_2 = \frac{3\pi\omega}{2} \tag{15.28}$$

where ω is the uncoupled resonant frequency in the dual resonant mode $\left(\omega = \frac{1}{\sqrt{L_1C_1}} = \frac{1}{\sqrt{L_2C_2}}\right)$ (Korioth et al. 1999; Shotts et al. 2005).

A transformer with a resonant frequency of 8 MHz for fast charging was designed in Petrella et al. (2016). It was a spiral-strip transformer, which consisted of a single-strip primary and multiple concentric strips for the secondary. The geometry of the primary (approximately 66 nH) was designed using the self-inductance formulas for a single-layer circular coil (Lundin 1985). The secondary had an approximate inductance of 2 μH. The primary capacitor bank was 6.6 nF. The load capacitor was formed out of a capacitor array (N4700, Murata) with a total capacitance of 200 pF. The coupling coefficient was estimated by measuring the self-inductance of the primary coil when the secondary coil was both shorted and not present. As a result, the coupling coefficient (k) was determined to be 0.54. Figure 15.22 shows typical waveforms for the primary voltage, secondary voltage, and secondary current. In the

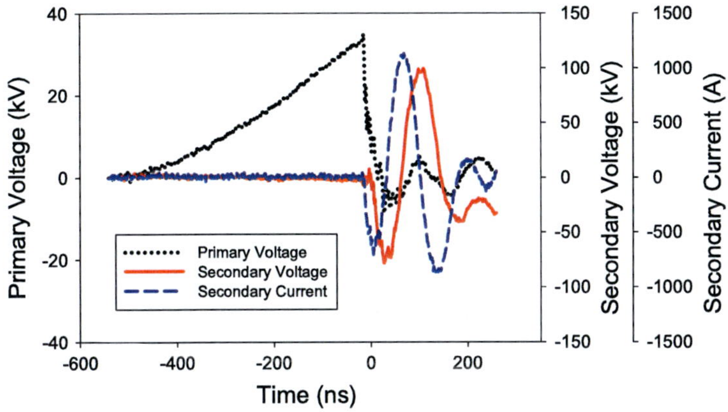

Fig. 15.22 An air-core, dual resonance transformer producing a 100 kV charging voltage (Petrella et al. 2016)

resonance condition, the electrical energy is converted to the magnetic energy or vice versa in a periodic fashion. An efficient transfer of energy occurs when the magnetic energy reaches its minimum while the electric energy simultaneously reaches its maximum. This can be clearly seen in the minimal current at ~ 120 ns, at which the secondary voltage reached its peak.

15.4.2 Marx Generator

Marx generators are the most popular high voltage pulse generation for output voltage that is higher than the charging voltage (Carey and Mayes 2002; Fitch 1971; Morrison and Smith 1972). The schematic of a typical Marx generator is shown in Fig. 15.23. Each stage consists of a capacitor, two resistors (or inductors), and a switch. The capacitors are charged in parallel through resistors and discharged in series through switches to attain high voltages. The switches can be turned on either for all the

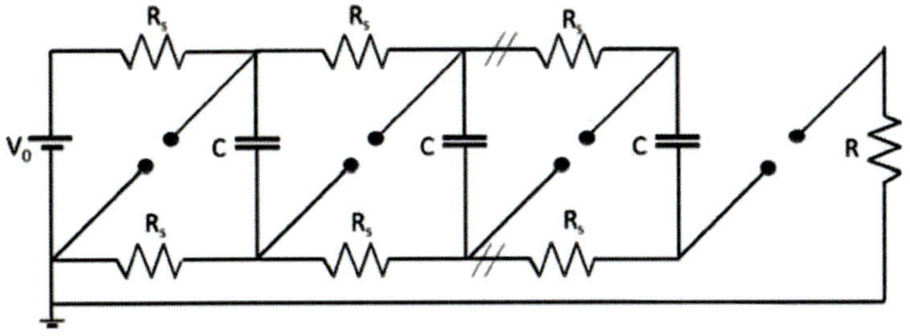

Fig. 15.23 A Marx generator consists of distributed resistors and capacitors

energy stored in the capacitors drained by a load (as in the spark gap switch case) or for a fixed period that equals the pulse duration (as in the solid-state switch case). The charging time constant of the last (N^{th}) capacitor is given by (Francis 1976; Heeren et al. 2007)

$$\tau_c = \alpha R_s C N^2$$

where $\alpha = 90\%$. When the Marx generator is fully charged, the lowest gap is allowed to fire by an external triggering. This makes the voltage across the second switch equal to twice the input voltage and results in self-closing. The remaining switches likewise experience overvoltage and sequentially fire, which is called the erection of the Marx generator. When the capacitors are discharging to the load resistance R, they also are discharging to the charging resistors (R_s). To make the Marx efficient, the time for discharging through R_s should be much longer than the pulse duration of the load.

The first order approximation of a Marx generator is an RLC circuit where $L_{eq} = N \times L_{stage}$ and $C_{eq} = \frac{C_{stage}}{N}$. L_{stage} and C_{stage} are the total inductance and capacitance of each stage. To avoid ringing, the resonance period $T = 2\pi\sqrt{L_{eq}C_{eq}}$ should be much longer than the pulse duration. Hence, the Marx operated in the critically damped condition produces a monopolar pulse whose duration is given by

$$t_{pulse} = \frac{C_{stage}}{N} \times R_{load} = C_{eq} \times R_{load} \tag{15.29}$$

For a spark-gap Marx, if the inter-stage coupling capacitance C_{and} the stray capacitance C_s to ground are significant, the capacitive coupling could occur between stages. As the first gap of the Marx is triggered, the charges among all the capacitances redistribute. The overvolted condition of other gaps may be disrupted and therefore breakdown delays can be expected, which makes the erection of Marx deviate from its ideal condition.

The spark-gap Marx possesses passive elements like resistors or inductors for charging the capacitors and preventing self-discharges of the capacitors. These elements increase the time required for the charging of the capacitors and limit the pulse repetition frequency. They also lower the efficiency if resistive charging is used. The charging elements can be replaced by active components such as solid-state switches. Furthermore, the spark gaps can be replaced by solid switches as a circuit show in Fig. 15.24 (Canacsinh et al. 2008).

The capacitors are charged/discharged via the power semiconductor switches (IGBTs in this case). The circuit consists of source module comprising of the DC power supply and an IGBT T0 with an anti-parallel diode and a diode D0, to makes sure that the power supply is disconnected from the circuit while the capacitors are discharging. Each stage of the circuit consists of five switches T_{ai}, T_{bi}, T_{ci}, T_{di} and T_{ei} (each with an anti-parallel diode), a diode D_{ei} and two energy storing capacitors C_n and C_{n+1}. The circuit is capable of generating negative pulses and positive pulses. In the charging phase, the switches T_0, D_0, T_{ai}, D_{bi}, T_{ei} and D_{ei} are on while the

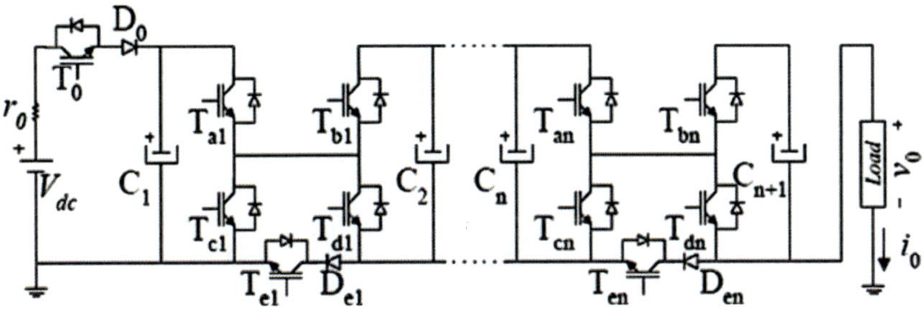

Fig. 15.24 A solid state Marx pulse generator (Canacsinh et al. 2008)

remaining switches T_{bi}, T_{ci}, T_{di} are off. The capacitors are charged in parallel and the voltage across the load is zero.

When the Marx fires to generate negative pulses, the switches T_{bi} and T_{ci} are on, whereas the remaining switches are off. All the capacitors (excluding C_1) discharge to the load. As a result, the load resistor sees a pulse with amplitude equal to

$$V_0 = nV_{dc} \tag{15.30}$$

with V_{dc} being the input voltage.

To generate positive pulses, the switches T_{ai}, and T_{di} are on and all the remaining switches are off. All the capacitors, except C_{n+1}, discharge to the load. As a result, the load resistor sees a pulse with amplitude shown in Eq. 15.30.

The advantage of this generator is that it enables the usage of the half-bridge semiconductor structure ensuring the voltage across each IGBT equals the voltage of each capacitor. Since the charging current is limited by the internal resistance of the elements, which is usually small, the circuit allows the kHz operation. Though the operation of the circuit looks simple, the difficulty lies in the control of the switches. Some switches are triggered from the high-side drivers, which need to be energized by a floating power supply. The switches must be synchronized with nanosecond resolution, and therefore the accuracy of the switch trigger circuits is important. There must be sufficient transition time when switches alternate from being on and off, otherwise there is a chance for the capacitors to discharge inside the stages, rather than through the load.

To produce subnanosecond pulses, a Marx generator is equipped with a peaking circuit that has a peaking capacitor and a peaking switch (Fig. 15.25) (Schoenbach et al. 2008b). The Marx charges the peaking capacitor (C_0) as a transition section before directly connects to a load. The peaking capacitor is generally at least an order of magnitude less than the total capacitance of the Marx (C_{eq}) (Camp 2012; Heeren et al. 2007). As a result, a resonance charging occurs through a stray output inductance and the voltage across C_0 is doubled. The peaking switch is overvolted to close and so a pulse with almost zero ristime can be created. Besides the short pulse, a longer pulse tail associated with discharging C_{eq} also appears to the load. A tail cut switch is thus needed to shorten the pulses to subnanosecond range.

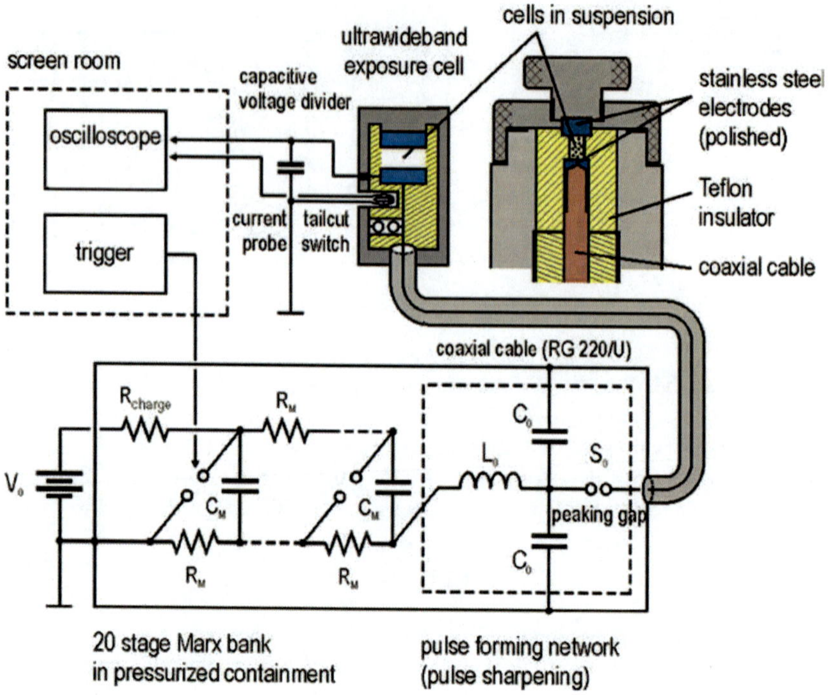

Fig. 15.25 A Marx generator with a peaking circuit (Schoenbach et al. 2008b)

References

Baker RJ, Johnson BP (1992) Stacking power MOSFETs for use in high speed instrumentation. Rev Sci Instrum 63:5799–5801

Baker RJ, Johnson BP (1993) Series operation of power MOSFETs for high speed, high voltage switching applications. Rev Sci Instrum 64:1655–1656

Barkhordarian V (1997) Application Note AN-1084 Power MOSFET Basics

Batarseh I (2011) Power electronics handbook. Butterworth-Heinemann, Burlington

Baum CE, Baker WL, Prather WD, Lehr JM, Loughlin JPO, Giri DV, Smith ID, Altes R, Fockler J, McLemore DM et al (2004) JOLT: a highly directive, very intensive, impulse-like radiator. Proc IEEE 92:1096–1109

Braginskii SI (1958) Theory of the development of a spark channel. Soviet Phys JETP 34:1548–1557

Brown D, Martin D (1987) Subnanosecond high-voltage pulse generator. Rev Sci Instrum 58:1523–1529

Camp JT (2012) Synergistic effect of subnanosecond pulsed electric fields and temperature on the viability of biological cells. Ph.D. thesis, Norfolk, VA: Old Dominion University

Canacsinh H, Redondo LM, Silva JF (2008) New solid-state Marx topology for bipolar repetitive high-voltage pulses. In: IEEE power electronics specialists conference, 15–19 June 2008, pp 791–795

Carey WJ, Mayes JR (2002) Marx generator design and performance. In: Conference Record of the Twenty-Fifth International Power Modulator Symposium and High-Voltage Workshop, 30 June-3 July 2002, pp. 625–628

E. Kuffel WSZaJK. 2000. High Voltage Engineering Fundamentals: Newnes.
Engel TG, Donaldson AL, Kristiansen M (1989) The pulsed discharge arc resistance and its functional behavior. IEEE Trans Plasma Sci 17:323–329
Fitch RA (1971) Marx - and Marx-Like—high-voltage generators. IEEE Trans Nucl Sci 18:190–198
Francis JF (1976) High voltage pulse techniques. Plasma Laboratory, Texas Tech University, Pulsed Power Lecture Series
Frey W, Sack M, Wuestner R, Mueller G (2009) Gas-insulated self-breakdown spark gaps: aspects on low-scattering and long-lifetime switching. Acta Physica Polonica A 115
Frey W, White JA, Price RO, Blackmore PF, Joshi RP, Nuccitelli R, Beebe SJ, Schoenbach KH, Kolb JF (2006) Plasma membrane voltage changes during nanosecond pulsed electric field exposure. Biophys J 90:3608–3615
Frost CA, Martin TH, Patterson PE, Rinehart LF, Rohwein GJ, Roose LD, Aurand JF (1993) Ultrafast Gas Switching Experiments. In: 9th IEEE international pulsed power conference, 21–23 June 1993, p 491
Godignon P, Soler V, Cabello M, Montserrat J, Rebollo J, Knoll L, Bianda E, Mihaila A (2017) New trends in high voltage MOSFET based on wide band gap materials. In: 2017 international semiconductor conference (CAS), 11–14 Oct 2017, pp 3–10
Grekhov IV, Korotkov SV, Stepaniants AL, Khristyuk DV, Voronkov VB, Aristov YV (2005) High-power semiconductor-based nano and subnanosecond pulse generator with a low delay time. IEEE Trans Plasma Sci 33:1240–1244
Heeren T, Camp JT, Kolb JF, Schoenbach KH, Katsuki S, Akiyama H (2007) 250 kV sub-nanosecond pulse generator with adjustable pulse-width. IEEE Trans Dielectr Electr Insul 14:884–888
Heffernan LK, Curry RD, McDonald KF (2005) A fast, 3MV marx generator for megavolt oil switch testing with an integrated abramyan network design. In: 2005 IEEE pulsed power conference, 13–15 June 2005, pp 596–599
Huiskamp T, Heesch EJMv, Pemen AJM (2015) Final implementation of a subnanosecond rise time, variable pulse duration, variable amplitude, repetitive, high-voltage pulse source. IEEE Trans Plasma Sci 43:444–451
Husain E, Nema RS (1982) Analysis of Paschen Curves for air, N2 and SF6 using the townsend breakdown equation. IEEE Trans Electr Insulation EI-17:350–353
Hussey TW, Davis KJ, Lehr JM, Roderick NF, Pate RC, Kunhardt E (1999) Dynamics of nanosecond spark-gap channels. In: 12th IEEE international pulsed power conference, 27–30 June 1999, pp 1171–1174
Ivanov BV, Smirnov AA, Shevchenko SA (2016) A study of charge losses in 4H-SiC drift step recovery diodes (DSRD). In: 2016 IEEE NW Russia young researchers in electrical and electronic engineering conference (EIConRusNW), pp 51–52, 2–3 Feb 2016
James Nilsson SR (2018) Electric circuits, 11th ed. Pearson
Katsuki S, Takano D, Namihira T, Akiyama H, Majima T (2001) Repetitive operation of water-filled Blumlein generator. Rev Sci Instrum 72:2759–2763
Kolb JF, Joshi RP, Xiao S, Schoenbach KH (2008) Streamers in water and other dielectric liquids. J Phys D Appl Phys 41:234007
Korioth JL, Copeland RP, Scholfield DW (1999) A novel super low inductance primary ring utilized in a pulsed dual resonant tuned transformer. In: 12th IEEE international pulsed power conference, vol 2, pp 811–814, 27–30 June 1999
Krishnaswamy P, Kuthi A, Vernier PT, Gundersen MA (2007) Compact Subnanosecond Pulse Generator Using Avalanche Transistors for Cell Electroperturbation Studies. IEEE Trans Dielectr Electr Insul 14:873–877
Lehr JM, Baum CE, Prather WD, Torres RJ (1998) Fundamental physical considerations for ultrafast spark gap switching. Ultra-Wideband Short-Pulse Electromagnetics 4 (IEEE Cat. No.98EX112), pp 11–20, 14–19 June 1999
Levinson S, Kunhardt EE, Kristiansen M, Guenther AH (1979) Simulation of inductive and electromagnetic effects associated with single and multichannel triggered spark gaps. In: 2nd IEEE international pulsed power conferences, United States, pp 433–436

Lim SW, Cho CH, Ryoo HJ, Kim JS, Rim GH, Jin YS (2011) Fabrication and operation testing of a dual resonance pulse transformer for PFL pulse charging. J Korean Phys Soc 59:3679–3682
Lundin R (1985) A handbook formula for the inductance of a single-layer circular coil. Proc IEEE 73:1428–1429
Martin JC (1992) Nanosecond pulse techniques. Proc IEEE 80:934–945
Martin TH, Seamen JF, Jobe DO (1993) Energy losses in switches. In: 9th IEEE pulsed power conference, Albuquerque, NM, USA, p 463, 21–23 June 1993
Miller SL, Ebers JJ (1955) Alloyed junction avalanche transistors. Bell Syst Tech J 34:883–902
Moran SL, Hardesty LW (1991) High-repetition-rate hydrogen spark gap. IEEE Trans Electron Devices 38:726–730
Morrison RW, Smith AM (1972) Overvoltage and breakdown patterns of fast marx generators. IEEE Trans Nucl Sci 19:20–31
Ng KK (2009) Complete guide to semiconductor devices. Wiley, New York
P. A. Pincosy PPaWRC (1992) High-pressure gas switch for a wideband source. In: 9th international conference on high-power particle beams. IEEE, Washington, DC
Persephonis P, Vlachos K, Georgiades C, Parthenios J (1992) The inductance of the discharge in a spark gap. J Appl Phys 71:4755–4762
Petrella RA, Xiao S, Katsuki S (2016) An air core pulse transformer with a linearly integrated primary capacitor bank to achieve ultrafast charging. IEEE Trans Dielectr Electr Insul 23:2443–2449
Raizer YP. 1997. Gas Discharge Physics Berlin: Springer.
Reed JL (1988) Greater voltage gain for Tesla-transformer accelerators. Rev Sci Instrum 59:2300–2301
Rein H, Zahn M (1975) Subnanosecond-pulse generator with variable pulsewidth using avalanche transistors. Electron Lett 11:21–23
Rohwein GJ (1979) A three megavolt transformer for PFL pulse charging. IEEE Trans Nucl Sci 26:4211–4213
Ryan HA, Hirakawa S, Yang E, Zhou C, Xiao S (2018) High-voltage, multiphasic, nanosecond pulses to modulate cellular responses. IEEE Trans Biomed Circuits Syst 12:338–350
Schoenbach K, Kolb J, Xiao S, Katsuki S, Minamitani Y, Joshi R (2008a) Electrical breakdown of water in microgaps. Plasma Sources Sci Technol 17:024010
Schoenbach K, Xiao S, Joshi R, Camp J, Heeren T, Kolb JF, Beebe S (2008b) The effect of intense subnanosecond electrical pulses on biological cells. IEEE Trans Plasma Sci 36:414–422
Shotts Z, Rose F, Merryman S, Kirby R (2005) Design methodology for dual resonance pulse transformers. In: IEEE pulsed power conference, Monterey, CA, USA, pp 1117–1120, 13–15 June 2005
Smith PW (2002) Transient electronics: pulsed circuit technology. Wiley, London
Sun Y, Xiao S, White JA, Kolb JF, Stacey M, Schoenbach KH (2007) Compact, nanosecond, high repetition rate, pulse generator for bioelectric studies. IEEE Trans Dielectr Electr Insul 14:863–870
Woodworth JR, Chalenski D, Sarkisov GS, Blickem JR (2005) 170-kV laser-triggered water switch experiments. IEEE Trans Plasma Sci 33:2051–2059
Xiao S, Zhou C, Yang E, Rajulapati SR (2018) Nanosecond bipolar pulse generators for bioelectrics. Bioelectrochemistry 123:77–87

Chapter 16
Pulse Delivery and Exposure Systems

Shu Xiao

Abstract This chapter discusses how electric pulses are delivered to a biological target using electrodes and a transmission line. The electric fields in the tissue near the electrodes can be determined through analytical approach or numerical calculation (Sect. 16.2). In the time domain, the pulses seen at the target are largely affected by the pulse duration and the electrical characteristics of the electrodes in the tissue, especially the associate capacitance (Sect. 16.3). It is also important to treat the electrode configuration as part of the transmission line for the delivery of ultrashort pulses when the exact waveform is determined. Two types of electrodes that can deliver ultrafast pulses for in vitro studies are discussed in Sect. 16.4.

16.1 Introduction

Electric pulses generated by a pulse generator can be delivered to a biological target by antennas noninvasively or by electrodes intrusively. Pulse radiating antennas tend to produce low-intensity electric fields inside a tissue, which can cause mild effects such as electrostimulation, although prolonged exposure can also cause strong effects such as hyperthermia. For effects such as electroporation or tissue ablation that require higher electric field intensity, electric pulses are delivered to the target by electrodes that are connected to the pulse generator. The intensity of the electric field and the field distribution are controlled by the electrode configuration and electrode shape. The electrodes that have high curvature, needles, e.g., are often chosen for easy penetration to the tissue. In addition, the enhanced field near the electrode surface relative to the average field allows for a localized treatment. When partially coated with an insulation layer, the electrodes can treat deep targets and spare shallow ones, e.g., when ablating the fat and meanwhile protecting the skin. A uniform electric field distribution for in-vitro studies on the other hand can be provided by parallel electrodes. With the ease of controlling the electric field, needle and parallel electrodes have been effective means to deliver electric pulses.

S. J. Beebe et al., *Ultrashort Electric Pulse Effects in Biology and Medicine*, Series in BioEngineering, https://doi.org/10.1007/978-981-10-5113-5_16

16.2 Tissue Properties and Electric Field Determination

To deliver pulses to a tissue, knowing the tissue electrical properties is beneficial for strategizing a specific endpoint application and maximizing the treatment efficiency. The tissue's impedance, for example, determines whether it affects the pulse waveshape and how much power it absorbs. More specifically, the response of a biological tissue to the pulse includes charge redistribution and associated energy losses, which are manifested through the dielectric permittivity ($\varepsilon_0\varepsilon_r$) and conductivity ($\sigma$). The relaxation time constant of the tissue determines whether it is viewed as being capacitive (a capacitor) or being resistive (a resistor), although in most cases it is a mixture of both. The time constant is defined as

$$t_{RC} = \varepsilon_0\varepsilon_r/\sigma \tag{16.1}$$

If a pulse is much longer than t_{RC}, the tissue can be treated as a resistive medium, in which the electric field distribution is primarily governed by the conductivity. If a pulse is shorter than t_{RC}, the tissue can be treated as a capacitive medium, in which the electric field distribution is determined by the dielectric permittivity.

The dielectric properties of a tissue are dispersive, i.e., frequency-dependent (Gabriel et al. 1996). For example, at 100 MHz, an adipose tissue has $\varepsilon_r \sim 11$ and $\sigma \sim 0.05$ S/m. At 1 MHz, $\varepsilon_r \sim 30$, $\sigma \sim 0.02$ S/m. The relationship of ε_r and σ is described in a complex permittivity:

$$\varepsilon^* = \varepsilon_0\varepsilon_r(\omega) - j\sigma(\omega)/\omega \tag{16.2}$$

ε^* can be alternately described by many models, such as the Debye model, the Cole–Cole model, and the Lorentz model (Cruciani et al. 2012; Grimnes and Martinsen 2000; Stoykov et al. 2003).

The electric field established by electrodes in principle can be measured with probes based on electro-optic effects, such as the Pockels effect. In a small electrode gap, however, the electric field is mostly calculated because of the lack of small profile, non-intrusive probes. The field equals the gradient of the scalar potential ϕ ($\mathbf{E} = -\nabla\phi$), which satisfies the Laplace's equation:

$$\nabla^2\phi = 0 \tag{16.3}$$

Solving Eq. 16.3 requires the boundary condition, which is the electrode potential difference, or the voltage, for a given electrode configuration. For a simple electrode geometry, analytical methods, such as a conformal transformation, can be used to solve the potential and field. For a complex geometry, numerical methods, such as finite element method (FEM), can be used to calculate the static field distribution by knowing the voltage.

Equation 16.3 applies to any single, homogeneous tissue, regardless of whether the tissue is dielectric or resistive. For a dielectric type, in which no free charges (or drift current) are present, it can be assumed that charges only exist on the electrodes. Outside the electrodes, the potential is governed by the Laplace equation. Alternatively, even though there is no drift current, the continuity of displacement current still requires $\nabla \cdot \mathbf{J} = 0$, where $\mathbf{J} = j\omega\varepsilon_0\varepsilon_r\mathbf{E}$. The zero divergence of $\mathbf{J}$ also leads to the Laplace's equation. For a resistive type, charge accumulation does not exist and only charge movement, i.e., the current flow, is found in the tissue: $\mathbf{J} = \sigma\mathbf{E}$. Again, since current flow is continuous, $\nabla \cdot \mathbf{J} = 0$ and hence, $\nabla \cdot (\sigma\mathbf{E}) = 0$, which also leads to Eq. 16.3. Therefore, it becomes unnecessary to distinguish the tissue type.

In the case that multiple tissue types are included, and the pulse duration is close to the relaxation time constant, separation of dielectric domain from resistive domain is not possible and Eq. 16.3 will no longer apply. An accurate determination of electric field relies on solving the time-dependent Maxwell equations. The Finite Difference in Time Domain (FDTD) method can be used. Besides the electric field and magnetic field, the electric flux density vector also needs to be solved. The complex permittivity (Eq. 16.2) relates the electric flux density vector and the electric field, adding one more equation to the FDTD iteration (Joseph et al. 1991).

16.3 Pulse Distortion

Because of tissue's inherent capacitance, the pulse in the tissue is not the same as that at the exit of the generator. The pulse risetime becomes slower as it charges the capacitance (Fig. 16.2). The falling phase also becomes longer as the capacitance is discharged to the resistive component at the end of the pulse. Meanwhile, the capacitance causes a reflection in the rise and fall phases because it produces a smaller impedance than that of the transmission line. A complete match between the tissue impedance and the line impedance is impossible in the transient phases. But in the steady phase of the pulse, the tissue resistance can match the line impedance. In some cases, a resistor is added to the tissue either in parallel or series to achieve the best matching condition.

For a pulse that is much longer than the capacitor charging time (and discharging time), the delayed transient response is not even observable on the oscilloscope. The impedance discontinuity in the transient phase poses little to no impact to the waveform. But if the rise or fall time is comparable to the pulse flat part, one need to be aware of the pulse distortion. For example, a 1 mm-gap electroporation cuvette with electrode area of 0.7 cm^2 has an approximate capacitance of 50 pF. If the transmission line that delivers the pulses is 10 Ω (five, 50 Ω cables in parallel), then the charging time constant is 500 ps. If the pulse duration is shorter than 500 ps, the actual pulse across the cuvette becomes slower in the rise time and the fall time, and its amplitude is smaller. For long pulses $\gg$ 1 ns, the actual pulse across the cuvette remains almost the same pulse shape as that from the generator. Note that in this case the resistive component of the cuvette medium is assumed to be 10 Ω, the same as

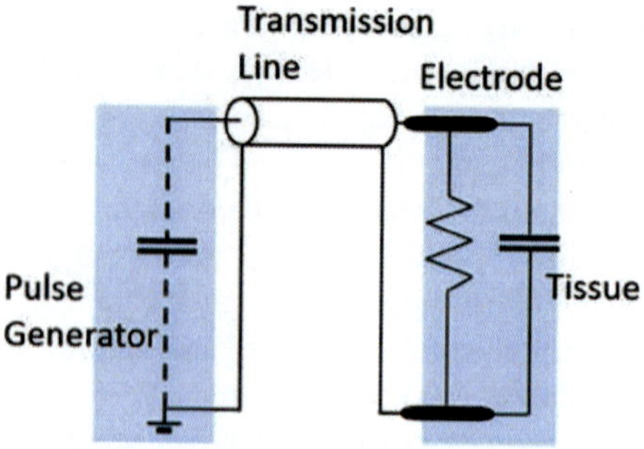

Fig. 16.1 The electric pulses generated from a pulsed generator are delivered to a biological target by a transmission line, which further connects to the electrodes inserted to a tissue or cell medium

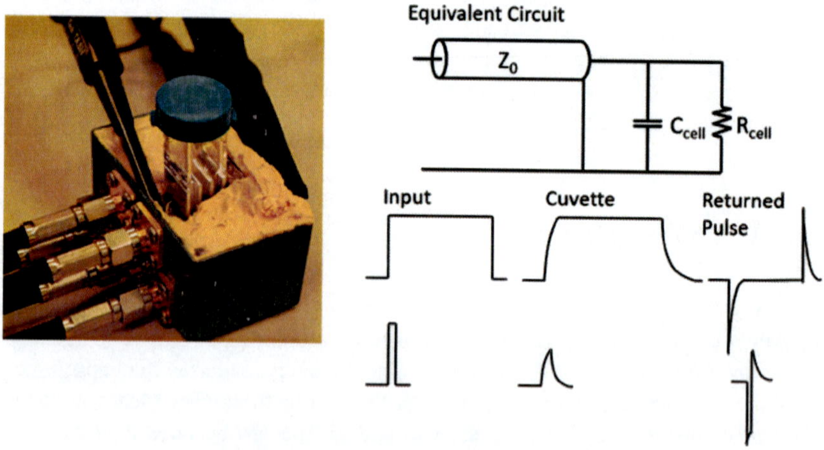

Fig. 16.2 (Left) An exposure system that works for pulses as short as 500 ps, (Right) The input and output of the system show the impact of the capacitance to the actual voltage across the cuvette

the transmission line impedance, so that the pulse energy is mostly absorbed by the medium despite reflections in the pulse transients.

For a short pulse (< 500 ps), if the amplitude needs to be retained, it is imperative to reduce the capacitance by reducing the cuvette contact area or increasing the cuvette gap distance. Doing so however increases the resistance of the cuvette, as the relaxation time constant $R_{cell}C_{cell}$ stays the same. An extra matching resistor may be placed in parallel so that the overall cuvette resistance can be lowered to match the transmission line. Integrating the matching resistor to the cuvette however is challenging as it may introduce extra distributed capacitance and inductance.

The return pulse from the tissue can put the generator at risk of overvoltage and failure. If the return pulse is not suppressed, it will experience multiple reflections and so the load will be exposed to multiple pulses even there is only one pulse coming out from the generator. To avoid these problems, one can design the generator's output impedance to be 50 Ω, and so any return pulse will be absorbed. A second option is to insert an impedance matching network (for example, a π network) midway between the source and the load. Both options suppress the pulse return from the load, but the pulse amplitude will be lowered.

16.4 Electrodes

There are many electrode shapes and configurations, either for surface application or tissue penetration, but cylindrical electrodes and parallel plates have been the primary electrode options, at least for in-vitro studies. Although the electric field in the cylindrical electrode gap is not uniform and the highest field is near the electrode surface, the field in the gap center projected from the electrodes is relatively homogeneous. This is useful for studying a small number of selected cells. Parallel plates on the other hand provide a large homogeneous electric field region between the electrodes and therefore allow a large volume of samples to be treated with the same electric field. Both electrode configurations can be made applicable for subnanosecond pulses.

16.4.1 Cylindrical Electrodes

In an in vitro exposure system (Xiao et al. 2017), two tungsten needle electrodes (diameter 100 μm, gap distance 170 μm) were positioned with a precision of a few μm by a micromanipulator above a cover slip under a microscope. The cells on the cover slip, positioned in the gap center between the electrodes, were exposed to the field from the electrodes. The electrodes were connected to the breakout of a coaxial cable (RG316). The connection was secured by applying epoxy glue, which also insulated the electrodes and prevented electrical breakdown (Fig. 16.3). There was no insulation on the other side of the tungsten rods over 3.5 mm for submerging in the cell medium. The coaxial cable with the electrodes was anchored on a brass rod and mounted on a micromanipulator for precise positioning above the cover slip. The angle of the electrodes relative to the cover slip was 45°. The intact side of the coaxial cable with the standard connector was connected to a π network.

The amplitude of the subnanosecond pulse in the coaxial cable exiting from the π network was kept constant. However, this was not the actual voltage across the electrodes as there were reflections due to the impedance discontinuities at the breakout of the coaxial cable (position 1, 2 in Fig. 16.3) and the change in insulation between tungsten electrodes from epoxy to a water medium (position 3 in Fig. 16.3). The actual electric field at the surface of a cover slip with attached cells was simulated by a 3-D time-domain electromagnetic solver, CST microwave studio (Computer Simulation Technology), taking these reflections into account.

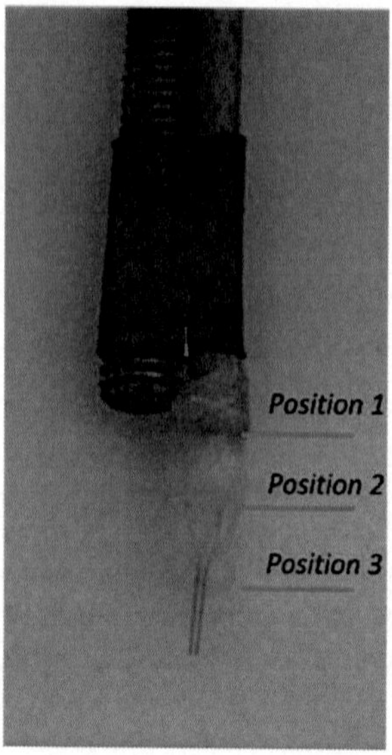

Fig. 16.3 The electrodes, consisting of two tungsten rods (diameter 100 μm, gap distance 170 μm), were connected to a coaxial cable (RG316), which was fixed on a brass rod and further mounted on a micromanipulator. The other side of the coaxial cable was connected to a π network. The *lines 1–3* indicate the locations where reflections occur before the pulse reaches the end of the electrodes

It was noticed that the full width at half maximum (FWHM) of the electric field pulse on the cover slip is broader than the pulse applied at the coaxial input (Fig. 16.4). This is due to the impedance discontinuity at the exit of the coaxial cable at the position close to the cover slip (Fig. 16.5a). Prior to 1 ns, the pulse already experiences an impedance discontinuity twice: one at the junction between the coaxial cable and the epoxy section that contains the rod electrodes and the other between the epoxy section and medium. At the end of the electrodes, one more reflection occurs at a time of about 1 ns. The reflected wave propagates along its original path but is reflected again at the junction between the water and epoxy section. Further, a reflection occurs once again at the junction between the coaxial cable and epoxy section. Despite these reflections, the total pulse width is still approximately 500 ps at the cell location. The pulse waveforms were found to be almost identical when the conductivity was varied from 0.5 to 1.5 S/m. The pulse broadening is mainly due to the water-submerged part of the tungsten rod electrodes, which have a length 3.5 mm. This contributes to a round-trip time of approximately 200 ps for pulses traveling in water. Hence, the accurate measurement of rod electrode length is important in characterizing the actual pulse width. An error of 1 mm length could result in a deviation of 58 ps of calculation from the actual pulse width seen at the cell location. On the other hand, the pulse broadening due to the epoxy section is negligible as the epoxy has a low dielectric constant (approximately 2.5). As such, the two conductors can be slightly off the axis of the coaxial cable and don't have to be strictly aligned with the axis

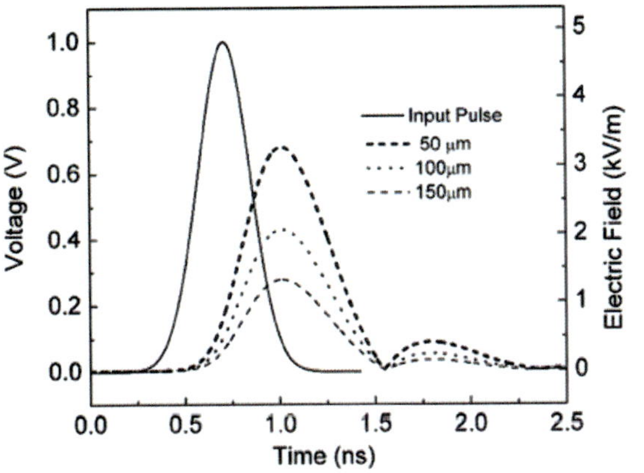

Fig. 16.4 A Gaussian pulse was used as the input of the coaxial cable in the field simulation. For an input voltage of 1 V, the peak electric fields on the cover slip at the center point beneath the electrodes were 3.2, 2.0 and 1.3 kV/m for the electrode tips positioned at 50, 100 and 150 μm above the coverslip

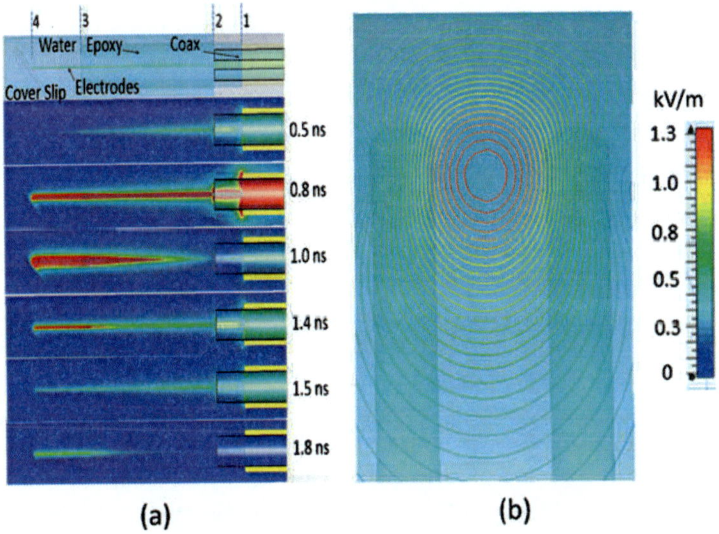

Fig. 16.5 **a** Side view of the coaxial cable in the simulation. Multiple impedance discontinuities cause reflections and broadening of the pulse: the breakout of the coaxial cable (indicated by *dash line 1*); the junction between the coaxial cable dielectrics and the epoxy section (*dash line 2*); and the junction between water and epoxy section (*dash line 3*). The electrodes, cover slip, and epoxy section are fully submerged in water. **b** The electric field distribution on the cover slip for the electrode height of 150 μm

of the coaxial cable. In general, the electrodes can be treated as two points, i.e., the edges of the tilted electrodes, so the electric field intensity on the cover slip is not sensitive to the tilt angle. Rather, it is determined by the distance between the two electrodes and the distance between the electrode bottom and the cover slip.

For a given input voltage, the electric field on the surface of the cover slip can be varied by raising the electrodes 50, 100 and 150 μm above the cover slip (Fig. 16.4). For all three electrode heights, the electric field in the center region is homogenous. The area covers approximately 50 μm × 70 μm for 50 μm, and 60 μm × 70 μm for both 100 and 150 μm (Fig. 16.5b). The cells in this region were used for study. For lower conductivities (0.5 and 1 S/m), the electric field at the center region is 25 and 10% higher.

16.4.2 Parallel-Plate Cuvette

A large volume of cell medium can be treated in a parallel-plate cuvette. For subnanosecond pulses, one needs to consider how to interface the transmission line with the electrodes. The terminal of a biconical transmission line can house the cuvette that connects the high voltage inner conductor to the grounding outer conductor (Fig. 16.6) (Baum 2007). This cuvette has a thin dielectric wall that does not perturb the field distribution. The medium in the biconical line and coaxial line is high pressure gas for high voltage operation.

The resistance and capacitance of the cuvette is calculated by Eqs. 16.4 and 16.5.

$$R_s = \frac{l}{\pi a^2 \sigma_s} \tag{16.4}$$

$$C_s = \varepsilon_s \varepsilon_0 \frac{\pi a^2}{l} \tag{16.5}$$

For a saline solution ($\varepsilon_s = 80$, $\sigma_s = 0.3$ S/m), $l = 2$ cm and $a = 2$ mm were chosen: R_S = 5.3 kΩ and C_S = 0.45 pF. When the pulse duration is less than the sample relaxation time, the sample looks capacitive to a 50 Ω source, producing a negative reflection and lowering the voltage across the sample. However, in this design the sample resistance is 5.3 kΩ, which is large compared to the line impedance, resulting in an approximately 100% reflection and almost doubling of the voltage across the sample, although the reflection of the pulses from the source end may cause ringing.

The subnanosecond pulse source can be placed at the other side of the transmission line, which has a pulse sharpening circuit (Fig. 16.7). The peaking switch's inductance, L_S, is approximately 1 nH for 1 mm gap. The fastest risetime can be estimated as 20 ps for Z_c that equals to 50 Ω. For such peaking switch, the risetime of about 100 ps have been achieved at voltages of over 100 kV (Baum et al. 2004). The wave launching section can be also designed as biconical shape as shown in Fig. 16.7. The characteristic impedance, Z_C, is determined by

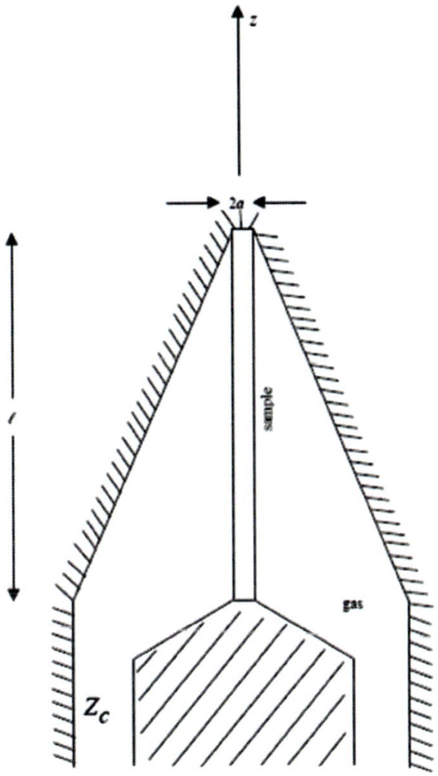

Fig. 16.6 A cuvette designed for small sample volume with low capacitance Baum (2007)

$$Z_C = 377 f_g \tag{16.6}$$

f_g is determined by two polar angles, Ψ_1 and Ψ_2:

$$f_g = \frac{1}{2\pi} \ln\left(\frac{\cot\left(\frac{\psi_1}{2}\right)}{\cot\left(\frac{\psi_1}{2}\right)}\right) \tag{16.7}$$

The biconical impedance Z_C should be matched to the middle section, i.e., the coaxial transmission line. From the wave propagation perspective, one needs to match the transit times for the various ray paths to the cylindrical section. As a result, Ψ_1 and Ψ_2 can be made adequately small. In the pulse source section prior to the peaking switch, the peaking capacitor, C_P, should contain low inductance. For the peaking switch to work, C_P will be charged up to the self-breakdown voltage. To minimize the propagation of the fast pulse back toward the high voltage source, a small inductance,

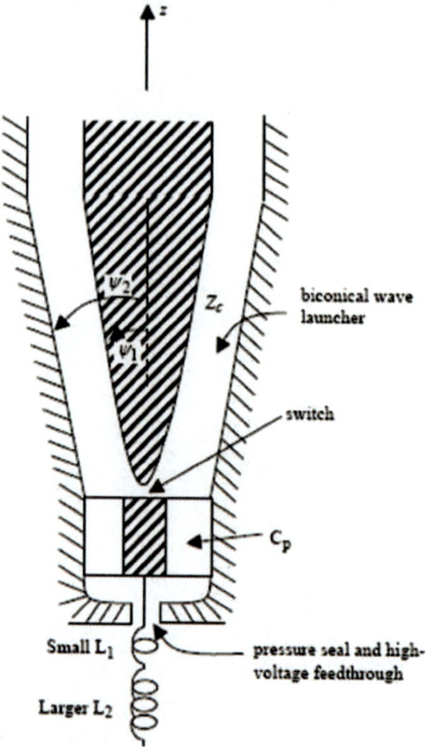

Fig. 16.7 Launching fast pulses toward biological sample Baum (2007)

L_1, for high frequency performance can be added. Next, L_1 is followed by a large inductance, L_2, in the charging line near C_P.

The parallel-plate cuvette in an ending of a biconical transmission line was constructed in (Camp et al. 2012). The line was adapted to an existing coaxial transmission line (Fig. 16.8). The radius of the sample is 2.4 mm and the gap distance is 1.8 mm. A conical waveguide was inserted between the coaxial cable and the cuvette to allow the radius transition. The impedance was maintained 50 Ω. The dielectric material between the inner conductor and the outer conductor is polypropylene with permittivity of 2.38. The conical waveguide allows a rather smooth transition from the transverse electric fields in the cable to the axial electric field in the sample. The cuvette's resistance was almost entirely matched to 50 Ω at 42 °C, the temperature at which the experiment was conducted. Still, the capacitance of the cuvette will slow the pulse risetime and lower the input pulse amplitude. Therefore, this cuvette is applicable to pulses longer than the pulse duration of 300 ps.

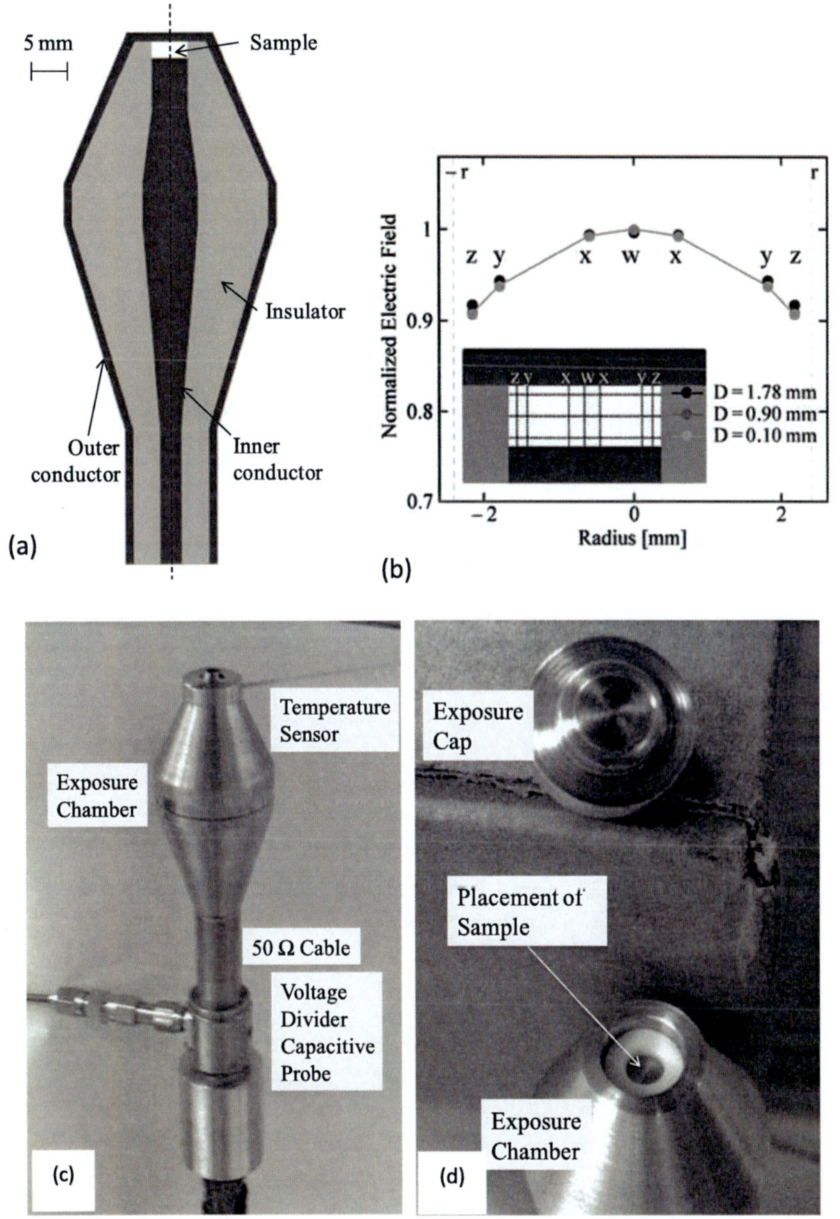

Fig. 16.8 **a** Cross-section view of exposure chamber. **b** Simulation results of the electric field strength in the sample. The insert represents the sample (white) between the two conductors (black). The data points show the normalized electric field values at the positions defined by the axial line, w, and the lines, x, y, and z, at various radial distances from the axis, at the intersections with the of lines at various axial distances, D. **c** Photograph of the exposure chamber and voltage probe. **d** Photograph of the interior of the exposure chamber (Camp et al. 2012)

References

Baum CE (2007) Subnanosecond sample holder. Measurement Note 60. https://ece-research.unm.edu/summa/notes/Measure/0060.pdf

Baum CE, Baker WL, Prather WD, Lehr JM, Loughlin JPO, Giri DV, Smith ID, Altes R, Fockler J, McLemore DM et al (2004) JOLT: a highly directive, very intensive, impulse-like radiator. Proc IEEE 92:1096–1109

Camp JT, Jing Y, Zhuang J, Kolb JF, Beebe SJ, Song J, Joshi RP, Xiao S, Schoenbach KH (2012) Cell death induced by subnanosecond pulsed electric fields at elevated temperatures. IEEE Trans Plasma Sci 40:2334–2347

Cruciani S, Santis VD, Feliziani M, Maradei F (2012) Cole-cole versus Debye models for the assessment of electromagnetic fields inside biological tissues produced by wideband EMF sources. Asia-Pac Symp Electromagn Compat 21–24:685–688

Gabriel S, Lau RW, Gabriel C (1996) The dielectric properties of biological tissues: II. Measurements in the frequency range 10 Hz to 20 GHz. Phys Med Biol 41:2251–2269

Grimnes S, Martinsen OG (2000) Bioimpedance and bioelectricity basics. Academic Press

Joseph RM, Hagness SC, Taflove A (1991) Direct time integration of Maxwell's equations in linear dispersive media with absorption for scattering and propagation of femtosecond electromagnetic pulses. Opt Lett 16:1412–1414

Stoykov NS, Kuiken TA, Lowery MM, Taflove A (2003) Finite-element time-domain algorithms for modeling linear Debye and Lorentz dielectric dispersions at low frequencies. IEEE Trans Biomed Eng 50:1100–1107

Xiao S, Semenov I, Petrella R, Pakhomov AG, Schoenbach KH (2017) A subnanosecond electric pulse exposure system for biological cells. Med Biol Eng Comput 55:1063–1072

Chapter 17
Pulse Voltage Measurement

Shu Xiao

Abstract Ultrafast pulses can be measured by a resistive divider or a capacitive divider (Sect. 17.1). In terms of the ease of construction and characterization, the resistive divider is a better choice. The divider is typically consisted of a high value resistor and a 50 Ω resistor, designed to match the impedance of a standard coaxial cable. The configuration and related issues of such a divider is discussed in Sect. 17.2. With robust resistors, the divider can be used to measure high voltage, nanosecond pulses. For shorter pulses, a coaxial structure that houses resistors with low inductance can be used. One such a resistor divider for subnanosecond pulses is shown in Sect. 17.3.

17.1 Introduction

In bioelectric applications, a real-time voltage monitoring is needed as high voltage pulses may cause a breakdown inside the tissue under treatment and the actual voltage becomes different from the output of the pulse generator. For some pulse generators that are sensitive to the load impedance, such as a PFL pulser, the actual pulse waveform at the load is susceptible to change if an impedance mismatch occurs. On one hand, the real-time voltage measurement allows the operator to determine the quality of the treatment. On the other hand, knowing the voltage across the electrodes allows the calculation of the electric field intensity inside the tissue, which helps interpretation of biological results.

Most oscilloscopes can receive pulse voltages up to 50–100 V if the input impedance is set as 1 MΩ. The input voltage is reduced to half for the input impedance of 50 Ω. Hence it is not feasible to measure or display high voltage pulses directly. A high voltage pulse must be converted to a low voltage pulse acceptable to the scope. The conversion can be achieved by a capacitor divider in which a small, high voltage capacitor is in series with a larger, low voltage capacitor. This method however is sensitive to stray inductances, which could cause pulse resonance and distortion. In addition, the stray capacitances of the capacitor divider could cause the inaccuracy of the voltage conversion ratio. Therefore, designing a high-performance capacitor divider is not trivial. A capacitor-resistor mixed divider can offer a better

S. J. Beebe et al., *Ultrashort Electric Pulse Effects in Biology and Medicine*, Series in BioEngineering, https://doi.org/10.1007/978-981-10-5113-5_17

performance in high impedance, broadband measurements, for which the capacitors take care of high frequency components and the resistors are responsible for low frequency to DC components, but the divider's design and characterization are demanding. In contrast, resistor dividers are easy to construct and characterize owing to nowadays the availability of high voltage, high performance resistors, which can even be fabricated with an electrolyte or by 3-D printing.

This chapter will discuss resistor dividers and attenuators for measurement of nanosecond and subnanosecond pulses. The performance is satisfactory in the range of a dozen of kilovolts. The attenuation ratio of these devices can be made to 1000:1 when they are used with standard, low voltage, and broadband attenuators.

17.2 Resistor Voltage Divider

A voltage divider consists of a couple of resistors, R_1 and R_2, as shown in Fig. 17.1a. The resistance of R_1 is usually much greater than that of R_2, i.e. $R_1 \gg R_2$ and is called the high voltage arm, which typically consists of a series of resistors. The value of R_1 is chosen such that the resistance draws very little current and hence shows no effect on the pulse generator. Whereas the smaller resistance R_2 is referred to as the low voltage arm, which is chosen to be equal to the cable characteristic impedance Z_0. Let V be the voltage to be measured and V_2 be the voltage observed on the oscilloscope. The relation between V and V_2 can be obtained as follows:

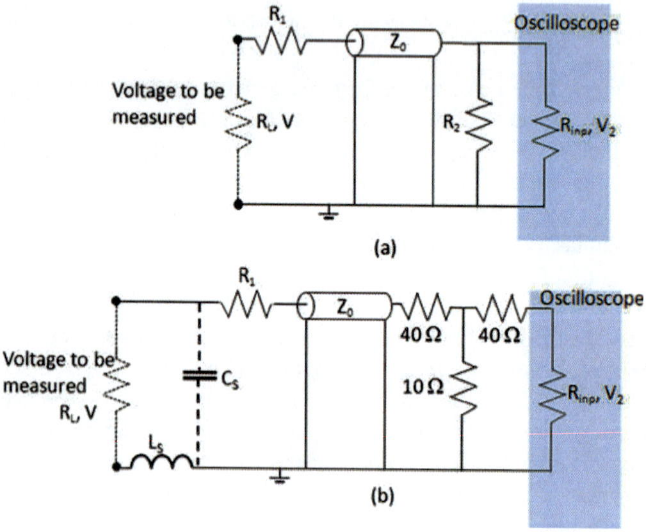

Fig. 17.1 A resistor divider to measure the voltages across the load (R_L). **a** A generic circuit; **b** A practical circuit

$$V = \frac{R_1 + R_2}{R_2} V_2 \cong \frac{R_1}{R_2} V_2 \tag{17.1}$$

The attenuation ratio $\left(\frac{R_1}{R_2}\right)$ should not be too large, as a heavily attenuated signal is susceptible to noise contamination. Electromagnetic noises in free space generated by power lines, power supplies, and (or) spark gaps in the pulse generators could easily enter the connecting cables through the cable leakage capacitance to the ground. They may cause pulse distortion or mask the original signal entirely, so it is rather important to keep the measured signal on the level of ~ 100 V, much higher than the unwanted signals. For example, if R_1 is chosen 5 kΩ, a 10 kV pulse that enters the cable ($Z_0 = 50$ Ω) is attenuated to 100 V. Before being fed to the oscilloscope, the signal should go through one or more attenuations. This becomes easy as standard coaxial attenuators can be used. So R_2 can be a resistance of a combination of standard attenuators and the input impedance of the oscilloscope, which can be either 50 Ω or 1 MΩ. For a 20 dB attenuator and 1 MΩ scope impedance (R_{inp}, Fig. 17.1b), the measured voltage V and the scope voltage V_2 are related by:

$$V = 5\frac{(R_1 + 50)}{50} V_2 \cong \frac{R_1}{10} V_2 \tag{17.2}$$

In practice, the circuit at the oscilloscope side is coaxially contained and therefore will not impact the performance of the divider. At the load (R_L) side, however, stray parameters must be taken into consideration, which affect the transients of the pulses arriving at the oscilloscope. The adherent inductances of R_L and R_1 as well as the lead inductance L_s are present and can easily amount to 10–100 nH. The capacitances of R_1 and the cable leakage capacitance C_0 can add up to a few pF. The stray capacitance and inductance can form an underdamped oscillation with a period up to a few ns, if they are energized by a pulse appeared across R_L. As a result, the pulse waveform on the oscilloscope becomes distorted after the oscillation is superimposed.

The stray parameters should be minimized to preserve the original pulse waveform. The reduction of stray capacitance is difficult, but it is generally estimated as a few pF. The stray inductance can be reduced by shortening the loop size, but ~ 20 nH is already a stretch as a rectangle loop of 2 cm × 2 cm has an inductance of ~ 36 nH. Besides reducing the loop inductance, the high voltage resistor (R_1) should be low-inductance. Wire-wound resistors can be made inductance-free by the 2-layer winding. One layer is wound in the opposite direction from the other layer so that the magnetic fluxes cancel each other and only resistance is exhibited (Creed and Collins 1963). Other none-inductance or low inductance resistors include metal film, carbon composition, and ceramic types.

The accuracy of the resistor divider ultimately relies on robust, linear resistors that exhibit the same resistance at different voltages. The resistance usually changes when a voltage is applied, which is defined as voltage coefficient of resistance (Barth 1988). Using resistors of high voltage coefficient of resistance causes inaccurate measurement. A 2-W carbon composition resistor was seen to decrease from 390 to 200 Ω during a pulse testing. While the issue of resistance varying with voltage persists, one can still build inexpensive and reliable voltage dividers with carbon composition resistors by combination of several resistors in parallel or series.

17.3 Coaxial Attenuators

A high-performance resistor attenuator in general provides a 50-Ω termination in both input and output when the 50 Ω coaxial cables are used for connection. An example is a 50 Ω, 20-dB attenuator, which is a T resistive network (40, 10, 40 Ω). In some cases, the attenuation ratio needs to be smaller than 20 dB and as a result a different resistor network is used, which however does not yield a 50 Ω output impedance. Such an attenuator can be used for lowering the output voltage from a pulse generator. This was shown in a π network for a subnanosecond pulse delivery system (Xiao et al. 2017). This attenuator acted as a 60-Ω absorber for pulses propagating in the 50-Ω coaxial cable. The attenuation for such an ideal network is 8 dB, and the return signal is − 18 dB.

When constructing a resistive network for the passage of subnanosecond pulses, the main challenge lies in minimizing stray parameters, such as the insertion inductance and shunt capacitance. In general, placing resistors in a coaxial setting can be a good option. The electric loop length of the resistive circuit is restricted to the same range as the diameter of the coaxial cable being used (~ 1 cm for RG217, e.g.) and the inductance is less than a few nH, which means that pulses with a rise time of one hundred picoseconds or less will be able to pass without distortion. For example, in a 50-Ω cable, if a loop inductance is 5 nH, the *L*/*Z* time is 100 ps. Besides reducing the loop length, the shunt capacitance should be reduced. Even if it is not, a compensating capacitance may be added to improve the high-frequency performance (Barth and Sarjeant 1981). Furthermore, the resistors placed in the coaxial configuration should cause only a minimum distortion of the radial electric field of the coaxial line. Otherwise, additional return loss will be generated, a consequence that is especially pronounced at high frequencies. Satisfying the conditions for minimum distortion therefore requires a very compact arrangement of the resistors in the π network.

In this design, the serial and parallel combinations of the resistors formed three equivalent resistances in the π network: 180, 64 and 180 Ω, as shown in Fig. 17.2. They were low-inductance carbon composition resistors housed in an aluminum hollow cylinder with the same diameter as the insulation layer of an RG217 cable. The overall length of the cylinder was 2 cm. The resistors were configured in a symmetrical layout, so that the resistance viewed from the left port was identical to that from the right port. To increase the withstand voltage of the network, silicon glue was filled among the resistors to prevent any high-voltage flashover.

To test the π network, a network analyzer (Keysight, N9926) was used to measure its reflection coefficient, S11 (shown in Fig. 17.3), a parameter that describes how much an electromagnetic wave is reflected by impedance discontinuities. Up to 3 GHz, the reflection coefficient was approximately − 17 dB, close to its design

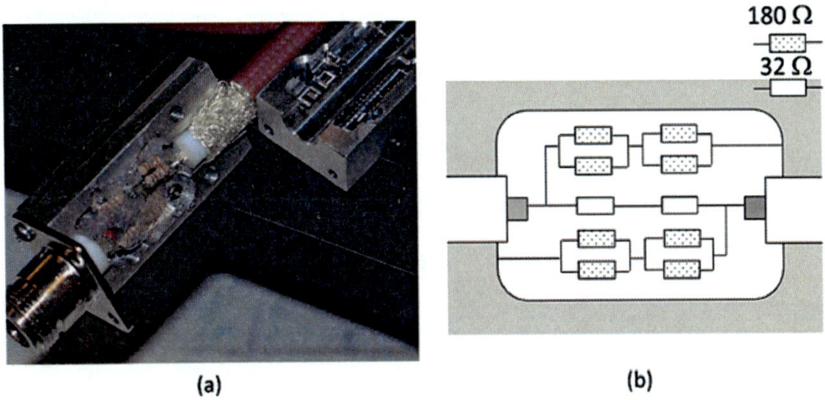

Fig. 17.2 A π resistor network served as an attenuator. **a** The photograph of the resistor network; **b** the detailed resistor arrangement

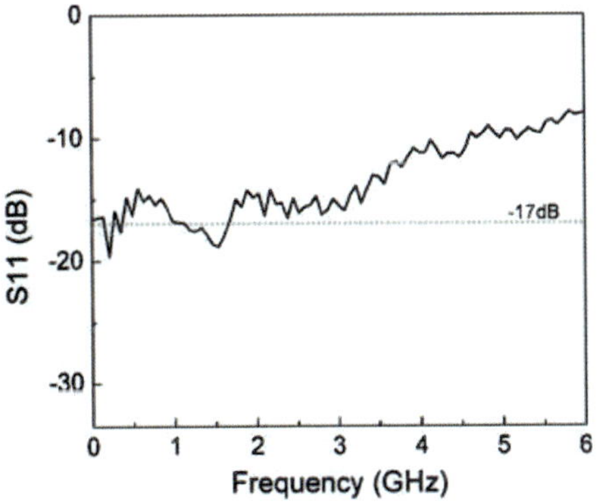

Fig. 17.3 Magnitude of S11 of the π resistor network. Up to 3 GHz, the reflection coefficient stayed relatively constant, approximately − 17 dB. Beyond 3 GHz, the reflection loss became larger and deviated from its low-frequency value

value of − 18 dB. Beyond 3 GHz, the reflection loss became greater. Using a Gaussian waveform as an example, the π network can relay a pulse of 270 ps without distortion. After characterizing the reflection coefficient in the frequency domain, a picosecond pulse generator (FPG 20-P, FID) was used as the source and a 20X attenuator (Barth 142-NMFP-26) as the load (Fig. 17.4) to test the time-domain performance of the π network. Further, an in-line V-dot probe (VDC-1–012, Farr Research) was inserted midway between the π network and the Barth attenuator to monitor the voltage. In total, three signals were obtained (1) from the resistor divider (Barth 20X attenuator, RD), (2) from the voltage probe installed in the FID pulser (Voltage Sensor 1, VS1), and (3) from the Farr V-dot probe (Voltage Sensor 2, VS2). Note that both VS1 and VS2 are capacitive probes. RD and VS2 measured the same

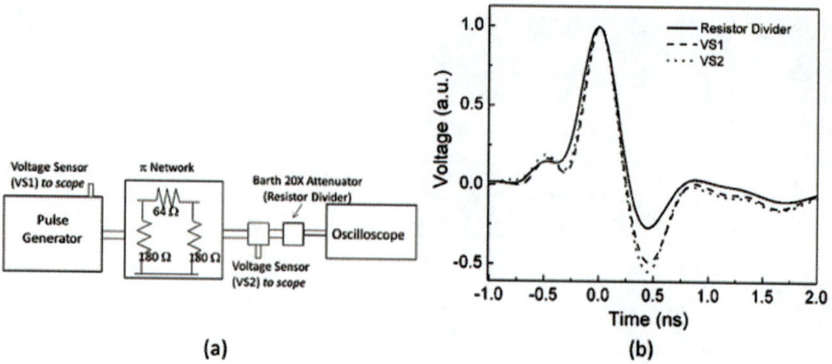

Fig. 17.4 The test setup for the π-network attenuator. A subnanosecond pulse generator was connected to the π network, which served as an attenuator as well as an impedance matching unit to avoid the generation of multiple pulses. **a** The overall schematic. Pulse waveforms were read from three sensors: the voltage from the Barth attenuator (Resistive Divider, 20X), the FID sensor (VS1), and the Farr V-dot sensor (VS2). **b** The pulses measured from the three sensors

high voltage signal after the π network, whereas VS1 measured the voltage before the π network. For easy comparison, the voltage amplitudes were normalized to unity. The three waveforms were almost identical in the first 1.2 ns, and each showed a pulse width of 300 ps (FWHM). However, after the main pulse, VS1 and VS2 showed a voltage reversal greater than 25%, whereas RD showed a negative swing of less than 25%. This result showed that the π network attenuator was satisfactory in the subnanosecond range. The maximum operable voltage for the attenuator was found to be 16 kV, which corresponds to an output of 6.2 kV.

References

Barth J (1988) Voltage coefficient of resistance application note. https://www.barthelectronics.com/pdf_files/Application%20note%201%20Voltage%20Coefficient%20Products_Pulse%20Page.pdf

Barth J, Sarjeant W (1981) Direct subnanosecond voltage monitors. In: 3rd IEEE pulsed power conference. Albuquerque, NM

Creed FC, Collins MMC (1963) The measurement of short-duration impulse voltages. IEEE Trans Commun Electron 82:621–630

Xiao S, Semenov I, Petrella R, Pakhomov AG, Schoenbach KH (2017) A subnanosecond electric pulse exposure system for biological cells. Med Biol Eng Comput 55:1063–1072

Chapter 18
usEPs in Pre-clinical Cancer Treatment

Stephen J. Beebe

Abstract One of the earliest possible medical applications for usEPs was to ablate tumors. Many laboratories carried out many studies to investigate the potential for usEPs to serve as a cancer therapy. Like simulations with cells in suspension, usEPs also passed through cells in tumor tissues forming high-density nanopores in all cell membranes as supraelectroporation, distinct from conventional electroporation. The first studies were smaller in scope, but showed proof of principle that usEPs could reduce fibrosarcoma tumor volume in mice. More extensive studies followed, showing that usEPs could treat B16f10 melanoma tumors in mice, although some tumors required more than one and as many as six 300 ns, 40 kV/cm treatment (6 × 1.2 Vs/cm). Later studies showed that as many as 5–6 Vs/cm was required to eliminate tumors completely. The blood supply to these melanoma tumors was also reduced, as was revascularization, as shown using endothelial markers. Other studies showed that usEPs ablated mouse liver tumors. While all of these studies investigated ectopic tumors within mice's flanks, studies also showed that usEPs could also completely ablate orthotopic rat liver tumors, using a 5 needle array with heterogeneous electric fields. Other studies demonstrated that usEPs could eliminate many different tumor types in mice, including some human tumors in immunodeficient mice. One study showed decreased tumor sizes in dog osteosarcoma.

18.1 Introduction

Apoptotic CD is one of the hallmarks of cancer (Hanahan and Weinberg 2000, 2011). These hallmarks were based on a cell-centric theory called the somatic mutation theory including eight essential alterations on normal cell physiology that constituted malignant growth, and evasion of programmed cell death (apoptosis) was one of these. Others include self-sufficiency in growth signals, insensitivity to growth-inhibitory (antigrowth) signals, limitless replicative potential, sustained angiogenesis, and tissue invasion, metastasis, reprogramming of energy metabolism,

S. J. Beebe et al., *Ultrashort Electric Pulse Effects in Biology and Medicine*,
Series in BioEngineering, https://doi.org/10.1007/978-981-10-5113-5_18

and evading immune destruction. In attempts to simplify cancer, the cancer hallmarks considered that all cell functions were based on standard biological mechanistic principles. They focused on several underlying principles that governed the transformation of normal human cells into malignant cancers.

Vogelstein and colleagues (2013) suggested a somewhat greater cancer complexity. They proposed that two to eight "driver genes" functioned in twelve signaling pathways regulating cell fate, survival, and genome maintenance. The authors used the gene mutation theory of cancer for both of these concepts. Basically, "driver gene" pathways expressed fundamental determinants and key cancer mutations that could be identified and treated. Coupled with these simple reductionist concepts was the theory that cancer followed a stepwise progression of central mutations and that mutations in cancer signaling pathways would establish causes of cancer. The molecular mechanistic foundations of these diseases could identify mutations that caused cancer. Knowing these molecular mechanisms, treatment strategies using mutation-specific small molecule drugs or targeted monoclonal antibodies could specifically target cancers as defined in cancer hallmarks, and driver gene pathways. These cancer hallmarks and driver gene concepts generated thousands of scientific articles using these basic uniting principles of malignancies (Weinberg 2014). These views served as a convenient basis for teaching and learning principles of molecular biology and cell signal transduction mechanisms within normal and diseased biological systems. However, therapeutic applications against these cancer principles have been less successful than hoped.

Many realized cancer obstructions make cancer hallmarks and driver gene pathways less relevant for therapeutic applications. These include the enormous number of cancer-causing mutations, overlaps among cancer hallmarks, genomic instability, structural and temporal clonal heterogeneity, cancer stem cells, epigenetic modifications, cancer evolution, and genome-centric instead of cell-centric considerations. While cancer hallmarks and driver genes underlie many of these cancer mechanisms, their compounded complexities present cancer differently regarding targeted cancer therapy. Intratumoral heterogeneity demands the use of multiple therapies. Combination therapy is the foundation of cancer treatments as drugs are used in succession, often when drug-resistance occurs with the initial treatment. More recently, combination therapies are used at the outset of treatment, and some may provide synergistic effects (Al-Lazikani et al. 2012; Gatzka 2018). However, tumor heterogeneity can generally normalize drug combinations, so the unique benefit of a particular drug on a specific subpopulation of cancer cells can sometimes be reduced (Zhao et al. 2014).

In contrast to the reductionist approaches taken in the somatic mutation theory, the **tissue organization theory** proposes that cancer is a tissue-based diseases and that proliferation and motility are default states of all cells (Sonnenschein and Soto 2014). The somatic mutation theory proposes that mutation occur in the genome of a single cell that cause the development of cancer. In contrast, the tissue organization field theory means that carcinogenesis is caused by changes in normal relationships between and among cells and their environments, such as interactions between stoma and epithelium where cancer is due to skewed developmental programs.

18.2 NPS for Cancer Therapy

For usEP therapy of cancer, it may not matter if cancer functions through the somatic reductionist theory or the tissue organization field theory. For the reductionist theories of gene and somatic mutation, it appears that usEPs can bypass cancer hallmarks associated with inhibiting apoptosis or regulated cell death (RCD), evasion of immune surveillance, sustained angiogenesis, as well as eliminate driver genes. For the tissue organization field theory that proposes that carcinogenesis is caused by changes in relationships between and among somatic, epithelia, and cancer cells and their environments, usEPs completely disrupts these cellular interactions in the TME, so their associations are no longer meaningful. Tumor cell heterogeneity is also of less significance because usEPs are not selective for any specifically given tumor cell dysfunction. It does not discriminate among them. Furthermore, it eliminates cancer cells and somatic cells, and this is generally a good response because all cells in the TME are mostly corrupt with somatic cells serving the functions of perverted tumor cells, so all these collaborations are terminated by usEP delivery to the TME.

As presented in the section on RCD, supra-electroporation with usEPs, considered less than 1 μs in duration, with high electric fields from 10 to 300 kV/cm, passes through the cell induces the high-density formation of pores that are less than 1 nm in all cell membranes. Supra-electroporation is distinct from conventional electroporation, considered greater than 100 μs with electric fields 0.1 1 kV/cm, which essentially goes around the cell and induces a lower density of larger pores primarily in the plasma membrane with some pores extending into the endoplasmic reticulum near the plasma membrane (Gowrishankar et al. 2006). While these simulations were with individual cells, the treatment of tumors raised how usEPs will affect multicellular structures. The same authors (Gowrishankar and Weaver 2006) published simulations of 300 ns usEPs used in the first extensive study of usEP treatments of melanoma tumors in mice. (Nuccitelli et al. 2006; see below). Briefly, unlike 100 μs, 1 kV/cm pulses, which were spatially heterogeneous, 300 ns pulses up to 80 kV/cm, caused homogeneous electroporation of plasma and nuclear membranes.

At the lowest electric fields, electroporation is limited to the plasma membranes. As the electric field strength increases from 10 to 80 kV/cm, all sections of the plasma membrane is electroporation as are the nuclear membranes and the tight junctions. The electric field are mostly uniform as indicated by the parallel equipotential lines. The authors indicated that there is "electrical transparency" So, considering the treatment of solid tumor tissue, it is interesting to see how usEPs affect tumor death when studied from a modeling viewpoint. Gowrishankar and Weaver (2006) used a transport lattice to consider the effects of usEPs on tumors. The method describes effects on complex tissues that are not always homogeneous based on electrical, chemical, and thermal activities. The model showed that supra-electroporation with a high density of small pores in all membranes is common with usEPs in cells and tissues. The studies also presented another difference between conventional electroporation and usEPs on tissues. The extent of cell electroporation is that tissues do not depend on the size, shape, or orientation of the tissue in the tissue landscape. There are homogeneous permeabilizations of intracellular membranes and plasma membranes, and efficacy should not be dependent on cell morphology. These homogeneous permeabilizations are an advantage because tumors can be exhibit quite diverse morphologies.

These pore size and density arguments consider that permeabilization is the most critical aspect of killing tumor cells with usEPs. Indeed, cell membrane permeability is inevitable, but other supplements usEPs take a toll on the cell. Based on previous work from the authors' laboratory (Beebe et al. 2012, 2013) and presented in the section on usEPs on mitochondria, an influx of Ca^{2+} is an essential factor because it is needed for the usEP-induced decrease in the mitochondrial membrane potential ($\Delta\Psi_m$), which comes about after opening the mitochondria permeability transition pore. However, decreases in the $\Delta\Psi_m$ can occur with high electric field charging impact, which probably permeabilizes the inner mitochondrial membrane (IMM). Nevertheless, an influx of Ca^{2+} alone is not necessarily sufficient to induce cell death. UsEPs affect mitochondria that appear to be independent of permeability changes in the IMM (Beebe et al. 2012, 2013). These considerations are extended in the section of mitochondria.

The applications of usEPs on tumors are applied using Blumlein line-based or MOSFET-based pulsed power technologies (see Chap. 15). The effects of usEPs can differ depending on the pulse durations, pulse rise-fall times, electric field intensities, and pulse repetition rate. The latter factor may induce temperature increases, which adds thermal effects to the electric field effect. In the absence of heat, it adds repetition to the intracellular effects increasing the charging impact as indicated by the $E\tau n$ formula for calculation Vs/cm.

Through the studies that use usEPs, the charging effects of this technology (Schoenbach et al. 2009), as measured by Vs/cm are considered essential. Since this pulsed power technology is deemed to be non-thermal, it seems less reasonable to use Joule heating units to define their effects. In some early studies, usEP conditions were varied by pulse duration. Electric fields were adjusted so that the energy density was held constant. Under these constant energy density conditions, cytochrome c release (see Fig. 9.5 in Chap. 9), caspase activation, phosphatidylserine externalization (Beebe et al. 2002, 2003), and protein kinase inactivation (Beebe 2015) were dependent on the pulse duration and Vs/cm. So, charging rather than energy density seems to be an important factor. In any event, usEP conditions are given, and the Vs/cm values for the given conditions.

18.3 An Initial Study for the Effects of usEPs on Fibrosarcoma Tumors in Mice

When it became evident that usEPs could kill bacteria, an immediate thought was to determine if usEPs could kill cancer cells and tumors. After usEP eliminated human

Table 18.1 UsEP reduce fibrosarcoma tumor size. Four 8-week-old C57BL/6 mice were inoculated subcutaneously in both flanks (one side for treatment and the other side as control) with 5×10^6 B10.2 Fibrosarcoma cells in 0.1 ml Hanks Balanced Salt Solution without Ca and Mg

Mouse fibrosarcoma tumor size in vivo					
Animal #	Total treatment (days)	# of days treated	Tumor size (mm²/day)		% decreased growth rate versus control
			Control	Treated	
1	20	10	24.6	15.2	37.2
2	12	7	30.4	14.1	53.6
3	15	9	8.2	3.2	61.0
4	12	8	11.7	4.7	59.8
(10 pulses @50 ns, 75 kV/cm)					52.8 ± 5.5%

When tumor masses were about 5–10 mm in diameter, dual needle electrodes were inserted into tumors and that were exposed to usEPs in vivo with 10 pulses, 50 ns and 75 kV/cm. The number of treatments and treatment days are indicated (Beebe et al. 2002)

Jurkat and HL-60 cancer cells (Beebe et al. 2002, 2003), we induced ectopic tumors of mouse B10.2 fibrosarcoma tumors and treated them with usEPs in C57BL/6 mice (Beebe et al. 2002). There was a significant decrease in tumor growth compared to sham-treated control tumors. Tumors treated twice with multiple 300 ns pulses at 75 kV/cm decreased in size and weight by about 60% eight days after treatment.

In another study (Table 18.1), tumors were treated multiple times with ten 50 ns pulses and 75 kV/cm (Beebe et al. 2002). With numerous treatments over several days, tumor sizes were decreased by about 50% compared to contralateral untreated control tumors.

These studies were the first to show that usEPs could decrease tumor growth. However, two improvements were included in future studies that were absent here. First, these initial studies did not surround the tumors with electric fields. Dual needle electrodes were inserted directly into the tumors, so not all tumor tissue was exposed to electric fields. These untreated tumor tissues will likely continue to grow. Second, it was later shown that greater pulse numbers and higher electric fields were needed to eliminate tumors. The authors were a bit hesitant treating tissues in live, albeit anesthetized, animals for the first time, not knowing what to expect. With proper lubricant, pulse breakdown was avoided; there were no muscle contractions, and animals recovered from anesthesia with no apparent side effects. In studies that followed, increased pulse numbers and multiple treatments were delivered until tumors were completely eliminated. Finally, it was determined that single treatments could eliminate tumors when electric fields and pulse numbers were sufficiently high.

Nevertheless, these early studies made a case for the possibility that usEPs could potentially be used as a cancer treatment. Studies that followed established that idea and extended it to include activation of host immune responses after tumor treatment (Lassiter et al. 2018; Guo et al. 2018), which will be discussed in a section on immunity (Chap. 19). Based on the studies discussed below, it is likely that some immune responses were present in these initial studies. However, the emphasis was on treating cancer by ablation and determining the mechanisms of cell death. During this time, the scientific community realized that cells died by a genetically programmed mechanism of cell death called apoptosis and not by what is now called accidental cell death (ACD) or membrane rupture or casual failure to maintain a membrane potential leading to necrosis. This idea was noted in the modeling studies presented above (Gowrishankar et al. 2006; Esser et al. 2009). The focus of usEP research was also fascinated by new applications of pulsed power technology, which had been used in for development of radar in World War II, particle accelerators, fusion research, and high-power pulsed lasers. Pulsed power could now be used to treat cancer. The transfer of technology from military to medical applications was incredibly attractive. The finding that pulse power technology could also induce immune responses was another leap of imagination about what this technology could do.

18.4 Murine Melanomas Were Shown to Self-destruct as usEPs Effectively Targeted Cancer

The first large-scale study of usEP treatment of cancer cells treated murine B16-F10 melanoma tumors SKH-1 immunocompetent, hairless, albino strain (Nuccitelli et al. 2006). This combination of cells and mice was not syngeneic. B16-F10 cells and hairless, albino mice were from different murine strains, so they were not immunologically compatible and could cause some immune reactions. Nevertheless, dark melanin in the tumors was easy to treat and analyze on an albino background. Two to four tumors were induced in each mouse, and they were treated with 300 ns pulse duration with a rise-fall time of 30 ns and electric fields greater than 20 kV/cm. Two electrode designs were used. One was a 5 needle array with a central positive electrode and 4 ground electrodes forming a square and spaced about 4 mm apart. The other design was stainless steel parallel plate electrodes. Tumor responses to usEPs with 5 needle array electrodes were dependent on pulse field strength and pulse number. However, for tumor responses to pulse number, a sufficient electric field must be achieved. When pulses of 10 kV/cm were applied, neither 10 pulses nor 0.03 Vs/cm nor 100 pulses or 0.3 Vs/cm had effects on tumor growth inhibition. However, when one hundred 300 ns duration pulses were 20 kV/cm 0.6 Vs/cm, there was a pulse number-dependent effect on tumor growth inhibition. Tumors decreased 75% in 8 days when with two treatment of 100 pulses with 300 ns durations and electric fields of 20 kV/cm or a total of 1.2 Vs/cm. When tumors were treated with 100 pulses and 40 kV/cm or 1.2 Vs/cm using parallel plate electrodes, black scabs appeared and remained for about two weeks on the stratum corneum before regeneration. With two separate 100 pulse treatments and 40 kV/cm or a total of 2.4 Vs/cm three days apart, tumors shrank 90% in two weeks; however, two weeks later, tumors began to regrow. Overall, it was found that using 100 pulses at 300 ns and 40 kV/cm complete tumor regression by day 65 could be achieved with treatments on day 0, 1,

2 and 21, 22, 23 or 7.2 Vs/cm (6 × 1.2 Vs/cm). Later studies further optimized these conditions.

UsEPs are non-thermal. One of the early concepts and proposed advantages for usEPs for ablation of tumor tissue was the absence of Joule heating effects. Although the electric field strength is most often quite high, the pulse duration of 100 ns or 300 ns used in tumor treatment were deemed too short to significantly heat tissue (Schoenbach et al. 2001). This would avoid heatsinks that are common with microwave and radiofrequency ablation (Kim 2018). In a tissue ablation environment, a heatsink is the passive diffusion or transfer of heat away from a heat generating device in a fluid medium such as the fluids that bathe tissues, which can cause collateral damage to tissues that were not intended to be heated. Using direct measurements and theoretical modeling and considering a pulse duration of 300 ns and an electric field strength of 2.2 kV/cm pulses with 2 s intervals, a maximum temperature was reached after 40 pulses or 0.03 Vs/cm. After 100 pulses (0.6 Vs/cm) a temperature of 29.4 °C was recorded (Pliquett and Nuccitelli 2014). However, as expected, if multiple pulses are delivered in rapid succession, heat will be generated, so the pulse repetition rate much be relatively low. With a pulse duration of 300 ns and an electric field of 40 kV/cm increased the temperature 3 K concluding that hyperthermia was not a major factor it usEP effects. In another study of temperature effects due to usEPs using 2000 pulses with a repetition rate of 5–7 pulse/s, pulse duration of 100 ns and electric field strength of 30 kV/cm or 8.0 Vs/cm did not raise temperatures above 40 °C (Nuccitelli et al. 2010). Temperatures above 43.5 °C for 35 min is considered hyperthermia (Borrelli et al. 1990). Thus, when pulse repetition rates are relatively low, usEPs can be considered to induce relatively pure electric field effects that are devoid of heating.

18.5 usEPs Eliminate Melanoma and Its Blood Supply

A second study by this group using the same SKH-1 hairless non-syngeneic mouse model found more optimal condition for treating these same melanoma tumors with usEPs (Nuccitelli et al. 2009) and investigated some of the cellular responses that might be responsible for tumor demise. Treating tumors with 300 pulses with durations of 300 ns (5 ns rise-fall time) and 40 kV/cm or 3.6 Vs/cm, 24% of tumors were eliminated with a single treatment, 59% showed complete remission with two treatments, and 18% required three treatments. Overall, either 1–3 applications of 300 or 600 usEPs with 300 ns durations and 40–50 kV/cm electric field strengths (70–90 A) or 3.6–9 Vs/cm per application eliminated 90% of the tumors. For each treatment, the temperature increase was about 3 °C.

While a significant focus of these early studies using usEPs for tumor treatment was tumor elimination, it was also of interest to observe other aspects of tumor treatment, including effects on tumor blood supply and cell death mechanisms. Transillumination imaging indicated that the tumor blood supply was greatly reduced within a day after tumor treatment. A 93% reduction in CD31 as a micro-vessel density marker corroborated these observations. DNA damage was evident as fragmented DNA in the in vitro comet assay, and the percent of DNA in the comet tails were proportional to the square root of the pulse number (see usEP effects on the nucleus). The antiapoptotic Bcl-2 levels were decreased 72%, consistent with apoptosis. However, this alone does not indicate apoptosis. It is the Bax (pro-apoptotic)/Bcl-2 (anti-apoptotic) ratio that is important, and Bax was not determined. It is also noteworthy that even when this ratio is increased, there are other downstream effectors that determine caspase-3 activation, which is the major apoptotic executioner. While this study showed that pro-apoptotic Bad was increased threefold, it is elevated de-phosphorylated Bad that is pro-apoptotic by heterodimerizing with anti-apoptotic Bcl-2 or Bcl-xl (Wang et al. 1999; Datta et al. 2002). It should be understood that these observations were made in the TME, which not only contained tumor cells but host somatic and immune cells, so the appearance of any of these pro- or anti-apoptotic factors does not specifically identify their presence in tumor cells. While this study and others that followed suggested that usEPs induced apoptosis in B16-F10 cells in vitro (Ford et al. 2010) and other tumors (Chen et al. 2012), this may not be the case because other did not observe caspase-3 activation in B16-F10 cells in vitro (Rossi et al. 2019). There were also increases in intracellular Ca^{2+} (see section on ER, Chap. 5) and in membrane conductance (see section on plasma membrane, Chap. 2). All these studies demonstrated that usEPs were effective in eliminating melanoma in mice, suggesting possible applications to treat melanoma in humans.

18.6 Optimization of usEP for Treating Melanoma

Continued studies with usEPs and melanoma determined to optimize conditions to eliminate B16-F10 melanoma stably expressing an enhanced green fluorescent protein (eGFP) in a single treatment using an athymic nude mouse (Nu/Nu) (Nuccitelli et al. 2010). This immunodeficient mouse lacks a thymus, so it produces $CD4^+$ and $CD8^+$ T-cells but is B-cell and NK cell competent. Because the previous studies were carried out in a non-syngeneic mouse model, there was the likelihood that some immune responses could have inhibited tumor growth and thereby affected the results. The pulses were delivered using a suction electrode that was compatible with human skin. Four different suction electrode configurations were tested. These were cups into which tumors would be sucked so that electrodes surrounded all tumors. The designs were 2-pole, 6-pole, and 6-pole dual designs and a 5 needle array with a positive center needle and four corner ground needles. While this was well-known from earlier studies yet not specifically discussed, this study also pointed out the importance of using an oil (olive oil here) to prevent sparking over the skin caused by air pockets between electrodes. Because of the low dielectric constant of air, flashing over the skin can occur as the flowing current ionizes the air gases. This sparking or flashing bypasses the tumor and burns the skin. This study showed that complete regression and absence of regrowth after treatment was achieved by delivering 2000 pulses with 100 ns durations, 15 ns rise-fall time, and electric fields of

30 kV/cm with a pulse repetition rate of 5–7 pulses (Hz), which delivered 6.0 Vs/cm to tumors. The frequency did not increase the temperature to over 40° C. Under these conditions, usEPs were likely enhanced by mild heating, which enhances electric field effects of irreversible electroporation (Edelblute et al. 2018). Considering all electrode configurations, the 6-pole dual-electrode was superior to the others. The suction design allowed the usEP effects to be highly localized to the tumor. The results demonstrated a pulse delivery device compatible with human skin that completely eliminated murine malignant melanomas with a single 6 min-long treatment.

Additional studies with B16-F10 melanoma tumors in the hairless, SKH-1 mouse model more closely investigated cellular and molecular mechanisms involved in usEP-induced tumor cell death that provided greater insight into exactly how usEPs could eliminate tumor growth (Chen et al. 2010). In these studies, 100 usEPs with 300 ns and 40 kV/cm or 1.2 Vs/cm were applied to tumors, and relative early responses (1–24 h) were observed to indicate how tumors might be dying. As stated above by earlier studies with the same model, these conditions would not completely eliminate tumors but decreased tumor size by about threefold lower than sham-treated tumors. Results demonstrated that the usEP-treated tumor microenvironment (TME) expressed anti-angiogenesis, apoptotic cell death, and DNA damage. Earlier studies indicated that usEPs decreased blood flow to B16-F10 tumors after treatment. This evidence was substantiated by the finding that there were reduced numbers of blood vessels, decreased blood vessel epithelial molecular markers including CD31, CD35, CD106, and reduced levels of vascular growth factors such as vascular endothelial growth factor (VEGF) and platelet-derived endothelial cell growth factor (PD-ECGF). There were also time-dependent increases in active effector caspases-3/-6/-7; however, it was not clear if these were apoptotic tumor cells or other cells in the TME. While there were increases in DNA damage markers, including TUNEL (terminal deoxynucleotidyl transferase dUTP nick end labeling), pyknotic nuclei, phospho-histone 2AX, and large DNA fragments on agarose gels, 180 bp fragmentation ladders typical of caspase-dependent DNA damage were not present. These results suggested that apoptosis may have been initiated but not completed. These findings could occur if other RCD mechanisms were induced that resulted in decreased ATP levels, which are needed to form the apoptosome with APAF-1, caspase-9, and deoxy-ATP, and/or incomplete caspase-mediated cleavage of PARP (poly-ADP-ribose polymerase). During apoptotic cell death, effector caspases cleave PARP, which is used in DNA repair because DNA repair is not needed, and it is specifically degraded during apoptotic cell death for efficient cell removal macrophages. When caspases do not inactivate PARP and there is excessive DNA damage, it becomes hyper-activated. It depletes ATP with excessive poly-adenylation of proteins and accumulation of PAR, which is associated with RCD by parthanatos or PARP-1 dependent cell death and the nuclear translocation of apoptosis-inducing factor (AIF) from mitochondria (Yu et al. 2006; David et al. 2009).

18.7 Additional Melanoma Studies with usEP

A more recent study with usEP treatment of B16-F10 melanoma used a syngeneic C57BL/6 mouse model. These studies also investigated cell death mechanisms in response to usEPs in vivo and in vitro (Rossi et al. 2019). Again, usEPs were shown to eliminate tumors, but the cell death mechanism was equivocal. UsEP treated B16-F10 tumors did not exhibit active caspase-3 in vivo or in vitro. In agreement with the literature (Koo et al. 2015), B16-F10 melanoma cells lacked RIP-3, so necroptosis was also not induced by usEPs in these tumors or cells. B16-F10 cells also did not undergo autophagy in a cell transfected with LC3/GFP. It was concluded that B16-F10 cells died by necrosis in response to usEPs; however, RCD mechanisms by parthanatos and pyroptosis were not determined. The definition of cell death by necrosis does not really indicate how cells die because necrosis is defined by cells that show necrosis morphology, which does not reveal a cell death process.

The finding that usEPs did not induce caspase activation and apoptosis in these cells is in conflict with a previous report indicating otherwise (Ford et al. 2010). Retrospectively, this study can be revisited. The study reported usEP-induced extrinsic apoptosis with some unique features. As expected through extrinsic apoptosis, cytochrome c was not released, although the mitochondrial membrane potential was decreased, and ATP levels were reduced. Atypical of apoptosis, annexin-V binding was absent, and SMAC/Diablo was not released from mitochondria, which facilitates apoptosis by blocking the caspase inhibitor IAP (inhibitor of apoptosis) and keeps active caspase in an inhibited state such as during a viral infection or with overly active caspases during normal cell function (Deveraux et al. 1997). If this result was present in vivo, this could explain DNA fragments without the presence of the typical apoptotic 180 bp DNA ladder (Chen et al. 2010). AIF was not released in response to usEPs, suggesting that parthanatos was not induced under these conditions. The presence of active caspases in this study was determined using FITC-VAD-fmk. In unpublished experiments, this caspase indicator showed usEP-induced active caspase when the well-characterized caspase catalytic activity assay was negative, casting doubt on the FITC-VAD-fmk as a valid caspase assay when cells are exposed to usEPs. Taking data from these three investigations (Ford et al. 2010; Chen et al. 2010; Rossi et al. 2019), the mechanisms of usEP-induced cell death in B16-F10 melanoma is in doubt. However, given that cell death did not occur immediately (Ford et al. 2010; Rossi et al. 2019), it is unlikely to be due to accidental cell death (ACD). Rossi et al. concluded that cells died by necrosis. However, this term is not informative given necrosis is defined as "dead." So, B16-F10 cells can undergo apoptosis because STS was effective. Still, necroptosis was apparently not functional given the absence of RIP-3, and cell viability was resistant to necroptosis as inducted by a cocktail of TNF-α, the pan-caspase inhibitor zVAD, and the Smac mimetic BV6 (Rossi et al. 2019). Since AIF was not released (Ford et al. 2010), parthanatos is an unlikely RCD mechanism. Because cell death decreased between 3 and 24 h, cells did not die immediately, so ACD can be excluded. However, pyroptosis, which is activated by DAMPs, PRAMPs, inflammation, and ferroptosis, is induced by ROS production lipid peroxidation (Tang et al. 2019), are still considerations. Pyroptosis is reported to occur rapidly, which may not allow apoptosis or other RCD mechanism time to occur completely. NPS does induce ROS, but it is not agreed upon whether it plays a significant role in cell death (Pakhomov et al. 2012; Nuccitelli et al. 2013). However, ROS effects on cell viability depend on the usEPs, and there may be cell-type differences. Regardless of the RCD mechanism, usEPs treated B16-F10 tumor self-destruction, and there is evidence for usEP-induced release of immunogenic factors (Nuccitelli et al. 2014; Guo et al. 2018; Rossi et al. 2019). More on this and other cancer models when we consider usEP-induced immunity.

18.8 usEP Eliminate Hepatocellular Carcinomas (HCCs)

The early studies that followed usEP effects on the B16-F10 melanoma non-syngeneic model investigated usEP effects on a mouse ectopic Hepa1-6 hepatocellular carcinoma (HCC) model (Chen et al. 2012). This study was designed to investigate the long term survival of HCC after usEP treatment and investigate what molecular and cellular events occurred in the TME after treatment. Morphological changes include cytoplasmic shrinkage and condensed cytoplas with disintegrating cytoskeletal structures. Nuclei exhibited greatly condensed chromatin that segregated into sharply defined bodies (nuclear pyknosis). Multi-angular cells exhibited decreased nuclear/cytoplasmic ratios and enlarged extracellular spaces. There were time-dependent decreases in tumor nuclei size to less than 20% of their original volume in 24 h, with the most significant reduction occurring within the first 3 h after treatment. Generally, the overall molecular and cellular conclusions of this study were similar to those with B16-F10 melanoma. There were time-dependent increases in TUNEL staining and caspase activation (caspase-3 > −7. −6) that peaked 3 h after treatment. There were time-dependent decreases in VEGF and a downstream VEGF respondent CD34, a standard endothelial micro-vessel density (MVD) marker, with the first three weeks after treatment. Like the melanoma study, active caspases and TUNEL staining could not be specifically assigned to Hepa1-6 tumor cells.

Two different electrode designs were tested against Hep1-6 HCC tumors. Electrodes with a 4 + 1 five needle array and needle-ring designs were compared. With all comparable usEP conditions, the 5-needle array was clearly superior.

In the five-needle array used in these studies, the electric field is heterogeneous within the TME as the treatment zone; more in the 4 + 1 five-needle array configuration, less in the needle-ring design (Fig. 18.2). The needle electrode was a 4 + 1 needle array with four grounded perimeter needles forming a square and a high voltage biased center needle. For the ring electrode, a coaxial ring applicator replaced the grounded needle electrodes. The inner ring diameter was 8 mm, the same as the diagonal distance of the needle array. Both electrode designs entirely surround tumors (0.3–0.5 cm).

The electric field intensity at the needle electrode surfaces reaches values as high as 200 kV/cm. The field decreases in the needle to ring electrode system to values on the order of 40 kV/cm (with lowest values close to the ring). In the 5-needle array the electric field decreases from the center electrode, reaches a minimum of about 40 kV/cm at two-thirds of the distance from center to the outer needles, and increases again to almost 90 kV/cm at the surface of the outer needle electrode. Of course, whereas the electric field distribution in the ring electrode is rotationally symmetric about the center electrode, in the 5-needle array it shows maxima at the radial positions, which are defined by the location of the outer electrodes.

Another aspect of these studies considered the hypothesis that shorter pulses with shorter rise-fall times were more effective due to intracellular effects. So, short rise-fall time pulses with 30 ns and 100 ns durations were compared with different electric fields and pulse numbers. These HCC tumors could also be successfully eliminated ($\geq$ 75%) with 900 usEPs (1 Hz) with durations of 100 ns and electric fields at 65–68 kV/cm and 900 pulses in a single treatment or three separate 300 pulse treatments on alternate days. This split treatment provided charging effects as high as 6.0 Vs/cm. This charging level was deemed necessary for successful ablation. When 300 or 500 pulses with charging values of 2.0 and 3.4 Vs/cm, respectively, ablation was ineffective (< 40%) for successful ablation. Decreasing the electric field from 68 kV/cm to 50 kV/cm (6.1–4.5 Vs/cm) decreased survival from about 75% to < 40%. Likewise, pulses with durations of 30 ns and electric fields as high as 68 kV/cm (1.8 Vs/cm) could not successfully ablate Hepa1-6 HCC. While 30 ns pulses may be capable to induce intracellular effects, the electric fields were not sufficiently high to induce RCD mechanisms. These studies and more recent studies with other tumor models suggest shorter pulse durations with short (or fast) rise-fall time, but electric fields and pulse numbers must be sufficient to eliminate tumors. As indicated in the ectopic mouse Hepa1-6, HCC charging effects greater than 4.5 and perhaps as high as 5.0 Vs/cm are necessary for tumor elimination. A continuation of these studies are discussed in the section on usEPs as possible immunotherapy. Table 18.1 in that section shows that mice with Hep1-6 tumors cleared are resistant to recurrence of these same tumors, which is the bases for that section on immunity.

In order to continue and extend usEP effects on tumors in vivo, it was important to begin to include orthotopic models, that is tumors initiated and treated in the tissues where they would normally grow. Since the post-ablation protective effect occurred in HCC tumors, an orthotopic N1-S1 HCC model was developed in the rat (Chen et al. 2014a), since a mouse model may be too fragile that required three surgeries with two of them with a week to ten days of each other. This model included a first surgery to inject tumor cells under the liver capsule, followed by a second surgery to treat the tumors when it was large enough to treat. This was usually around 7 days. Then 7 weeks later, a third surgery would include the orthotopic challenge injection of live N1-S1 cells to see if the protective vaccine-like effect was present. This turned out to be an excellent model in most all respects for the initial intended purposes, which were to confirm the protective vaccine-like effect in a orthotopic HCC model and then to define the usEP-induced host immune mechanisms that defined the in vivo vaccination.

Orthotopic rat N-S1 HCC tumors were first treated in rat livers with conditions that were shown to eliminate tumors in the mouse melanoma and HCC models.

This included 1000 pulse with 100 ns durations and 50 kV/cm. These conditions were successfully eliminated > 80% of all N1-S1 tumors treated. To determine if fewer pulse numbers could be effective, 100, 300 and 500 pulses. (0.5, 1.5, 2.5 Vs/cm, respectively) were used under the same 100 ns 50 kV/cm conditions. While these lower pulse numbers effectively reduced tumor size, they were not sufficient to completely eliminate tumors without regrowth. Thus, in all models tested as many as 1000 pulses at 100 ns durations and 50 kV/cm (5.0 Vs/cm) successfully eliminated mouse B16F10 melanoma, mouse Hepa1-6 HCC and rat N1-S1 HCC tumor models. In all of these experiments, pulses were delivered at 1–3 pulses per second to avoid heat build-up and thermal effects. Thus, usEP-induced elimination were solely due to electric field effects. While 1000 pulses seem like a large number of electric fields to deliver, the total time tumor cells experience these electric fields is 0.1 ms. That is shorter than the time of a single heart-beat; however, the power levels are extremely high—around 10–20 MW. Considering these conditions, this usEP treatment can be considered as pulsed power ablation.

After we had tested the limits of usEP conditions for tumor elimination, we carried out a series of studies to eliminate tumors and then orthotopically challenge tumor-free rats with live N1-S1 tumor cells to determine if a protective vaccine-like effect was present as an indication of immunity. This turned out to be the case. These studies are discussed, and this section continued in the section on usEPs as possible immunotherapy (Fig. 18.1).

18.9 Moderate Heat Enhances usEP Ablation of Ectopic Tumors

The avoidance of heat with usEP treatment was a factor because we wanted to determine only electric field effects to prevent contamination with this second factor of Joule heating. However, it was hypothesized that moderate heating would increase the efficacy of usEP effects and possibly produce a synergistic effect by decreasing tumor impedance and increasing the conductivity of the tumor (Edelblute et al. 2018). Using a 980 nm wavelength infrared laser, a heating system with a temperature sensor was designed that was programmable and automatically adjustable to reach and maintain tissue temperatures of 43 °C. The results indicated that heat acted synergistically with usEPs with a reduction of the voltage of about 27% to obtain an equivalent level of tumor regression and overall survival. Using a lower voltage of 9.8 kV for tumor treatment compared to untreated mice and mice treated with usEP alone or heat alone, a 53% reduction in voltage could achieve an equivalent effect on tumor regression and overall survival.

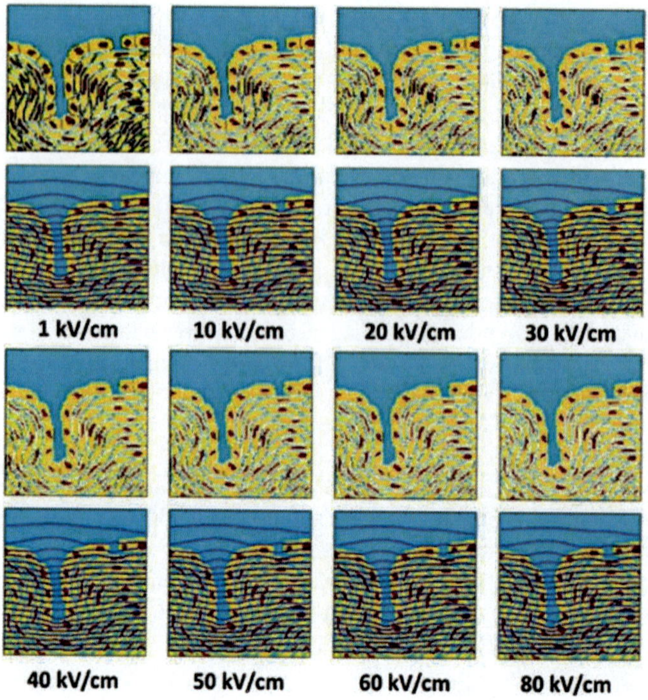

Fig. 18.1 Responses of multicellular tissues with usEPs of 300 ns and increasing electric fields. A multicellular tissue model shows pairs of panels with the top panels showing electroporation in white and bottom panels showing equipotential lines in blue. 300 ns pulses are shown in all panels with the electric field strength indicated below (Gowrishankar and Weaver 2006)

18.10 usEPs Demonstrate Efficacy Against Spontaneous Canine Osteosarcoma

Osteosarcoma (OS) accounts for more than 80% of bone tumors and is the most common primary tumor in dogs. Most OS occur in the forelimbs than in hindlimbs (64%), ribs or skull (28.5%), or in viscera (7.5%). The principal treatment is by limb-sparing or amputation, which increases survival. Chemotherapy with carboplatin, doxorubicin, cisplatin, or combinations has been used. The median survival for dogs with limb OS after diagnosis is about one year; about 15–30% of dogs with OS will live about 2 years. Alternative treatments include more or different drugs. Most dogs with osteosarcoma die of metastasis (Szewczyk et al. 2015).

However, in the only reported study in dogs, usEPs were used to treat four dogs in a controlled study compared to untreated controls (n = 5) and amputation (n = 3) (Chen et al. 2017). A dual needle puncture electrode was made from medullo-puncture needles. Dogs were treated with 500 usEPs with durations of 100 ns and electric field strength of 40 kV/cm or 2 Vs/cm. With a 6 month follow up, all control

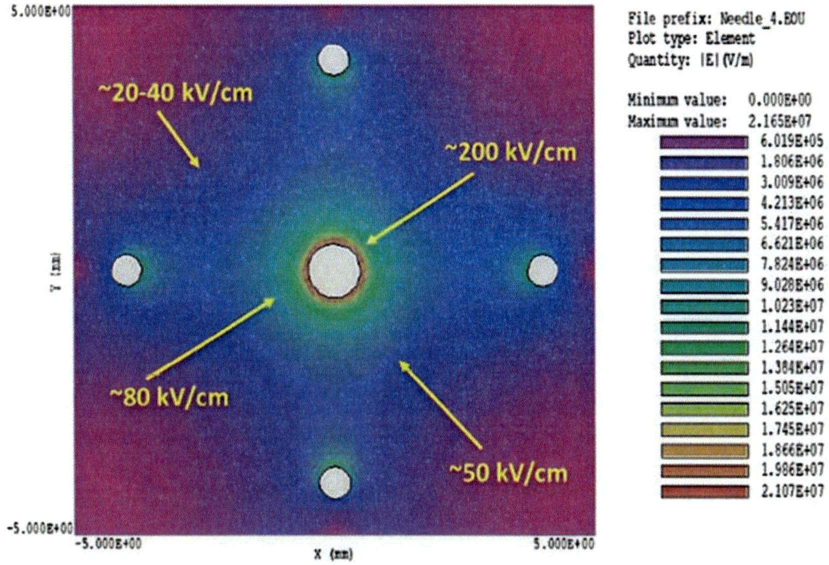

Fig. 18.2 Five-needle electrode array and electric field strength delivery. The indicated usEP conditions are 60 ns 50 kV/cm (25 kV/ 0.5 cm) delivers heterogeneous electric field amplitudes with the highest field near the center positive electrode with increases around the four ground electrodes

animals had metastasis, while only one of the usEP treated dogs showed metastasis. No dogs treated by amputation had metastatic disease. After 6 months, one of the five control dogs, three of the four usEP treated dogs, and all three of the amputees survived.

The study demonstrated that the usEP treatment extended overall survival for six months after treatment, and tumor growth was significantly inhibited. Alkaline phosphatase levels, which are elevated in dogs with OS, were reduced. usEPs could provide an alternative to treated OS without disfiguring amputation. Nevertheless, the authors' proposition that usEP treatment prevents metastasis is not fully substantiated. It cannot be ruled out that the tumors were cleared before metastasis happened, much as amputation likely occurred before metastasis arose.

18.11 usEPs Treatments of Human Cancer Xenografts

All studies with usEP treatment of tumors up to this point have been tumors in mice, although not all have been syngeneic. The ultimately dissimilar tumor models are xenografts where the tumors type and the animal are not from the same species. The most common xenografts are using human tumors in immuno-compromised mice, most often nude mice. These mice lack a thymus and do not express $CD4^+$ or CD8 $^+$ T-cells, so they are immune-compromised. They express natural killer cells, so the innate system is present, and the mice do express B-cells. These mice also lack hair and, therefore, the nickname "nude" mice. Since they accept grafts from other species, they are especially valuable for studies that test new cancer treatments. There have been several xenograft models that have tested the effects of usEPs on human tumors.

18.11.1 Human Melanomas

Melanoma was one of the first major projects undertaken for cancer treatment in part because melanoma skin tumors would be more redily accessible for treatment with usEP electrodes. Even so, melanoma is a significant systemic disease that readily metastasizes to other areas on the skin and to other organs. This is a dangerous disease that has attracted a sizable number of treatment strategies, including meaningful success with immune checkpoint inhibitors. So, there is substantial competition to demonstrate the therapeutic success that exceeds the present treatment efficacy. Perhaps the most precarious aspect of melanoma is the potential for metastasis. Part of the checkpoint inhibitors' success is the capacity to activate the patient's immune system, which is the foremost "job" of the checkpoint inhibitors. As we will discuss later, usEPs can act as immunotherapy and induce the host immune system in two well-characterized models. Nevertheless, as adjuncts and follow up studies of mouse melanoma models, studies addressing usEP treatment of human melanoma xenografts in nude mice are presented here.

A375 cells, a human melanoma cell line, were transfected with lentivirus expressing eGFP driven by the cytomegalovirus promoter (lentiviral particles (Pslv-CMV-eGFP) and injecting in immune-compromised nude mice for melanoma tumor formation (Guo et al. 2014). Using stainless steel plate electrodes, the tumors were treated with 1000 usEP with durations of 300 ns (35 ns rise-fall time) and electric field strength of 20 kV/cm or 6 Vs/cm. The tumors were decreased tenfold in the first five days as determined by eGFP fluorescence, and all tumors were eliminated in a single treatment. These studies essentially confirmed in human melanoma what had been observed in mouse melanoma studies, except a single time point at 5 days is used instead of time courses. Five days post-treatment, in the TME, there were around 50-fold increases in TUNEL staining, decreases in CD31microvascular staining, VEGF, and PCNA. The authors also demonstrated twofold decreases in Bcl-2 and twofold increases in Bax. Much like the finding in the B16-F10 melanoma study, the results suggest usEP-induced apoptosis induction and decreased angiogenesis, which identify conditions that inhibit angiogenesis.

As in the mouse melanoma studies and all those studies investigating events in the TME with specific identification of tumor cells, tumor microenvironment, it is not clear that tumor cells are demonstrating these apoptosis markers are tumor cells. On the other hand, the PCNA marker confirms that most cells in the TME cease dividing and the MVD present on endothelial cells such as those in the vasculature suggest the absence of vessel formation, which is confirmed by a decrease in the upstream driver for angiogenesis, VEGF. The studies confirm that usEPs can eliminate human melanoma in nude mice.

18.11.2 Human Hepatocellular Carcinomas

Human hepatocellular carcinomas have also been extensively used in usEP treatment. In two studies, human metastatic HCC HCCLM3 tumors were generated by implanting tumor tissue blocks into nude mice (Yin et al. 2014; Chen et al. 2014b, 2017). To get tissue blocks, HCCLM3 cells were injected into mice to form tumors. When tumors formed, small tumor tissue blocks were implanted into mice to be treated with usEPs. The electrode design consisted of a needle as the anode placed in the tumor center and a semi-circular ring as the cathode positioned on the tumor periphery. For this treatment, "one half of the tumor was treated with the initial semi-circular ring placement, and then the ring was rotated 180° to complete the tumor treatment." (Yin et al. 2014). Mice were either treated a single time or single dose (SD) with 300 usEPs with durations of 100 ns and electric field strength of 40 kV/cm or 1.2 Vs/cm or treated with three doses or the multi-fractionated dose (MFD) at 48 h interval with 0.4 Vs/cm for a total of 1.2 Vs/cm. However, from the explanation of the electrode placement and the electrodes' rotation, it may be that each tumor half or side is only exposed to half the total treatment dose.

Nevertheless, compared to the control group on days 7 and 14 days after treatment, tumor growth in SD and MFD groups was significantly reduced. Mice treated a single time with the SD 1.2 Vs/cm protocol showed a 40% reduced growth compared to control 16 days after treatment. After 60 days, about a 40% survival rate characterized the SD treatment. In contrast, the MFD group exhibited no tumor growth than its original size yet appear to be stable after 16 days. After 60 days, about a 90% survival rate characterized the MFD treatment. Using these protocols usEPs slow tumor growth, downgrade tumor burden, and increase overall survival. The more significant effects of the MFD were substantiated by a much greater macrophage infiltration into the TME recognized by MAC387 (Chen et al. 2014b). MAC387 is an S-100-like Ca^{2+} binding protein associated with myeloid differentiation (Goebeler et al. 1994; Roth et al. 1993).

Since the HCCLM3 is metastatic, the author showed that all 7 of the control untreated mice exhibited metastasis to the lungs. In contrast, 2 of 7 mice in the SD group presented with metastatic lungs, while none of the 7 mice in the MFD group showed metastasis in the lung (Yin et al. 2014). In the group that had tumor resection, one of the 7 mice exhibited metastasis. The authors suggested that usEP reduced pulmonary metastasis to the lungs in the model. While this may be the case, a more likely possibility is that in the absence of metastasis, the tumors were effectively treated before they could metastasize. The author did indicate that non-lethal usEPs enhanced the numbers of HCC cells phagocytized by differentiated THP1 macrophages; however, they did reveal what the non-lethal usEP treatment was. Based on the data present here, it is not sufficiently demonstrate that usEPs reduce metastasis other than successfully killing a sufficient number of HCC cells before they can metastasize. Nevertheless, the study demonstrates that usEPs effectively control human HCC growth in a nude animal model.

This study was similar to a study with usEP treatment in a mouse Hep1-6 HCC model. In that study, treatments comparing a single 19.1 Vs/cm treatment with three treatments with 2 Vs/cm per treatment made little difference to treatment success (Chen et al. 2012). However, the present study used 1.8 Vs/cm usEP conditions that were much lower. Using lower usEP conditions, it is likely that sub-threshold conditions are more likely to see differences than when higher threshold conditions are used. In the usEP studies reviewed so far and those upcoming, tumor elimination requires higher conditions than the 1.8 Vs/cm used here. If these studies compared 6 Vs/cm in one treatment or in three equally dived treatments, differences might not be evident. Nevertheless, using sub-threshold condition, which are lower than what has been shown to induce tumor elimination, it was possible to show that overall, it may be better to divide usEP treatments over three applications instead of a single treatment.

18.11.3 Human Breast Cancer

UsEPs were also effective in human MCF-7 breast cancer in nude mouse models (Wu et al. 2014). Pulses were applied through clamp electrodes with three treatment of 240 usEP at 4 pulses per second (Hz) with a duration of 100 ns and electric field intensities of 30 kV/cm or 0.72 Vs/cm per treatment for a total of 2.16 Vs/cm over the three treatments. This breast cancer model confirms what had been shown in human melanoma and human HCC that usEPs cause tumor shrinkage, TUNEL staining, decreases in Bcl-2, and decreases in vessel formation and angiogenesis markers. By day 14, after treatment, the tumor weighed 79% less than control tumors. While these tumors grew more slowly, the usEP conditions were not sufficient to completely eliminate tumors. As indicated in previous studies, these conditions are at least twofold lower than required to completely eradicate solid tumors.

18.11.4 Pancreatic Cancer

The earliest human tumor xenografts treated with usEPs were conducted using human AsPC-1 pancreatic tumor cells in matrigel injected into a female athymic nude mouse model using pulse durations of 7, 10, and 20 ns (Garon et al. 2007). In these studies, using a five-needle electrode array electrode delivering three 75 pulse bursts at 20 Hz separated by 1 min with 20 ns durations and 40–45 kV/cm, half of the tumors were cleared; however, efficacy was considered only 4 days after treatment and only regression and not clearance, could be declared. Considering the Vs/cm charging effects in these studies were no greater than 0.13 Vs/cm, these treatments were 50–60 fold lower than effects on melanoma and HCC tumors in mice, so transient effects on tumor regression could be expected. However, the Vs/cm measure does not include a factor for repetition rate, and these studies were carried out at 20 Hz or 20 pulses per second. While these pulse durations are 15–40 times shorter than the 300 ns pulse durations that were found to increase temperatures to 40 °C, their electric fields are similar, yet the temperature was not determined, so it is not clear that these usEP effects were due to electric field alone. Nevertheless, given that 30 ns pulses with charging effects greater than these were not sufficient, it is likely that higher electric fields and/or greater pulse numbers are needed unless the high repetition rate in the presence or absence of heat has effects to overcome the duration or pulse numbers factors.

18.11.5 Human Squamous Carcinomas

Another study using pulses at the lowest nanosecond level used 50, 200, and 400 pulses at 50 Hz, a 5 needle array that delivered 7 and 14 ns durations with electric fields of 31 and 40 kV/cm to treat cutaneous papillomas and squamous carcinomas in SENCAR (SENsitivity to CARcinogenesis) mice. To develop tumors, mice were treated with methyl-N'-nitro-N-nitrosoguanidine and tumor promoter 12-O-tetradecanoylphorbol-13-acetate (TPA), which developed tumors in 20–30 weeks (Yin et al. 2012). These tumors were eliminated for at least one week using one (85%) or two (15%) treatments with 14 ns pulse duration. It would be important to know if these tumors regrew in the following weeks to evaluate it as an effective therapy. The pulse duration of 7 ns was less effective when using 400 pulses at 50 Hz. Although the pulse durations are short, these high repetition rates could increase temperatures, which were not determined.

18.11.6 Human Glioma Xenograft in a Vascularized Avian Chorioallanttoic Membrane (CAM)

Bardet and colleagues (2016) determined effects of a single 10 ns usEPs Japanese quail (Coturnix coturnix japonica) chorioallanttoic membrane (CAM) implanted with human U87glioblastoma cells stably expression a fluc2 and green fluorescent protein (eGFP) driven by the cytomegalovirus (CMV) promotor. The cells were engrafted into a developmental window of developing quail eggs, which acted as a host promoting the growth of the tumor. This established a glioblastoma tumor organoid adjacent to the developing bird embryo. This approach provided growth of fluorescently labeled millimeter-sized spheroid tumors that were treated with usEPs while using multiphoton microscopy. This gave a bird's eye view (pun intended) of the developing vasculature of the tumor and the embryo. The tumor grew in such a way that easily distinguished from the CAM. Numerical simulations estimated that the electric field amplitude was 35–45 kV/cm for single 10 ns usEPs or 3.5–4.5 × 10^{-4} Vs/cm. Even at such low charging effects, the usEPs collapsed the perfusion neovasculature and affected the diameter of capillaries and larger vessels in normal tissue. There was a pronounce effect on capillaries. Since the CAM blood network is not innervated, the authors suggested that the usEP effects were directly on endothelial cells, perhaps at their junctions. Mathematical models estimated that the electric fields were 40% higher in the endothelial cells than the rest of the tissues (Sersa et al. 2008).

18.11.7 Human Triple Negative Breast Cancer (TNBC) Patient-Derived Xenograft (PDX) in NOD-Scid Gamma (NGS) and NSGM3 Mice

This cancer model is derived from Jackson Labs. The PDX came in an NSG mouse, which was then removed and engrafted into other mice for treatment. The NSG mice were developed and marketed by Jackson Labs. They carry the strain NOD . Cg-*Prkdc*^{scid} *Il2rg*^{tm1Wjl}/SzJ. They are the most immune deficient mice available lacking T-cells, B-cells, NK cells and are deficient in multiple cytokine signaling pathways. They are also defective in many innate immune mechanisms (Shultz et al. 1995, 2005).

Human TNBC-PDXs were developed in NSG immunocompromised mice (NOD . Cg-*Prkdc*^{scid} *Il2rg*^{tm1Wjl}/SzJ) by implanting tumor sections about 2 mm^3 in size into the flank. Tumors grew for approximately 6 weeks until reaching a treatable size of about 350–600 mm^3. Tumors were treated with usEP conditions previously successful in complete tumor elimination in mouse and rat models: 1000 pulse (~ 3 Hz) 100 ns durations and 50 kV/cm delivering 5 Vs/cm to each tumor via pinch electrodes. Decreases in growth were evident within days and were completely eliminated in 3–4 weeks post treatment. The 6 usEP-treated tumors in Fig. 18.3 represent

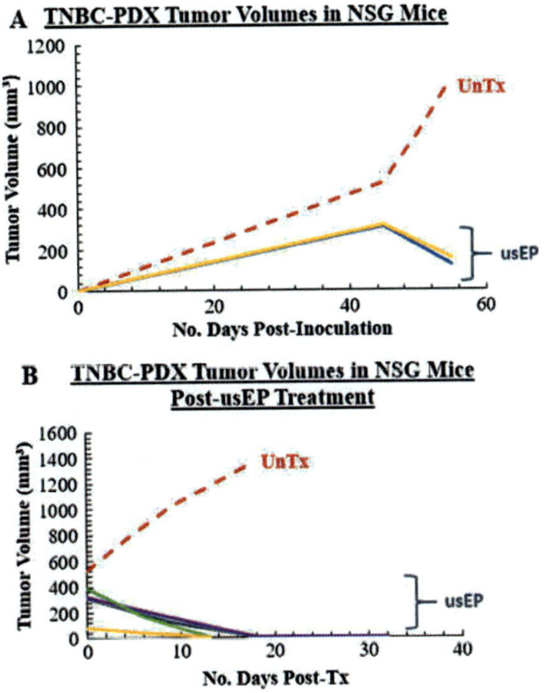

Fig. 18.3 UsEP eliminate human TNBC PDX in NSG mice. Human tumor tissue sections (~ 2 × 2 mm) were implanted into NSG immunodeficient mice **a** Implanted tumors grewing for about 6 weeks and then treated with usEPs, 1000 pulses (3 Hz) at 100 ns, 50 kV/cm, 5 Vs/cm (usEPTx) or sham treated. 45 days after implantation Tumor grown was followed for 10 days. **b** Tumors were grown for about 6 weeks (not shown) and then treated with usEP conditions as in panel A (red) or shame treated (black). Tumor growth was flowed for about 30 days. (Ruedlinger, Guo, and Beebe unpublished)

as many as 15 and all were eliminated by the treatment protocol used here.

Using the same tumor model (TNBC-PDX), tumors were implanted in NSG-SGM3 mice (NOD . Cg-*Prkdc*^{scid} *Il2rg*^{tm1Wjl} Tg(CMV-IL3, CSF2, KITLG)1Eav/MloySzJ). These mice were developed from the same background as NSG mice and therefore also lack mouse T-cells, B-cells, and functional NK cells with deficiencies in cytokine signaling. NSG-SGM3 mice contain three co-expressed transgenes regulated by the human cytomegalovirus promoter/enhancer sequence. The three transgenes (SGM3) include human stem cell factor (*SCF*) gene (S), which is required for maintenance of normal basal hematopoiesis by acting on hematopoietic stem cells to facilitate their entry into the cell cycle; human granulocyte/macrophage-colony stimulating factor 2 (*GM-CSF*) (G), which increases the numbers and functions of neutrophil, eosinophil; and monocyte/macrophage numbers and functions; and human interleukin-3 (*IL-3*), which stimulates multipotent hematopoietic stem cells to differentiate into myeloid progenitor cells and stimulates all myeloid cells,

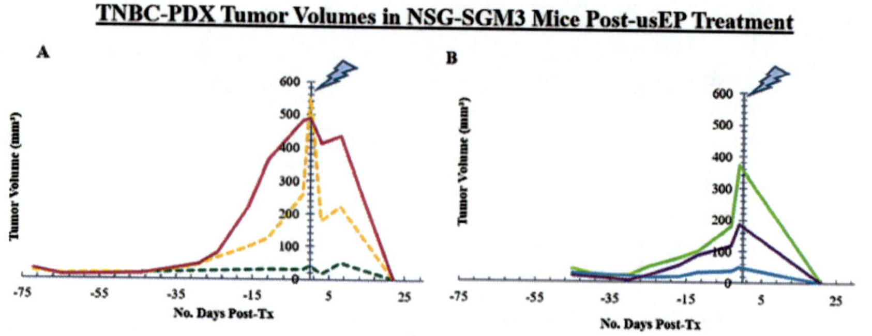

Fig. 18.4 UsEP eliminate human TNBC PDX in NSG-MSG3 mice. Panels A and B represent two individual study trials. **a** Three tumors were engrafted, two of which required about 75 days to reach treatable sizes of 400–600 mm^3. One of the three tumors failed to grow in the time frame. Tumors were treated with 1000 pulses with 100 ns durations and 50 kV/cm or 5 Vs/cm. **b** Three tumors were engrafted, two of which reached a treatable size of about 200–400 mm^3 in about 40 days. One tumor filed to grow in the time frame. These tumors were treated as in panel A (Ruedlinger, Guo, and Beebe, unpublished)

which includes granulocytes, monocytes, and dendritic cells, to proliferate and function with other cytokines. The NSG-SGM3 mouse was developed to improve engraftment and growth of human leukemias but was used here in the development of humanized mice, which is not reported here.

Like the TNBC tumors grown in Fig. 18.3, usEPs were effective to eliminate these tumors with delivery of 5 Vs/cm within about 3 weeks. The studies in Figs. 18.3 and 18.4 indicate that usEPs under these conditions were sufficient to eliminated TNBC PDXs without the need of immune system intervention.

Another approach was to use human breast cancer cell lines in immunodeficient mice. We used the human MDA-MB-231 VIM-RFP clone, which are MDA-MB-231 clone expressing vimentin labeled red fluorescent protein. The MDA-MB-231 cell line is human mesenchymal epithelial, basal B breast cancer cell line isolated from a pleural effusion of a 51-year-old Caucasian female with a metastatic mammary adenocarcinoma. It is estrogen receptor negative, fibroblast-like and highly invasive (Lacroix et al. 2004).

When MDA-MB-231 VIM-RFP cells were treated with usEPs delivering 50 kV/cm in NSG-mice (Fig. 18.5a) or NSG-SGM3-mice (Fig. 18.5b), tumors were not eliminated and continued to grow in both immunodeficient mouse models. This is the only tumor model investigated by our laboratory, including human, mouse, and rat models, in which usEP treatment failed to eliminate tumors.

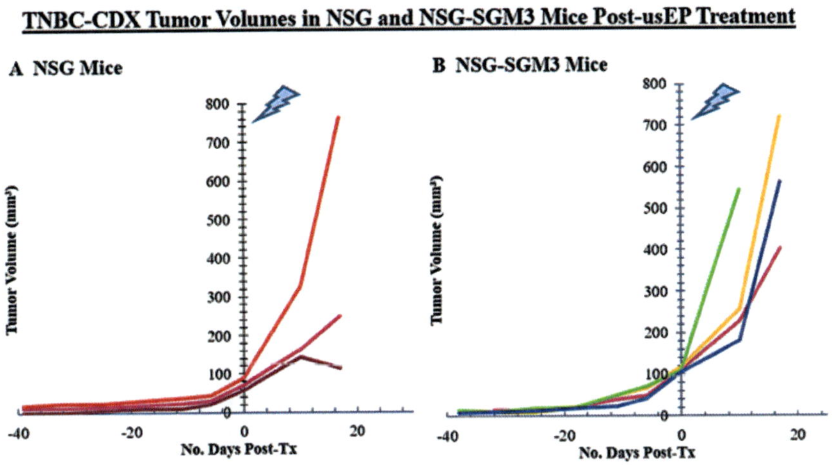

Fig. 18.5 UsEP treatment of cell derived xenografts (CDX) tumors in immunodeficient mice. Human TNBC MDS-MD-231-VIM-RFP cells were grown as defined at ATCC until 50–60% confluent and injected into the flanks of NSG mice (**a**) and NSG-SGM3 mice (**b**) (Ruedlinger, Guo, and Beebe, unpublished)

18.11.8 Summary of usEPs in tumor treatment

In general, the studies with human tumors in immune-deficient mice confirmed findings for similar studies with the same mouse tumors in immune competent mice. There are significant effects on vasculature, indicating neovascularization and angiogenesis are impeded, suggesting that this will minimize metastasis. However, the data presented does not rule out the possibility that the primary tumor was eliminated before metastasis could begin. Nevertheless, some data provide more convincing arguments that usEPs may prevent metastasis, but this is based on the presence of in situ vaccination and the presence of immune responses that prevented tumor recurrence (Guo et al. 2018) (see the section on immunity). Yet, with all the studies carried out, it is still unclear what regulated cell death mechanism(s) are responsible for tumor elimination, although it is likely to be cancer type-specific. While some conclusions have been reached from in vitro data, as presented above and elsewhere, specific RCD mechanisms in the TME are unknown because they have not been co-localized with the cancer cells in the treatment zone. Furthermore, as indicated in Figs. 9.9 and 9.10 in the section on Regulated Cell Death, showing necroptosis and parthanatos in TNBC cells, respectively, there may be more than one RCD mechanism in a given cell type. The focus on usEP-induced RCD is more than academic. It is hypothesized that RCD is one determinant of immunity, especially is they lead to the release of immunogenic cell death (ICD) factors like calreticulin, HMGB1, and ATP, among others. Given that such a description of immune induction for cancer therapy is a determining factor for designing immunotherapy, discovering as much as possible about these questions is essential for reconstructing these complex processes.

The xenograft studies on pancreatic cancer and squamous cell carcinoma with pulse duration $\leq$ 20 ns suggest that there may be limits to usEP durations that can effectively treat tumors. This limit is most likely determined by achieving electric fields that are sufficiently intense when electric fields are so short. It is well known that a decrease in pulse duration can be compensated by an increase in the electric field. This may also be offset by increasing the pulse number. Overall, these studies suggest that cells must be exposed to pulses short in duration and rise-fall time on the order of the plasma membrane charging time constant. These usEPs must deliver charging with adequate intensity and duration that provide sufficient charging of cell membranes and likely other structures, especially in mitochondria. This suggests an intensity threshold for appropriate short durations and rise-fall times and that this must be delivered enough times to afford cells no time to reinforce survival mechanisms. Based on available data, charging values on the order of 4–6 Vs/cm appear necessary for tumor elimination. As discussed in the section on immunity, there may be a distinct threshold for an adequate immune response, although this likely depends on the immunogenicity of the cancer model providing greater immunity with higher immunogenic models. Surprisingly, usEPs did not eliminate MDA-MB-231 VIM-RFP tumors in immunodeficient mice (Fig. 18.5), suggesting immunity may be necessary in this models that may be less immunogenic.

However, to reiterate, there are limits to the weight of the charging values for treatment efficacy and immunity. For example, there is no factor for the rise-fall time in the formula, which is important for usEP-induced intracellular effects unique to this technology. With some limited data among different cancer models, a 100 ns pulse with a 5–10 ns rise-fall time may be more effective for immune induction than a 200 ns pulse with a 40–50 ns rise-fall time (Ruedlinger, Guo, and Beebe et al. unpublished). However, this requires a larger study and likely depends on the cancer model's immunogenicity. Since there are three factors involved in the formula $E\tau n$—pulse duration (τ), electric field (E), and number (n)—the charging value can be reached by manipulating any of these conditions. So, consider the pulse duration of 100 ns with 5–10 ns rise time as adequate for the hypothesized short duration factors. A charging value of say 5 Vs/cm could be reached by using an electric field strength that could be below the presumed intensity threshold, but the Vs/cm value could be reached by increasing the pulse number. Again, this presumes that an electric field threshold is required instead of a sufficient charging requirement reached by delivering the charges enough times. Besides, there is a possibility for some temperature effects if the pulse repetition rate is fast enough. Modeling studies indicate small temperature changes (Esser et al. 2009), and experimental data indicate that a repletion rate of 5–7 Hz raises temperatures to ~ 40 °C but not enough for thermal injury (Nuccitelli et al. 2010). This cannot be specifically considered non-thermal. It was also shown that such moderate temperatures can enhance efficacy, and there is no factor for temperature in the charging formula. However, usEPs are delivered with no thermal effects so the results depend on charging and not thermal. Nevertheless, the charging formula near 5.0 Vs/cm appears to hold for the available data from the authors' labs. Using this formula can be helpful when considering effects of conditions that are and are not effective in tumor elimination.

References

Bardet SM, Carr L, Soueid M, Arnaud-Cormos D, Leveque P, O'Connor RP (2016) Multiphoton imaging reveals that nanosecond pulsed electric fields collapse tumor and normal vascular perfusion in human glioblastoma xenografts. Sci Rep 6:34443

Beebe SJ (2015) Considering effects of nanosecond pulsed electric fields on proteins. Bioelectrochemistry 103:52–59

Beebe SJ, Fox PM, Rec LH, Buescher ES, Somers K, Schoenbach KH (2002) Nanosecond pulsed electric field (nsPEF) effects on cells and tissues: apoptosis induction and tumor growth inhibition. IEEE Trans Plasma Sci 30:286–292

Beebe SJ, Fox PM, Rec LJ, Willis EL, Schoenbach KH (2003) Nanosecond, high-intensity pulsed electric fields induce apoptosis in human cells. FASEB J 17:1493–1495

Beebe SJ, Chen YJ, Sain NM, Schoenbach KH, Xiao S (2012) Transient features in nanosecond pulsed electric fields differentially modulate mitochondria and viability. PLoS ONE 7(12):e51349

Beebe SJ, Sain NM, Ren W (2013) Induction of cell death mechanisms and apoptosis by nanosecond pulsed electric fields (nsPEFs). Cells 2:136–162

Borrelli MJ, Thompson LL, Cain CA, Dewey WC (1990) Time-temperature analysis of cell killing of BHK cells heated at temperatures in the range of 43.5–57.0 °C. Int J Radiat Oncol Biol Phys 19(2):389–399

Chen X, Kolb JF, Swanson RJ, Schoenbach KH, Beebe SJ (2010) Apoptosis initiation and angiogenesis inhibition: melanoma targets for nanosecond pulsed electric fields. Pigment Cell Melanoma Res 23:554–563

Chen X, Zhuang J, Kolb JF, Schoenbach KH, Beebe SJ (2012) Long term survival of mice with hepatocellular carcinoma after pulse power ablation with nanosecond pulsed electric fields. Technol Cancer Res Treat 11:83–93

Chen R, Sain NM, Harlow KT, Chen YJ, Shires PK, Heller R, Beebe SJ (2014a) A protective effect after clearance of orthotopic rat hepatocellular carcinoma by nanosecond pulsed electric fields. Eur J Cancer 50:2705–2713

Chen X, Yin S, Hu C, Chen X, Jiang K, Ye S, Feng X, Fan S, Xie H, Zhou L, Zheng S (2014b) Comparative study of nanosecond electric fields in vitro and in vivo on hepatocellular carcinoma indicate macrophage infiltration contribute to tumor ablation in vivo. PLoS ONE 9(1):e86421

Chen X, Chen Y, Jiang J, Wu L, Yin S, Miao X, Swanson RJ, Zheng S (2017) Nano-pulse stimulation (NPS) ablate tumors and inhibit lung metastasis on both canine spontaneous osteosarcoma and murine transplanted hepatocellular carcinoma with high metastatic potential. Oncotarget 8(27):44032–44039

Datta SR, Ranger AM, Lin MZ, Sturgill JF, Ma YC, Cowan CW, Dikkes P, Korsmeyer SJ, Greenberg ME (2002) Survival factor-mediated BAD phosphorylation raises the mitochondrial threshold for apoptosis. Dev Cell 3:631–643

David KK, Andrabi SA, Dawson TM, Dawson VL (2009) Parthanatos, a messenger of death. Front Biosci 14:1116–1128

Deveraux QL, Takahashi R, Salvesen GS, Reed JC (1997) X-linked IAP is a direct inhibitor of cell-death proteases. Nature 388:300–304

Edelblute CM, Guo S, Hornef J, Yang E, Jiang C, Schoenbach K, Heller R (2018) Moderate heat application enhances the efficacy of nanosecond pulse stimulation for the treatment of squamous cell carcinoma. Technol Cancer Res Treat 17:1533033818802305

Esser AT, Smith KC, Gowrishankar TR, Weaver JC (2009) Towards solid tumor treatment by nanosecond pulsed electric fields. Technol Cancer Res Treat 8(4):289–306

Ford WE, Ren W, Blackmore PF, Schoenbach KH, Beebe SJ (2010) Nanosecond pulsed electric fields stimulate apoptosis without release of pro-apoptotic factors from mitochondria in B16f10 melanoma. Arch Biochem Biophys 497(1–2):82–89

Garon EB, Sawcer D, Vernier PT, Tang T, Sun Y, Marcu L, Gundersen MA, Koeffler HP (2007) In vitro and in vivo evaluation and a case report of intense nanosecond pulsed electric field as a local therapy for human malignancies. Int J Cancer 121(3):675–682

Gatzka MV (2018) Targeted tumor therapy remixed-an update on the use of small-molecule drugs in combination therapies. Cancers (Basel) 10 pii:E155. https://doi.org/10.3390/cancers10060155

Goebeler M, Roth J, Teigelkamp S, Sorg C (1994) The monoclonal antibody MAC387 detects an epitope on the calcium-binding protein MRP14. J Leukoc Biol 55(2):259–261

Gowrishankar TR, Weaver JC (2006) Electrical behavior and pore accumulation in a multicellular model for conventional and supra-electroporation. Biochem Biophys Res Commun 349(2):643–653

Gowrishankar TR, Esser AT, Vasilkoski Z, Smith KC, Weaver JC (2006) Microdosimetry for conventional and supra-electroporation in cells with organelles. Biochem Biophys Res Commun a341(4):1266–1276

Guo F, Yao C, Li C, Mi Y, Peng Q, Tang J (2014) In vivo evidences of nanosecond pulsed electric fields for melanoma malignancy treatment on tumor-bearing BALB/c nude mice. Technol Cancer Res Treat 13(4):337–344

Guo S, Jing Y, Burcus NI, Lassiter BP, Tanaz R, Heller R, Beebe SJ (2018) Nano-pulse stimulation induces potent immune responses, eradicating local breast cancer while reducing distant metastases. Int J Cancer 142(3):629–640

Hanahan D, Weinberg RA (2000) The hallmarks of cancer. Cell 100:57–70

Hanahan D, Weinberg RA (2011) Hallmarks of cancer: the next generation. Cell 144:646–674

Kim C (2018) Understanding the nuances of microwave ablation for more accurate post-treatment assessment. Review Future Oncol 14(17):1755–1764

Koo GB, Morgan MJ, Lee DG, Kim WJ, Yoon JH, Koo JS, Kim SI, Kim SJ, Son MK, Hong SS, Levy JM, Pollyea DA, Jordan CT, Yan P, Frankhouser D, Nicolet D, Maharry K, Marcucci G, Choi KS, Cho H, Thorburn A, Kim YS (2015) Methylation-dependent loss of RIP3 expression in cancer represses programmed necrosis in response to chemotherapeutics. Cell Res 25:707–725

Lacroix M, Haibe-Kains B, Hennuy B, Laes JF, Lallemand F, Gonze I et al (2004) Gene regulation by phorbol 12-myristate 13-acetate in MCF-7 and MDA-MB-231, two breast cancer cell lines exhibiting highly different phenotypes. Oncol Rep 12:701–707

Lassiter BP, Guo S, Beebe SJ (2018) Nano-pulse stimulation ablates orthotopic rat hepatocellular carcinoma and induces innate and adaptive memory Immune mechanisms that prevent recurrence. Cancers (Basel) 10(3):69

Lazikani B, Banerji U, Workman P (2012) Combinatorial drug therapy for cancer in the post-genomic era. Nat Biotechnol 30(7):679–92. https://doi.org/10.1038/nbt.2284

Nuccitelli R, Pliquett U, Chen X, Ford W, James Swanson R, Beebe SJ, Kolb JF, Schoenbach KH (2006) Nanosecond pulsed electric fields cause melanomas to self-destruct. Biochem Biophys Res Commun 343:351–360

Nuccitelli R, Chen X, Pakhomov AG, Baldwin WH, Sheikh S, Pomicter JL, Ren W, Osgood C, Swanson RJ, Kolb JF, Beebe SJ, Schoenbach KH (2009) A new pulsed electric field therapy for melanoma disrupts the tumor's blood supply and causes complete remission without recurrence. Int J Cancer 125:438–445

Nuccitelli R, Tran K, Sheikh S, Athos B, Kreis M, Nuccitelli P (2010) Optimized nanosecond pulsed electric field therapy can cause murine malignant melanomas to self-destruct with a single treatment. Int J Cancer 127:1727–1736

Nuccitelli R, Lui K, Kreis M, Athos B, Nuccitelli P (2013) Nanosecond pulsed electric field stimulation of reactive oxygen species in human pancreatic cancer cells is $Ca^{(2+)}$-dependent. Biochem Biophys Res Commun 435(4):580–585

Nuccitelli R, Wood R, Kreis M, Athos B, Huynh J, Lui K, Nuccitelli P, Epstein EH Jr (2014) First-in-human trial of nanoelectroablation therapy for basal cell carcinoma: proof of method. Exp Dermatol 23:135–137. https://doi.org/10.1111/exd.12303

Pakhomova ON, Khorokhorina VA, Bowman AM, Rodaitė-Riševičienė R, Saulis G, Xiao S, Pakhomov AG (2012) Oxidative effects of nanosecond pulsed electric field exposure in cells and cell-free media. Arch Biochem Biophys 527(1):55–64

Pliquett U, Nuccitelli R (2014) Measurement and simulation of Joule heating during treatment of B-16 melanoma tumors in mice with nanosecond pulsed electric fields. Bioelectrochemistry 100:62–68

Rossi A, Pakhomova ON, Pakhomov AG, Weygandt S, Bulysheva AA, Murray LE, Mollica PA, Muratori C (2019) Mechanisms and immunogenicity of nsPEF-induced cell death in B16F10 melanoma tumors. Sci Rep 9:431

Roth J, Burwinkel F, van den Bos C, Goebeler M, Vollmer E, Sorg C (1993) MRP8 and MRP14, S-100-like proteins associated with myeloid differentiation, are translocated to plasma membrane and intermediate filaments in a calcium-dependent manner. Blood 82(6):18751883

Schoenbach KH, Beebe SJ, Buescher ES (2001) Intracellular effect of ultrashort electrical pulses. Bioelectromagnetics 22(6):440–448

Schoenbach KH, Joshi RP, Beebe SJ, Baum CE (2009) A scaling law for membrane permeabilization with nanopulses. IEEE Trans Dielectr Electr Insul 16:1224–1235

Sersa G, Jarm T, Kotnik T, Coer A, Podkrajsek M, Sentjurc M, Miklavcic D, Kadivec M, Kranjc S, Secerov A, Cemazar M (2008) Vascular disrupting action of electroporation and electrochemotherapy with bleomycin in murine sarcoma. Br J Cancer 98:388–398

Shultz LD, Schweitzer PA, Christianson SW et al (1995) Multiple defects in innate and adaptive immunologic function in NOD/LtSz-scid mice . J Immunol 154(1):180–191

Shultz LD, Lyons BL, Burzenski LM et al (2005) Human lymphoid and myeloid cell development in NOD/LtSz-scid IL2R gamma null mice engrafted with mobilized human hemopoietic stem cells. J Immunol 174(10):6477–6489

Sonnenschein C, Soto AM (2014) Cancer and the Elusive Cancer Genes. Med Sci (paris) 30(6–7):688–692

Szcwczyk M, Lechowski R, Zabielska K (2015) What do we know about canine osteosarcoma treatment? Review Vet Res Commun 39(1):61–67

Tang D, Kang R, Berghe TV, Vandenabeele P, Kroemer G (2019) The molecular machinery of regulated cell death. Cell Res 29(5):347–364

Vogelstein B, Papadopoulos N, Velculescu VE, Zhou S, Diaz LA Jr, Kinzler KW (2013) Cancer genome landscapes. Science 339:1546–1558

Wang HG, Pathan N, Ethell IM, Krajewski S, Yamaguchi Y, Shibasaki F, McKeon F, Bobo T, Franke TF, Reed JC (1999) Ca^{2+}-induced apoptosis through calcineurin dephosphorylation of BAD. Science 284:339–343

Weinberg RA (2014) Coming full circle-from endless complexity to simplicity and back again. Cell 157:267–271. https://doi.org/10.1016/j.cell.2014.03.004

Wu S, Wang Y, Guo J, Chen Q, Zhang J, Fang J (2014) Nanosecond pulsed electric fields as a novel drug free therapy for breast cancer: an in vivo study. Cancer Lett 343(2):268–274

Yin D, Yang WG, Weissberg J, Goff CB, Chen W, Kuwayama Y, Leiter A, Xing H, Meixel A, Gaut D, Kirkbir F, Sawcer D, Vernier PT, Said JW, Gundersen MA, Koeffler HP (2012) Cutaneous papilloma and squamous cell carcinoma therapy utilizing nanosecond pulsed electric fields (nsPEF). PLoS ONE 7(8):e43891

Yin S, Chen X, Hu C, Zhang X, Hu Z, Yu J, Feng X, Jiang K, Ye S, Shen K, Xie H, Zhou L, James Swanson R, Zheng S (2014) Nanosecond pulsed electric field (nsPEF) treatment for hepatocellular carcinoma: a novel locoregional ablation decreasing lung metastasis. Cancer Lett 346(2):285–291

Yu SW, Andrabi SA, Wang H, Kim NS, Poirier GG, Dawson TM, Dawson VL (2006) Apoptosis-inducing factor mediates poly(ADP-ribose) (PAR) polymer-induced cell death. Proc Natl Acad Sci USA 103:18314–18319

Zhao B, Hemann MT, Lauffenburger DA (2014) Intratumor heterogeneity alters most effective drugs in designed combinations. Proc Natl Acad Sci U S A 111:10773–10778. https://doi.org/10.1073/pnas.1323934111

Chapter 19
usEPs as a Possible Immunotherapy

Stephen J. Beebe

Abstract That usEPs could ablate tumors was a significant finding in the use of this technology for cancer therapy. The finding that usEPs could also induce immunity was a bonus for this treatment as possible immunotherapy. However, the data to support this immune induction by usEPs had several different sets of suggestive evidence. One approach showed slower tumor growth in immunocompetent mice vs. growth in immunodeficient mice. A second approach showed the slower growth of a secondary tumor after ablation of a primary tumor, suggesting that the primary treatment caused an immune response that slowed secondary tumor growth. The presence of CD4+ T-cells in the primary treated tumors and CD4+ cells in the untreated secondary tumor was used as evidence. However, the CD4+ cells require further characterization to differentiate CD4+ CD25+ Foxp3+ T-regulatory immunosuppressor cells (Tregs) from CD4+ CD44+ with the presence or absence of CD62L+ as T-central or T effector memory cells, respectively. The strongest evidence for usEP-induced immunity indicated the complete absence of secondary tumor growth after primary tumor treatment. Such responses were present in cancer models in the ectopic mouse liver, orthotopic mouse breast, and rat liver cancers. The absence of secondary tumor growth is called vaccine effects or in situ vaccination. Thus, the treatment of the primary tumor induces immunity and vaccinates the animals by the usEP treatment. The latter two cancers exhibited early decreases in immunosuppressor Tregs and myeloid-derived suppressor cells (MDSC), which resolve suppression of immune responses, and increases in dendritic cells (DCs) in the TME that could identify antigens and induced immunity. The rat liver cancer model also showed activation of the innate immune natural killer (NK) cells with specific activation markers in its TME and the presence of effector and central memory cells in the mouse breast and rat liver (TME), which were cytotoxic. An ectopic mouse pancreatic cancer model that did not show a vaccine effect failed to show a decrease in Tregs and MDSC in the TME and blood and did not show activated T-cells, suggesting immunosuppression prevented an immune response. Continued studies will determine immunity, cell death mechanisms, and ICD factors (calreticulin, ATP, and HMGB1) in other immunogenic cancer models.

S. J. Beebe et al., *Ultrashort Electric Pulse Effects in Biology and Medicine*, Series in BioEngineering, https://doi.org/10.1007/978-981-10-5113-5_19

19.1 Introduction

There are many different scientific approaches that provide various levels of evidence for the presence of immunity. Some of this evidence is indirect, and some of it is direct. There are also many facets, qualities, and facades that lend support or establish immunity. The complexity of immune mechanisms is sometimes glossed over, presenting more commentary and less valid evidence. There appears to be a greater need for a more strict, immunological confirmation for some of the findings that merely suggest immunity instead of specifically identifying that an immune response is present.

A general immunological problem is the differentiation between immune cell defense against immunogenic cell death in response to pathogens versus silent or tolerogenic self-cell death (Green et al. 2009). Adaptive antitumor immunity may be enhanced by dying in the right way. Cancer cells and oncogenic viruses can inhibit or promote cell death by one or more of several RCD mechanisms. Immunotherapy can stimulate specific immunogenic cell death (ICD) mechanisms and promote cancer immunogenicity, by inhibiting immune checkpoints, signaling through dendritic cell (DC) receptors, which can cross-present tumor antigens to T-cells and thereby improving adaptive immunity and enhancing therapeutic efficacy (Kroemer et al. 2013; Karaman et al. 2018; Morrison et al. 2013; Guo et al. 2014). Cells undergoing ICD express different cell surface receptors and release factors that enhance immunogenicity. These factors recruit and activate DCs to identify antigens, phagocytose them and process them before presenting them to T-cells. ICDs include the presence of the most abundant endoplasmic reticulum (ER) protein, calreticulin; the most abundant intracellular metabolite, ATP; and most abundant non-histone chromatin binding protein HMGB-1. These and others (Hou et al. 2013) are collectively called danger associated molecular patterns (DAMPs) that have dominant effects on immunogenicity when they are released and appear from dying cells (Tesniere et al. 2008a, b; Kepp et al. 2011; Inoue and Tani 2014). The finding that usEPs have intracellular effects on the ER, DNA, the nucleus, mitochondria, and other subnuclear structures provide possible mechanisms for releasing these factors that induce ICD. It appears likely that different RCD mechanisms can lead to release of these factors and induce ICD. Some of this is due to shifts in focus from apoptosis to autophagy and more diverse forms of RCD such as necroptosis, pyroptosis, and pathanatos (Inoue and Tani 2014). Autophagy also appears to enhance the release of DAMPS and the release of DAMPs appears to enhance autophagy. Since autophagy is primarily a survival mechanism, the presence of autophagy in dying cells that autophagy cannot save, identifies the least fit tumor cells releasing DAMPs (Hou et al. 2013).

As a general overview, consider that immune mechanisms are influenced by activated cytotoxic immune cells, which attack cancer cells, and activated immune cells that suppress immune mechanisms. These immune suppressor cells not only prevent immune cell activation, but thereby support cancer progression. Immune suppressor cells include T-regulatory cells (Tregs), myeloid-derived suppressor cells (MDSCs), and tumor-associated macrophages (TAMs), among others. Immune cell suppression is a significant barrier for cancer immunotherapy and for any cancer therapy, including usEPs (Escors 2014). This immunosuppressive cellular barrier must be destroyed so innate and adaptive immune cells can become cytotoxic and attack cancer cells. Immune suppressor cells in the TME are generally dominant by their numbers and percentages with specific suppressor immune cell phenotypes based on cell surface receptors. More convincingly, immune suppressor cells can be more directly demonstrated by their specific functional suppressor activity against T-effector cells. On the other side of immunity, activated adaptive immunity includes cells that are CD3+ and also express CD4+ or CD8+ phenotypes. However, the presence of these CD4+ or CD8+ cell phenotypes does not mean that these cells are active against tumors. CD4+ and CD8+ cells can be present, but if immunosuppressor cells are also present and dominant, they can be inactivated or be anergic, meaning they fail to respond to their specific antigen. Alternatively, T-cells could be exhausted, meaning they cannot react to their antigens because of chronic antigen stimulation (Blank et al. 2019; Schwartzberg et al. 2019). It is also possible that the presence of CD4+ cells could mean the presence of T-reg cells, which also express CD25 and the transcription factor FoxP3. Therefore, to identify activated T-cells, additional phenotypes should be demonstrated to show cells are activated by expressing CD44 as effector memory T-cells (CD62L− CCR7−) and central memory T-cells (CD62L+ CCR7+). Once again, activated T-cells can be demonstrated at a more direct level by isolating their phenotypes and showing their functional activity by secretion of cytokines such as IFNγ or by showing isolated T-cells directly induce apoptosis in cancer cells. In several studies, usEP-induced immunity has been demonstrated at all of these levels to various degrees.

19.2 usEPs Induce Immunity

The earliest hint that there was an immune response with usEP-induce immunity was indirectly shown in ectopic mouse Hepa1-6 HCC tumors (Beebe et al. 2011) and in a later broader study in orthotopic rat N1-S1 HCC (Chen et al. 2014). These studies were then followed by specific inspection at immunity in the N1-S1 HCC model at the immune cell phenotype levels predictive of immunity (Lassiter et al. 2018). In these studies, tumors are treated and eliminated by usEPs. When tumors have vanished for about seven weeks after treatment, and no additional growths were observed from the treatment zone or elsewhere, the animals are challenged with another cell injection of the same cells that formed the tumor that was eliminated by usEPs.

As examples for the search for immunity in response to usEPs, a brief study in the mouse ectopic Hepa 1–6, more in-depth with a rat orthotopic N1-S1 HCC models in a mouse orthotopic 4T1-luc breast cancer model will be primarily highlighted. The rat HCC and the mouse breast cancer models provided essential data to support immune-mediated protective vaccine effects in both cancer models. Data from the studies of other investigators in similar models will be presented, leading to common conclusions that usEPs can effectively induce the development of host immune responses that can facilitate the elimination of tumors after usEP treatment. In the rat HCC and the mouse breast cancer models, we show that the host mounts a robust immune response after treatment and a more apparent immune response when the HCC model is challenged with possible recurrences of the same HCC cancer.

Table 19.1 shows a series of studies that were carried out with ectopic Hepa1-6 mouse HCC tumors in C57Bl/6 mice treated with usEPs. Table 19.1 describes results from an experiment when 16 one-month old mice were divided randomly into an untreated, control group of 8 mice and a treated group of 8 mice. The HCC tumors were initiated in all mice with 1×10^6 cells in all mice. When the tumors reached about 0.4 cm, the control group was sham-treated, and the other group was treated with 900 pulses (1–2 Hz) with 100 ns durations and electric fields intensity at 55 kV/cm (5.0 Vs/cm). In 6 of the 8 usEP treated mice, tumors were eliminated for as long as 60 days. The other two treated mice and all of the control group were euthanized according to our IACUC protocol. When the 6 successfully treated mice were tumor free for 60 day, tumors were initiated in the opposite flank as before. None of these mice grew tumors for as long as 49 days before the experiment was terminated. In a control group of naïve mice, which had never had tumors, these same injected cells readily grew. While these studies showed no direct evidence of an immune response, a highly likely explanation was that an immune response was induced, since naïve mice grew tumors and usEP-treated mice did not.

This small ectopic mouse HCC study was extended to a rat orthotopic N1-S1 HCC model (Chen et al. 2014), which provided a larger rat liver than a mouse liver and a larger animal model that could withstand experimental protocol requirements for three surgeries with two of them close together. One surgery is required to initiate tumors, a second surgery 7–10 days later to treat tumors and a third surgery 7 weeks later to challenge the liver with another tumor cell injection like the initial one.

Table 19.1 UsEP treatment of Hepa1-6 HCC may provide host immunity in C57Bl/6 mice (Beebe et al. 2011)

Group	# of mice (1 Mo)	Tumor diameter 1st treatment	1st treatment	Survival ratio	Survival days	Growth after 2nd injection	Tumor free (days)
Treated	8	0.4 cm	100 ns 55 kV/cm 900 pulse	6/8	60	0/6	49
Control	8	0.4 cm	No nsPEFs	0/8	13 ± 5.3	–	–

19.2.1 *usEP Treatment Induces an Immune-Mediated Vaccine Effect in Orthotopic N1-S1 HCC Tumors*

When orthotopic rat N1-S1 hepatocellular carcinoma tumors were treated with usEPs, greater than 82–90% were eliminated entirely. Two tumors showed partial responses, one tumor was stable, and three others continued to grow, albeit slowly without a complete response. Sham-treated tumors (88.5%) readily grew, eventually requiring euthanasia due to tumor burden (Chen et al. 2014). Since then, scores of tumors have been eliminated with virtually 100% efficacy when entire tumors are exposed to sufficient conditions of 1000 pulses with electric fields strengths of 50 kV/cm and durations of 100 ns (5.0 Vs/cm). When tumors were treated with 300 or 500 pulses (1.5 and 2.5 Vs/cm, respectively), tumors grew more slowly but were not eliminated.

Figure 19.1 (Chen et al. 2014) shows a Kaplan–Meier plot from an extensive survival and challenge study with the rat N1-S1 HCC model. The black arrow (bottom left) shows tumor initiation. Twenty sham-treated tumor-bearing rats required euthanasia within 40 days due to tumor burden (green line). Twenty-one of 23 (91.3%) tumor-bearing rats were treated with 1000 pulses with 100 ns, 50 kV/cm, and at 1 Hz became tumor-free for 7 weeks. All surviving rats that were then challenged with a second identical injection (black arrow, top) in the same lobe or different lobe bearing the initial tumor failed to grow tumors and were tumor-free for as long as 20 weeks (red line). In contrast, 24 naïve, age-matched rats receiving the same injections required euthanasia between 4–5 weeks after injection due to tumor burden (blue line). This study demonstrated a protective, vaccine-like effect, suggesting an immune response. Because this was not a routine vaccination, we initially referred to

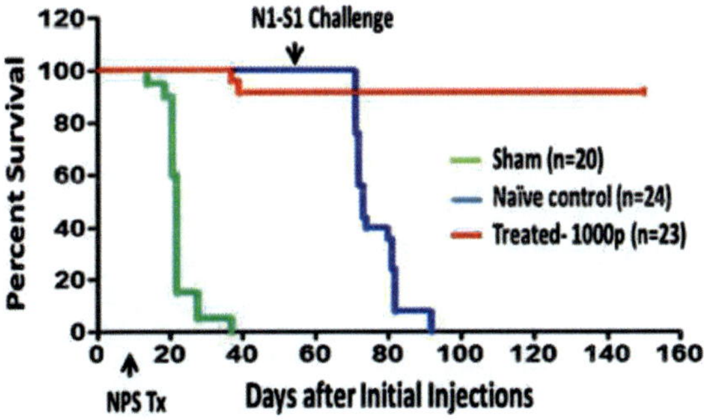

Fig. 19.1 usEPs eliminate N1-S1 tumors and provide a protective vaccine effect—N1-S1 tumors were initiated by injection of 1 × 10^6 cells under the liver capsule. One-week later tumors were treated with 1000 pulses at 1 Hz and 100 ns durations and 50 kV/cm (5.0 Vs/cm (arrow bottom left). 91% (21 of 23) of treated rats were tumor free for 7 weeks after treatment (red line). Rats were then challenged with a second identical injection of N1-S1 cells in the same or adjacent liver lobe (arrow top) in treated rats (red line) and in age-matched naïve control rats (blue line). Rats were followed for 150 days after initial tumor initiation when the study ended (from Chen et al. 2014)

this as a vaccine-like effect. Furthermore, there was no direct evidence that adaptive immune memory cells were present or decreased immunosuppressive cells. However, there was no other explanation for these findings, so an immune response had to be assumed until additional evidence was presented.

19.2.2 N1-S1 Tumor Microenvironment Exhibits Time-Dependent Increases in Granzyme B After usEP Treatment

Figure 19.2 shows initial studies to determine if an immune response was present by demonstrating the time-dependent increase in granzyme B in the TME after usEP treatment of N1-S1 tumors (Chen et al. 2014). One of the mechanisms of the innate and adaptive immune cells is to physically interact with tumor cells and utilize their perforin/granzyme pathway to induce apoptosis in their target cells Cytotoxic natural killer cells, and T-lymphocytes express the serine protease granzyme B (GzB) in specialized cytotoxic granules in secretory lysosomes. Upon contact with the target cancer cell, the granules degranulate, releasing perforin, which disrupts the plasma membrane, and GzB protease initiates apoptosis of cancer cells. The granules are stored in a way that protects the killer cell from its own killer molecules. Furthermore, GzB and perforin expression occur at lower levels until the killer cell is activated when GzB expression is increased.

Once inside the cells, GzB can directly activate caspase-3, and other pro-caspases. However, for full activity, GzB must act through the mitochondria, so it cleaves Bid, leading to the release of cytochrome c and other mitochondria-stored

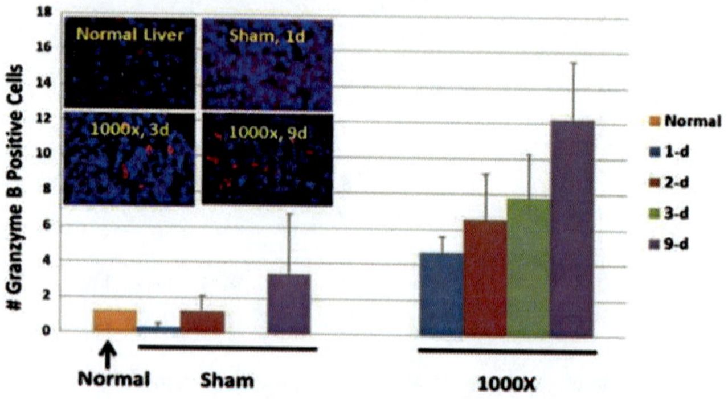

Fig. 19.2 usEPs induces granzyme B (GzB) expression after treatment of N1-S1 HCC tumors—N1-S1 tumor were treated with 1000 usEPs (1 Hz) with durations of 100 ns and electric field strengths of 50 kV/cm or they were sham treated. On days 1, 2, 3, and 9 after treatment, tumor sections were prepared and stained with an antibody to GzB. Typical images are shown in the inset and quantitative data are shown in bar graph format. A section is also shown of normal liver (from Chen et al. 2014)

factors such as the second mitochondria-derived activator of caspases or SMAC and inhibitor of caspase-activated DNase (ICAD). SMAC is also known as DIABLO. SMAC/DIABLO impedes the inhibitor of apoptosis proteins (IAPs) and thereby releases active caspases and reverses caspase inhibition (Vucic et al. 2002). ICAD release from mitochondria (Enari et al. 1998) leads to fragmenting DNA (Trapani 1995; Barry et al. 2000; Lord et al. 2003; Lieberman 2010).

The N1-S1 HCC study shows a time-dependent increase in GzB staining in the TME after treatment. The increased presence of GzB with time is most likely due to an increased presence of cytotoxic cells after treatment and the increase in GzB as T-cells and NK cells become more active. The data demonstrating resistance to the recurrence of the N1-S1 HCC cancer upon the challenge injections and the presence of GzB provides invisible evidence (the resistance) and the appearance of effector molecules that can induce apoptosis in target cells. However, T-regulatory cells use perforin and GzB to prevent T-cell-mediated tumor clearance (Cao et al. 2007). So, it is not clear if the GzB is expressed in cytotoxic T-cells and NK cells or T-reg suppressor cells.

19.2.3 *Increased Numbers of CD4+ and CD8+ Tumor-Infiltrating Lymphocytes (TILs) Permeate the N1-S1 HCC Tumor Microenvironment (TME)*

Figure 19.3 shows that one week after usEP treatment of N1-S1 HCC tumors, immunofluorescent microscopy demonstrates increases in CD4+ and CD8+ TILs in the TME. CD4+ and CD8+ co-receptor serve as recognition and adhesion molecules, help stabilize the T-cell receptor (TCR) interactions with target cells, and enhance T-cell activation signaling. CD8 is expressed by cytotoxic T-cells, but not all CD8+ cells are necessarily cytotoxic. They have to be activated first, as indicated by the expression of effector or central memory cell receptors. CD4 is expressed by helper T-cells, and CD8+ is expressed by cytotoxic T-cells. CD8+ TILs serve to kill tumors using GzB, as shown above. CD4 T-cells perform several functions, including activation of innate immune cells, T-cells, and B-cells. Although not as readily recognized, CD4+ T-cells can also be cytotoxic (Haabeth et al. 2014).

However, CD4+ helper T-cells are best known for their influence on other immune cells by secreting cytokines that will be most effective to the host's immune needs. By their production of cytokines and chemokines, they enhance B-cell antibody production, recruit phagocytes to infection sites, and enhance T-cell functions. By their differentiation into CD4+ cell subsets, including Th1, Th2, Th17, and iTregs, they secrete many different cytokines that enrich host cell immune function (Zhu and Paul 2008). CD4+ T-cells can also induce T-regulatory cell immunosuppressive functions by producing GzB. So, while the increased presence of T-cell as indicated by CD4+ and CD8+ T-cells is expected as indicators of immunity, the CD4 and CD8 receptors by themselves do not prove that they are serving immune functions.

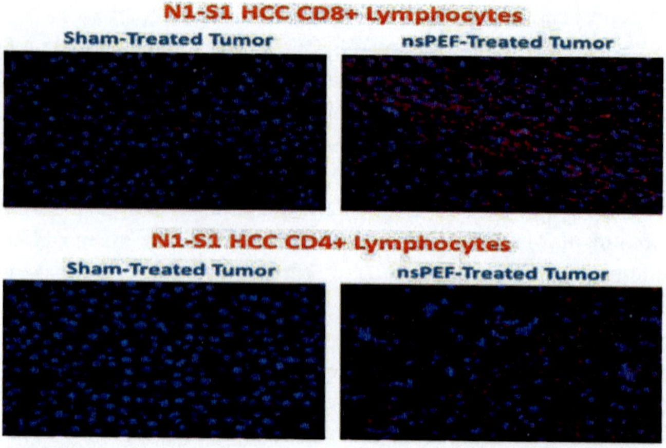

Fig. 19.3 Increased presence of CD8+ and CD4+ lymphocytes after treatment of N1-S1- HCC tumors—N1-S1 tumors were sham-treated (left panels) or treated with 1000 usEPs (1 Hz) with durations of 100 ns and electric field strengths of 50 kV/cm (5.0 Vs/cm; right panels). One week after treatment, receding tumors were removed, fixed in formalin, and prepared on slides for immunofluorescent microscopy with antibodies to CD8 (top panels) and CD4 (bottom panels). Slides were back-stained with DAPI. (Chen R, Sain N, and Beebe SJ, unpublished)

As indicated above, other distinguishing markers or phenotypes are necessary to demonstrate the presence of immunity. If the TME remains immunosuppressive, the CD4+ and CD8+ T-cells will not be able to carry out their helper and cytotoxic cell functions.

19.2.4 usEP Treatment Enriches the Presence of Memory T-Cells in the Spleens of Rats with N1-S1 Regressing Tumors

To determine if NPS induced an immune response after treatment of rat orthotopic N1-S1 HCC tumors, CD4+ and CD8+ effector memory T-cells (Tem: CD44+ CD62L−) and central memory T-cells (Tcm: CD44 + CD62L+) were analyzed in the spleens (Fig. 19.4) and blood (Lassiter et al. 2018). During the 7-day post-treatment study compared to pre-treated, tumor-bearing rats, CD4+ Tem and Tcm lymphocytes increased as much as 2.0- and 5.2-fold, respectively, while blood CD8+ Tem and Tcm cells increased 1.8 and 4.8-fold, respectively. Generally, these cells remained elevated by the 7th day after treatment. Over the 7 days study in the spleen, compared to tumor-bearing, untreated rats, CD4+ Tem cells increased 1.6-fold by day 7, but CD4+ Tcm cells were not increased. CD8+ Tem and Tcm cells increased 1.4- and 2.0-fold, respectively, but were below untreated rats by day 7.

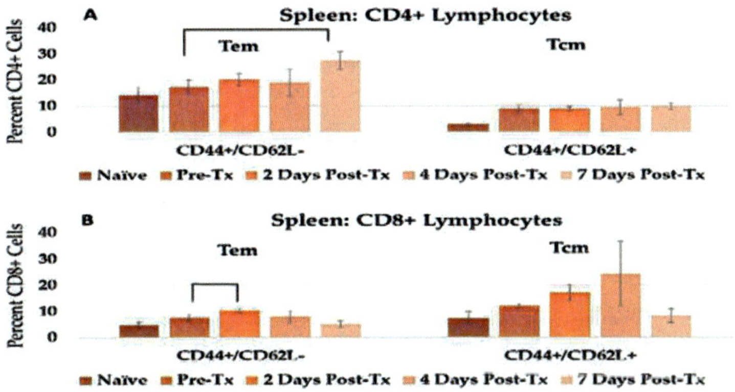

Fig. 19.4 usEPs induce increases in CD4+ Tem and CD8+ Tem and Tcm in spleens after treatment. Lymphocytes were isolated from spleens various times after usEP post-treatment then stained with antibodies to CD4, CD8, CD44 and CD62L; analysis was done on CD4+ cells (**a**) and CD8+ cells (**b**). Error bars = SEM; Naïve, 2 Days, 4 Days n = 4; Pre-Tx n = 2; 7 Days n = 5. **a** CD44+/CD62L- 7 Days Post-Tx: $p = 0.02$. **b** CD44+/CD62L+ 2 Days Post-Tx: $p = 0.004$; connecting lines $p = 0.05$ (modified from Lassiter et al. 2018)

Perhaps the most critical location for immunity information is in the usEP treated TME. Generally, the TME is the most difficult to analyze by flow cytometry, especially after treatment with usEPs, for several reasons. First, many cell types and autofluorescence and dead and dying cells present a lot of debris to search through. For solid tumors, the digestion process can also be challenging and sometimes incomplete. Debris is a significant problem. For orthotopic HCC tumors, digestion was not necessary. However, analysis of the HCC TME by flow cytometry was not especially helpful in determining immunity mechanisms in HCC for yet another reason. The liver is not an isolated organ like the breast or the pancreas or a solid tumor. The liver receives the animal's full blood supply, 50–60% of which arrives through the hepatic portal vein from the gut carrying remnants of gut bacteria. So, it was reasonable to consider immune cell phenotypes in the spleen and blood as alternative tissues for the identification of immunity.

As indicated above, CD4+ T-cells serve valuable functions as helper cells modifying other immune cells' functions. CD8 T-cells are lymphocytes that carry out cytotoxic activity against tumor cells. These lethal functions are adjunct activity to the usEP cytotoxic function. Differentiation of these CD8+ T-cells, as well as CD4+ T-cells into memory cells, is vital to immune responses like those seen in the N1-S1 HCC model in response to usEPs. Simply, memory cells must respond to specific antigens that are descriptive of the tumor cells and must exhibit long-term endurance. While effector cells and memory cells share some molecular and functional similarities, effector cells contract after their initial cytotoxic activities while memory cells persist log-term and respond to specific antigens' recurrence with robust proliferation (Badovinac et al. 2002, 2005). Also, memory T-cells are more numerous, proliferate rapidly, execute a lethal function, and secrete cytokines, populate multiple organs, and exhibit different transcription, metabolic, epigenetic, and functional states compared to effector T-cells (Martin and Badovinac 2018). Memory cell phenotypes can be heterogeneous that can depend on the time since antigen stimulation, numbers of repeated antigen exposures, and the genetic background and history of exposures to various antigens. The history of antigen exposures is less an issue in clean animal facilities, which raises questions about whether studies with "dirty" or outbred mice may be more representative of the human condition (Martin and Badovinac 2018).

Generally, Tem cells are more cytotoxic and express receptors (integrin and chemokine) that localize them to inflammatory sites, while Tcm cells express receptors that enhance their homing to secondary lymphoid organs. To define effector memory cells (Tem), we used CD44+ and CD62L- phenotype and CD44+ CD62L+ to define Tcm cells. Many different receptors can be used to define these T-cell memory subsets (Martin and Badovinac 2018). Like CD62L+ cells, the CCR7+ receptor is also used to define Tcm cells as lymphoid homing receptors. CD62L is the L selectin adhesion molecule that recognizes sialylated carbohydrate groups, and CCR7 is a chemokine receptor that binds CCL19 and CCL21. So, both CD62L+ and CCR7+ cells define Tcm while the absence of these markers indicates Tem cells when other appropriate markers are present. In addition to CD44+ T-cells, other makers also are used for defining these subclasses or memory T-cells. For example, Cx3Cr1+ CD27− cells represent Tem cells while Cx3Cr1− CD27+ cells represent Tcm cells. Other subsets of memory cells are not included here (Martin and Badovinac 2018).

19.2.5 usEPs Decrease the Percentages of Tregs in the N1-S1 HCC TME

As mentioned earlier, one of the most significant hurdles for immunotherapy is the resolution of the immunosuppressive TME, where the hosts immune cells and host somatic cells collaborate to promote tumor progression. Given that more than 80% of tumors are cleared with a single treatment, usEPs significantly disrupts the TME. However, for an immune response to be useful as a mechanism against cancer, immunotherapy must significantly reduce immune suppressive cells such as the T-regulatory cells (T-regs).

In the naïve rat liver, there are very few T-regs ulatory cells (Fig. 19.5). However, the tumor-bearing, pre-treated rat HCC tumors represent about 6% of the tumors' CD4+ cells. This number slightly increases, but not significantly by the second day after usEP treatment. It is not until the 4th day after treatment that Tregs decreased by twofold in the TME and then decreased to levels similar to that of tumor-free naïve rats by day 7 post-treatment. This slight increase in Tregs soon after usEP treatment seems to be shared in other models. In recent studies in 4T1-luc mammary cancer tumors, Tregs increased in the first days after usEP treatment but decreased in the following days after treatment (Nanajian, Beebe, and Guo, unpublished). However, it appears critical that the Tregs and other immunosuppressive cells exit the treatment

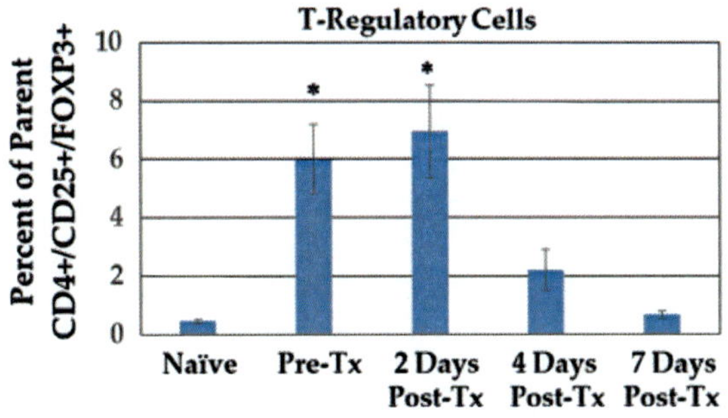

Fig. 19.5 usEPs cause dissolution of immunosuppressive Tregs in the TME. Lymphocytes were isolated from the liver of naïve rats, pre-treated rats, and from rats at various times after treating N1-S1 HCC tumors with 1000 usEPs with durations of 100 ns and electric field of 50 kV/cm (5 Vs/cm). Tumors were processed without enzymatic digestions, stained with indicated antibodies and analyzed by flow cytometry CD4+ cells were gated and analyzed for CD25+ and Foxp3+ cells. Foxp3/CD25/CD4 (naïve n = 4, others n = 2) error bars = SEM. * $p = 0.05$. From Lassiter et al. (2018)

zone to establish an immune-mediated vaccine effect and long-lasting immunity in this N1-S1 HCC model.

19.2.6 usEPs Activate the Innate Immune System

Another approach was used to consider signatures of immunity in the HCC TME that touched on both the adaptive and innate immune systems. The adaptive immune system is an evolutionary improvement on an older and less sophisticated innate immune system. This memory system comprises T- and B-lymphocytes that express a limitless catalog or repertoire of antigen receptors produced based on specific antigen structures and sequences by site-specific somatic recombination in the T-cell and B-cell receptors (TCR, BCR). These cells undertake cell divisions and maturations before expressing their effector functions. This allows them to recognize remembered, specific antigens for rapid and robust responses. Their memory functions are based on processes in three well-defined phases. The first is expanding specific clones based on antigen recognition with a major histocompatibility complex (MHC) framework. This is followed by contraction when most effector cells undergo apoptosis. A relatively small number of cells survive and enter the third memory phase when they endure, self-renew, and wait for antigen recurrence (Luckey et al. 2006; Williams and Bevan 2007; Vivier et al. 2011; Gabrielli et al. 2016).

The innate immune system consists of myeloid and lymphoid cells. Of specific interests here are natural killer T– (NKT) cells and innate lymphoid cells (ILC), which include a most essential subset called natural killer (NK) cells (Hazenberg and Spits 2014). In contrast to the TCRs' recombination features, NK and NKT cells have limited sets of germline-encoded receptors that lack antigen specificity but exhibit rapid effector responses upon activation. So, NK cells exhibit cytotoxic and cytokine secreting functions with prior recognition or sensitization. NKs function alongside and are most like CD8+ T-cells using perforin, GzB, and TNF family death receptors for target cell killing (Sun and Lanier 2011). They function by balancing the expression of inhibitory and activating receptors (Moretta et al. 2001, 2008; Sun et al. 2009; Vivier et al. 2011; Gabrielli et al. 2016). Some inhibitory receptors recognize HMC I, expressed on all virtually all healthy cells, thereby preventing NK cells from attacking self-cells. Some tumor transformed cells delete expression of MHC I, which protects them from CD8+ T-cells; however, this leads to NK cell activation in the so-called "missing-self hypothesis" (Ljunggren and Kärre 1990). Two important activating receptors in NK and NKT-cells are the NKG2D and CD161 (NK1.1) receptors (Fergusson et al. 2011; Lanier 2015). The NKG2D receptor recognized "stress ligands" that are often expressed on cancer cells marked by DNA damage (Lanier 2015). An NK cell that identified stress ligands on cancer cells, which do not express MHC I, will kill that cell by releasing perforin and GzB, inducing apoptosis. Likewise, the CD161 receptor can also cause cancer cell death by promoting perforin and GzB and releasing IFNγ. Tumor cell killing can also occur by sequential release of IFNγ by NKT-cells and release of perforin and GzB by NK cells (Smyth et al. 2002).

Recent revelations of some innate cell functions have muddled the distinctions between innate and adaptive immunity. "The high degree of cooperation and interdependence between immunological mechanisms makes a clear-cut distinction between the two concepts increasingly difficult" (Gabrielli et al. 2016). The recent disclosures that NK cells exhibit attributes of immunological memory make these distinctions more clouded. These considerations made it increasingly attractive to investigate the role of innate immunity in N1-S1 HCC cells treated with usEPs. One approach was to analyze immune cells in the TME before and after treatment with usEPs using antibodies to CD3 and CD161 (NK1.1) and CD56 (Lassiter et al. 2018). In Fig. 19.6, NK, NKT, and T-cells were differentiated on flow cytometry because NK cells do

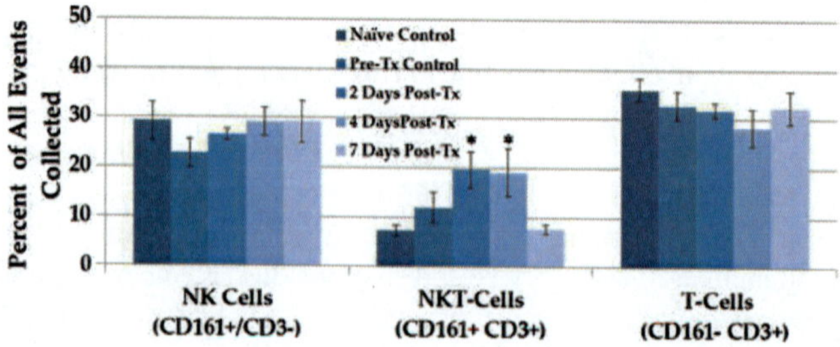

Fig. 19.6 Expression of NK, NKT and T-cells in N1-S1 HCC tumors before and after usEP treatment. Tumors were treated with 1000 usEPs (1 Hz) with 100 ns durations and 50 kV/cm (5.0 Vs/cm) or sham treated. Tumors were removed from livers and processed without enzymatic digestions, stained with indicated antibodies and analyzed by flow cytometry. Error bars = standard error of the mean (SEM); Naïve n = 4, Pre-Tx n = 5; others n = 6. * $p = 0.05$ (from Lassiter et al. 2018)

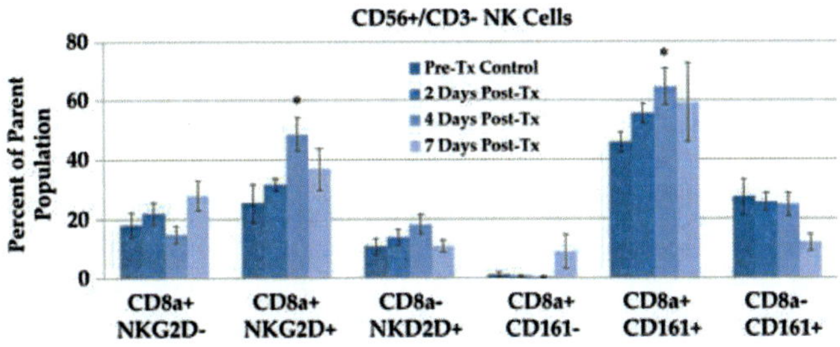

Fig. 19.7 usEPs induce increases in subsets of NK cells after treatment with usEPs. N1-S1 tumors were sham treated or treated with usEPs as described in Fig. 19.6. Immune cells were stained with antibodies for the indicated phenotypes. Cells were gated on CD3- CD161+ and further analyzed with antibodies for CD8, CD161 and/or CD314 (NKG2D) and analyzed by flow cytometry. Naïve n = 4, Pre-Tx n = 5; others n = 6; error bars = SEM. * $p = 0.05$ (modified from Lassiter et al. 2018)

not express CD3, T-cells do not express CD161 or CD56, and NKT-cells express all of them—CD3, CD161, and CD56.

There were no significant changes in the levels of NK cells or T-cells in naïve liver tissues, pre-treated control tumors, or in tumors in the seven days after usEP treatment. In contrast, there were significant increases in CD161+ CD3+ NKT-cells 2 and 4 days after treatment (Fig. 19.6) and in CD56+ CD3+ NKT-cells 4 days after treatment (Lassiter et al. 2018). However, 7 days after treatment, all NKT-cell phenotypes were back to control levels.

Although there were no significant changes in NK cells before and after treatment with usEPs using these markers, a closer look at NK and NKT-cells presented a different and interesting picture of these innate immune cells' roles in usEP-induced immunity in the N1-S1 HCC model. Figure 19.7 shows results for NK cells, which were closely mimicked by changes in NKT-cells (Lassiter et al. 2018).

Figure 19.7 shows that NK cells that were CD8+ and CD314+ (NKG2D+) or CD8+ CD161+ were increased in the TME in a time-dependent manner with significant changes on day 4 post treatment. NK cells that did not express CD8 or cells that expressed CD8 but did not express CD161or NKG2D were low and not increased. UsEP specifically induced expression of NK cells that expressed CD8 and one or the other or both (not shown) of two different activation receptors NKG2D and CD161. Similar results were observed in the same NKT-cell phenotypes (Lassiter et al. 2018). That NK cells and NKT-cells with these activation phenotypes were not present in normal liver tissue or in tumors before treatment with usEPs, suggests that the usEP treatment induced expression of ligand that were recognized and upregulated in cells with these two activation receptors. These observation suggest that they played valuable roles in the anti-tumor immunity and the usEP immunotherapeutic effects.

So, in a summary of the immune-mediated events that establish usEP immunotherapy, the initial protective vaccine effect provided "invisible" evidence that immunity was highly likely. However, there were no specific tangible immune characteristics other than the in situ vaccination by the usEP treatment itself, which was by no means trivial. Yet, a developing immune response after treatment was supported by several features typical of immunity in time-dependent manners in the days following treatment. That the vaccine effect was immune-mediated was supported by the increase in GzB in the TME; the increase in CD4+ and CD8+ T-cells in the TME, albeit without distinction by flow cytometry; the increases in CD4+ and CD8+ Tem and Tcm cells in the spleen and blood; the decrease in Tregs in the usEP treated TME; and the increases in specific subsets of NK and NKT-cells with specific activation receptors. These features defined phenotypic characteristics of immune cells after the usEP treatment of N1-S1 HCC in rat liver.

19.2.7 *usEP Treatment Vaccinates Mice Against N1-S1 HCC*

While those data defined the developing immune response, perhaps more meaningful demonstrations of immunity occurred in immune responses when tumor-free animals are challenged with the tumor cells cleared by usEP treatment. At the time of this writing, an analysis of the immune response shaped by the in situ vaccination has only been investigated in this orthotopic N1-S1 HCC model discussed here. While the immune phenotypes and functions need to be more carefully studied in broader perspectives, the response to the challenge with live tumor cells differs from the response during treatment. While the adaptive immune responses are more readily associated with immunity in general, we will first consider the innate immune response after challenge before we discuss the adaptive immune responses. Part of this presentation strategy is because we just discussed the innate immune response after usEP treatment and because there were surprises in the observations.

Figure 19.8 shows NK cells in the rat liver after usEP treated, tumor-free rats were challenged with N1-S1 cells in the same liver lobe that carried the primary tumor or another lobe. Comparing the liver NK cell responses following usEP treatment in Fig. 19.7 with the NK cell responses after challenge in Fig. 19.8 demonstrates

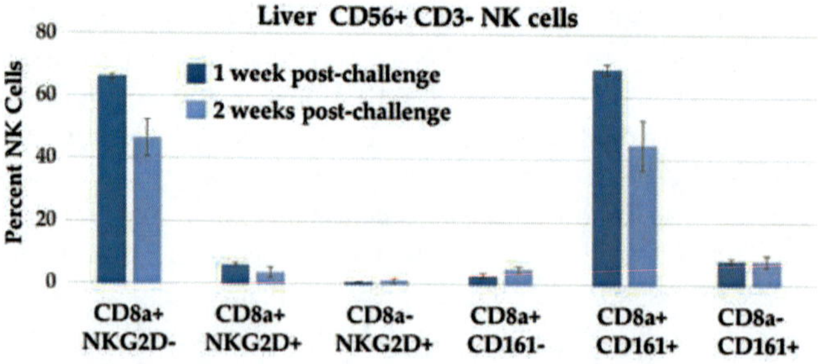

Fig. 19.8 Liver NK cell subset phenotypic analysis post-challenge shows a vaccine-like effect. Lymphocytes were isolated from liver at 1 and 2 weeks post challenge and then stained with CD56, CD3, CD8a, NKG2D (CD314) and CD161. Analysis done on CD56+/CD3− cells. N = 3; error bars = SEM. abridged from Lassiter et al. 2018

some interesting differences in how immunity is initiated and how it responds to a recurrence. The most apparent differences are the absence of CD8+ NKG2D NK phenotypes and the return of CD8+ CD161 NK cells after the challenge injections. Although not shown here, the NKG2D receptor is not present on CD161 cells like after usEP treatment (Lassiter et al. 2018). Another difference was a much more significant CD8+ cell presence without the NKG2D receptor, yet an absence of CD8+ cells unless they expressed the CD161 receptor. Similar results occurred in the spleen, except there were greater percentages of CD8+ CD161− NK cells (15–20%) and smaller percentages of CD8+ CD161+ NK cells (12–15%) (Lassiter et al. 2018). The absence of the NKG2D receptor after challenge suggests that this receptor is specific for a stress ligand expressed after N1-S1 HCC cells are treated with usEPs, but not in untreated HCC cells that do not apparently express that ligand. Another intriguing finding is the continued presence or the return of the CD161 receptor on CD8+ NK cells in the liver. This suggests that these CD8+ CD161+ NK cells could be memory cells, a unique feature of innate immune cells. NK cells (Cerwenka and Lanier 2016; Sun et al. 2009; Vivier et al. 2011) and other innate immune cells (Sun et al. 2014; Netea et al. 2016) are capable of immunological memory. CD161 can also act as a co-receptor defining a specific T-cell subset among T-cell lineages (Fergusson et al. 2014) that secrete IL-17, among other functions, recruits and activates granulocytes to promote inflammation (Maggi et al. 2010). It is highly likely that innate immune cells in the liver play an important role in HCC tumor elimination after usEP treatment, at least in part, dependent on usEP-specific activation of subsets of NKs and NKT-cells that express activation receptors NKG2D-CD314 and CD161.

The possible presence of innate immunity in this anti-tumor context is fascinating because innate immunity or trained immunity is best characterized in response to bacterial or viral infective agents instead of reactions to tumors antigens or ligands, as suggested here. The memory mechanisms in these innate cells is distinct from T-cell mediated memory (Boraschi and Italiani 2018). First, there is a lack of gene rearrangements typical of T- and B-cell receptors; instead, there are transcription and epigenetic re-programming involving histone acetylation and DNA methylation. There are also microRNA or miRNA roles, which require RNA silencing and post-transcriptional regulation of gene expression. Basically, a specific mRNA will be degraded faster and translated less, thereby extinguishing the translation of a specified protein (Ambros 2004; Bartel 2004, 2018). Innate immunity is initiated by DAMPs and PAMPs recognized by pattern recognition receptors (PRRs) (Andersson et al. 2001; Scaffidi et al. 2002; Obeid et al. 2007; Kazama et al. 2008; Panaretakis et al. 2008; Elliott et al. 2009; Chiba et al. 2012; Hou et al. 2013). The responses are carried out in NK cells by inhibitory and stimulatory or activating receptors and cytokines. The environment shapes innate memory involving metabolic changes in innate cells (Gardiner and Finlay 2017; Poznanski et al. 2018; O'Brien and Finlay 2019). Innate immunity appears to include only monocytes and macrophages, precisely localize to specific organs, and short-lived. However, the example of CD161 presence in the described challenge studies occurred 7 weeks are the treatment of the N1-S1 HCC tumors. Although we did not analyze immune mechanisms, the protective vaccine

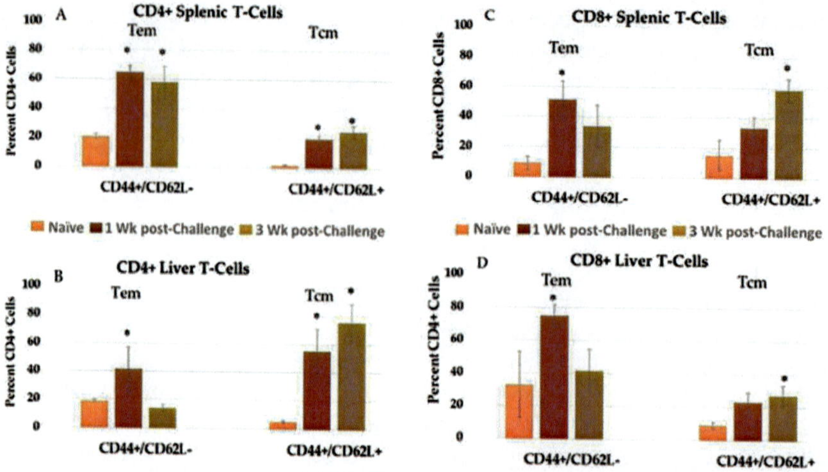

Fig. 19.9 Immune characteristics in the spleen and liver after orthotopically challenge of usEP treated, tumor-free rats with live N1-S1 HCC cells—Seven weeks after rats were treated with usEPs or about 6 weeks after treated tumors were not evident, 1×10^6 viable N1-S1 HCC cells were injected under the liver capsule in the same manner as the primary tumors were initiated. One week and three weeks after challenge injections, rats were euthanized and the spleens (panels **a** and **c**) and liver tissues at the injection site (panels **b** and **d**)were analyzed for indicators of Tem and Tcm cells by flow cytometry using antibodies CD4, CD8, CD44, CD62L. Naïve and 3 weeks n = 2, 1 week n = 6. n = 5; error bars = SEM. * $p = 0.05$ (abridged from Lassiter et al. 2018)

effect lasted for 8 months in several rats, the last time tested (Lassiter et al. 2018) (Fig. 19.9).

For analyses of the immune cell phenotypes in the challenged rat spleen and liver, we determine the presence of Tem cells with CD44+ and CD60L− T-cells and for Tcm cells with CD44+ CD62L+ T-cells. Unlike the liver TME after treatment with usEPs, the challenged environment's study was less complicated, challenging, and productive. In both the spleens and livers, there were significant increases in CD4+ and CD8+ Tem and Tcm at one and/or two weeks after challenge injections. The most significant increases were 10–15-fold in CD4 Tcm cells in the spleen and liver for up to 3 weeks after treatment. CD8+ Tcm cells were also elevated 2–fourfold in the spleen and liver up to 3 weeks after challenge. CD4+ Tem in the liver tended to only be significantly increased in the first week but remained elevated in the spleen for the entire post-challenge study. Thus, there were possibilities for robust responses from both CD4+ and CD8+ Tem and Tcm cells in the first three weeks after challenging rats that were in situ vaccinated against N1-S1 HCC by usEPs.

usEPs induce adaptive immune responses in blood, spleen, and liver that lasts for months in rats treated for HCC. Robust innate immune responses are explicitly induced by usEPs in liver NK and NKT-cells expressing dominant activation receptors CD8+ NKG2D- (CD314-) and CD161. The CD161, which is present in some memory cells, is expressed on CD8+ NK and NKT-cells, suggesting that some innate memory immune responses may be present in addition to adaptive memory responses.

19.2.8 Splenic CD8+ Lymphocytes Cytotoxicity After usEP Treatment Demonstrates Active Immunity

These immune marker responses are meaningful and provide substantial evidence that usEPs induced host immunity, leading to in situ vaccination and thereby affords protection against recurrence of the same cancer. However, all of the data presented so far to support usEP-induced immunity in the host are based on phenotypic changes in host immune cells after treatment and after challenge with usEPs. To enhance the quality of data for immunity, it was essential to show some functional data demonstrating T-cells' direct cytotoxic function. Therefore, we isolated CD4+ and CD8+ T-cell by negative selection from spleens of rats two weeks after challenge and incubated them with Cell Tracker-violet labeled N1-S1 cells. N1-S1 cells were monitored for apoptosis using active caspases and annexin-V binding as apoptosis markers in Fig. 19.10.

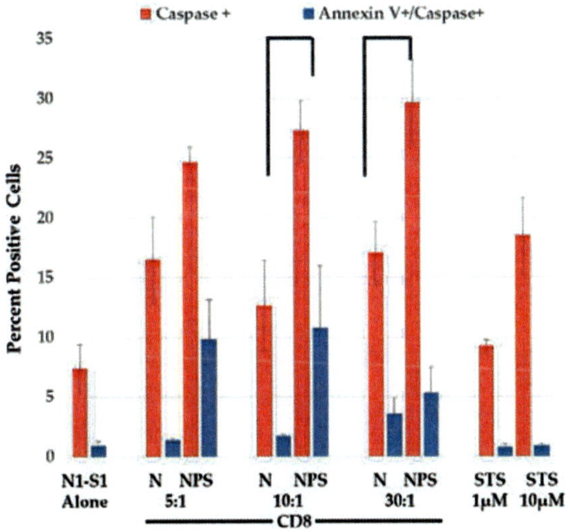

Fig. 19.10 Functional cytotoxicity of splenocyte CD8+ T-cells from in situ vaccinated rats against N1-S1 HCC cells – CD8+ T-cells isolated from spleens of naïve rats (N) and rats 2 weeks post-NPS* by negative selection (NPS). CD8+ T-cells were incubated with N1-S1 cells in ratios of 5:1, 10:1 and 30:1 for 16hrs and prepared for flow cytometry. N1-S1 cells were labeled with Cell tracker-violet and specifically analyzed for apoptosis with caspase activity with FITC-VAD-fmk and Annexin-V binding as marker for phosphatidylserine externalization. Staurosporine (1 and 10 M) were used as positive controls. N = 3 and error bars = SEM. Connecting lines $p = 0.05$ (Abridged from Lassiter et al. 2018). NPS, NanoPulse Stimulation, is another term for usEPs

CD4+ lymphocytes were 92% pure, and CD8+ lymphocytes were 96% pure, with only 2.2 and 1.4% cross-contamination, respectively (not shown). Studies with CD4+ cells showed that they were not cytotoxic so that data are not shown here (Lassiter et al. 2018). N1-S1 cell percentage for active caspases (red bars) and caspases plus Annexin-V (blue bars) when lymphocytes to N1-S1 ratios were 5:1, 10:1, and 30:1. Compared to naive CD8+ T-lymphocytes, CD8+ lymphocytes from NPS treated rats exhibited 1.5-, 2.2- and 1.7-fold more active caspase positive cells, respectively, and 7.3, 6.3 and 1.5-fold more caspase—Annexin-V double-positive cells, respectively. However, statistically significant cytotoxicity was only observed in CD8+ T-cells for increases in caspase activity when lymphocytes: N1-S1 cell ratios were 10:1 and 30:1. These results indicate that not only were CD8+ Tem and Tcm elevated post-challenge compared to control CD8+ T-lymphocytes, as shown in Fig. 19.10, they were also more cytotoxic than control CD8 cells from naïve control rats. It also demonstrated that although some data support CD4+ T-lymphocyte cytotoxicity (Haabeth et al. 2014), they were not cytotoxic in the usEP treated N1-S1- HCC cells.

Overall, studies for effects of usEPs in the N1-S1 HCC model clearly demonstrated immune-mediated vaccine effects that coincided with elevated levels of Tem and Tcm lymphocytes in the blood, spleen, and rats' liver compared to sham-treated controls. These same subsets were also elevated in the spleens and livers after challenge injections. Notably, the CD8+ T-lymphocytes from spleens were cytotoxic towards N1-S1 cells, demonstrating both phenotypic and functional immune mechanisms induced by usEPs in this model.

Also, specific subsets of CD8+ NKG2D+ and CD8+ CD161+ NK and NKT-cells were elevated in the TME after usEP treatment. The NK and NKT-cell responses during treatment and after challenge were different in that only the CD8+ CD161+ NKs and NKT-cell subsets were elevated after challenge; CD8+ NKG2D+ cells were absent or at deficient levels. These data suggest that the CD8+ NKG2D+ subset recognizes ligands that are present after release from tumors by usEP treatment. In contrast, the CD8+ CD161+ subset appears to have been specifically activated by ligand(s) recognized on the usEP treated and non-treated N1-S1 HCC cells. This suggests that the CD8+ CD161+ NK and NKT-cells may represent innate memory cells. While this needs to be verified by additional studies, it could be one of the few possible examples of innate memory from NK and NKT-cells towards cancer antigens or ligands. Essentially, all innate memory data is with innate cells toward bacterial or viral ligands. The data clearly demonstrate adaptive and innate immunity after treatment and after challenge with both phenotypic and functional evidence of immunity to support the vaccine effect and in situ vaccination by usEPs.

19.2.9 usEPs Induce Immune Responses, Eradicating Breast Cancer and Reducing Distant Metastases

The mouse 4T1 model is a well-recognized pre-clinical breast cancer model because it occurs with high spontaneous metastasis to distant organs like the spleen, liver, lung, and bone (Wu et al. 2001; Tao et al. 2008) like human breast cancers do (Yoneda et al. 2000; Tao et al. 2008). The 4T1-Luc model in the studies described here expresses a luciferase reporter gene, which produces light when given its luciferin substrate. Therefore, it is possible to monitor the primary tumor's cessation and verify distant metastasis (Guo et al. 2018; Beebe et al. 2018). The studies described here are presented in some detail because they corroborate and complement and the HCC studies and add a breadth of data that clearly supports usEP-induced immune response in another orthotopic model of breast cancer. The HCC and the 4T1-luc model described usEP-induced immunity in two orthotopic models in two rodent species. In contrast to the most often ectopic cancer models, orthotopic models present a more typical TME for a given cancer. Given that the immunosuppressive TME is a primary hurdle for immunotherapy or any therapy, an orthotopic model has advantages because of the natural tissue of origin. Based on the challenges confronting immunotherapy (Chai et al. 2020), the immunosuppressive mechanisms in the TME are the most challenging to incapacitate and overpower.

The most potent and discriminating aspects of the usEP-induced immunity in this metastatic 4T1-luc breast cancer model were the potential for avoidance of metastasis. Mice with usEP-treated 4T1-luc tumors experienced complete tumor regression with significantly decreased metastases to spleen, liver, and lungs. Like the usEP-treated HCC animals, these mice rejected a second tumor induction when challenged 3–7 weeks after primary tumors were eliminated. The destruction of the TME and immunity induction resulted from 1000 usEPs with durations of 100 ns and 50 kV/cm or 5.0 Vs/cm. When 300 or 500 pulses were applied or 1.5 and 3.0 Vs/cm, tumors were not cleared. Again, charging values around 5.0 Vs/cm were sufficient to eliminate the 4T1-luc tumors like the N1-S1 HCC tumors. But the incomplete treatment of 4T1-luc tumors provided an opportunity to compare metastasis from control mice and incompletely treated mice with equivalent tumor sizes.

So, 4T1-luc tumors were incompletely treated with 600 pulses or 3.0 Vs/cm. Under these conditions, 50% of mice were tumor-free as determine by luciferase imaging and caliper measurements. In the other half, tumors began to grow 2 weeks later, and median survival significantly increased 25 days compared to sham-treated control mice. For tumors that continued to grow, tumor doubling-times were about 50% longer than other control animals. To analyze metastasis, organs from control, and incompletely treated mice with the same tumor size were observed for metastasis by luciferase activity. For control mice, metastasis was observed on days 20, 37, and 38. For incompletely treated mice with similar tumor sizes, metastasis was analyzed on days 52, 62, 84, and 111. Metastases were observed in the spleen, lung, and/or liver in 82% (9/11) of control mice. In contrast, six of seven mice (86%) that were incompletely treated presented no detectable metastases in the same tissues by luciferase activity by ex vivo imaging of isolated organs. The most probable cause of this inhibition of metastasis was antitumor immunity induced by usEPs. It should be informative to trace events that likely led to the usEP-induced prevention of metastasis and the immunity against 4T1-luc breast cancer.

As discussed above, DAMPs released from dead and dying cells present signals that activate antigen-presenting cells (APC) like dendritic cells (DCs). When 4T1-luc cells were treated with usEPs in vitro (10 usEP at 1 Hz with durations of 60 ns and electric fields 20 kV/cm or 3.8×10^{-3} Vs/cm), calreticulin (~5-fold), HMGB1 (~2-fold) and ATP (~3-fold) were released into the supernatant 4 and/or 24 h later. Thus, data support the hypothesis that usEPs destroy the TME and induce the release of ICD markers that activate DCs. These in vitro experiments also showed that these same charging strengths usEPs induced co-stimulatory molecules CD40 and CD86 on bone marrow-derived mouse DCs. The pulsing conditions are much different in vitro and cannot be directly compared with pulses in vivo, but they demonstrate the potential for DAMPs to be released from 4T1-luc cells in the TME. While usEPs that kill cancer cells in the TME are much more robust, conditions similar to these in vitro conditions might occur peripherally to the lethal treatment zone and could activate DCs in those niches. Nevertheless, the release of ICD factors from usEP treated 4T1-luc tumors is consistent with the initiation of immunity responsible for the resistance to 4T1-luc challenge injections or the so-called protective vaccine effect.

The investigation for hallmarks of immunity took place in the spleen and blood of naïve, tumor-bearing, and tumor-free mice. Tumor-bearing mice expressed significantly fewer effector (Tem, CD44+ CD62L-) and central memory (Tcm, CD44+ CD62L+) CD8+ and CD4+ T cells in the spleen and blood compared to naïve mice. However, CD4+ and CD8+ T memory cells in the blood or spleen of treated, tumor-free mice were improved or exceeded levels of control mice, especially central memory CD4+ and CD8+ T cells. Tumor-free mice exhibited a 15–20-fold CD4+ and CD8+ memory T-cells over tumor-bearing mice. A similar dynamic occurred after restimulation with an anti-CD3 antibody for IFNγ producing CD4+ and CD8+ T cells from tumor-bearing mice that were decreased compared to naive control mice but was significantly increased compared to tumor-bearing and naïve mice. Tumor-specific IFNγ production was also increased when splenocytes were co-culture with tumor lysate. The significant reduction in cytotoxic T cells in tumor-bearing mice can be explained by the common immunosuppressive nature associated with 4T1-luc breast cancer.

The recovery of memory CD4+ and CD8+ T-cells after usEPs treatment over the suppression in the tumor-bearing mice and the prevention of spontaneous metastasis were primarily due to the destruction of the TME and the withdrawal of the immunosuppressive nature of the malignant status of 4T1-luc breast cancer. This immune suppression is perpetrated by the highly metastatic potential of the formidable presence of T-regulatory cells (Tregs) and myeloid-derived suppressor cells (MDSCs), which are crucial for angiogenesis (Murdoch et al. 2008) and a well-known co-cancer hallmark for invasion and metastasis (Hanahan and Weinberg 2000, 2011). After usEP treatment, Treg levels in the blood of tumor-bearing mice decreased from 78% of the CD4+ cell population to about 10%, which is similar to that in naïve mice. The MDSC levels in the blood of tumor-bearing mice decreased 15–20-fold after treatment. Also, the MDSC levels were correlated with recovery from tumor burden. For example, animals with regressing tumors exhibited only 14.0% MDSCs in the blood than about 38% MDSCs in the blood of animals with growing tumors. MDSCs are significant barriers to the success of adoptive immunotherapy by suppressing T-cell immunity (Bunt et al. 2006; Gabrilovich and Nagaraj 2009). At the same time, MDSCs were decreasing tumor associated F4/80+ CD11c+ macrophage dendritic cells (M-DCs) increased about 7-fold. The coincident decrease in suppressor cells and the increase in APC revises TME for activation of M-DCs, which was observed by the up-regulation of the activation markers CD40.

These events provide insight into how usEPs resolved the immunosuppression in the TME and create an environment that fostered an infiltration of adaptive and innate immune cells that ultimately provided protective immunity. usEPs induce the release of DAMPs that enhance chances for antigen recognition as suppressor cells decrease and activated APC increase in the TME (Nuccitelli et al. 2017; Guo et al. 2018). It is also possible that usEPs themselves activate DCs that reside in tissue peripheral to the treatment zone. Activated DCs present antigen to T-cells, and effector cytotoxic and memory T cells are generated to eliminate residual cancer cells. The presence of the immune response in the context of the mouse 4T1-luc model experimental protocol appears to be sufficient to decrease the potential for metastasis in the well-characterized metastatic breast cancer model.

Using these two orthotopic rodent models of HCC in the rat and breast cancer in the mouse, it was possible to show that the innate and adaptive immune system can contribute to the elimination tumors if the entire tumors are exposed to full usEP treatment. Immunity is demonstrated by the vaccine effect in two animal models that have been in situ vaccinated with usEPs.

However, two issues should be discussed involving usEP-induced immunity. One consists of the concept of immunogenicity, and the other concerns the "size" of the tumor load identified by the immune system. Immunogenic cancers are more likely to induce immune responses or express an immune response to a non-self-antigen, which could more easily be revealed by the treatment, identified by the APC, and presented to T-cells. Some cancers and animal cancer models are more immunogenic than others. The breast and liver cancers models were sufficiently immunogenic to induce a vaccine effect and an in situ vaccination. Tumor antigens were recognized as dead and dying cancer cells after treatment, and they were readily identified as untreated cancer cells injected in the challenge studies. As far as immunogenicity to usEPs is concerned, the rat liver and mouse breast cancer models were more immunogenic than other cancer models.

The second factor is the size of the tumor presented to the usEP-induced immune system when some immunity has already been produced. In the case of a challenge injection, the tumor cells can be identified by immune cells before a TME is formed. Thus, the tumor load is relatively low, and the tumor cells can be eliminated before they form an immunosuppressive TME. Likewise, it is probable that if an immune response were initiated soon after the usEP treatment and/or if the metastatic cells were released late from the primary tumor, it may be expected that the migrating tumor cells will not have formed an immunosuppressive TME by the time tumor cells can be recognized by cytotoxic T-cells. The primary problem of cancer and malignancy is the formation of the immunosuppressive TME, which, once formed, it becomes a formidable opponent that can preclude effective immunotherapy. The best way to avoid the immunosuppressive TME is, unfortunately, to prevent the formation of a tumor. Another study from the authors' group addresses immunogenicity and the formation of an immunosuppressive TME.

19.2.10 Other Evidence that usEPs Induce Immune Responses

19.2.10.1 UV-Induced Melanoma and Immunity

The B16f10 melanoma is a commonly used cancer model and a popular one for treating cancer with usEPs because of the relative ease of treatments with a needle or pinch electrodes. Another melanoma model for usEP treatment is UV induced cancer in a C57/BL6-HGF/SF transgenic mouse (Nuccitelli et al. 2012). This mouse overexpresses the hepatocyte growth factor/scatter factor (HGF/SF) driven by the metallothionein gene promoter. The HGF/SF mice exhibit subcutaneous melanoma in response to a single exposure to an erythemal dose of ultraviolet radiation shortly after birth, meaning an amount of radiation that temporarily turns the exposed skin red. The melanoma is this model resembles that seen in human skin (Takayama et al. 1997). With an application of 2000 usEPs with 100 ns durations, 20 ns rise-fall time, and electric fields of 30 kV/cm or 6.0 Vs/cm, tumors in 79% (11/14) of mice were eliminated with one treatment. Three other mice were tumor-free after a second treatment. The mice's skin is significantly less pigmented three weeks after treatment; however, pigmentation was normal over the next months and was a normal color three months after treatment. Although the authors suggest that usEP treatment "locally disrupts normal melanogenesis," this is not supported by the findings. It is more correct to suggest usEP treatment under these conditions eliminate tumors and transiently disrupt normal melanocyte pigment production but does not affect melanogenesis permanently because skin pigmentation returns to normal. This indicates that stem cells survived the usEP treatment, or if they did not, stem cells migrated into the treatment zone, and melanogenesis at the treatment site was preserved.

In this study by Nuccitelli et al. (2012), they report that usEPs induces apoptosis in these UV-induce melanomas. It has become a habit for reporting apoptosis as a consequence after usEP treatment. This may be due to the initial finding that usEPs induced apoptosis in Jurkat cells (Beebe et al. 2002, 2003) and the finding that immunofluorescent microscopy found active caspases in the TME of mouse ectopic melanoma and HCC (Chen et al. 2010, 2012) and orthotopic rat HCC (Chen et al. 2014). However, as indicated earlier, these findings did not show that cancer cells exhibited active caspases. So, if correct that usEPs induce apoptosis in the TME after treatment; however, it is not necessarily in the cancer cells. As indicated earlier, both orthotopic N1-S1 HCC and 4T1-luc cause vaccine effects, yet they do not undergo apoptosis in response to usEPs in vitro. While there are reports that apoptosis induces immunogenic factors or apoptosis (Scheffer et al. 2003; Krysko et al. 2006), apoptosis is most known for "silent" cell death, which as discussed earlier is more appropriately call programmed cell death, such that inflammation is not induced into the physiological environment. Tang et al. (2016) indicate evidence that virtually all RCD mechanisms have been associated with ICD mechanisms. As indicated in the case here, there is the misconception that DNA fragmentation and TUNEL staining, in 30% of cells, suggests the presence of apoptosis. Apoptotic DNA fragmentation is very specific because caspases activate endonucleases that fragment the chromatin into units defined by nucleosomes, so the DNA fragments are multiples of about 180-bp oligomers. Because the nucleosomes DNA degradation only allows nucleotide substrates every 180 bps (Matassov et al. 2004). These fragments are identified by gel electrophoreses as the so-called DNA ladder separated by 180 bp on the gel. DNA by fragmentation by TUNEL does not supply this specificity because there is no requirement for TUNEL staining to define such fragments. TUNEL does not provide the specialized 180 bp fragmentation (Grasl-Kraupp et al. 1995). The authors also use the definition of pyknosis for apoptosis. However, pyknosis represents a condensation of nuclear chromatin but does not distinguish between cells undergoing necrosis or apoptosis. So, neither description of TUNEL nor pyknosis correctly identify apoptosis. Under these definitions, the mechanism(s) of RCD that describes these UV-induce melanoma tumors tumor cell deaths occurred.

In reality, there is not adequate evidence for immunity to appear in the title. An initial complication in this strategy is the use of allograft models for immune studies. An allograft introduces a tissue graft from a donor into a recipient that are not genetically identical. Melanoma tumors from B16-F10eGFP melanoma cells were injected in the flanks of SKH-1 (immunocompetent) and 12 Nu/Nu (immunodeficient) mice. The introduction of B16f10-eGFP tumors into an SKH-1 mice host could be expected to present an immune response because they are not syngeneic and therefore not immunologically compatible. Of course, the Nu/Nu mice could not induce an immune response because they lack T-cells. To determine if usEPs induced an immune response, determine the presence or absence of a vaccine effect. After fourteen or 28 days of treating primary tumors in one flank with usEPs or surgically resecting them, SKH-1 mice and Nu/Nu mice were challenged with a secondary tumor in the opposite side.

First, tumors grew significantly faster (3-fold) in immunodeficient mice than in immunocompetent mice. When second tumors were implanted 14 and 28 days after treatment, they grew more slowly than primary tumors in the SKH-1 mice, but those infused 28 days grew more slowly than those implanted 14 days after treatment. In contrast, second tumors in the Nu/Nu immunodeficient mice grew at the same rate as the primary tumors. However, second tumors grew more slowly in mice with primary tumors resected, but these tumors continued pigment and GFP. The reduced growth of secondary tumors after resection is due to the presence of an immune response that is independent of usEP treatment. This could be due to the allograftic nature of this tumor model. The author concluded that since the slower growth of the secondary tumor was not present in the immunocompromised model, the growth inhibition in the usEP-treated immunocompetent model was due to an immune response that was better than that observed in the resected model.

Suspecting an immune response, tumors were analyzed on various days after implantation. CD4+ cells were observed in secondary tumors from mice that had primary tumors treated with usEPs. With an absence of CD4+ cells in untreated tumors, CD4+ cells were observed in treated primary tumors between 14 and 19 days after usEP treatment. Notably, CD4+ cells were observed in untreated secondary tumors removed from mice with primary tumors treated with usEPs. These results were reported as long as another tumor in that mouse had been treated with usEPs; however, they did not specifically report the presence or absence of CD4+ cells in secondary tumors when the primary tumor was resected.

The study concluded that the usEP treatment of a primary tumor induced a protective immune response that was superior to tumor excision for slowing the growth of a secondary challenge tumor injection. That such a slowed tumor growth was not present in the immunocompromised model, provides additional evidence to support an immune response. CD4+ cells in the primary and notably in the secondary tumor further suggested an immune response. However, CD4+ cells should be further classified since helper T-cells, and suppressive Tregs are CD4+.

The slower growth of secondary tumors in the usEP treated model than in the excision model supports this conclusion that an immune response may have been present. However, this slower secondary tumor growth was also present after excision of the primary tumor. This suggests that the usEP treatment may have enhanced an immune response that was naturally present in this allograft model. The time-dependent increase in CD4+ cells was not trivial but actually falls short of supporting immune response. However, CD4+ cells are generally not cytotoxic but serve helper functions. Also, CD4+ cells can be effector cells with immune functions, but they can also be exhibit immunosuppressive functions if they also express CD25 and Foxp3. Without an extended analysis of these CD4+ cells, we assume they provided some helper immune functions that may have contributed to slower growth of secondary tumors, but immunosuppressive Tregs may have contributed to preventing growth.

It is notable that in this allograftic model, usEPs did not induce a distinct vaccine effect, but only slowed the growth of challenge tumor injections. There are several possible reasons for this that are open to debate and further experimentation. First, this model may be less immunogenic than the N1-S1 HCC and the 4T1-luc breast cancer model. This could be due to the tumor itself, the animal model, or a combination of these two interacting systems. Second, a second reason for this less effective assumed immune response could be due to the usEP treatment conditions. As the authors stated and as has been published (Schoenbach et al. 2001, 2004, 2009), the pulse duration and rise-fall time should be shorter (or faster) than the time for the redistribution of charges over the plasma membrane that would prevent electric fields to enter the cells.

While the pulse duration and rise-fall time are short, the electric field may be above the threshold for tumor ablation, but below a threshold for immune induction. This would suggest that increasing the pulse number to 2000 pulses could compensate for an alleged sub-threshold electric field of 30 kV/cm (6 Vs/cm). Yet increasing the charging value to 6.0 Vs/cm, may not be effective to achieve a sufficient electric field strength threshold for immune induction. This study's charging value is greater than that of 5.0 Vs/cm in the HCC and breast cancer model. This suggests that the charging value may not be valid alone as an indicator of immunity unless there is an optimal value, below which these conditions are ineffective. For one thing, the charging value does not include a factor for rise-fall time.

This optimal charging value concept implicates an effective mechanism(s) of cell death that is immunogenic, that is induction of immunogenic factors such as ATP, HMDB1, and calreticulin (Tesniere et al. 2008a, 2008b; Krysko et al. 2012) among other less well characterize ICD factors (Hou et al. 2013). For example, it could be possible that overcharging tumor cells with a value of 6.0 Vs/cm could trigger a less effective cell death than 5.0 Vs/cm. Nevertheless, as discussed earlier, the authors' suggestion that apoptosis is the RCD mechanism in this study is not sufficiently supported by their data. In any event, there is not enough data to resolve these issues until additional data is presented that adequately addresses these issues.

19.2.10.2 B16f10 Melanoma and Immunity

A similar strategy to determine if usEPs induced a vaccine effect was carried out using the B16f10 melanoma model. In this study, B16f10 tumors were treated with 750 usEPs with 200 ns durations and 25 kV/cm or 3.75 Vs/cm (Rossi et al. 2019). An in vitro analysis of RCD mechanisms did not demonstrate caspase activation or PARP cleavage and failed to express necroptosis and autophagy in response to usEPs. The author confirmed the literature (Koo et al. 2015) that B16F10 cells lack RIP3 expression, as measured by RT-PCR and immunoblot analysis. Given the absence of apoptosis, necroptosis, and autophagy, mechanisms that induced ICD, the authors concluded the cell died by necrosis. Once again, necrosis does not tell us how the cell died; it means they are dead. If necrosis is equated with accidental cell death (ACD), this is also not likely correct because the cells did not die immediately. Cell death was similar at 3 h and 24 h over a series of increasing electric fields with increased cell death. Since necrosis is not really a cell death mechanism but a statement of death, it is not clear how these cells died.

Focusing on ICD as mechanisms that induce immune responses, the authors relied on data finding that apoptosis (Casares et al. 2005; Obeid et al. 2007; Garg et al. 2012), necroptosis (Yang et al. 2016), and autophagy (Michaud et al. 2011) induced RCD and that necrosis does not induce ICD (Scheffer et al. 2003; Gamrekelashvili et al. 2012). However, they did not determine whether usEPs induced ICD markers in B16f10 cells. usEPs generated ICD markers in several different cell types that also caused caspase activation in vitro (Nuccitelli et al. 2017). usEPs induced ICD markers in vitro in 4T1-luc breast cancer cells that did not induce apoptosis (Guo et al. 2018; Beebe et al. 2018). The data are consistent with the likelihood that the expression of ICD depends on the death-inducing stimulus, the intensity or dose of the trigger, and the cell type. It is perhaps noteworthy to recall that the only verified and stable examples of usEP-induced ICD that induced immunity occurred on rat N1-S1 and mouse 4T1-luc breast cancer models, and neither of them appeared to induced apoptosis.

Rossi et al. (2019) report an 85% efficacy without tumor relapse for usEPs and 80% efficacy following surgical removal. The mice were challenged with second B16f10 tumor cells 7 weeks after usEP treatment to determine if a vaccine effect was present. A vaccine effect occurred in 33% and 28.6% of mice in the usEP and surgery group, respectively. Considering that natural immunity was responsible for the surgery excision group, usEPs did not appear to significantly affect the natural host immunity. However, the expression of natural immunity seems unusual given that B16f10 melanoma is considered poorly immunogenic.

Explanations for the absence of a significant boost in host immunity rest on the arguments presented in other models. In this case, likely, the usEP conditions were not optimal. Again, it could be argued that 100 ns is better than a 200 ns duration based on the plasma membrane charging time constant. Although the rise-fall time was not reported, the pulser used in these studies has a 40–50 ns rise-fall time, which is slower than the 4–5 ns rise time for the 100 ns pulser in the N1-S1 HCC and the 4T1-luc breast cancer models. The pulsing conditions delivered 3.5 V/cm, which is lower than the 5.0 Vs/cm in the successful vaccine effects. This low value is influenced by the relatively low 30 K/cm and the lower pulse number of 750 despite the longer pulse duration. These 3.5 V/cm indicate values near the threshold for tumor ablation but below a threshold for immune induction if that is possible in the model. This raises perhaps a more important question concerning the B16f10 model's immunogenicity, which is likely lower than the rat HCC and the mouse breast model. So, there seems to be an agreement in the literature that the mechanism for cell death induction. However, I do not think there is a universal answer given the variabilities of cancer mechanisms that inhibit cell RCD mechanisms.

In addition to sufficient usEP conditions and the cell type, some differences determine ICD makers' release *in vitro* and usEP conditions in vivo. In the study by Rossi et al. (2019), the differences 12.5–125-fold greater in vitro, as might be expected. For the N1-S1 and 4T1-luc models, the differences are 15–20 fold. So, it is difficult to predict how these conditions impact the cells in suspension versus the impact of tissue matrices. Another issue that has not been raised is the role played by the TME as those tissue matrices. The roles of the dynamic interactions between tumor cells, TME niches, and immunomodulators in the TME can positively or negatively impact tumor progression and immune responses (Tang et al. 2016). It may not be coincidental that both effective immune responses have occurred in orthotopic models.

19.2.10.3 Another B16f10 Melanoma Study

In this study B16f10 tumors were treated with 100, 200, and 300 usEPs with 100 ns durations and 30 kV/cm or 0.3, 0.6, and 0.9 Vs/cm, respectively (Zhang et al. 2019). There were a pulse number-dependent decreases in tumor size for 100 and 200 pulses correlated with tumor ablation efficacy, "within the tolerance range of the mice." This was stated given a blatant and unexplained statement that "The mice with 300 pulses were dead". Under normal usEP circumstances, mice with ectopic or orthotopic tumors are treated with us EP conditions at least five times this Vs/cm conditions that survive as tumor-free mice. Although the authors do not indicate the tumor sizes before treatment, the images of mice-bearing tumors suggest that tumors are significantly larger than would be allowed by AAALAC standards practiced worldwide. Another term used in this manuscript was inappropriate when referring to euthanasia as execution ("executed the mice"). These issues cast an uncomfortable light on these finding's value because the animals do not appear to have been treated humanely by AAALAC standards.

Nevertheless, 200 pulses with a 0.6 Vs/cm treatment inhibited tumor growth were used to analyze T-cell phenotypes and function in a four-day snapshot after usEP treatment. Differences between CD4+ and CD8+ T-cells before versus after usEPs treatments were relatively small, with increases of 3.2% and 0.8%, respectively. T-regs decreased by almost 2%, and MDSC decreased most impressively by about 20%. Although the T-cell changes exhibited unimportant meaning, decreases in immunosuppressive cells were more considerable. Given the immunosuppressive barrier consequence in the TME, these results appear to be leaning toward a more favorable TME after usEP treatment. The study also shows that compared to untreated control mice, IL-2, and TNF-α increased and IL-10 decreased significantly, while IFN-γ, TGF-β increased without statistical difference. The IL-2, TNF-α, and IFN-γ would be expected to enhance T-cells and NK cells' cytotoxicity. The results suggest some immune system enhancement during tumor growth inhibition even at low Vs/cm conditions.

19.2.10.4 A Vaccination Approach Demonstrates UsEP-Induced Immunity in Colon Carcinoma and Lymphoma Cells

Although usEP induces immune responses and vaccinations in situ in the mouse breast and rat HCC models, this is not the traditional way to vaccinate animals. A more classical approach for vaccinating and demonstrating immunity is to test a vaccine made from usEP treated cells like that done with CT-26 colon carcinoma cells and EL-4 lymphoma cells (Rossi et al. 2019). The authors focused on the induction of ER stress as the most likely mechanism to induce ICD. Morotomi-Yano et al. (2012) nicely demonstrated that usEPs induced novel cellular stress mechanisms in HeLaS3, which was discussed in the RCD section. Rossi and colleagues showed that their chosen cell line responded slightly differently to usEPs in vitro. CT-26 cells were more resistant to treatment requiring about 3-fold higher 300 pulse conditions or 0.42 Vs/cm vs. 100 pulses or 0.14 Vs/cm with 200 ns durations at 7 kV/cm for equivalent cell death (60%). CT-26 cells exhibited a more generalized stress response responding through both IRE1ά and XBP-1 expression and PERK-mediated eIRF2ά phosphorylation while EL-4 responded only by PERK-mediate eIRF2ά phosphorylation without XBP-1 expression. With the phosphorylation of eRF2ά, both cell lines responded to usEPs with the integrated stress response (IRS). While both cell lines exhibited caspase-3 activation, calreticulin (CRT) externalization, and HMBG-1 release, only CT-26 cells released ATP.

To induce vaccinations, the approaches used somewhat fewer usEP-treated cells to use as a possible vaccine than is generally used (0.6 instead of 1.0 × 106), 1/10th the number for CT-26, and 1/30th the number for EL-4 to determine immunity by challenge injections. After usEP treatment, there were still 15% and 25% live cells for CT26 and EL-4, respectively. Yet, about 60% (9/15) and 25% (6/25) of injections for CT26 and EL-1, respectively, did not develop tumors at the injection sites. In mice that did not develop tumors, a protective vaccine developed in 78% (7/9) mice for CT26 and 80% (8/10) mice for EL-4. The authors also reported that for mice that did develop tumors at the vaccination site, most (5/6) developed tumors at the challenge site, but with significantly slower growth.

Generally, these studies demonstrate immune responses generated with usEPs, but demonstrations of immunity required that standard conditions needed to be reduced for vaccination and challenge injections. While immunogenicity could be induced with an incomplete stress response from EL-4 cells, this cell line more readily responded to usEPs than CT-26 cells yet less readily produced vaccinations. However, based on the expression/release of ICD factors, it is likely that CRT was the only assayed ICD factor that played a role in immunogenicity because, after treatment, the cells were centrifuged and resuspended to concentrate them. UsEP treatment makes cells much more vulnerable to centrifugation, likely bursting many cells, so, likely, the ATP and HMGB-1 were not present in the vaccination injection. It could be argued that the absence of the IRE1ά response could have made them a somewhat less likely candidate for vaccination. If this hypothesis is correct, there may be other factors that play roles in immunization, in addition to the classically considered CRT, ATP, and HMGB-1.

References

Ambros V (2004) The functions of animal microRNAs. Nature 431(7006):350–355

Andersson U, Wang H, Palmblad K, Aveberger AC, Bloom O, Erlandsson-Harris H, Janson A, Kokkola R, Zhang M, Yang H, Tracey KJ (2001) High mobility group 1 protein (HMG-1) stimulates proinflammatory cytokine synthesis in human monocytes. J Exp Med 192(4):565–570

Badovinac VP, Porter BB, Harty JT (2002) Programmed contraction of CD8(+) T cells after infection. Nat Immunol 3(7):619–626

Badovinac VP, Messingham KA, Jabbari A, Haring JS, Harty JT (2005) Accelerated CD8+ T-cell memory and prime-boost response after dendritic-cell vaccination. Nat Med 11(7):748–756

Bartel DP (2004) MicroRNAs: genomics, biogenesis, mechanism, and function. Cell 116(2):281–297

Bartel DP (2018) Metazoan microRNAs. Cell 173(1):20–51

Barry M, Heibein JA, Pinkoski MJ, Lee SF, Moyer RW, Green DR, Bleackley RC (2000) Mol Granzyme B short-circuits the need for caspase 8 activity during granule-mediated cytotoxic T-lymphocyte killing by directly cleaving Bid. Cell Biol 20(11):3781–3794

Beebe SJ, Fox PM, Rec LH, Buescher ES, Somers K, Schoenbach KH (2002) Nanosecond pulsed electric field (nsPEF) effects on cells and tissues: apoptosis induction and tumor growth inhibition. IEEE Trans Plasma Sci 30:286–292

Beebe SJ, Fox PM, Rec LJ, Willis EL, Schoenbach KH (2003) Nanosecond, high-intensity pulsed electric fields induce apoptosis in human cells. FASEB J 17:1493–2145

Beebe SJ, Ford WE, Ren W, Chen X (2011) Pulse power ablation of melanoma with nanosecond pulsed electric fields. In: Morton R (ed) Treatment of metastatic melanoma. In Tech Croatia, pp 231–268. ISBN 978-953-307-574-7. https://doi.org/10.5772/22850

Beebe SJ, Lassiter BP, Guo S (2018) Nanopulse stimulation (NPS) induces tumor ablation and immunity in orthotopic 4T1 mouse breast cancer: a review. Cancers (basel) 10(4):97

Blank CU, Haining WN, Held W, Hogan PG, Kallies A, Lugli E, Lynn RC, Philip M, Rao A, Restifo NP, Schietinger A, Schumacher TN, Schwartzberg PL, Sharpe AH, Speiser DE, Wherry EJ, Youngblood BA, Zehn D (2019) Defining 'T cell exhaustion'. Nat Rev Immunol 11:665–674

Boraschi D, Italiani P (2018) Innate immune memory: time for adopting a correct terminology. Front Immunol 9:799

Bunt SK, Sinha P, Clements VK et al (2006) Inflammation induces myeloid-derived suppressor cells that facilitate tumor progression. J Immunol 176:284–290

Casares N, Pequignot MO, Tesniere A, Ghiringhelli F, Roux S, Chaput N, Schmitt E, Hamai A, Hervas-Stubbs S, Obeid M, Coutant F, Métivier D, Pichard E, Aucouturier P, Pierron G, Garrido C, Zitvogel L, Kroemer G (2005) Caspase-dependent immunogenicity of doxorubicin-induced tumor cell death. J Exp Med 202(12):1691–1701

Cerwenka A, Lanier LL (2016) Natural killer cell memory in infection, inflammation and cancer. Nat Rev Immunol 16(2):112–123

Chai LF, Prince E, Pillarisetty VG, Katz SC (2020) Challenges in assessing solid tumor responses to immunotherapy. Cancer Gene Ther 27(7–8):528–538

Cao X, Cai SF, Fehniger TA, Song J, Collins LI, Piwnica-Worms DR, Ley TJ (2007) Granzyme B and perforin are important for regulatory T cell-mediated suppression of tumor clearance. Immunity 27(4):635–646

Chen R, Sain NM, Harlow KT, Chen YJ, Shires PK, Heller R, Beebe SJ (2014) A protective effect after clearance of orthotopic rat hepatocellular carcinoma by nanosecond pulsed electric fields. Eur J Cancer 50(15):2705–2713

Chen X, Kolb JF, Swanson RJ, Schoenbach KH, Beebe SJ (2010) Apoptosis initiation and angiogenesis inhibition: melanoma targets for nanosecond pulsed electric fields. Pigment Cell Melanoma Res 23:554–563

Chen X, Zhuang J, Kolb JF, Schoenbach KH, Beebe SJ (2012) Long term survival of mice with hepatocellular carcinoma after pulse power ablation with nanosecond pulsed electric fields. Technol Cancer Res Treat 11:83–93

Chiba S, Baghdadi M, Akiba H, Yoshiyama H, Kinoshita I, Dosaka-Akita H, Fujioka Y, Ohba Y, Gorman JV, Colgan JD, Hirashima M, Uede T, Takaoka A, Yagita H, Jinushi M (2012) Tumor-infiltrating DCs suppress nucleic acid-mediated innate immune responses through interactions between the receptor TIM-3 and the alarmin HMGB1. Nat Immunol 13(9):832–842

Elliott MR, Chekeni FB, Trampont PC, Lazarowski ER, Kadl A, Walk SF, Park D, Woodson RI, Ostankovich M, Sharma P, Lysiak JJ, Harden TK, Leitinger N, Ravichandran KS (2009) Nucleotides released by apoptotic cells act as a find-me signal to promote phagocytic clearance. Nature 461(7261):282–286

Enari M, Sakahira H, Yokoyama H, Okawa K, Iwamatsu A, Nagata S (1998) A caspase-activated DNase that degrades DNA during apoptosis, and its inhibitor ICAD. Nature 391(6662):43–50

Escors D (2014) Tumour immunogenicity, antigen presentation and immunological barriers in cancer immunotherapy. New J Sci 2014:734515
Fergusson JR, Fleming VM, Klenerman P (2011) CD161-Expressing hHuman T cells. Front Immunol 2:36
Fergusson JR, Smith KE, Fleming VM, Rajoriya N, Newell EW, Simmons R, Marchi E, Björkander S, Kang YH, Swadling L, Kurioka A, Sahgal N, Lockstone H, Baban D, Freeman GJ, Sverremark-Ekström E, Davis MM, Davenport MP, Venturi V, Ussher JE, Willberg CB, Klenerman P (2014) CD161 defines a transcriptional and functional phenotype across distinct human T cell lineages. Cell Rep 9(3):1075–1088
Gabrielli S, Ortolani C, Del Zotto G, Luchetti F, Canonico B, Buccella F, Artico M, Papa S, Zamai L (2016) The memories of NK cells: innate-adaptive immune intrinsic crosstalk. J Immunol Res 2016:1376595
Gabrilovich DI, Nagaraj S (2009) Myeloid-derived suppressor cells as regulators of the immune system. Nat Rev Immunol 9:162–174
Gamrekelashvili J, Ormandy LA, Heimesaat MM, Kirschning CJ, Manns MP, Korangy F, Greten TF (2012) Primary sterile necrotic cells fail to cross-prime CD8(+) T cells. Oncoimmunology 1(7):1017–1026
Gardiner CM, Finlay DK (2017) What fuels natural killers? metabolism and NK cell responses. Front Immunol 8:367
Garg AD, Krysko DV, Vandenabeele P, Agostinis P (2012) The emergence of phox-ER stress induced immunogenic apoptosis. Oncoimmunology 1(5):786–788
Grasl-Kraupp B, Ruttkay-Nedecky B, Koudelka H, Bukowska K, Bursch W (1995) Schulte-Hermann R In situ detection of fragmented DNA (TUNEL assay) fails to discriminate among apoptosis, necrosis, and autolytic cell death: a cautionary note. Hepatology 21(5):1465–1468
Green DR, Ferguson T, Zitvogel L, Kroemer G (2009) Immunogenic and tolerogenic cell death. Nat Rev Immunol 9:353–363
Guo S, Jing Y, Burcus NI, Lassiter BP, Tanaz R, Heller R, Beebe SJ (2018) Nano-pulse stimulation induces potent immune responses, eradicating local breast cancer while reducing distant metastases. Int J Cancer 142(3):629–640
Guo ZS, Liu Z, Bartlett DL (2014) Oncolytic immunotherapy: dying the right way is a key to eliciting potent antitumor immunity. Front Oncol 4:74
Haabeth OA, Tveita AA, Fauskanger M, Schjesvold F, Lorvik KB, Hofgaard PO, Omholt H, Munthe LA, Dembic Z, Corthay A, Bogen B (2014) How do CD4(+) T cells detect and eliminate tumor cells that either lack or express MHC class II molecules? Front Immunol 5:174
Hanahan D, Weinberg RA (2000) The hallmarks of cancer. Cell 100:57–70
Hanahan D, Weinberg RA (2011) Hallmarks of cancer: the next generation. Cell 144:646–674
Hazenberg MD, Spits H (2014) Human innate lymphoid cells. Blood 124(5):700–709
Hou W, Zhang Q, Yan Z, Chen R, Zeh HJ III, Kang R, Lotze MT, Tang D (2013) Strange attractors: DAMPs and autophagy link tumor cell death and immunity. Cell Death Dis 4(12):e966
Inoue H, Tani K (2014) Multimodal immunogenic cancer cell death as a consequence of anticancer cytotoxic treatments. Cell Death Differ 21:39–49
Kroemer G, Galluzzi L, Kepp O, Zitvogel L (2013) Immunogenic cell death in cancer therapy. Annu Rev Immunol 31:51–72
Krysko DV, D'Herde K, Vandenabeele P (2006) Clearance of apoptotic and necrotic cells and its immunological consequences. Apoptosis 11(10):1709–1726
Karaman MW, Herrgard S, Treiber DK, Gallant P, Atteridge CE, Campbell BT, Chan KW, Ciceri P, Davis MI, Edeen PT, Faraoni R, Floyd M, Hunt JP, Lockhart DJ, Milanov ZV, Martin MD, Badovinac VP (2018) Defining memory CD8 T cell. Front Immunol 9:2692
Kazama H, Ricci JE, Herndon JM, Hoppe G, Green DR, Ferguson TA (2008) Induction of immunological tolerance by apoptotic cells requires caspase-dependent oxidation of high-mobility group box-1 protein. Immunity 29(1):21–32

Kepp O, Galluzzi L, Martins I, Schlemmer F, Adjemian S, Michaud M, Sukkurwala AQ, Menger L, Zitvogel L, Kroemer G (2011) Molecular determinants of immunogenic cell death elicited by anticancer chemotherapy. Cancer Metastasis Rev 30(1):61–69

Koo GB, Morgan MJ, Lee DG, Kim WJ, Yoon JH, Koo JS, Kim SI, Kim SJ, Son MK, Hong SS, Levy JM, Pollyea DA, Jordan CT, Yan P, Frankhouser D, Nicolet D, Maharry K, Marcucci G, Choi KS, Cho H, Thorburn A, Kim YS (2015) Methylation-dependent loss of RIP3 expression in cancer represses programmed necrosis in response to chemotherapeutics. Cell Res 25:707–725

Krysko DV, Garg AD, Kaczmarek A, Krysko O, Agostinis P, Vandenabeele P (2012) Immunogenic cell death and DAMPs in cancer therapy. Nat Rev Cancer 12:860–875

Lanier LL (2015) NKG2D Receptor and its ligands in host defense. Cancer Immunol Res 3(6):575–582

Lassiter BP, Guo S, Beebe SJ (2018) Nano-pulse stimulation ablates orthotopic rat hepatocellular carcinoma and induces innate and adaptive memory immune mechanisms that prevent recurrence. Cancers (basel) 10(3):69

Lieberman J (2010) Anatomy of a murder: how cytotoxic T cells and NK cells are activated, develop, and eliminate their targets. Immunol Rev 235:5–9

Ljunggren HG, Kärre K (1990) In search of the 'missing self': MHC molecules and NK cell recognition. Immunol Today 11(7):237–244

Lord SJ, Rajotte RV, Korbutt GS, Bleackley RC (2003) Granzyme B: a natural born killer. Immunol Rev 193:31–38

Luckey CJ, Bhattacharya D, Goldrath AW, Weissman IL, Benoist C, Mathis D (2006) Memory T and memory B cells share a transcriptional program of self-renewal with long-term hematopoietic stem cells. Proc Natl Acad Sci USA 103(9):3304–3309

Maggi L, Santarlasci V, Capone M, Peired A, Frosali F, Crome SQ, Querci V, Fambrini M, Liotta F, Levings MK, Maggi E, Cosmi L, Romagnani S, Annunziato F (2010) CD161 is a marker of all human IL-17-producing T-cell subsets and is induced by RORC. Eur J Immunol 40(8):2174–2181

Matassov D, Kagan T, Leblanc J, Sikorska M, Zakeri Z (2004) Measurement of apoptosis by DNA fragmentation. Methods Mol Biol 282:1–17

Martin MD, Badovinac VP (2018) Defining Memory CD8 T Cell. Front Immunol 9:2692

Michaud M, Martins I, Sukkurwala AQ, Adjemian S, Ma Y, Pellegatti P, Shen S, Kepp O, Scoazec M, Mignot G, Rello-Varona S, Tailler M, Menger L, Vacchelli E, Galluzzi L, Ghiringhelli F, di Virgilio F, Zitvogel L, Kroemer G (2011) Autophagy-dependent anticancer immune responses induced by chemotherapeutic agents in mice. Science 334(6062):1573–1577

Moretta A, Bottino C, Vitale M, Pende D, Cantoni C, Mingari MC, Biassoni R, Moretta L (2001) Activating receptors and coreceptors involved in human natural killer cell-mediated cytolysis. Annu Rev Immunol 19:197–223

Moretta A, Marcenaro E, Parolini S, Ferlazzo G, Moretta L (2008) NK cells at the interface between innate and adaptive immunity. Cell Death Differ 15(2):226–233

Morotomi-Yano K, Oyadomari S, Akiyama H, Yano K (2012) Nanosecond pulsed electric fields act as a novel cellular stress that induces translational suppression accompanied by eIF2α phosphorylation and 4E-BP1 dephosphorylation. Exp Cell Res. 318(14):1733–1744

Morrison MJ, Pallares G, Patel HK, Kroemer G1, Galluzzi L, Kepp O, Zitvogel L (2013) Immunogenic cell death in cancer therapy. Annu Rev Immunol 31:51–72

Murdoch C, Muthana M, Coffelt SB et al (2008) The role of myeloid cells in the promotion of tumour angiogenesis. Nat Rev Cancer 8:618–631

Netea MG, Joosten LA, Latz E, Mills K, Natoli G, Stunnenberg HG, O'Neill LA, Xavier RJ (2016) Trained immunity: a program of innate immune memory in health and disease. Science 352:aaf1098

Nuccitelli R, Tran K, Lui K, Huynh J, Athos B, Kreis M, Nuccitelli P, De Fabo EC (2012) Non-thermal nanoelectroablation of UV-induced murine melanomas stimulates an immune response. Pigment Cell Melanoma Res 25(5):618–629

Nuccitelli R, McDaniel A, Anand S, Cha J, Mallon Z, Berridge JC, Uecker D (2017) Nano-Pulse Stimulation is a physical modality that can trigger immunogenic tumor cell death. J Immunother Cancer 5:32

Obeid M, Tesniere A, Ghiringhelli F, Fimia GM, Apetoh L, Perfettini JL, Castedo M, Mignot G, Panaretakis T, Casares N, Métivier D, Larochette N, van Endert P, Ciccosanti F, Piacentini M, Zitvogel L, Kroemer G (2007) Calreticulin exposure dictates the immunogenicity of cancer cell death. Nat Med 13(1):54–61

O'Brien KL, Finlay DK (2019) Immunometabolism and natural killer cell responses. Nat Rev Immunol 19(5):282–290

Panaretakis T, Joza N, Modjtahedi N, Tesniere A, Vitale I, Durchschlag M, Fimia GM, Kepp O, Piacentini M, Froehlich KU, van Endert P, Zitvogel L, Madeo F, Kroemer G (2008) The co-translocation of ERp57 and calreticulin determines the immunogenicity of cell death. Cell Death Differ 15(9):1499–1509

Rossi A, Pakhomova ON, Pakhomov AG, Weygandt S, Bulysheva AA, Murray LE, Mollica PA, Muratori C (2019) Mechanisms and immunogenicity of nsPEF-induced cell death in B16F10 melanoma tumors. Sci Rep 9:431

Poznanski SM, Barra NG, Ashkar AA, Schertzer JD (2018) Immunometabolism of T cells and NK cells: metabolic control of effector and regulatory function. Inflamm Res 67(10):813–828

Scaffidi P, Misteli T, Bianchi ME (2002) Release of chromatin protein HMGB1 by necrotic cells triggers inflammation. Nature 418(6894):191–195

Scheffer SR, Nave H, Korangy F, Schlote K, Pabst R, Jaffee EM, Manns MP, Greten TF (2003) Apoptotic, but not necrotic, tumor cell vaccines induce a potent immune response in vivo. Int J Cancer 103(2):205–211

Schoenbach KH, Beebe SJ, Buescher ES (2001) Intracellular effect of ultrashort electrical pulses. Bioelectromagnetics 22(6):440–448

Schoenbach KH, Joshi RP, Kolb JF, Chen N, Stacey M, Buescher ES, Beebe SJ, Blackmon P (2004) Ultrashort electrical pulses open a new gateway into biological cells. Proc IEEE 92:1122–1137

Schoenbach KH, Joshi RP, Beebe SJ, Baum CE (2009) A Scaling Law for Membrane Permeabilization with Nanopulses. IEEE Trans Dielectrics Electr Insul 16:1224–1235

Schwartzberg PL, Sharpe AH, Speiser DE, Wherry EJ, Youngblood BA, Zehn D.Defining T cell exhaustion'.Nat Rev Immunol 19(11):665–674

Smyth MJ, Crowe NY, Pellicci DG, Kyparissoudis K, Kelly JM, Takeda K, Yagita H, Godfrey DI. Sequential production of interferon-gamma by NK1.1(+) T cells and natural killer cells is essential for the antimetastatic effect of alpha-galactosylceramide. Blood 99(4):1259–1266

Sun J, Beilke J, Lanier L (2009) Adaptive immune features of natural killer cells. Nature 457:557–561

Sun JC, Lanier LL (2011) NK cell development, homeostasis, and function: parallels with CD8+ T cells. Nat Rev Immunol. 11(10):645–657

Sun JC, Ugolini S, Vivier E (2014) Immunological memory within the innate immune system. EMBO J 17:1295–1303

Takayama H, LaRochelle WJ, Sharp R, Otsuka T, Kriebel P, Anver M, Aaronson SA, Merlino G (1997) Diverse tumorigenesis associated with aberrant development in mice overexpressing hepatocyte growth factor/scatter factor. Proc Natl Acad Sci U S A. 94(2):701–706

Tang H, Qiao J, Fu YX (2016) Immunotherapy and tumor microenvironment. Cancer Lett 370(1):85–90

Tao K, Fang M, Alroy J, Sahagian GG (2008) Imagable 4T1 model for the study of late stage breast cancer. BMC Cancer 8:228

Tesniere A, Apetoh L, Ghiringhelli F, Joza N, Panaretakis T, Kepp O, Schlemmer F, Zitvogel L, Kroemer G (2008a) Immunogenic cancer cell death: a key-lock paradigm. Curr Opin Immunol 20(5):504–511

Tesniere A, Panaretakis T, Kepp O, Apetoh L, Ghiringhelli F, Zitvogel L et al (2008b) Molecular characteristics of immunogenic cancer cell death. Cell Death Differ 15:3–12

Trapani JA (1995) Target cell apoptosis induced by cytotoxic T cells and natural killer cells involves synergy between the pore-forming protein, perforin, and the serine protease, granzyme B. Aust N Z J Med 25(6):793–799

Vivier E, Raulet DH, Moretta A, Caligiuri MA, Zitvogel L, Lanier LL, Yokoyama WM, Ugolini S (2011) Innate or adaptive immunity? The example of natural killer cells. Science 331(6013):44–49

Vucic D, Deshayes K, Ackerly H, Pisabarro MT, Kadkhodayan S, Fairbrother WJ, Dixit VM (2002) SMAC negatively regulates the anti-apoptotic activity of melanoma inhibitor of apoptosis (ML-IAP). J Biol Chem 277(14):12275–12279

Williams MA, Bevan MJ (2007) Effector and memory CTL differentiation. Annu Rev Immunol 25:171–192

Wu RS, Kobie JJ, Besselsen DG, Fong TC, Mack VD, McEarchern JA, Akporiaye ET. Comparative analysis of IFN-gamma B7.1 and antisense TGF-beta gene transfer on the tumorigenicity of a poorly immunogenic metastatic mammary carcinoma. Cancer Immunol Immunother 50:229–240

Yoneda T, Michigami T, Yi B, Williams PJ, Niewolna M, Hiraga T (2000) Actions of bisphosphonate on bone metastasis in animal models of breast carcinoma. Cancer 88(Suppl 12):2979–2988

Yang H, Ma Y, Chen G, Zhou H, Yamazaki T, Klein C, Pietrocola F, Vacchelli E, Souquere S, Sauvat A, Zitvogel L, Kepp O, Kroemer G (2016) Contribution of RIP3 and MLKL to immunogenic cell death signaling in cancer chemotherapy. Oncoimmunology 5(6):e1149673

Zhang X, Zhang Y, Chen J, Wu Y, Zhang J, Wang J (2019) Nanosecond pulsed electric field inhibits malignant melanoma growth by inducing the change of systemic immunity. Med Oral Patol Oral Cir Bucal 24(4):e555–e561

Zhu J, Paul WE (2008) CD4 T cells: fates, functions, and faults. Blood 112(5):1557–1569